Bioactive Compounds from Fruit and Vegetable Waste: Extraction and Possible Utilization

Bioactive Compounds from Fruit and Vegetable Waste: Extraction and Possible Utilization

Editors

Noelia Castillejo Montoya
Lorena Martínez-Zamora

MDPI

Basel • Beijing • Wuhan • Barcelona • Belgrade • Novi Sad • Cluj • Manchester

Editors
Noelia Castillejo Montoya
Università degli Studi
di Foggia
Foggia
Italy

Lorena Martínez-Zamora
University of Murcia
Murcia
Spain

Editorial Office
MDPI AG
Grosspeteranlage 5
4052 Basel, Switzerland

This is a reprint of articles from the Special Issue published online in the open access journal *Foods* (ISSN 2304-8158) (available at: https://www.mdpi.com/journal/foods/special_issues/47KGJ458Y7).

For citation purposes, cite each article independently as indicated on the article page online and as indicated below:

Lastname, A.A.; Lastname, B.B. Article Title. *Journal Name* **Year**, *Volume Number*, Page Range.

ISBN 978-3-7258-1775-7 (Hbk)
ISBN 978-3-7258-1776-4 (PDF)
doi.org/10.3390/books978-3-7258-1776-4

Contents

About the Editors

Noelia Castillejo Montoya

Noelia Castillejo Montoya received her PhD from Universidad Politécnica de Cartagena (Spain) in 2022. In this period, she had the opportunity to focus deeply on abiotic stresses to increase bioactive compounds in fruits and vegetables, and on revalorization of food waste to produce new sustainable foods. This was thanks to a research project funded by Seneca Foundation-Science and Technology Agency of the Region of Murcia (Spain) to study the development and minimal fresh processing of high-quality sprouts using eco-sustainable techniques. She had done a postdoctoral researcher for one year at the Università degli studi di Foggia (Italy), where she worked on postharvest issues related to fruits and vegetables. She was dedicated to fruit and vegetable postharvest research, with emphasis on how hyperspectral imaging can give us valuable information to predict or identify some postharvest damages or even the content of bioactive compounds in fruits and vegetables. She is currently doing a two-year postdoctoral research stay at the Universitat Politècnica de València (Spain), thanks to a Ministry gran (JDC2022-049432-I) funded by MCIN/AEI/10.13039/501100011033 and by the European Union "NextGenerationEU"/PRTR. Her expertise in this field is evident in her extensive publication record, which includes 35 papers in prestigious international journals such as Postharvest Biology and Technology, Journal of the Science of Food and Agriculture, Plant Physiology and Biochemistry, LWT, among others.

Lorena Martínez-Zamora

Lorena Martínez-Zamora received her PhD from the University of Murcia in 2019. Now, she is working as a postdoctoral researcher at Universidad Politécnica de Cartagena (Spain), thanks to her contract financed by the Re-qualification program of the Spanish University System, funded by the EU NextGeneration, Margarita Salas modality by the University of Murcia. With a PhD in Food Technology, Nutrition and Bromatology and a degree in Nutrition and Dietetics from the University of Murcia, Lorena Martínez Zamora has been able to focus her incipient research career on improving the healthiness of plant and animal foods using novel technologies and natural extracts, obtained mainly from the recovery of food discards. Currently, she is involved in a project about the revalorization of food by-products obtained from food losses. So far, she has published 37 scientific papers and 11 scientific reviews in popular JCR journals, such as Current Opinion in Food Science, Postharvest Biology and Technology, LWT, Plant Physiology and Biochemistry, Journal of the Science of Food and Agriculture, Antioxidants, Foods, etc. Despite her youth, she hopes (and works for it) that her scientific career will continue to be full of achievements such as the publication of her first Special Issue Reprint.

Preface

Fruit and vegetable waste is a current problem in our society, as one third of the world's production is lost or wasted at some point in the food chain. Within the Sustainable Development Goals (SDGs) to reduce food waste by 50%, we would like to contribute with this Special Issue, where several by-products are presented, and how to incorporate them into a food matrix as raw material.

This Special Issue is for the entire scientific community, especially for the younger researchers who must continue this path, proposing sustainable and innovative ways to take advantage of the by-products present in the agri-food industry.

Finally, we would like to thank all those who have made this Special Issue possible. Starting with all the leading researchers from all over the world who have done their bit to produce a very comprehensive Special Issue, of course to the editors and staff of the prestigious journal Foods for all the help received and finally to the future readers who will make it possible for these papers to be the basis for many others.

Noelia Castillejo Montoya and Lorena Martínez-Zamora
Editors

foods

MDPI

Editorial

Bioactive Compounds from Fruit and Vegetable Waste: Extraction and Possible Utilization

Noelia Castillejo [1,*] and Lorena Martínez-Zamora [2,3,*]

1 Department of Agricultural Sciences, Food, Natural Resources and Engineering, University of Foggia, Via Napoli 25, I-71122 Foggia, Italy
2 Department of Food Technology, Nutrition and Food Science, Faculty of Veterinary Sciences, University of Murcia, Espinardo, 30071 Murcia, Spain
3 Postharvest and Refrigeration Group, Department of Agronomical Engineering and Institute of Plant Biotechnology, Universidad Politécnica de Cartagena, Cartagena, 30203 Murcia, Spain
* Correspondence: noelia.castillejomontoya@unifg.it (N.C.); lorena.martinez23@um.es (L.M.-Z.)

1. Introduction

Globally, there is a serious problem with fruit and vegetable waste, which can result from improper food handling or storage techniques or from the disposal of inedible portions of produce. Approximately one-third of worldwide production is lost or wasted at some point in the food chain, according to reports from the Food and Agriculture Organization (FAO) [1]. One of the Sustainable Development Goals (SDGs), reducing food waste by 50%, is one of the FAO's major challenges for 2050. Waste is considered the raw material for a circular economy, which is based on the idea of a "zero waste" economy driven by society [2].

By-products from fruits and vegetables can be employed as innovative ingredients or food fortifiers because they are high in bioactive components [3]. Just as we may cut down on the quantity of wasted fruit and vegetable matter, an ideal extraction of these health-promoting components will enable efficient utilization of these compounds [4]. Additionally, these by-products support a circular economy because they have a wide range of possible uses in various industries [5]. By doing so, we can enhance food security and nutrition while creating more sustainable farming systems.

Therefore, the purpose of this Special Issue is to showcase current developments in cutting-edge green technology for the extraction of bioactive substances from fruit and vegetable by-products, along with their possible applications. To achieve this purpose, leading researchers from all over the world contributed to this Special Issue by providing new methods to address this problem.

Citation: Castillejo, N.; Martínez-Zamora, L. Bioactive Compounds from Fruit and Vegetable Waste: Extraction and Possible Utilization. *Foods* **2024**, *13*, 775. https://doi.org/10.3390/foods13050775

Received: 23 February 2024
Accepted: 27 February 2024
Published: 1 March 2024

2. An Overview of the Published Articles

Thirteen scientific works are included in this Special Issue, with them focusing on the optimization of the extraction of bioactive compounds, the characterization of several by-products, the development of new food products, and the evaluation of the functionality of these extracts.

Reis et al. [contribution 1] optimized the extraction of *Passiflora cincinnata*, which belongs to the same genus as passion fruit, to obtain several bioproducts such as oil rich in linoleic acid, antioxidant extracts, and oil microparticles with antioxidant extracts. The seeds were pressed, and the best extraction condition (solid–liquid ratio 1:48 with 58% ethanol at 74 °C) produced an extract rich in lignans, with high antioxidant capacity and antimicrobial activity against Gram-positive and Gram-negative bacteria. In addition, the oil was microencapsulated and enriched with *Passiflora cincinnata* antioxidant extracts, improving the oxidation stability of the product by up to 30%.

In this line of research, Sai-Ut et al. [contribution 2] optimized the extraction of phenolic compounds from lychee and longan seeds using response surface methodology (RSM).

The best condition for maximizing extraction yield, phenolic content, and antioxidant activity using a 1:20 solid–liquid ratio was 41% and 53% ethanol at 51 °C and 58 °C for 139 min and 220 min for lychee and longan seeds, respectively.

Moreover, Afifi et al. [contribution 3] characterized the main metabolites present in orange peels (albedo and flavedo) and juices from different countries. Sixty-six metabolites were detected through the use of nano-liquid chromatography coupled to a high-resolution electrospray-ionization quadrupole time-of-flight mass spectrometer. In *Citrus sinensis*, eleven metabolites were discovered for the first time. Certain flavonoids were significantly more abundant in the citrus peel than in the juice, suggesting that the peel had greater potential for its use in industry and the circular economy. The huge number of metabolites found in the peel, which is the main by-product during fruit processing, means that it should be considered for its use as a new ingredient in the food industry.

Within the development of new food products, Salas-Millán et al. [contribution 4] developed a melon-based sparkling wine with 12% *v*/*v* ethanol. Within the volatile profile, the compounds isoamyl acetate, ethyl decanoate, 3,6-nonadienyl acetate, and (E,Z)-nonadien-1-ol contributed to the sweet, fruity, banana, tropical, nutty, and melon aroma. The sensory evaluation of this melon-based wine at the end of its shelf-life scored 92 points, with 100 being the maximum. This study shows how by-product revalorization can yield new products with good sensory acceptance, market potential, and a distinctive aroma, such as the aforementioned innovative sparkling wine.

Following this topic, Pinho et al. [contribution 5] focused their research on the extraction of carotenoids from guarana peels, and their encapsulation via spray drying to improve their stability. The 1:2 ratio (concentrated ethanolic extract:gum arabic) was added to oatmeal paste. The minimum temperature applied was 70 °C, with the temperature reaching up to 90 °C. The oatmeal paste enriched with carotenoid microparticles showed lower viscosity because the presence of the microparticles reduced the accessible water. However, pigment loss decreased with encapsulation. These findings imply that the production of stable and useful products may be possible with the addition of encapsulated carotenoids to cooked foods at higher temperatures.

Also, in this line of research, Pérez et al. [contribution 6] studied the revalorization of broccoli and carrot by-products, formulating a new vegetable drink. Although three types of preservative treatments were applied, namely pasteurization, ultrasound, and high hydrostatic pressure, only the last treatment process was able to show the best results, maintaining a high concentration of carotenoids and sulfur compounds, which contributed to prolonging its shelf-life, increasing its antioxidant capacity, and reducing its microbial load, respectively.

Besides the determination and quantification of the main bioactive compounds in the by-products, it is crucial to determine the effect of these compounds on health. Hence, Olennikov et al. [contribution 7] characterized the waste generated by the lingonberry processing industry and its potential functionality. The lipid profile and antioxidant status of hamsters fed with a high-fat diet ended up being normalized by polysaccharides from lingonberry, according to the results of in vitro and in vivo experiments. These results provide credence to the idea that food processing waste can be used to produce hypolipidemic and antioxidant substances in the pharmaceutical sector.

Additionally, in this research line, Han et al. [contribution 8] studied the hepatoprotective effects of tamarind shell extract using zebrafish and chicken embryos as in vitro models. The use of tamarind shell extracts made it possible to successfully reverse the ethanol-induced pathological alterations, liver dysfunction, and ethanol–metabolic enzyme malfunction in the chick embryo liver, zebrafish, and HepG2 cells. Tamarind extract suppressed the level of excess reactive oxygen species in zebrafish and HepG2 cells and restored the altered potential of the mitochondrial membrane. This suggested that tamarind shell extract attenuates alcohol liver disease by activating NRF2 to suppress ethanol-induced oxidative stress.

Furthermore, Coelho et al. [contribution 9] studied a potential prebiotic ingredient consisting of tomato flour (seed and peel) rich in phenolic compounds and carotenoids extracted via ohmic and conventional extraction. For this purpose, the flour was subjected to simulated gastrointestinal digestion, and its fermentability potential and impact on gut microbiota were determined. The probiotic strain SFCONV favored the growth of *Bifidobacterium animalis*, while SFOH favored the growth of *Bifidobacterium longum*, probably due to the different carbohydrate profiles of the flours. Overall, the flours used were able to function as a direct substrate to favor the potential prebiotic growth of *Bifidobacterium longum*. The results of the fecal fermentation model showed the highest growth of Bacteroidetes with SFOH and the highest values of Bacteroides with SFCONV.

In this line of research, Zhou et al. [contribution 10] also studied the functionality of walnut kernel extracts (defatted or whole), measuring antioxidant activity in vitro and in vivo. The defatted walnut kernel extract showed the highest antioxidant activity under extraction conditions with 58% ethanol at 48 °C for 77 min.

In addition, similar to this topic, Phumat et al. [contribution 11] studied the functionality of *Benincasa hispida* peel extracts from consumer residues, with the extracts obtained from 95% ethanol showing the best results in terms of anti-aging and antioxidant activity. They highlighted the absence of toxicity and the irritant effect of this extract, which is promising for the development of new products.

In addition, two scientific reviews are included in this collection. The characterization of the main bioactive compounds found in several by-products is important to understand how they can be re-utilized. In this sense, as editors of this Special Issue, Martínez-Zamora et al. [contribution 12] presented a systematic review of lemon by-products as new flavonoid-rich ingredients, analyzing their extraction using green technologies and the incorporation of these flavonoid-rich extracts into new foods. This analysis reported that 89% of the papers studied used green technologies and solvents for extraction and 18 papers were related to the reuse of these extracts, although only 35% of these works evaluated the functionality of such incorporation.

Lastly, Chetrariu and Dabija [contribution 13] presented an updated review on spent grain and its use as an ingredient in various food products. This food is high in fats, fiber, protein, vitamins, and minerals. Because of its high moisture content and microbial sensitivity, it needs to be processed using an appropriate, economical, and ecologically friendly recovery approach. This by-product, due to its nutritional properties, is utilized as a raw material to make a variety of food products, including frankfurters, fruit drinks, pasta, pastries, muffins, waffles, yogurt and plant-based yogurt substitutes, and other snacks. The circular economy finds opportunities in the regeneration and recycling of waste materials and energy that become inputs to other food processes.

Therefore, the extraction and potential utilization of bioactive compounds from fruit and vegetable waste have emerged as a promising and relevant research field in agro-food waste management [6]. This area provides a unique opportunity to address both food loss reduction and the implementation of more sustainable and environmentally friendly practices [7]. Bioactive compounds derived from these waste materials can contribute to human nutrition and health and facilitate the creation of value-added products, thus promoting circular economy principles [8].

3. Conclusions and Future Perspectives

In summary, exploring bioactive compounds from fruit and vegetable waste presents a promising avenue in agro-food waste management, offering opportunities to reduce food loss and adopt sustainable practices. These compounds, with potential benefits for human nutrition and health, open doors for the development of value-added products. The ongoing exploration and development of new extraction techniques and innovative applications by young researchers are crucial to maximize the potential of this research avenue, with a view to a prosperous future in the areas of enhanced sustainability and comprehensive waste management in the food industry.

Author Contributions: N.C. and L.M.-Z. contributed equally to the writing and editing of this Editorial. All authors have read and agreed to the published version of the manuscript.

Funding: Project PID2021-123857OB-I00, financed by the Spanish Ministry of Science and Innovation, the Spanish State Research Agency/10.13039/501100011033/, and the FEDER. This work was also financed by the Autonomous Community of the Region of Murcia through the Seneca Foundation and the European program NextGenerationEU throughout the AGRO-ALNEXT project.

Data Availability Statement: Not applicable.

Acknowledgments: L.M.-Z.'s contract was financed by the Program for the Re-qualification of the Spanish University System funded by the EU NextGeneration, Margarita Salas modality, by the University of Murcia.

Conflicts of Interest: The authors declare no conflicts of interest.

List of Contributions:

1. Reis, C.C.; Freitas, S.P.; Lorentino, C.M.A.; Fagundes, T.d.S.F.; da Matta, V.M.; dos Santos, A.L.S.; Moreira, D.d.L.; Kunigami, C.N.; Jung, E.P.; Ribeiro, L.d.O. Bioproducts from *Passiflora cincinnata* Seeds: The Brazilian Caatinga Passion Fruit. *Foods* **2023**, *12*, 2525. https://doi.org/10.3390/foods12132525.
2. Sai-Ut, S.; Kingwascharapong, P.; Mazumder, M.A.R.; Rawdkuen, S. Optimization of ethanolic extraction of phenolic antioxidants from lychee and longan seeds using response surface methodology. *Foods* **2023**, *12*, 2827. https://doi.org/10.3390/foods12152827.
3. Afifi, S.M.; Kabbash, E.M.; Berger, R.G.; Krings, U.; Esatbeyoglu, T. Comparative untargeted metabolic profiling of different parts of Citrus sinensis fruits via liquid chromatography–mass spectrometry coupled with multivariate data analyses to unravel authenticity. *Foods* **2023**, *12*, 579. https://doi.org/10.3390/foods12030579.
4. Salas-Millán, J.Á.; Aguayo, E.; Conesa-Bueno, A.; Aznar, A. Revalorization of Melon By-Product to Obtain a Novel Sparkling Fruity-Based Wine. *Foods* **2023**, *12*, 491. https://doi.org/10.3390/foods12030491.
5. Pinho, L.S.; Patel, B.K.; Campanella, O.H.; Rodrigues, C.E.D.C.; Favaro-Trindade, C.S. Microencapsulation of Carotenoid-Rich Extract from Guaraná Peels and Study of Microparticle Functionality through Incorporation into an Oatmeal Paste. *Foods* **2023**, *12*, 1170. https://doi.org/10.3390/foods12061170.
6. Pérez, P.; Hashemi, S.; Cano-Lamadrid, M.; Martínez-Zamora, L.; Gómez, P.A.; Artés-Hernández, F. Effect of Ultrasound and High Hydrostatic Pressure Processing on Quality and Bioactive Compounds during the Shelf Life of a Broccoli and Carrot By-Products Beverage. *Foods* **2023**, *12*, 3808. https://doi.org/10.3390/foods12203808.
7. Olennikov, D.N.; Chemposov, V.V.; Chirikova, N.K. Polymeric compounds of lingonberry waste: Characterization of antioxidant and hypolipidemic polysaccharides and polyphenol-polysaccharide conjugates from Vaccinium vitis-idaea press cake. *Foods* **2022**, *11*, 2801. https://doi.org/10.3390/foods11182801.
8. Han, S.-C.; Huang, R.-P.; Zhang, Q.-Y.; Yan, C.-Y.; Li, X.-Y.; Li, Y.-F.; He, R.-R.; Li, W.-X. Antialcohol and Hepatoprotective Effects of Tamarind Shell Extract on Ethanol-Induced Damage to HepG2 Cells and Animal Models. *Foods* **2023**, *12*, 1078. https://doi.org/10.3390/foods12051078.
9. Coelho, M.C.; Costa, C.; Roupar, D.; Silva, S.; Rodrigues, A.S.; Teixeira, J.A.; Pintado, M.E. Modulation of the gut microbiota by tomato flours obtained after conventional and ohmic heating extraction and its prebiotic properties. *Foods* **2023**, *12*, 1920. https://doi.org/10.3390/foods12091920.
10. Zhou, X.; Gong, X.; Li, X.; An, N.; He, J.; Zhou, X.; Zhao, C. The Antioxidant Activities In Vitro and In Vivo and Extraction Conditions Optimization of Defatted Walnut Kernel Extract. *Foods* **2023**, *12*, 3417. https://doi.org/10.3390/foods12183417.
11. Phumat, P.; Chaichit, S.; Potprommanee, S.; Preedalikit, W.; Sainakham, M.; Poomanee, W.; Chaiyana, W.; Kiattisin, K. Influence of Benincasa hispida Peel Extracts on Antioxidant and Anti-Aging Activities, including Molecular Docking Simulation. *Foods* **2023**, *12*, 3555. https://doi.org/10.3390/foods12193555.
12. Martínez-Zamora, L.; Cano-Lamadrid, M.; Artés-Hernández, F.; Castillejo, N. Flavonoid Extracts from Lemon By-Products as a Functional Ingredient for New Foods: A Systematic Review. *Foods* **2023**, *12*, 3687. https://doi.org/10.3390/foods12193687.

13. Chetrariu, A.; Dabija, A. Spent Grain: A Functional Ingredient for Food Applications. *Foods* **2023**, *12*, 1533. https://doi.org/10.3390/foods12071533.

References

1. FAO. *The State of Food and Agriculture: Moving Forward on Food Loss and Waste Reduction*; FAO: Rome, Italy, 2019; ISBN 9789251317891.
2. Blasi, A.; Verardi, A.; Sangiorgio, P. The zero-waste economy: From food waste to industry. In *Membrane Engineering in the Circular Economy*; Elsevier: Amsterdam, The Netherlands, 2022; pp. 63–100. [CrossRef]
3. Gómez-García, R.; Campos, D.A.; Aguilar, C.N.; Madureira, A.R.; Pintado, M. Valorisation of food agro-industrial by-products: From the past to the present and perspectives. *J. Environ. Manag.* **2021**, *299*, 113571. [CrossRef] [PubMed]
4. Cano-Lamadrid, M.; Martínez-Zamora, L.; Castillejo, N.; Artés-Hernández, F. From Pomegranate Byproducts Waste to Worth: A Review of Extraction Techniques and Potential Applications for Their Revalorization. *Foods* **2022**, *11*, 2596. [CrossRef] [PubMed]
5. Campos, D.A.; Gómez-García, R.; Vilas-Boas, A.A.; Madureira, A.R.; Pintado, M.M. Management of fruit industrial by-products—A case study on circular economy approach. *Molecules* **2020**, *25*, 320. [CrossRef] [PubMed]
6. Ben-Othman, S.; Jõudu, I.; Bhat, R. Bioactives from agri-food wastes: Present insights and future challenges. *Molecules* **2020**, *25*, 510. [CrossRef] [PubMed]
7. Dora, M.; Biswas, S.; Choudhary, S.; Nayak, R.; Irani, Z. A system-wide interdisciplinary conceptual framework for food loss and waste mitigation strategies in the supply chain. *Ind. Mark. Manag.* **2021**, *93*, 492–508. [CrossRef]
8. Liu, Z.; de Souza, T.S.; Holland, B.; Dunshea, F.; Barrow, C.; Suleria, H.A. Valorization of food waste to produce value-added products based on its bioactive compounds. *Processes* **2023**, *11*, 840. [CrossRef]

foods

MDPI

Article

Bioproducts from *Passiflora cincinnata* Seeds: The Brazilian Caatinga Passion Fruit

Carolina Cruzeiro Reis [1], Suely Pereira Freitas [1], Carolline Margot Albanez Lorentino [2], Thayssa da Silva Ferreira Fagundes [3], Virgínia Martins da Matta [4], André Luis Souza dos Santos [2], Davyson de Lima Moreira [3,5,*], Claudete Norie Kunigami [6], Eliane Przytyk Jung [6] and Leilson de Oliveira Ribeiro [6,*]

1 Laboratory of Vegetable Oil, Federal University of Rio de Janeiro, Rio de Janeiro 21941-909, Brazil; carolinacruzeiror@gmail.com (C.C.R.); freitasp@eq.ufrj.br (S.P.F.)

2 Laboratory for Advanced Studies of Emerging and Resistant Microorganisms, Federal University of Rio de Janeiro, Rio de Janeiro 21941-902, Brazil; carolline.margot@gmail.com (C.M.A.L.); andre@micro.ufrj.br (A.L.S.d.S.)

3 Laboratory of Natural Products, Rio de Janeiro Botanical Garden Research Institute, Rio de Janeiro 22460-030, Brazil; thayssafagundes@id.uff.br

4 Food Engineering Department, Embrapa Agroindústria de Alimentos, Rio de Janeiro 23020-470, Brazil; virginia.matta@embrapa.br

5 Post-Graduation Program in Translational Drugs and Medicines, Institute of Technology in Medicines, Oswaldo Cruz Foundation, Rio de Janeiro 21040-900, Brazil

6 Laboratory of Organic and Inorganic Chemical Analysis, National Institute of Technology, Rio de Janeiro 20081-312, Brazil; claudete.kunigami@int.gov.br (C.N.K.); eliane.jung@int.gov.br (E.P.J.)

* Correspondence: davysonmoreira@jbrj.gov.br (D.d.L.M.); leilson.oliveira@int.gov.br (L.d.O.R.)

Abstract: The present work aimed to obtain bioproducts from *Passiflora cincinnata* seeds, the Brazilian Caatinga passion fruit, as well as to determine their physical, chemical and biological properties. The seeds were pressed in a continuous press to obtain the oil, which showed an oxidative stability of 5.37 h and a fatty profile rich in linoleic acid. The defatted seeds were evaluated for the recovery of antioxidant compounds by a central rotation experimental design, varying temperature (32–74 °C), ethanol (13–97%) and solid–liquid ratio (1:10–1:60 m/v). The best operational condition (74 °C, 58% ethanol, 1:48) yielded an extract composed mainly of lignans, which showed antioxidant capacity and antimicrobial activity against Gram-positive and Gram-negative bacteria. The microencapsulation of linoleic acid-rich oil through spray drying has proven to be an effective method for protecting the oil. Furthermore, the addition of the antioxidant extract to the formulation increased the oxidative stability of the product to 30% (6.97 h), compared to microencapsulated oil without the addition of the antioxidant extract (5.27 h). The microparticles also exhibited favorable technological characteristics, such as low hygroscopicity and high water solubility. Thus, it was possible to obtain three bioproducts from the Brazilian Caatinga passion fruit seeds: the oil rich in linoleic acid (an essential fatty acid), antioxidant extract from the defatted seeds and the oil microparticles added from the antioxidant extract.

Keywords: vegetable oil; antioxidant extract; spray dryer; microparticles; antibacterial activity

Citation: Reis, C.C.; Freitas, S.P.; Lorentino, C.M.A.; Fagundes, T.d.S.F.; da Matta, V.M.; dos Santos, A.L.S.; Moreira, D.d.L.; Kunigami, C.N.; Jung, E.P.; Ribeiro, L.d.O. Bioproducts from *Passiflora cincinnata* Seeds: The Brazilian Caatinga Passion Fruit. *Foods* **2023**, *12*, 2525. https://doi.org/10.3390/foods12132525

Academic Editors: Noelia Castillejo Montoya and Theodoros Varzakas

Received: 22 May 2023
Revised: 9 June 2023
Accepted: 26 June 2023
Published: 29 June 2023

1. Introduction

The passion fruit is native to Latin America. The most used commercial species are the yellow passion fruit (*Passiflora edulis* Curtis) and the purple passion fruit (*Passiflora edulis* Sims). However, although less commercially explored, species such as *P. alata*, *P. quadrangularis* L., *P. cincinnata* Mast., and *P. mollisima* Bailey present unique sensory, chemical, and technological characteristics, increasing the fruit's potential for edible and cosmetic purposes. Passion fruit juice and pulp are the main products derived from its processing. They can be used as ingredient in ice cream, jams, or even consumed *in natura* [1–3]. According to the Brazilian

Institute of Geography and Statistics (IBGE), the global production of passion fruit was estimated to be 1.5 million tons, with Brazil being the largest producer, having produced about 1 million tons in 2021 [4].

Passiflora cincinnata is a native species to Brazil, cultivated in the North, Northeast and Southeast regions of the country. It is also found in other countries such as Argentina, Bolivia, Colombia, Paraguay, and Venezuela. This species is known as "maracujá-do-mato" or "maracujá-da-caatinga" and it has been mainly exploited by rural families'cooperatives in Bahia State (Northeast) in an extractive way. Its fruit is composed of 34% pulp, 26% seed and 40% peel [5]. Thus, its processing can generate 66% residues. As *Passiflora cincinnata* exhibits higher tolerance to water stress and pests when compared to other *Passiflora* species, this has motivated the Brazilian Agricultural Research Corporation, Embrapa, to develop cultivars with higher productivity and yield, and which contribute to the strengthening of the passion fruit agro-industrial chain in Brazil [6]. Thus, an increase in the generation of waste is envisioned, requiring studies that evaluate its potential in obtaining bioproducts with higher added value. Herein, it is important to highlight that studies with the residue of this species are still scarce in the literature, being restricted to the extraction of oil from seeds.

The use of passion fruit seeds to obtain oil has been the subject of much research since it represents a good way to add value to the agro chain of this fruit, as reported by Reis et al. [7]. Authors obtained oil of *Passiflora setacea*, *Passiflora alata*, and *Passiflora tenuifila* with increased oxidative stability by adding a hydroethanolic extract from fruit seeds. However, an optimization study to recover antioxidant compounds from defatted seeds has not been performed. Furthermore, both the composition of the extracts and the fatty acid profile of the oils have not been evaluated.

Products based on passion fruit seed oil are commercialized with functional appeal, which enhances its use and justifies the improvement of the extraction process for different *Passiflora* species and the application of alternative technologies to retain the oxidative stability of the oil and expand its use in the food and cosmetics industries [2,7].

Passion fruit seed oil contains polyunsaturated fatty acids, which are of interest to the pharmaceutical, cosmetic, and food industries. Moreover, bioactive compounds, mainly phenolics with antioxidant and anti-inflammatory properties, have been identified in oils of various *Passiflora* species [8,9]. Passion fruit seed oil contains 60–70% unsaturated fatty acids, consisting mainly of linoleic acid (C18:2), an essential fatty acid involved in cellular functions and in the formation of other acids in human metabolism [10,11]. In this way, it has a high nutritional value. On the other hand, linoleic acid is very susceptible to oxidation due to the low oxidative stability of polyunsaturated fatty acids.

Microencapsulation by spray drying is an already consolidated technique in food preservation and has been recently evaluated in the processing of edible oils. Due to its flexibility, using a spray dryer for microencapsulation of oils shows advantages when operating on a large scale. Additionally, the wall material used for emulsion preparation protects the encapsulated oil from light, oxygen, and moisture, for example [12,13]. In their previous study, Reis et al. [7] obtained antioxidant emulsions consisting of a mixture of oils from different passion fruit seeds (*Passiflora setacea*, *Passiflora alata*, and *Passiflora tenuifila*) and antioxidant extracts from their defatted seeds. Although these emulsions have shown higher oxidative stability when compared to pure oil, they are still very susceptible to oxidation. Thus, the use of microencapsulation can guarantee the chemical stability of both oil and antioxidant compounds. In addition to extending the shelf-life of the obtained products, the elaborated powders take up less space during storage when compared to liquid products such as emulsions and extracts, thus being a current approach to obtain bioproducts from *Passiflora cincinnata* seeds [13]. To our knowledge, this approach has not been evaluated for the biorefinery of seeds of this passion fruit species.

Therefore, the present work aimed to obtain bioproducts from passion fruit seeds (*P. cincinnata*) by lipid extraction, recovery of antioxidant compounds from defatted seeds and microencapsulation of seeds oil with antioxidant extract, as well as to determine the physical, chemical and biological properties of the obtained bioproducts.

2. Materials and Methods

2.1. Material

The seeds of *P. cincinnata* used in this work were obtained from fruits grown in experimental fields of Embrapa Semiarid (Petrolina, Brazil) after depulping. Seeds were stored at −20 °C until their processing. After thawing, the seeds were washed in a sieve under running water to remove the arils. Then, they were autoclaved at 120 °C for 20 min to reduce the microbial load.

2.2. Seeds Oil Extraction

The seeds were dried to reach about 10% moisture. After that, the raw material was ground in a knife mill (MA-048, Marconi, Piracicaba, SP, Brazil). Crushed passion fruit seeds were processed in continuous press (CA59G, Oekotec, Mönchengladbach, NRW, Germany) to obtain the first-extraction crude oil decanted for 24 h in a dark cabinet. The oil was centrifuged at 3000 rpm (CR22N, Hitachi Koki, Minato, Tokyo, Japan) for 15 min to separate very fine particles. Then, the clarified seeds oil was stored at −20 °C until use, and the pressed cake (defatted seeds) was reserved for antioxidant compound extraction.

The oil extraction yield was calculated from the ratio between the mass of oil recovered by cold pressing (m) and the mass of oil extracted by Soxhlet (M).

$$Y\ (\%) = \frac{m}{M} * 100 \quad (1)$$

2.2.1. Fatty Acid Composition

The fatty acid composition of *P. cincinnata* seeds oil samples was analyzed using gas chromatograph–mass spectrometry (GC–MS) (6890-5975, Agilent, Santa Clara, CA, USA). Fatty acid methyl esters were prepared and analyzed in accordance with Jung et al. [14], and the samples were injected into a Carbowax DB23 (60 m × 0.25 mm i.d. × 250 μm film) column; helium as carrier gas flowed at 1 mL min^{-1}. The inlet temperature was 230 °C, the injection volume was 1 μL with a split ratio of 50:1, and the oven temperature was 200 °C for 40 min (isothermal). The transfer line temperature was 250 °C. The detected compounds were identified by matching their mass spectra with the reference spectra in the Willey7Nist05 library. The results were expressed in percentages based on the area of the chromatogram peaks without correction.

2.2.2. Oxidative Stability

The oxidative stability of *P. cincinnata* seeds oil was measured using the Rancimat® 743 (743, Metrohm, Riverview, FL, USA) equipment, passing a stream of purified air of 10 L h^{-1} at 110 °C through the oil sample (3 g). Results were expressed as an induction period representing a time interval until the sample reaches a high oxidation level.

2.3. Antioxidant Compound Recovery from Defatted Seeds

The *P. cincinnata* defatted seeds evaluated in the extraction of the antioxidant compounds were ground in a knife mill coupled to a 1 mm diameter circular mesh sieve. A rotational central composite design with eight factorials, six axials and three central points was employed to select the best extraction parameters. The selected independent variables were the concentration of ethanol (Pershy Chemical's, São Gonçalo, RJ, Brazil) as solvent, temperature, and solid–liquid ratio, according to Table 1. The temperature was limited to 74 °C to avoid loss of solvent. The extraction of antioxidant compounds was performed by agitated solvent extraction, using 125 mL glass flasks properly covered and heated for 1 h under constant stirring of 150 rpm, conditions based on preliminary studies and literature data [15,16]. The experimental data were analyzed by response surface methodology using a second order polynomial equation. Analysis of variance (ANOVA) to test for the lack of fit and coefficient of determination (R^2) were used to verify the model's significance.

Table 1. Total phenolic compounds content and antioxidant capacity of the hydroethanolic extract of *Passiflora cincinnata* defatted seeds obtained under different experimental conditions.

Trials	Temperature (°C)	Ethanol (%)	Solid/Liquid Ratio (g mL^{-1})	TPC [1]	ABTS$^{\bullet+}$ [2]	DPPH$^{\bullet}$ [2]	FRAP [3]
1	40 (−1)	30 (−1)	1:20 (−1)	845 ± 25	43 ± 2	111 ± 3	223 ± 8
2	40 (−1)	30 (−1)	1:50 (+1)	912 ± 31	51 ± 1	107 ± 5	216 ± 4
3	40 (−1)	80 (+1)	1:20 (−1)	1016 ± 27	64 ± 3	148 ± 6	284 ± 4
4	40 (−1)	80 (+1)	1:50 (+1)	1220 ± 25	69 ± 1	153 ± 6	304 ± 6
5	65 (+1)	30 (−1)	1:20 (−1)	1371 ± 57	80 ± 2	182 ± 8	395 ± 8
6	65 (+1)	30 (−1)	1:50 (+1)	1799 ± 119	105 ± 2	214 ± 12	462 ± 19
7	65 (+1)	80 (+1)	1:20 (−1)	1870 ± 35	114 ± 2	280 ± 11	567 ± 18
8	65 (+1)	80 (+1)	1:50 (+1)	2302 ± 31	130 ± 2	292 ± 10	614 ± 17
9	32 (−1.68)	55 (0)	1:35 (0)	991 ± 42	63 ± 4	128 ± 7	303 ± 6
10	74 (+1.68)	55 (0)	1:35 (0)	2538 ± 141	178 ± 6	370 ± 20	795 ± 5
11	53 (0)	13 (−1.68)	1:35 (0)	720 ± 18	45 ± 1	92 ± 5	211 ± 2
12	53 (0)	97 (+1.68)	1:35 (0)	377 ± 17	21 ± 1	41 ± 0	103 ± 1
13	53 (0)	55 (0)	1:10 (−1.68)	1451 ± 69	99 ± 4	210 ± 5	474 ± 17
14	53 (0)	55 (0)	1:60 (+1.68)	1809 ± 80	111 ± 8	226 ± 5	462 ± 16
15	53 (0)	55 (0)	1:35 (0)	1810 ± 124	121 ± 1	235 ± 1	515 ± 6
16	53 (0)	55 (0)	1:35 (0)	1616 ± 57	117 ± 1	222 ± 4	527 ± 1
17	53 (0)	55 (0)	1:35 (0)	1637 ± 64	110 ± 3	234 ± 3	504 ± 3

The coded values of the independent variables are in parentheses. Results expressed as mean ± standard deviation (coefficient of variation < 10%). TPC—Total phenolic compounds; [1] Results expressed as mg GAE 100 g^{-1} of sample; [2] Results expressed as μmol Troloxg^{-1} of sample; [3] Results expressed as μmol Fe^{2+} g^{-1} of sample.

To determine the best condition for extraction of the antioxidant compounds from *P. cincinnata* defatted seeds, the technique of simultaneous optimization of independent variables (desirability) was used. The desirability function is based on the conversion of each response in an individual desirability (d). After that, they were combined into an overall desirability (D), using the geometric mean. The D value ranges from zero (0) to one (1), in which the value of 1 corresponds to the most desirable response [17]. Under the best operational condition, more assays were performed and observed results were compared with those predicted in order to validate the chosen operational condition.

2.3.1. Total Phenolic Content

Total phenolic content (TPC) was measured according to the method described by Singleton and Rossi [18]. For the assays, 250 μL of each diluted sample was mixed with 1250 μL of 10% Folin–Ciocalteu reagent (Imbralab, Ribeirão Preto, Brazil) for 2 min after 1000 μL of sodium carbonate solution (7.5% *w*/*v*–Dinâmica, Indaiatuba, Brazil) was added. The mixture was incubated for 15 min at 50 °C. Then, the absorbance was measured at 760 nm in a spectrophotometer (5100, Metash, Songjiang District, Shanghai China) vs. a blank prepared with distilled water. Gallic acid (Sigma-Aldrich, Saint Louis, MO, USA) was used as a standard, and the results were expressed as mg gallic acid equivalents per 100 g of sample (mg GAE 100 g^{-1}).

2.3.2. Evaluation of the Antioxidant Capacity

- ABTS$^{\bullet+}$

The ABTS$^{\bullet+}$ antiradical activity (2,2-azino-bis-(3-ethylbenzthiazoline-6-sulfonic acid-Sigma-Aldrich, Saint Louis, MO, USA) was determined according to methodology reported by Gião et al. [19]. For the reactions, 30 μL of each sample was added to 3 mL of ABTS$^{\bullet+}$. The absorbance was measured at 734 nm in a spectrophotometer (5100, Metash, Songjiang District, Shanghai, China) after 6 min of reaction, using ultra-pure water as blank. The results were expressed as μmol of Trolox g^{-1}of sample (Sigma-Aldrich, Buchs, Switzerland).

- DPPH•

The DPPH• radical (Sigma-Aldrich, Steinheim, Germany) scavenging activity was determined according to Hidalgo et al. [20]. Briefly, 100 μL of each sample was added to 2.9 mL of DPPH• solution (6×10^{-5} M in methanol–Vetec, Rio de Janeiro, Brazil) for 30 min The absorbance was measured in a spectrophotometer (5100, Metash, Songjiang District, Shanghai, China) at 517 nm using methanol as a blank. The DPPH• radical scavenging activity was calculated using Trolox solution (Sigma-Aldrich, Buchs, Switzerland) as a standard, and the results were expressed as μmol of Trolox g^{-1} of sample.

- Ferric reducing/antioxidant power (FRAP)

The FRAP assay was performed according to Benzie and Strain [21] with slight modifications. Briefly, three stock solutions were made, 300 mM acetate buffer (pH 3.6), 10 mM TPTZ (Sigma-Aldrich, Buchs, Switzerland) in 40 mM HCl (Isofar, Duque de Caxias, Brazil) and 20 mM $FeCl_3$ (Neon, São Paulo, Brazil); for the analysis 25 mL of acetate buffer, 2.5 mL of TPTZ solution, and 2.5 mL of $FeCl_3$ were mixed. Then, 100 μL of each sample was reacted with 3 mL of FRAP at 37 °C for 30 min. The absorbance was measured in a spectrophotometer (5100, Metash, Songjiang District, Shanghai, China) at 593 nm using ultra-pure water as a blank. The ferric ion reducing ability was calculated using $FeSO_4 \cdot 7H_2O$ (CRQ, Diadema, Brazil) as standard, and the results were expressed as μmol Fe^{2+} $g^{-1}$of sample.

2.3.3. BiocompoundsProfile by UPLC–MS/MS

The defatted seeds extract was analyzed by UPLC–MS/MS using a Nexera UFLC system (Shimadzu) coupled to a high-resolution mass spectrometer with electron spray ionization (ESI), QTOF–MS Impact (Bruker, Billerica, MA, USA). A 1 μL sample (5 mg mL^{-1}-HPLC grade methanol, Tedia, São Paulo, Brazil) was injected into an Acquity BEH C18 column (100 mm × 2.1 mm i.d., 1.7 μm particle size) at 40 °C with a constant flow rate of 0.3 mL min^{-1}. Mobile phases A (0.1% formic acid in ultrapure water, MilliQ, Merck, Darmstadt, Germany) and B (acetonitrile with 0.1% formic acid) were used in a gradient elution system as follows: the eluent was maintained at 5% B for the first 5 min, followed by a gradient from 5 to 100% B until 39 min, and maintained at 100% B until 44 min. Mass spectra were acquired in positive and negative ion modes, and the analyte was monitored in the m/z range of 100 to 2000.

The ion source parameters were adjusted under the following conditions: capillary voltage of 4500 V, nebulizer gas pressure of 4.0 bar, desolvation gas flow rate of 10.0 L min^{-1}, transfer capillary temperature at 200 °C, and an entrance voltage of −500 V applied to the spectrometer (end plate offset). MS^2 spectra were acquired in auto MS/MS mode. Data processing was performed using Data Analysis 4.2 software (Bruker). Compound analysis was performed considering the pseudomolecular ion (negative or positive) and mass fragmentation pattern (MS^2) in manual dereplication and/or comparison with databases (MassBank).

2.3.4. Antimicrobial Assay

In this set of experiments, Gram-negative (*Acinetobacter baumannii* ATCC 19606, *Escherichia coli* ATCC 25922, *Klebsiella pneumoniae* ATCC 13883, and *Pseudomonas aeruginosa* ATCC 27853) and Gram-positive (*Staphylococcus aureus* ATCC 29213, *Staphylococcus epidermidis* ATCC 12228 and *Bacillus subtilis* 168 LMD 74.6) bacteria were grown in Mueller–Hinton agar (Difco, Franklin Lakes, NJ, USA) for 24 h at 35 ± 2 °C. Sabouraud–dextrose agar (Difco, Franklin Lakes, NJ, USA) was used for the cultivation of yeasts, *Candida albicans* ATCC 90028 and *Candida tropicalis* ATCC 750 for 24 h at 35 ± 2 °C. Antimicrobial activity was evaluated using the broth microdilution method with 96-well polystyrene plates, standardized according to documents M07-A9 (for bacterial assays) and M27-A3 (for fungal assays). To determine the minimum bactericidal and fungicidal concentration (MBC and MFC, respectively), 10 μL of the wells that had no visible microbial growth were inoculated in Mueller–Hinton culture medium and Sabouraud-dextrose Agar for 24 h at 37 °C. The MBC and MFC were considered to be the lowest concentration capable of completely inhibiting microbial growth on the agar surface. To eliminate the interference of ethanol in the results, the hydroalcoholic antioxidant extract was subjected to vacuum evaporation. The results were expressed as mg GAE mL^{-1} of ethanol-free extract.

2.4. *Microencapsulation by Spray Dryer*

To evaluate the properties of the microparticles obtained by spray drying, the assays were carried out with three different formulations: (i) oil dispersed in the aqueous phase containing wall material (ii) a mixture of oil and defatted seeds' ethanol-free antioxidant extract dispersed in the aqueous phase containing wall material and (iii) wall material dispersed in the aqueous phase only.

An oil-in-water emulsion was prepared using 7.5% pressed oil, 7.5% maltodextrin (10DE, MOR-REX 1910, Ingredion, São Paulo, SP, Brazil) and 22.5% modified starch (Capsul®06560101CE, Ingredion, São Paulo, SP, Brazil). Modified starch and maltodextrin 10DE were used as wall material in the ratio of 3:1 (*w*/*w*), respectively. The emulsion was homogenized in a blender for 2 min. These operational conditions were adapted from James et al. [22].

The spray dryer was operated under the following conditions: chamber inlet temperature and air velocity of 170 °C and 3 m s^{-1}, respectively, emulsion inlet flow rate of 485 mL h^{-1} and 0.7 mm diameter atomization nozzle.

2.4.1. Microparticles' Oxidative Stability

The oxidative stability of microparticles was measured using a Rancimat® 743 (743, Metrohm, USA), by passing a stream of purified air of 10 L h^{-1} at 110 °C through the oil sample (3 g). Results were expressed as an induction period representing a time interval until the sample reaches a high oxidation level.

2.4.2. X-ray Diffraction

X-ray diffraction (XDR) was performed using a diffractometer (Miniflex, Rigaku, Japan). Samples of the wall material, microencapsulated oil, containing antioxidant extract from defatted seeds or not, were directly analyzed using the equipment. Cu Kα radiation was employed with 2θ ranging from 10° to 80°, with a scan speed of 0.06° 2θ s^{-1}.

2.4.3. Morphology of Microparticles

The morphology of microparticles was analyzed by scanning electron microscopy (SEM) (TM303 plus, Hitachi, Japan). The powder samples were mounted on a specimen holder using double scotch tape under vacuum. The microscope was operated at 15 kV.

2.4.4. Fourier Transform Infrared Spectroscopy (FTIR)

The analysis of the chemical structures was performed by FTIR (Nexus 470, Themo Nicolet, USA). The sample powder was blended with spectroscopy-grade KBr (Vetec, Rio de Janeiro, Brazil) and pressed into pellets, and the oil was dripped in a KBr window. The FTIR spectrum of samples was measured from 4000 to 400 cm^{-1}, with a resolution of 4 cm^{-1} and 32 accumulations.

2.4.5. Particle Size

A particle size analyzer was used to determine the size distribution of microparticles using laser diffraction (1064, Cilas, France) coupled with ultrasound to increase the dispersibility of the sample. A small amount of powder was suspended in isopropanol and submitted to five particle size distribution readings. The spread of particle sizes was calculated as the scattering according to Equation (2):

$$\text{span} = \left(\frac{D_{90} - D_{10}}{D_{50}}\right) \quad (2)$$

where D_{10}, D_{50}, D_{90} are defined as diameters corresponding to the 10, 50, and 90% cumulative volumes, respectively. The results were expressed in micrometers.

2.4.6. Moisture Content

The moisture content of microparticles was determined by the gravimetric method at 105 °C. The results were expressed as a percentage [23].

2.4.7. Hygroscopicity

Hygroscopicity was determined according to Cai and Corke [24]. Approximately 1 g of sample was placed at 25 ± 2 °C in a desiccator with a saturated NaCl solution (73% relative humidity), and the weight gain due to moisture absorption was measured after one week. The results were expressed in absorbed moisture percentage.

2.4.8. Solubility

The solubility of the microparticles in water was evaluated according to Cano-Chauca et al. [25]. One gram of sample was added to a beaker containing 100 mL distilled water; which was stirred at high speed for 5 min, followed by centrifugation at 3000× *g* for 15 min. After that, a 25 mL aliquot of the supernatant was dried at 105 °C until constant weight. The solubility was calculated by the weight difference and expressed as a percentage.

2.5. Statistical Analysis

All analytical determinations were carried out at least in triplicate, except the RANCIMAT and the UPLC-MS/MS trials. Analysis of variance (ANOVA) followed by Fisher's LSD test was performed using Statistica® v.13.0.

3. Results and Discussion

3.1. Oil Extraction and Characterization

3.1.1. Yield

The oil content obtained in a lipid extractor using petroleum ether as solvent was about 15%. This result was used as the basis for calculating the efficiency of the process. Lopes et al. [26] found values between 16.7 and 19.2% for lipid content of *P. cincinnata* seeds. The trials were performed with fruits from different accessions, and the oil extraction was conducted in a lipid extractor after moisture removal. The lipid content in *P. cincinnata* was significantly different compared to other passion fruit species. The value obtained was lower when compared with data reported by De Paula et al. [27] for *P. setacea* (31 to 34% of oil/93% efficiency of pressing) and *P. alata* (23% of oil/84% efficiency of pressing) species. Extraction of *P. cincinnata* seeds oil by pressing showed an efficiency of 79%, using seeds with 11% moisture content. The moisture content was selected from preliminary experiments and literature data [28].

3.1.2. Fatty Acid Profile

The fatty acid profile of *P. cincinnata* seeds oil is shown in Table 2. The seeds oil is rich in unsaturated fatty acids, represented mainly by linoleic acid. The observed profile is similar to those reported in the literature by Araujo et al. [5] and Lopes et al. [26]. Linoleic acid is an essential fatty acid since human metabolism does not synthesize it. Therefore, it must be obtained through food intake. It is important to stress that the consumption of polyunsaturated fatty acids may decrease the risk of developing cardiovascular diseases [29]. However, due to the unsaturation in their structure, polyunsaturated fatty acids are very susceptible to oxidation.

Table 2. Fatty acid profile of *Passiflora cincinnata* seeds oil expressed as percentage (%).

Fatty Acids	Present Work	Lopes et al. [25]	Araújo et al. [5]
Palmitic (C16:0)	12.14 ± 1.00	10.2	9.2
Stearic (C 18:0)	1.09 ± 0.26	2.9	3.0
Oleic (C 18:1)	8.43 ± 1.32	11.3	15.4
Linoleic (C 18:2)	78.34 ± 2.22	74.3	70.3
Linolenic (C 18:3)	-	0.6	0.6

3.2. Antioxidant Compound Recovery from Defatted Seeds

3.2.1. Evaluation of the Extraction Process

The TPC content and antioxidant capacity of extracts from *P. cincinnata* defatted seeds are shown in Table 1. Trial 10 performed at 74 °C, 55% ethanol as solvent, and 1:35 g mL^{-1} solid–liquid ratio, presented the highest values for TPC, ABTS$^{\bullet+}$, DPPH$^{\bullet}$, and FRAP
(2538 mg GAE 100 g^{-1} of sample, 178 µmol Trolox g^{-1} of sample, 370 µmol Trolox g^{-1} of sample, and 795 µmol Fe^{2+} g^{-1} of sample, respectively). The lowest values were found in trial 12 at 53 °C, 97% ethanol in the same solid–liquid ratio (377 mg GAE 100 g^{-1}of sample, 21 µmol Trolox g^{-1} of sample, 41 µmol Trolox g^{-1} of sample, 103 µmol Fe^{2+} g^{-1} of sample, respectively). Thus, the results obtained in trial 10 were 6.7, 8.5, 9 and 7.7 times higher than those observed in trial 12, showing that the responses were strongly influenced by temperature and the percentage of ethanol employed in the extraction. Additionally, as can be seen in Table S1, the results of the experimental design showed a positive correlation ($p < 0.05$), indicating that the increase in the concentration of total phenolic compounds produces an extract with higher antioxidant capacity as measured by different methods.

The temperature was the most relevant parameter, as shown in the Pareto diagram (Figure 1). Its linear effect was significant and positive. In this way, higher temperatures favor the solubility of phenolic compounds and decrease the viscosity of the extraction system, leading to a better extraction yield [16,30]. Notably, this behavior was also observed for the antioxidant capacity of the extracts regardless of the method used (Table 1, Figure S1).

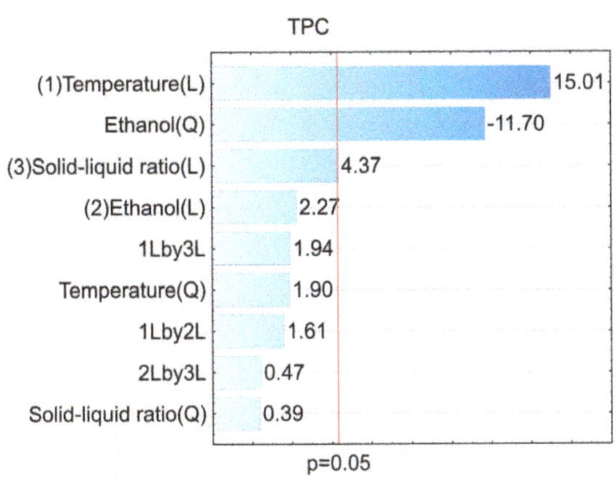

Figure 1. Influence of the independent variables on the total phenolic compounds in the extract of *Passiflora cincinnata* defatted seeds.

Leal et al. [31] obtained extract of *P. cincinnata* seeds with a total phenolic content of about 2528 mg mg GAE 100 g^{-1}, when the assays were conducted at room temperature, 95% ethanol, for 72 h. In the current study, the value found at 74°C, 55% ethanol, solid–liquid ratio of 1:35 for 1 h was similar (2538 mg mg GAE 100 g^{-1} 100 g^{-1} of sample). However, the selected operational conditions present as advantages a shorter extraction time and a lower ethanol concentration in the extraction solution.

Table 1 shows that extraction solutions with intermediate polarity (between 55 to 80% ethanol) were more suitable for extracting antioxidant compounds from defatted seeds. This occurs as a function of the polarity of the recovered compounds. The quadratic effect of this factor was significant (Figure 1), corroborating that there is a maximum value of ethanol concentration which promotes higher attainment of phenolic compounds and improves the antioxidant capacity of the extracts, which can also be seen in the Pareto diagrams for $ABTS^{\bullet+}$, $DPPH^{\bullet}$ and FRAP assays (Figure S1). Shi et al. [32], when extracting phenolic compounds from grape seeds, reported that the best ethanol concentration in the extraction solution was between 55% and 65%. Alcântara et al. [33] evaluated the extraction of phenolic compounds in chia seeds using water, ethanol and acetone in different proportions. In this study, the authors observed that the polarity of the extraction solution significantly influenced the recovery of phenolic compounds. The best solvent mixture was prepared with 17% water, 17% ethanol and 66% acetone.

The solid–liquid ratio was significant for TPC and FRAP responses (Figures 1 and S1). However, this factor made a low contribution to the results. From Table 1, it is possible to verify that results obtained by varying only the solid–liquid ratio, as in the trials 1 and 2, 3 and 4, 5 and 6, 7 and 8 and 13 and 14, the increment in the total phenolic compounds concentration ranged from 1.1 to 1.3 times only (trials 1 and 2; 5 and 6).

All models were significant for predicting the behavior of the responses in relation to the independent variables, as the calculated F-values were higher than the listed F-values ($F_{9,7} = 3.68$) at $p = 0.05$. The calculated F-values for TPC, $ABTS^{\bullet+}$, $DPPH^{\bullet}$ and FRAP responses were 10.2, 11.3, 8.5, and 11.8, respectively. However, some lack of fit was observed for $DPPH^{\bullet}$ and FRAP responses as it showed p-value < 0.05, and the calculated F-values were lower than the listed F-value for this parameter. The R^2 values of the fitted models were 0.93, 0.94, 0.91, 0.94 for TPC, $ABTS^{\bullet+}$, $DPPH^{\bullet}$ and FRAP responses, respectively, showing that the models accounted for at least 91% of data variability obtained by this experimental design. The R^2 of the adjusted models was superior to 0.81, reinforcing the good fit of the data, although the $DPPH^{\bullet}$ and FRAP responses have shown some lack of fit. As can be seen in Figure S2, which describe the values observed and those predicted, there is a good agreement between them, making it possible to use these responses to choose the best operational condition to recover antioxidant compounds from the residue.

The desirability tool was employed to obtain the best operational condition for recovering antioxidant compounds from defatted seeds. In this case, the overall desirability was 0.9985 (Figure 2); the closer to 1, the more statistically reliable the result. It indicated as the best operational condition 74 °C, 58% hydroethanolic solution and a solid–liquid ratio of 1:54. However, in order to save solvent in the process, we selected a solid–liquid ratio lower than that estimated by the statistical tool (1:54), since a solid–liquid ratio higher than 1:48 does not significantly increase the recovery of antioxidant compounds from residue. In the interval of the present work (1:48–1:60/Figure 2), there was no significant increase in the recovery of antioxidant compounds from defatted seeds that would justify the increase in solvent in the process. Therefore, the solid–liquid ratio of 1:48 was adopted in the current study. Thus, the antioxidant compound-rich extract constitutes another bioproduct obtained from the *P. cincinnata* seeds.

In this operational condition, observed values of TPC (2868 mg GAE 100 g^{-1} of sample) and antioxidant capacity by $ABTS^{\bullet+}$ (195 μmol Trolox g^{-1} of sample), $DPPH^{\bullet}$ (368 μmol Trolox g^{-1} of sample), and FRAP assays (694 μmol Fe^{2+} g^{-1} of sample) were close to the predicted values by the models as follows: TPC (2952 mg GAE 100 g^{-1} of sample) and antioxidant capacity by $ABTS^{\bullet+}$ (177 μmol Trolox g^{-1} of sample), $DPPH^{\bullet}$ (383 μmol Trolox g^{-1} of sample) and FRAP assays (795 μmol Fe^{2+} g^{-1} of sample) with coefficients of variation less than 10%. The dataset reinforces the operational condition selected for the recovery of antioxidant compounds from *P. cincinnata* defatted seeds.

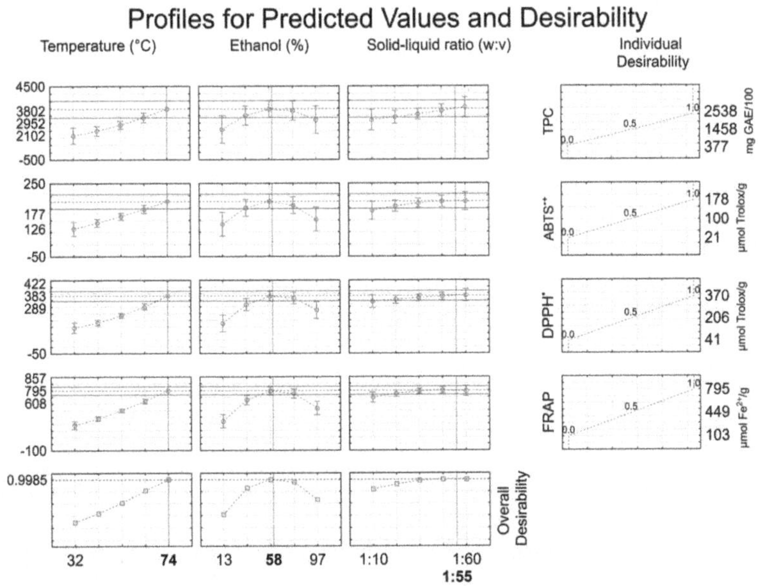

Figure 2. Profiles for predicted values and overall desirability for independent variables temperature, ethanol concentration and solid–liquid ratio employed in the extraction of antioxidant compounds from *Passiflora cincinnata* defatted seeds.

3.2.2. Biocompound Profile

The chemical composition of the crude extract from *P. cincinnata* seeds was analyzed by UPLC–MS/MS (Table 3). The compound identification was performed by manual dereplication, allowing the putative annotation of eight new lignans in the genus *Passiflora* (**1**–**8**). The base peak chromatographic profile in negative ionization mode can be seen in Figure 3. The MS/MS fragmentation spectra of the major compounds showed the main product ions m/z 165, m/z 147 and m/z 135, suggesting the presence of lignans analogous to secoisolariciresinol and berchemol [34,35]. These two compounds have already been described in the genus *Passiflora* [36,37]. The MS/MS data, combined with the exact mass data of the pseudomolecular ions $[M - H]^-$ suggested 3-demethoxy derived compounds.

The main fragment m/z 165 ($C_9H_9O_3^-$) results from the cleavage of the C8-C8′-carbon, followed by loss of CH_4. The two major lignans (m/z 327.1280 and m/z 329.1437) were identified as tetrahydrofuranolignans (**2** and **3**). The other six lignans comprise glycosylated structures, including one tetrahydrofurano lignan (**1**) and five butanediol lignans (**4**–**8**). In addition to lignans, two glycosylated flavonoids were identified in positive ionization mode. Comparison with MS/MS fragmentation databases allowed the annotation of isovitexin 2″-O-arabinoside (**9**, m/z 565.1555, $[M + H]^+$) and adonivernith (**10**, m/z 581.1507, $[M + H]^+$) (Figure 4). The elimination of 132 e 252 u revealed the pentose and hexose moiety; however, the glycan stereochemistry could not be determined by LC–MS. Different glycosylated flavonoids have already been reported for *P. cincinnata* [31]. All the annotated compounds have phenolic hydroxyl groups which may contribute to the antioxidant capacity of the extract [38].

Table 3. Compounds annotated by UPLC–MS/MS (ESI positive and negative modes) in *Passiflora cincinnata* defatted seeds extract.

Compound Name	T_R (min)	Molecular Formula	Adduct Ion	Experimental *m/z*	Main MS[2]Fragment Ions	Relative Percentage (%)
3-demethoxy-8-dehydroxy-berchemol 4-*O*-glucoside (**1**)	11.8	$C_{25}H_{32}O_{10}$	$[M-H]^-$	491.2162	165.0786, 147.0655, 135.0650	5.9
3-demethoxy-8-dehydroxy-berchemol (**2**)	13.2	$C_{19}H_{22}O_5$	$[M-H]^-$	329.1431	165.0793, 147.0660, 146.0572, 135.0659, 129.0540	10.7
2-(1,3-benzodioxol-5-yl)tetrahydro-4-[(4-hydroxyphenyl)methyl]-3-furanmethanol (**3**)	14.8	$C_{19}H_{20}O_5$	$[M-H]^-$	327.1280	163.0624, 162.0538, 147.0657, 135.0649,	8.5
Secoisolariciresinol 4-*O*-xylopyranoside (**4**)	15.0	$C_{25}H_{34}O_{10}$	$[M-H]^-$	493.2113	329.1427, 299.1287, 179.0591, 165.0777, 147.0656, 137.0443	10.6
3-demethoxy-secoisolariciresinol 4-*O*-glucoside (**5**)	15.5	$C_{25}H_{34}O_{10}$	$[M-H]^-$	493.2108	329.1429, 299.1292, 165.0779, 147.0661, 137.0445	9.8
2-methoxy-bisdemethoxy-secoisolariciresinol 4-*O*-glucoside (**6**)	16.0	$C_{25}H_{34}O_{10}$	$[M-H]^-$	493.2107	165.0781, 147.0658, 135.0650	4.9
2′-methoxy-bisdemethoxy-secoisolariciresinol 4-*O*-glucoside (**7**)	16.5	$C_{25}H_{34}O_{10}$	$[M-H]^-$	493.2102	165.0780, 147.0657, 135.0651	2.6
3-demethoxy-secoisolariciresinol 4-*O*-glucosil glucoside (**8**)	14.8	$C_{31}H_{44}O_{15}$	$[M-H]^-$	655.2615	165.0778, 163.0622, 162.0543	7.2
Isovitexin 2″-*O*-arabinoside (**9**)	11.5	$C_{26}H_{28}O_{14}$	$[M+H]^+$	565.1555	283.0595, 313.0702, 397.0917, 415.1010, 433.1253	<0.5
Adonivernith (**10**)	10.8	$C_{26}H_{28}O_{15}$	$[M+H]^+$	581.1507	299.0564, 329.0659, 413.0844, 431.1016, 449.1073	<0.5

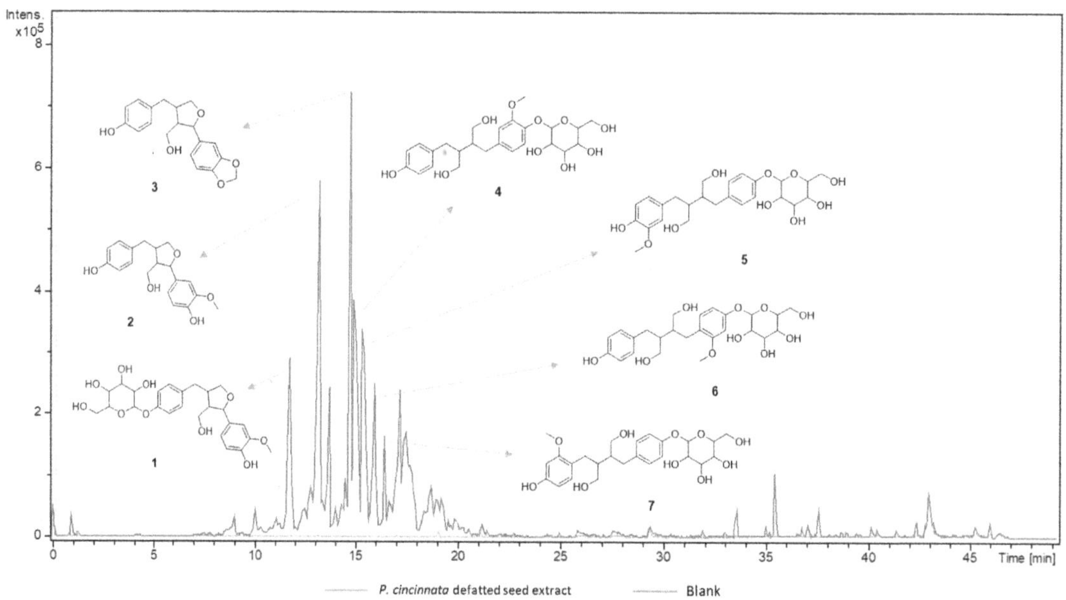

Figure 3. Base peak chromatogram (UPLC–MS/MS) in negative ionization mode of *Passiflora cincinnata* defatted seeds extract and the respective lignans (**1–7**) annotated for base peaks. The diglycosylated lignan **8** (Figure 4) is overlapped with the major ion signal at *m/z* 327.1280.

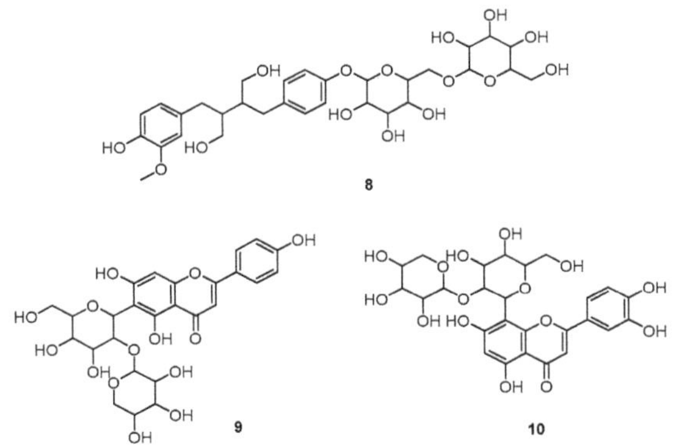

Figure 4. Minor compounds annotated in the *Passiflora cincinnata* defatted seeds extract.

3.2.3. Antimicrobial Action

The antimicrobial activity of the *P. cincinnata* defatted seeds extract (obtained in better conditions) was tested against various microorganisms, including yeasts and both Gram-positive and Gram-negative bacteria (Table 4). Initially, we performed the MIC assay. However, the extract interfered with the reading due to its high turbidity. Based on this, we decided to plate 10 μL of each system (untreated and treated with different concentrations of the extract) onto the surface of solid media to observe the bactericidal (MBC)/fungicidal (MFC) effects. The results showed that the *P. cincinnata* defatted seeds extract exhibited bactericidal effects against all tested Gram-positive bacteria (*S. aureus*, *S. epidermidis* and *B. subtilis*), with MBC values ranging from 0.302 to 0.602 mg GAE mL^{-1} (Table 4). The extract also inhibited the growth of Gram-negative bacteria, except for *P. aeruginosa*, with an MBC value of 0.602 mg GAE mL^{-1}, while the growth of fungal cells was not affected by the extract under the employed experimental conditions.

Table 4. Antimicrobial activity of *Passiflora cincinnata* defatted seeds extract.

Microorganisms	MBC/MFC [mg GAE mL^{-1}] [a]
Gram-positive bacteria	
Bacillus subtilis 168 LMD 74.6	0.602
Staphylococcus aureus ATCC 29213	0.302
Staphylococcus epidermidis ATCC 12228	0.302
Gram-negative bacteria	
Acinetobacter baumannii ATCC 19606	0.602
Escherichia coli ATCC 25922	0.602
Klebsiella pneumoniae ATCC13883	0.602
Psedomonas aeruginosa ATCC 27853	ND
Fungi	
Candida albicans ATCC 90028	ND
Candida tropicalis ATCC 750	ND

ND—not determined. [a] Results expressed as mg gallic acid equivalent/mL of ethanol-free extract. MBC—minimum bactericidal concentration. MFC—minimal fungicidal concentration.

Several studies have attributed the inhibitory effect of seed extracts against different bacteria to their phenolic compounds [15]. These compounds have the ability to bind with the bacterial cell wall and then inhibit bacterial growth. Additionally, phenolic compounds may precipitate proteins and inhibit the enzymes of microorganisms. Siebra et al. [3] studied the effect of a hydroethanolic extract of different parts of *P. cincinnata* Mast against *S. aureus* and *E. coli* and did not observe antimicrobial activity. However, the authors combined the extract of each part with an antimicrobial to reduce the resistance of these microorganisms to the antibiotics, and the results were successful for this type of application.

The antibacterial activity observed in the current work may be related to the phytochemical profile of the extract, as revealed by LC–MS/MS (lignan-rich) (Table 3) [39].

3.3. Microparticle Characterization

3.3.1. Oxidative Stability of Oil Microparticles

The induction times of *P. cincinnata* pure oil and microencapsulated *P. cincinnata* oil with and without antioxidant extract are shown in Table 5. An oxidative stability of 5.37 h for pure oil was found. This value falls within the range reported by Reis et al. [7], who evaluated the oxidative stability of *Passiflora* species seeds oils recovered by continuous pressing, including *P. alata*, *P. tenuifila*, and *P. setacea* (3.5–7.3 h). These values are typical for passion fruit seeds oils, which are rich in polyunsaturated fatty acids.

Table 5. Induction time of *P. cincinnata* seeds oil and its microparticles.

Samples	Induction Time (Hours)
Pure oil	5.37± 0.18 [b]
Oil + antioxidant extract microencapsulated	6.97± 0.73 [a]
Oil microencapsulated	5.27± 0.53 [b]

Different letters indicate significant differences.

It can also be observed that pure oil had oxidative stability increased by about 30% when it was microencapsulated after addition of the antioxidant extract. These results confirm the positive effects of bioactive compounds contained in defatted seeds on the oxidative stability of the microparticles, acting together with the wall material to protect the nutritional and chemical quality of the *P. cincinnata* seeds oil. It is essential to highlight that oil microparticles, with added (or not) antioxidant compounds from defatted seeds, are bioproducts of interest for the food and pharmaceutical industries. Therefore, their preparation can add value to the Brazilian passion fruit agro chain.

3.3.2. DRX

Figure 5 shows the diffractograms obtained for the wall material and *P. cincinnata* oil microencapsulated with and without antioxidant extract.

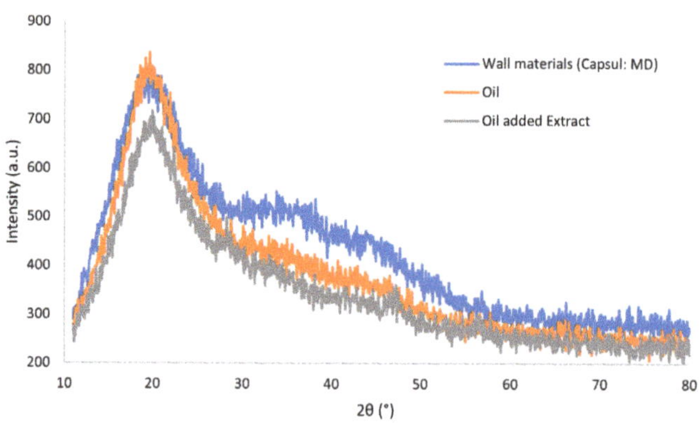

Figure 5. Diffractogram of the microparticles containing *Passiflora cincinnata* seeds oil (orange); microparticles containing *Passiflora cincinnata* seeds oil and antioxidant extract (gray); microparticles of wall material only (blue).

An amorphous profile was observed for all samples, with non-crystalline character, with only one characteristic signal near 20°. This verifies that incorporating *P. cincinnata* seeds oil and antioxidant extract in the microparticles did not influence the amorphous character typical of polysaccharides [40]. According to Pereira et al. [40], amorphous systems are better for microencapsulation, as they can form a glassy structure by removing water. Additionally, these systems dissolve more easily than those that contain crystalline components since the dissolution of the crystals occurs only on the surface exposed to the solvent. Microparticles of fish oil encapsulated with inulin, isolated whey protein, and maltodextrin by Botrel et al. [41] also exhibited an amorphous structure with a minimum of organization, based on the occurrence of significant diffuse peaks near 20°.

3.3.3. Morphology and Particle Size

As shown in Figure 6, most microparticles were spherical without cracks or fissures, ensuring better protection of the bioactive compounds (oil and phenolic compounds). Particle size distribution at different feed compositions is shown in Table 6. In general, adding antioxidant extract to feed emulsions resulted in the largest particle size ($p < 0.05$), although the distribution was more homogeneous (lowest span values, $p < 0.05$). The microparticles formed by emulsions containing oil and wall material and those containing only wall material were smaller and more heterogeneous than microparticles from emulsions containing oil, antioxidant compounds and wall material. This can be explained as a function of the emulsion's viscosity. More viscous emulsions give rise to larger droplets, favoring obtaining larger microparticles. This pattern was also reported by Tonon et al. [42], when evaluating the effect of air temperature and feed composition on the particle size distribution of flaxseed oil microparticles.

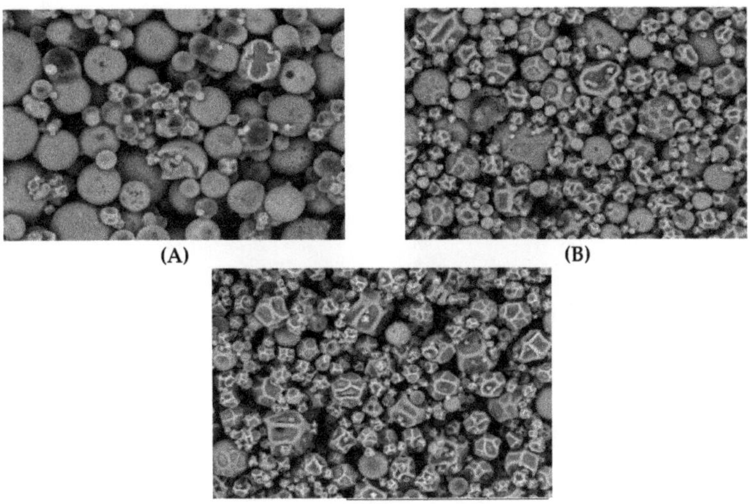

Figure 6. Micrograph of microparticles containing *Passiflora* cincinnata seeds oil (**A**); microparticles containing *Passiflora* cincinnata seeds oil plus antioxidant extract (**B**); microparticles of wall material only (**C**), visualized by MEV.

Table 6. Particle size parameters and moisture, hygroscopicity and solubility values of *Passiflora cincinnata* seeds oil microparticles.

Samples	Average Particle Size (μm)	Span Value	Moisture (%)	Hygroscopicity (%)	Solubility (%)
Oil, phenolic extract and wall material microencapsulated	20.63 ±0.81 b	1.46 ±0.11 b	4.83 ±0.13 b	7.17 ± 0.10 b	76.97 ±0.21 a
Oil and wall material microencapsulated	16.40 ±0.54 a	2.37 ±0.16 a	4.10 ±0.06 c	7.53 ±0.04 b	77.03 ±0.36 a
Wall material microencapsulated	15.50 ±0.05 a	2.22 ±0.06 a	6.16 ±0.12 a	10.75 ±0.77 a	72.87 ±1.74 b

Different letters in the same column indicate significant differences.

3.3.4. FTIR

The absorption spectra from FTIR analysis of pure *P. cincinnata* seeds oil, oil microparticles with and without antioxidant extract and wall material are shown in Figure 7. The *P. cincinnata* seeds oil spectra revealed a strong vibrational mode associated with 3009 cm^{-1} which referred to C-H sp^2 stretching, and intense bands at 2854 and 2925 cm^{-1} which were attributed to the symmetric and asymmetric axial deformation (stretching) of C-H bonds of the methyl (CH_3) and methylene (CH_2) of the fatty acid in triacylglycerol. The band at 1747 cm^{-1} corresponds to vibrations of stretching of the C=O group. The band near 1163 cm^{-1} corresponds to the C-O ester group. Bands at 1667–1640 cm^{-1} can be assigned to overlapping of the olefinic C=C stretching and O-H (water). The wall material presents a prominent band at 3384 cm^{-1} related to stretching O-H. The phenolic compounds were in small quantities, so they could not be identified in the analysis, which may be due to the overlapping of their characteristic bands common to other substances in the formulation. However, they were confirmed by LC–MS analysis (Table 3). The absorption spectra for microencapsulated samples, the broad band between 3650 and 3100 cm^{-1} was observed and refers to the wall material used. The characteristic signals of the *P. cincinnata* seeds oil was also observed, characterizing its incorporation throughout the structure of the microparticles. The signals observed were the stretching band of saturated alkanes near 2925 cm^{-1}, and carbonyls groups near 1747 cm^{-1}, confirming the presence of fatty acids esters. The absorption band at 720 cm^{-1} is attributed to the symmetric stretching vibration of $(CH_2)_n$ groups of n greater than four, indicating the presence of long hydrocarbon linear chains from *P. cincinnata* seeds oil. Thus, the incorporation of the oil into the microparticles can be inferred, given its presence being confirmed by the characteristic peaks of the oil in the absorption spectrum of the microparticles [43,44].

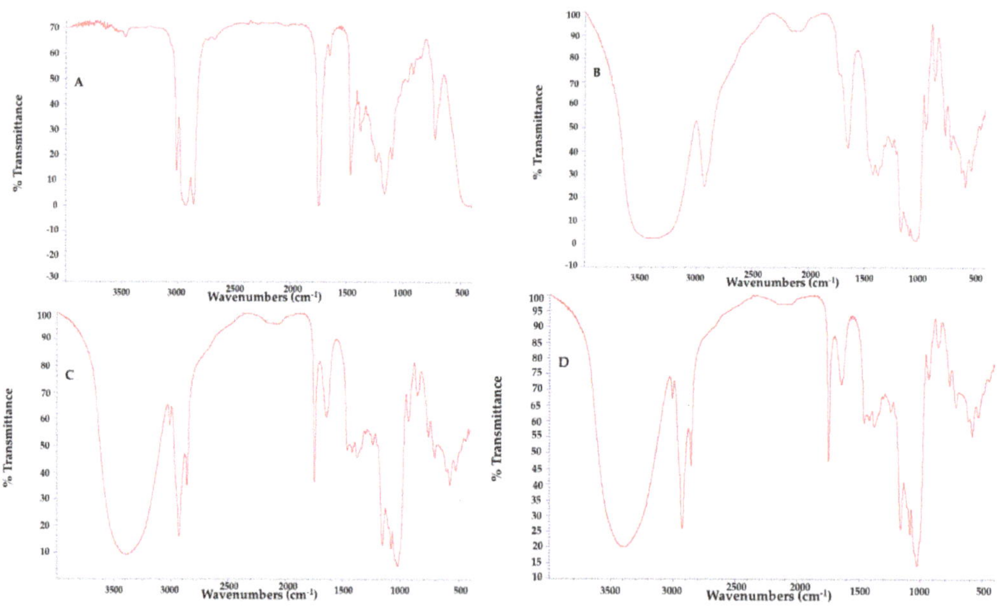

Figure 7. FTIR spectra of *Passiflora cincinnata* seeds oil (**A**) wall material (**B**) microparticles containing *Passiflora cincinnata* seeds oil and antioxidant extract (**C**) microparticles containing *Passiflora cincinnata* seeds oil (**D**).

3.3.5. Moisture, Solubility and Hygroscopicity

The moisture, solubility and hygroscopicity of *P. cincinnata* seeds oil microparticles are presented in Table 6. The moisture values ranged from 4.10% to 6.16%. Moisture data are essential since high moisture could favor microparticle oxidation and reduce their stability. Hijo et al. [45] reported that for dry products such as microparticles used in the food industry, the ideal moisture range is between 3 and 4%, where deterioration by microorganisms is reduced. Drying operational conditions, soluble solid content, and oil concentration influence microparticles' moisture, which explains the data observed in the present work. In this way, the mixture containing wall material and water forms a more hygroscopic solution when compared to oil/water/wall material emulsions with or without the addition of the antioxidant extract. Thus, the probability of the microparticles containing only the wall material, retaining more water during the process is higher, as they tend to come into equilibrium with the ambient humidity more easily.

The hygroscopicity values of the microparticles ranges from 7.17 to 10.75% (Table 6). As oil and water are immiscible, the oil microparticles presented smaller hygroscopicity values than those from wall material–water mixture. According to Nurhadi et al. [46], microparticles with hygroscopicity higher than 20% can be considered very hygroscopic; therefore, they are not stable, being more susceptible to water absorption and, to the stickiness of the material and oxidation of the target compounds [13].

The solubility was higher than 76% for oil microparticles with or without antioxidant compounds (Table 6). The addition of antioxidant extract did not affect the solubility of samples ($p < 0.05$). These results are according to de Oliveira et al. [15], who elaborated microparticles of buriti oil by freeze–drying using carbohydrates as wall material, whose highest value reported was 71%. Botrel et al. [41] reported 79% solubility for microparticles of fish oil atomized in spray dried using different wall materials, where the microparticles were found to have good solubility. In this way, *P. cincinnata* seeds oil microparticles demonstrate good solubility in water and, therefore, have potential for application in aqueous systems.

4. Conclusions

This study was successful in obtaining *P. cincinnata* seeds oil by pressing and recovering antioxidant compounds from the defatted seeds. The best operational conditionsto obtain an antioxidant extract of defatted seeds was 74 °C, 58% ethanol as solvent, and a solid–liquid ratio of 1:48. The main compounds identified by UPLC MS–MS in the extract were lignans that may contribute to antioxidant and antimicrobial activities. The microencapsulation was adequate to preserve *P. cincinnata* oil, and the addition of antioxidant extract proved to be a great method to increase the oxidative stability. Thus, the present work may contribute to adding value to the Brazilian Caatinga passion fruit agro chain by obtaining three bioproducts: pure oil, antioxidant extract and oil microparticles with antioxidant extract.

Supplementary Materials: The following supporting information can be downloaded at: https://www.mdpi.com/article/10.3390/foods12132525/s1, Figure S1: Influence of the independent variables on the antioxidant capacity of the extract of *Passiflora cincinnata* defatted seeds. Figure S2: Precicted versus observed values for responses of the experimental design (TPC—Total phenolic compounds and antioxidant capacity by ABTS$^{•+}$, DPPH$^{•}$ and FRAP assays). Table S1: Pearson' Correlation for results of experimental design.

Author Contributions: Conceptualization, S.P.F. and L.d.O.R.; methodology, C.C.R., V.M.d.M., C.N.K. and E.P.J.; formal analysis, C.C.R., D.d.L.M., T.d.S.F.F. and C.M.A.L.; investigation, C.C.R., C.N.K. and E.P.J.; resources, V.M.d.M. and S.P.F.; data curation, A.L.S.d.S., D.d.L.M., C.N.K. and L.d.O.R.; writing—original draft preparation, C.C.R., T.d.S.F.F. and C.M.A.L.; writing—review and editing, V.M.d.M., S.P.F., D.d.L.M., A.L.S.d.S. and L.d.O.R.; supervision, S.P.F., E.P.J. and L.d.O.R. All authors have read and agreed to the published version of the manuscript.

Funding: This research was funded by COORDENAÇÃO DE APERFEIÇOAMENTO DE PESSOAL DE NPIVEL SUPERIOR–BRASIL (CAPES)–Finance Code 01, grant number 01.

Data Availability Statement: The data used to support the findings of this study can be made available by the corresponding author upon request.

Acknowledgments: The authors acknowledge to Universidade Federal do Rio de Janeiro (UFRJ), Instituto Nacional de Tecnologia (LATEP, LACAT/INT), Empresa Brasileira de Pesquisa Agropecuária (EMBRAPA). The authors are also grateful for financial support from FAPERJ through project E-26/210.694/2021 and the Oswaldo Cruz Foundation Analytical Platform by UPLC–MS/MS analysis. The authors also acknowledge Francisco Pinheiro de Araújo, from Embrapa Semiarid for the supply of passion fruit.

Conflicts of Interest: The authors declare no conflict of interest.

References

1. Brandão, L.E.M.; Nôga, D.A.M.F.; Dierschnabel, A.L.; Campêlo, C.L.D.C.; Meurer, Y.D.S.R.; Lima, R.H.; Engelberth, R.C.G.J.G.J.; Cavalcante, J.S.; Lima, C.A.; Marchioro, M.; et al. *Passiflora cincinnata* Extract Delays the Development of Motor Signs and Prevents Dopaminergic Loss in a Mice Model of Parkinson's Disease. *Evid. Based Complement. Altern. Med.* **2017**, *2017*, 8429290. [CrossRef]
2. de Lavor, É.M.; Leal, A.E.B.P.; Fernandes, A.W.C.; de Almeida Ribeiro, F.P.R.; de Menezes Barbosa, J.; e Silva, M.G.; de Andrade Teles, R.B.; da Silva Oliveira, L.F.; Silva, J.C.; Rolim, L.A.; et al. Ethanolic Extract of the Aerial Parts of *Passiflora cincinnata* Mast. (Passifloraceae) Reduces Nociceptive and Inflammatory Events in Mice. *Phytomedicine* **2018**, *47*, 58–68. [CrossRef] [PubMed]
3. Siebra, A.L.A.; Oliveira, L.R.; Martins, A.O.B.P.B.; Siebra, D.C.; Albuquerque, R.S.; Lemos, I.C.S.; Delmondes, G.A.; Tintino, S.R.; Figueredo, F.G.; da Costa, J.G.M.; et al. Potentiation of Antibiotic Activity by *Passiflora cincinnata* Mast. Front of Strains *Staphylococcus aureus* and *Escherichia coli*. *Saudi J. Biol. Sci.* **2018**, *25*, 37–43. [CrossRef] [PubMed]
4. IBGE. *PAM—Produção Agrícola Municipal*; Instituto Brasileiro de Geografia e Estatística: Rio de Janeiro, Brazil, 2021.
5. Araújo, A.J.B.; Santos, N.C.; Barros, S.L.; Vilar, S.B.O.; Schmidt, F.L.; Araujo, F.P.; Azevedo, L.C. Caracterização Físico—Química e Perfil Lipídico Da Semente de Maracujá Do Mato (*Passiflora cincinnata* Mast.). *Cad. Pesqui. Ciência Inovação* **2019**, *2*, 14–22.
6. Faleiro, F.G.; Tadeu, N.; Junqueira, V.; Braga, M.F.; deOliveira, E.J.; Peixoto, J.R.; Costa, A.M. Germoplasma e Melhoramento Genético Do Maracujazeiro: Histórico e Perspectivas. *Planaltina Embrapa Cerrados* **2011**, 38. Available online: https://ainfo.cnptia.embrapa.br/digital/bitstream/item/76032/1/doc-307.pdf (accessed on 24 June 2023).
7. Reis, C.C.; Mamede, A.M.G.N.; Soares, A.; Freitas, S.P. Production of Lipids and Natural Antioxidants from Passion Fruit Seeds. *Grasas Aceites* **2020**, *71*, e385. [CrossRef]
8. Lucarini, M.; Durazzo, A.; Raffo, A.; Giovannini, A.; Kiefer, J. Passion Fruit (*Passiflora* spp.) Seed Oil. In *Fruit Oils: Chemistry and Functionality*; Springer: Cham, Switzerland, 2019; pp. 577–603. [CrossRef]
9. Ribeiro, D.N.; Alves, F.M.S.; dos Santos Ramos, V.H.; Alves, P.; Narain, N.; Vedoy, D.R.L.; Cardozo-Filho, L.; de Jesus, E. Extraction of Passion Fruit (*Passiflora cincinnata* Mast.) Pulp Oil Using Pressurized Ethanol and Ultrasound: Antioxidant Activity and Kinetics. *J. Supercrit. Fluids* **2020**, *165*, 104944. [CrossRef]
10. Malacrida, C.R.; Jorge, N. Yellow Passion Fruit Seed Oil (*Passiflora edulis* f. Flavicarpa): Physical and Chemical Characteristics. *Braz. Arch. Biol. Technol.* **2012**, *55*, 127–134. [CrossRef]
11. Moreira, N.X.; Curi, R.; Mancini Filho, J. Ácidos Graxos: Uma Revisão. *Nutr. Rev. Soc. Bras. Aliment. Nutr.* **2002**, *24*, 105–123.
12. Geranpour, M.; Assadpour, E.; Jafari, S.M. Recent Advances in the Spray Drying Encapsulation of Essential Fatty Acids and Functional Oils. *Trends Food Sci. Technol.* **2020**, *102*, 71–90. [CrossRef]
13. De Oliveira Ribeiro, L.; Freitas, S.P.; da Matta, V.M.; Jung, E.P.; Kunigami, C.N. Microencapsulation of the Extract from *Euterpe edulis* Co-Product: An Alternative to Add Value to Fruit Agro-Chain. *Waste Biomass Valoriz.* **2021**, *12*, 1803–1814. [CrossRef]
14. Jung, E.P.; Ribeiro, L.O.; Kunigami, C.N.; Figueiredo, E.S.; Nascimento, F.S. Ripe Banana Peel Flour: A Raw Material for the Food Industry. *Rev. Virtual Quim.* **2019**, *11*, 1712–1724. [CrossRef]
15. Freitas de Oliveira, C.; Giordani, D.; Lutckemier, R.; Gurak, P.D.; Cladera-Olivera, F.; Ferreira Marczak, L.D. Extraction of Pectin from Passion Fruit Peel Assisted by Ultrasound. *LWT-Food Sci. Technol.* **2016**, *71*, 110–115. [CrossRef]
16. Ribeiro, L.d.O.; de Freitas, B.P.; Lorentino, C.M.A.; Frota, H.F.; dos Santos, A.L.S.; Moreira, D.d.L.; do Amaral, B.S.; Jung, E.P.; Kunigami, C.N. Umbu Fruit Peel as Source of Antioxidant, Antimicrobial and α-Amylase Inhibitor Compounds. *Molecules* **2022**, *27*, 410. [CrossRef] [PubMed]
17. Derringer, G.; Suich, R. Simultaneous optimization of several response variables. *J. Qual. Technol.* **1980**, *12*, 214–219. [CrossRef]
18. Singleton, V.L.; Rossi, J.A. Colorimetry of Total Phenolics with Phosphomolybdic-Phosphotungstic Acid Reagents. *Am. J. Enol. Vitic.* **1965**, *16*, 144–168. [CrossRef]
19. Gião, M.S.; González-Sanjosé, M.L.; Rivero-Pérez, M.D.; Pereira, C.I.; Pintado, M.E.; Malcata, F.X. Infusions of Portuguese Medicinal Plants: Dependence of Final Antioxidant Capacity and Phenol Content on Extraction Features. *J. Sci. Food Agric.* **2007**, *87*, 2638–2647. [CrossRef]
20. Hidalgo, M.; Sánchez-Moreno, C.; de Pascual-Teresa, S. Flavonoid–Flavonoid Interaction and Its Effect on Their Antioxidant Activity. *Food Chem.* **2010**, *121*, 691–696. [CrossRef]
21. Benzie, I.F.F.; Strain, J.J. The Ferric Reducing Ability of Plasma (FRAP) as a Measure of "Antioxidant Power": The FRAP Assay. *Anal. Biochem.* **1996**, *239*, 70–76. [CrossRef]
22. Da Silva James, N.K.; Castro, L.P.S.; Freitas, S.P.; Nogueira, R.I. Increasing Energy Efficiency in Microencapsulation of Soybean Oil by Spray Drying. *Braz. J. Dev.* **2019**, *5*, 8082–8095. [CrossRef]
23. AOAC. *Official Methods of Analysis of the Association of Analytical Chemists International*; AOAC: Gaithersburg, MD, USA, 2005.

24. Cai, Y.Z.; Corke, H. Production and Properties of Spray-Dried Amaranthus Betacyanin Pigments. *J. Food Sci.* **2000**, *65*, 1248–1252. [CrossRef]
25. Cano-Chauca, M.; Stringheta, P.C.; Ramos, A.M.; Cal-Vidal, J. Effect of the Carriers on the Microstructure of Mango Powder Obtained by Spray Drying and Its Functional Characterization. *Innov. Food Sci. Emerg. Technol.* **2005**, *6*, 420–428. [CrossRef]
26. Lopes, R.M.; Sevilha, A.C.; Faleiro, F.G.; Silva, D.B.D.; Vieira, R.F.; Agostini-Costa, T.D.S. Estudo Comparativo Do Perfil de Ácidos Graxos Em Semente de Passifloras Nativas Do Cerrado Brasileiro. *Rev. Bras. Frutic.* **2010**, *32*, 498–506. [CrossRef]
27. De Paula, R.C.M.; Soares, A.G.; Freitas, S.P. Volatile Coumponds in Passion Fruit Seed Oil (*Passiflora setacea* BRS Pérola Do Cerrado and *Passiflora alata* BRS Doce Mel). *Chem. Eng. Trans.* **2015**, *44*, 103–108. [CrossRef]
28. Pradhan, R.C.; Mishra, S.; Naik, S.N.; Bhatnagar, N.; Vijay, V.K. Oil Expression from Jatropha Seeds Using a Screw Press Expeller. *Biosyst. Eng.* **2011**, *109*, 158–166. [CrossRef]
29. Astrup, A.V.; Bazinet, R.; Brenna, J.T.; Calder, P.C.; Crawford, M.A.; Dangour, A.; Donahoo, W.T.; Elmadfa, I.; Galli, C.; Gerber, M.; et al. Fats and Fatty Acids in Human Nutrition. Report of an Expert Consultation. *FAO Food Nutr. Pap.* **2010**, *91*, 1–166.
30. Cacace, J.E.; Mazza, G. Mass Transfer Process during Extraction of Phenolic Compounds from Milled Berries. *J. Food Eng.* **2003**, *59*, 379–389. [CrossRef]
31. Leal, A.E.B.P.; de Oliveira, A.P.; dos Santos, R.F.; Soares, J.M.D.; de Lavor, E.M.; Pontes, M.C.; de Lima, J.T.; Santos, A.D.d.C.; Tomaz, J.C.; de Oliveira, G.G.; et al. Determination of Phenolic Compounds, in Vitro Antioxidant Activity and Characterization of Secondary Metabolites in Different Parts of *Passiflora cincinnata* by HPLC-DAD-MS/MS Analysis. *Nat. Prod. Res.* **2020**, *34*, 995–1001. [CrossRef]
32. Shi, J.; Yu, J.; Pohorly, J.; Young, J.C.; Bryan, M.; Wu, Y. Optimization of the Extraction of Polyphenols from Grape Seed Meal by Aqueous Ethanol Solution. *J. Food Agric. Environ.* **2003**, *1*, 42–47.
33. Alcântara, M.A.; de Lima Brito Polari, I.; de Albuquerque Meireles, B.R.L.; de Lima, A.E.A.; da Silva Junior, J.C.; de Andrade Vieira, É.; dos Santos, N.A.; de Magalhães Cordeiro, A.M.T. Effect of the Solvent Composition on the Profile of Phenolic Compounds Extracted from Chia Seeds. *Food Chem.* **2019**, *275*, 489–496. [CrossRef]
34. Hanhineva, K.; Rogachev, I.; Aura, A.-M.; Aharoni, A.; Poutanen, K.; Mykkänen, H. Identification of Novel Lignans in the Whole Grain Rye Bran by Non-Targeted LC–MS Metabolite Profiling. *Metabolomics* **2012**, *8*, 399–409. [CrossRef]
35. Eklund, P.C.; Backman, M.J.; Kronberg, L.Å.; Smeds, A.I.; Sjöholm, R.E. Identification of Lignans by Liquid Chromatography-Electrospray Ionization Ion-Trap Mass Spectrometry. *J. Mass Spectrom.* **2007**, *43*, 97–107. [CrossRef] [PubMed]
36. Kuhnle, G.G.C.; Dell'Aquila, C.; Aspinall, S.M.; Runswick, S.A.; Joosen, A.M.C.P.; Mulligan, A.A.; Bingham, S.A. Phytoestrogen Content of Fruits and Vegetables Commonly Consumed in the UK Based on LC–MS and 13C-Labelled Standards. *Food Chem.* **2009**, *116*, 542–554. [CrossRef]
37. Van Linh, N.; Trung Tuong, N.; Xuan Phong, P.; Trang, D.T.; Nhiem, N.X.; Hoai An, D.; Huu Tai, B. New Phenylethanoid and Other Compounds From *Passiflora foetida* L., with Their Nitric Oxide Inhibitory Activities. *Nat. Prod. Commun.* **2022**, *17*, 1934578X2211411. [CrossRef]
38. Favela-Hernández, J.M.J.; García, A.; Garza-González, E.; Rivas-Galindo, V.M.; Camacho-Corona, M.R. Antibacterial and Antimycobacterial Lignans and Flavonoids from *Larrea tridentata*. *Phyther. Res.* **2012**, *26*, 1957–1960. [CrossRef] [PubMed]
39. Wu, G.; Chang, C.; Hong, C.; Zhang, H.; Huang, J.; Jin, Q.; Wang, X. Phenolic Compounds as Stabilizers of Oils and Antioxidative Mechanisms under Frying Conditions: A Comprehensive Review. *Trends Food Sci. Technol.* **2019**, *92*, 33–45. [CrossRef]
40. Pereira de Oliveira, J.; Almeida, O.P.; Campelo, P.H.; Carneiro, G.; de Oliveira Ferreira Rocha, L.; Santos, J.H.P.M.; Gomes da Costa, J.M. Tailoring the Physicochemical Properties of Freeze-Dried Buriti Oil Microparticles by Combining Inulin and Gum Arabic as Encapsulation Agents. *LWT-Food Sci. Technol.* **2022**, *161*, 113372. [CrossRef]
41. Botrel, D.A.; De Barros Fernandes, R.V.; Borges, S.V.; Yoshida, M.I. Influence of Wall Matrix Systems on the Properties of Spray-Dried Microparticles Containing Fish Oil. *Food Res. Int.* **2014**, *62*, 344–352. [CrossRef]
42. Tonon, R.V.; Grosso, C.R.F.F.; Hubinger, M.D. Influence of Emulsion Composition and Inlet Air Temperature on the Microencapsulation of Flaxseed Oil by Spray Drying. *Food Res. Int.* **2011**, *44*, 282–289. [CrossRef]
43. Silverstein, R.M.; Webster, F.X.; Kiemle, D.J. *Identificação Espectrométrica de Compostos Orgânicos*; livros técnicos e científicos: Rio de Janeiro, Brazil, 2013; ISBN 978-85-216-3637-3.
44. Zhang, L.; Zeng, X.; Fu, N.; Tang, X.; Sun, Y.; Lin, L. Maltodextrin: A Consummate Carrier for Spray-Drying of Xylooligosaccharides. *Food Res. Int.* **2018**, *106*, 383–393. [CrossRef]
45. Hijo, A.A.C.T.; da Costa, J.M.G.; Silva, E.K.; Azevedo, V.M.; Yoshida, M.I.; Borges, S.V. Physical and Thermal Properties of Oregano (*Origanum vulgare* L.) Essential Oil Microparticles: Oregano Essential Oil Microparticles. *J. Food Process. Eng.* **2015**, *38*, 1–10. [CrossRef]
46. Nurhadi, B.; Andoyo, R.; Mahani; Indiarto, R. Study the Properties of Honey Powder Produced from Spray Drying and Vacuum Drying Method. *Int. Food Res. J.* **2012**, *19*, 907–912.

foods

MDPI

Article

Optimization of Ethanolic Extraction of Phenolic Antioxidants from Lychee and Longan Seeds Using Response Surface Methodology

Samart Sai-Ut [1], Passakorn Kingwascharapong [2], Md. Anisur Rahman Mazumder [3,4] and Saroat Rawdkuen [3,5,*]

1 Department of Food Science, Faculty of Science, Burapha University, Chonburi 20131, Thailand; samarts@go.buu.ac.th
2 Department of Fishery Products, Faculty of Fisheries, Kasetsart University, Bangkok 10900, Thailand; passakorn.ki@ku.th
3 Food Science and Technology Program, School of Agro-Industry, Mae Fah Luang University, Chiang Rai 57100, Thailand; anisur.research@mfu.ac.th
4 Department of Food Technology and Rural Industries, Bangladesh Agricultural University, Mymensingh 2202, Bangladesh
5 Unit of Innovative Food Packaging and Biomaterials, School of Agro-Industry, Mae Fah Luang University, Chiang Rai 57100, Thailand
* Correspondence: saroat@mfu.ac.th; Tel.: +66-5391-6739; Fax: +66-5391-6737

Abstract: Lychee seeds (LS) and longan seeds (LoS) are excellent sources of phenolic compounds (PCs) with strong antioxidant activity (AOA). The aim of this study was to optimize the extraction conditions regarding extraction yield (EY), extractable phenolic compound (EPC), and AOA from LS and LoS using surface response methodology (RSM). Solvent concentration, extraction temperature, time, and solid to liquid ratio were optimized using RSM. Increasing the solid to solvent ratio from 1:05 to 1:40 (w/v), increased EY for LoS, however, EY did not change from 1:20 to 1:40 for LS. Solid–liquid ratio 1:20 was chosen for this study. Increasing the quantity of solvent leads to higher EPC and FRAP. The results showed that LoS exhibited higher AOA than LS measured as DPPH, ABTS, and FRAP, respectively. Ethanol concentrations and temperatures significantly ($p < 0.05$) affect EY, EPC, and AOA. The results ($R^2 > 0.85$) demonstrated a good fit to the suggested models and a strong correlation between the extraction conditions and the phenolic antioxidant responses. The ethanol concentrations of 41 and 53%, temperatures of 51 and 58 °C, and the corresponding times of 139 and 220 min were the optimal conditions that maximized the EY, EPC, and AOA from LS and LoS.

Keywords: antioxidant activity; extraction yield; extractable phenolic compound; DPPH; ABTS; FRAP solvent concentration

Citation: Sai-Ut, S.; Kingwascharapong, P.; Mazumder, M.A.R.; Rawdkuen, S. Optimization of Ethanolic Extraction of Phenolic Antioxidants from Lychee and Longan Seeds Using Response Surface Methodology. *Foods* **2023**, *12*, 2827. https://doi.org/10.3390/foods12152827

Academic Editors: Noelia Castillejo Montoya and Lorena Martínez-Zamora

Received: 6 July 2023
Revised: 17 July 2023
Accepted: 22 July 2023
Published: 25 July 2023

1. Introduction

Lychee and longan have been grown commercially in Thailand for more than 100 years, which is concentrated in the upper Northern provinces. These fruits are the most important subtropical fruits among the top fruit crops grown in Thailand, which is currently the biggest producer in the world. Thailand produced 43.92 thousand metric tons of lychees in 2022, with the northern area producing the highest, at 40.8 thousand metric ton. For longan, Thailand produced 1700 thousand metric tons in the year 2022, with the northern area producing the highest, at about 1040 thousand metric tons [1]. There is an increase in plantations with good prospects for exporting these crops. Longan production is also considered to be an economically important fruit for export. With the technology to control flowering and yield, the lychee and longan are planned to grow more in other regions in Thailand. Most fresh lychee and longan are consumed locally. Thai longan is well-known for its quality; there is a demand in international markets. Lychee is currently popular among customers because of its flavorful and eye-catching appearance. However, the majority of the by-products from longans and lychees are the pericarp and seeds, which are wasted [2,3]. Both longan and lychee are consumed fresh as well as processed by drying or preserving in syrup because of their sweet and unique flavor. In any event, seed and peel are thrown away as residue. In addition, the contemporarily generated waste from lychee, including 10 to 20% from seeds and pericarp 15%, depending on the variety, has caused serious issues for the environment and the economic viability of the processing industries [4,5]. During the processing of lychee, 30–40% of the by-product is generated, which is typically thrown away by the processing industry [6]. For the food industries, this massive seed generation during lychee processing is an overwhelming challenge. However, seeds are used to cure a number of ailments in traditional Chinese medicine, such as pathogenic colds, stagnant humor, orchitis, neuralgia, testicular enlargement, hernias, and stomach problems [7,8]. Consequently, lychee seeds (LS) have the potential to be an active component in a variety of pharmaceutical and food preparations.

Longan seed (LoS) includes a variety of nutrients, most notably carbohydrates (75.57%) [9]. In addition, it contains moisture (7.40%), crude fiber (7.89%), ash (1.73%), protein (7.17%), and fat (0.23%) [9]. The LoS also includes a number of phenolic acids, including butanoic acid, caffeic acid, geraniin, isomallotinic acid, chebulagic acid, and flavogallonic acid [10,11]. A few research studies showed that high concentrations of polyphenolic compounds, such as gallic acid, corilagin, and ellagic acid [3,12], as well as ethyl gallate, 1-O-galloyl-d-glucopyranose, brevifolin, methyl brevifolin carboxylate, and 4-O-l-rhamnopyranosyl-ellagic acid [13], were found in the LoS. Starch polysaccharides (40.7%), proteins (4.93%), crude fibers (24.5%), lipids (3.2%), and minerals such as 0.28% Mg and 0.21% Ca are all abundant in LSs [14,15]. A significant amount of tocopherol, or vitamin E (9.4%) as well as unsaturated fatty acids was found in LS oil [16]. All of the essential amino acids were available in the LS, with the exception of two unique amino acids named -methylenecyclopropylglycine (-MCPG) and hypoglycin A (HGA), which restrict the usage of lychee seeds in food due to their hypoglycemic impact on human health [17]. The ethaolic extract of LS contains five major phenolic compounds, including some bioactive compounds: gallic acid, epicatechin-3-gallate, (-)-gallocatechin, (-)-epicatechin, and procyanidin B_2 along with a few minor phenolics such as protocatechuic aldehyde, 5-O-coumaroyl methyl quinate, protocatechuic acid, and daucosterol [18,19]. The bioactive compounds found in LS act as powerful antioxidants and are liable to a variety of biological processes, making it a viable option for use in food formulations.

In addition to being successful, an extraction technique for bioactives from plants should be safe, affordable, and environmentally beneficial. The method and solvent selection play an important role. A crucial step in separating these bioactive chemicals is extraction. The effectiveness of the extraction can be considerably influenced by a variety of parameters, including solvent composition, extraction duration, temperature, solvent to solid ratio, and extraction pressure [20–22]. Polyphenols from LoS have been isolated and extracted using a variety of techniques [23]. Sudjaroen et al. [24] used chromatography techniques to separate polyphenols from methanol extracts of LoS. Zheng et al. [13] separated 95% ethanol extract of LoS powder with various solvent. Li et al. [25] identified more feasible methods to isolate bioactive compounds from LoS using alkaline extraction and acid precipitation methods. To find the ideal extract conditions, Potisate and Pintha [26] investigated the effect of solvent and extraction techniques on antioxidant activity of LS. They divided LS into three types: whole seeds, pulp seeds, and pericarp seeds. Conventional and microwave extraction methods were used with different solvent such as distilled water, ethanol, citric acid, and baking soda. Paliga et al. [27] investigated the effect of pressurized solvent (n-butane) on the extraction yield (EY), chemical composition and antioxidant activity (AOA) of LS extract using 7–100 bar and temperature in the range of 25 °C to 70 °C. Type and concentration of solvent, extraction temperature and time, solid-to-liquid ration, pressure, and pH may strongly affect the extraction process. It is necessary to optimize the extraction process before extraction of phenolic compounds [28].

Solvent extraction techniques are regarded as the simplest and most straightforward way to identify antioxidants in plant components. Although modern extraction technology, such as supercritical fluid extraction, enzyme assisted extraction, ultrasonic assisted extraction, microwave assisted extraction as well as some new extraction solvents such as ionic liquid, low eutectic solvent and glycerol have been used by few researchers. The most important controlling variables are typically solvent polarity, solvent concentration, extraction time, and temperature [29,30]. Soong and Barlow [31] utilized solvent extraction methods to extract phenolic compounds and compare AOA between seeds and edible portion of jackfruit, avocado, longan, mango, and tamarind [31]. Nonetheless, Eberhardt et al. [32] claimed that it would be challenging to develop a universally optimal extraction strategy because of the complex internal matrix and variety of antioxidant chemicals found in natural sources. This study tries to use the simplest and traditional solvent extraction method to optimize the ethanolic extraction of phenolic antioxidants from LS and LoS using response surface methodology (RSM).

In general, empirical or statistical approaches can be used to optimize a process, with the former having restrictions on full optimization. Evaluation of the impacts of various process factors and their interactions on response variables is possible using the response surface methodology (RSM) [33]. The RSM is well known for optimizing extraction process. Additionally, RSM is a superb statistical method for maximizing variables, and when applied correctly, it reveals the ideal circumstances for process optimization. Since interactions between the various elements cannot be identified by taking each one into account separately, it is beneficial for identifying the effect of individual or combination of independent variables on the process. The goal of this study was to use RSM to maximize the phenolic antioxidants found in LS and LoS.

2. Materials and Methods

2.1. Chemicals

Absolute ethanol was supplied by Merck (Darmstadt, Germany). Fluka (Steinheim, Germany) provided the 2,2-azinobis (3-ethyl-benzothiazoline-6-sulfonic acid) (ABTS), 2,4,6-tripyridyl-s-triazine (TPTZ), and Folin–Ciocalteu phenol reagent. Sigma chemical Co. (St. Louis, MO, USA) supplied the 2,2′-diphenyl-picrylhydrazyl (DPPH) and gallic acid.

2.2. Seeds Preparation

The LS and LoS were collected from Fanginterfoods Co., Ltd. Chiang Mai, Thailand. Pericarp tissues were removed from the seeds. The seeds were washed by running. Extra water was removed by air drying until all water droplets were entirely evaporated. The seeds were dried using a tray drier (BP-80, KN Thai TwoOp, Bangkok, Thailand) at 50 °C for 24 h. A hammer mill (Model CMC-20) was used to grind the dry seeds into a fine powder. The pulverized seeds were passed through a sieve with a mesh size of 20 (0.84 mm), followed being kept in a high-density polyethylene zipper (Ziploc®, San Diego, CA, USA) and stored at −20 °C until further use [34,35].

2.3. Extraction of Phenolic Compounds

The impact of the solid-to-liquid ratio on extraction was determined using six ratios (1:5, 1:10, 1:15, 1:20, 1:30, and 1:40 (*w*/*v*)). The solid–liquid ratio was selected based on previous study of Rawdkuen et al. [35]. The entire extraction volume was made of up to 30 mL of ethanol and distilled water (50:50, *v*/*v*) solution. Ethanol is used to extract phenolic antioxidant in compliance with the standards governing the use of food-grade solvents [34]. The suspensions were stirred at 150 rpm for 4 h at room temperature (28 ± 2 °C). The extraction solution was centrifuged (MPW-352R, MPW. MED Instruments, Warszawa, Poland) for 15 min at 8000 rpm [35] The supernatant was filtered using a Buckner funnel and Whatman No. 4 solvent-resistant filter paper. The ratio that produced the highest EPC and AAO value was selected for RSM.

2.4. Response Surface Methodology

An experimental plan based on a three-factor/five-level design known as a rotatable central composite design, which included 17 experimental runs, including three replicates at the center point, was used to optimize the extraction of PCs from the LS and LoS by RSM. The ethanol concentration (X_1; 40–80%, *v*/*v* ethanol/water), extraction temperature (X_2; 40–80 °C), and extraction time (X_3; 60–180 min) were the independent variables [35]. Five levels of independent variables were reported in coded and uncoded forms (Table 1). The regression coefficients were obtained by multiple linear regressions after the experimental data were fitted to a 2nd-order polynomial model (Equation (1)). Using Minitab statistical software, the desirability function approach was used to determine the optimal extraction conditions. The STATISTICA Kernel Release 7.0.61.0 EN (StatSoft Inc., Tulsa, OK, USA) was used to develop the response surface plots.

$$Y = \beta_0 + \sum_{i=1}^{3} \beta_i X_i + \sum_{i=1}^{3} \beta_{ii} X_i^2 + \sum_{i-1}^{2} \sum_{j=2}^{3} \beta_{ij} X_i X_j \quad (1)$$

where X_1, X_2, and X_3 are the independent variables that influence the response Y's; and β_0 = intercept; β_i (i = 1, 2, 3) = linear; β_{ii} (i = 1, 2, 3) = quadratic, and β_{ij} (i = 1, 2, 3; j = 2, 3) = cross-product terms, respectively.

Table 1. Independent variables for optimization with coded and actual values.

Independent Variables	Units	Symbols	Code Levels				
			−α	−1	0	+1	+α
Ethanol concentration	%, *v*/*v*	X_1	26.36	40	60	80	93.64
Temperature	°C	X_2	26.36	40	60	80	93.64
Time	min	X_3	19.09	60	120	180	220.9

2.5. Validation of the Model

The prediction RSM equations were used to determine the optimal conditions for extracting the phenolic antioxidants from the LS and LoS based on the ethanol concentration, extraction temperature and time (Table 2). Following the determination of the 2nd-order model prediction and the multifactor analysis of variance, the desire function technique was used to determine the optimal extraction conditions.

Table 2. Three-factor, three-level face-centered cube design for response surface methodology.

Standard Order [a]	Run Order [b]	X_1	X_2	X_3
		Ethanol Concentration (%)	Temperature (°C)	Time (min)
1	15	40 (−1)	40 (−1)	60 (−1)
2	11	80 (+1)	40 (−1)	60 (−1)
3	4	40 (−1)	80 (+1)	60 (−1)
4	13	80 (+1)	80 (+1)	60.00(−1)
5	1	40 (−1)	40 (−1)	180 (+1)
6	3	80 (+1)	40 (−1)	180 (+1)
7	16	40 (−1)	80 (+1)	180 (+1)
8	6	80 (+1)	80 (+1)	180 (+1)
9	12	26.36 (−α)	60 (0)	120 (0)
10	2	93.64 (+α)	60 (0)	120 (0)
11	14	60 (0)	26.36 (−α)	120 (0)
12	7	60 (0)	93.63 (+α)	120 (0)
13	9	60 (0)	60 (0)	19.09 (−α)
14	5	60 (0)	60 (0)	220.90 (+α)
15	8	60 (0)	60 (0)	120 (0)
16	10	60 (0)	60 (0)	120 (0)
17	17	60 (0)	60 (0)	120 (0)

[a] No randomized; [b] Randomized.

2.6. Determination of EPC

The EPC was measured by following the modified method of Swain and Hillis [36] using Folin-Ciocalteu technique. A quartz vial containing 50 μL of the extract, 200 μL of deionized water, and 50 μL of the Folin-Ciocalteu reagent was filled, and the contents were then well agitated using a Vortex. After allowing the combination to react for 6 min, 500 μL of a 7% (*w*/*v*) Na_2CO_3 solution was added, and it was well mixed. For 90 min, the solution was incubated at room temperature in the dark. A UV spectrophotometer (Bio-chrom/Libra S22, Waterbeach, Cambridge, UK) was used to detect the absorbance at 760 nm, and the data were represented in gallic acid equivalents (GAE; mg/100 g dry weight (DW) using a reference curve for gallic acid (0–200 μg/mL). If the observed absorbance value was higher than the linear range of the standard curve, further dilution would be carried out.

2.7. Determination of DPPH Radical Scavenging Activity

The method used to measure DPPH radical scavenging activity was somewhat modified from that published by Brand-Williams et al. [37]. Six hundred (600) μL samples were added with 600 μL of 0.20 mM DPPH in 95% ethanol. After vigorous mixing, the mixture was left to remain at room temperature in the dark for 30 min. A UV spectrophotometer (Biochrom/Libra S22, Waterbeach, Cambridge, UK) was used to measure the absorbance of the resultant solution at 520 nm. The sample blank was made in the same way for each concentration, except that ethanol was used in place of the DPPH solution. The gallic acid standard curve was a logarithm between 2 and 25 μg/mL. Results are presented as mg/100 g DW of gallic acid equivalents (GAE). If the observed DPPH value was higher than the linear range of the standard curve, more dilution would be required.

2.8. Determination of Ferric Reducing Antioxidant Powder (FRAP)

The FRAP assay was evaluated according to the modified method of Benzie and Strain [38]. Stock solutions were prepared with 20 mM $FeCl_3.6H_2O$ solution, 10 mM TPTZ solution, and 300 mM acetate buffer (pH 3.6). 25 mL of acetate buffer, 2.5 mL of TPTZ solution, and 2.5 mL of $FeCl_3.6H_2O$ solution were mixed to prepare a workable solution. The mixture, known as the FRAP solution, was incubated for 30 min at 37 °C in a water bath (Mem-mert, D-91126, Schwabach, Germany). 810 μL of FRAP solution was mixed with 90 μL of sample (concentration range of 0.5 to 10 mg/L) and left at room temperature for 30 min in the dark. The ferrous tripyridyltriazine complex (colored product) was quantified at 595 nm using a UV spectrophotometer (Biochrom/Libra S22, Waterbeach, Cambridge, UK). A blank sample for each concentration was prepared by using distilled water instead of $FeCl_3$ in the FRAP solution. Gallic acid concentrations between 10 and 100 μg/mL were used to prepare the standard curve. The activity was measured in terms of mg/100 g DW of gallic acid equivalents (GAE).

2.9. Determination of ABTS Antioxidant Activity

2,2′-azino-bis-(3-ethylbenzothiazoline-6-sulfonic) acid (ABTS) radical scavenging activity was measured using modified described by Arnao et al. [39]. The stock solutions were prepared by mixing 2.6 mM potassium persulfate solution and 7.4 mM ABTS solution. The two stock solutions were mixed in equal parts to prepare the working solution, which was then left to react for 12–16 h at room temperature and in complete darkness. The solution was then diluted by mixing 5 mL of ABTS solution with 50 mL of methanol. This resulted in an absorbance of 1.1 ± 0.02 units at 734 nm measured by a UV spectrophotometer (Biochrom/Libra S22, Waterbeach, Cambridge, UK). A new ABTS solution was prepared for each test. 950 μL of ABTS solution was mixed with 50 μL of sample (concentration range of 0.5 to 10 mg/L) and left at room temperature for 120 min in the dark. The same procedure was followed to create a sample blank at each concentration; with the exception that methanol was used in place of the ABTS solution. The absorbance was quantified at 734 nm using a UV spectrophotometer (Biochrom/Libra S22, Waterbeach, Cambridge, UK). A gallic acid standard curve was prepared, with concentrations ranging from 2 to 50 μg/mL. The activity was measured in terms of mg/100 g DW of gallic acid equivalents (GAE).

2.10. Statistical Analysis

All of the analysis was performed in triplicate and statistical analysis was carried out by analysis of variance (ANOVA). Duncan's multiple-range test was used to compare means. SPSS software (SPSS 10.0 for Windows, SPSS Inc., Chicago, IL, USA) was used to conduct the analysis.

3. Results and Discussion

3.1. Selection of Solid-to-Liquid Ratio

Six ratios (1:05, 1:10, 1:15, 1:20, 1:30 and 1:40; *w*/*v*) were tested for their effects on the extraction of PCs from LS and LoS over the course of 4 h at 25 °C and using a 50% (*v*/*v*) ethanol solution as the solvent. The findings demonstrate that the extraction of PCs mainly depends on the solid-to-liquid ratio (Figure 1). Increasing the solid-to-liquid ratio, increased the EPC, and FRAP. This suggests that a higher solid–liquid ratio can increase the solvent–plant material interaction surface area. This allows a higher mass transfer of soluble chemicals from the substance to the solvent. This might also be due to a higher proportion of solvent providing reaction with matrix [40].

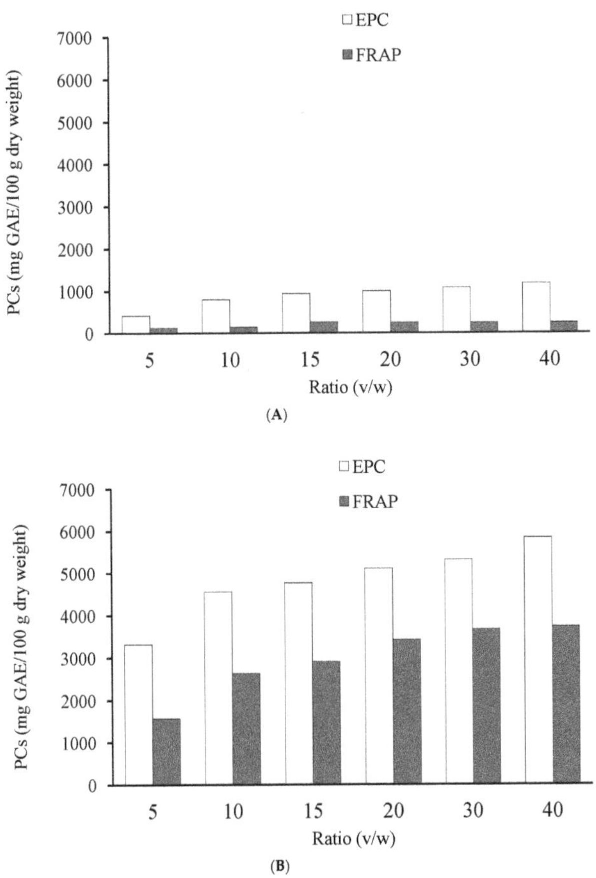

Figure 1. Effect of the solid-to-liquid ratio on the extraction of ferric reducing antioxidant power (FRAP) and extractable phenolic content (EPC) from LS (**A**) and LoS (**B**) using aqueous ethanol (50%, *v*/*v*) at 25 °C for 4 h. GAE = gallic acid equivalents; small letters represents the statistical analysis for EPC and capital letters represent the statistical analysis for FRAP. Bar represents the standard deviation.

The EY from LoS increased significantly ($p < 0.05$) when the amount of solvent was increased (Table 3). The extraction yield from LS did not change significantly ($p > 0.05$) between the ratios of 1:20 to 1:40 (w/v) (Table 3). Exceeding the solvent concentration of a certain value, leading to the changes polarity of the solvent, resulted in a reduction in the rate of phenolic chemical extraction [41]. The solubility and equilibrium constants of the process are affected by changes in the solid-to-solvent ratio [35]. According to FRAP, the optimal solid–liquid ratio for extracting PCs from LS is between the ratios of 1:20 and 1:30 (w/v) (Figure 1B). It appears that increasing the quantity of solvent leads to higher EY, EPC, and FRAP antioxidant activity. This might be the reason for the poor solubility of these compounds in the solvent as well as the dependence of isolated compound yields on the predicted dielectric constant of the extraction solvent [42]. The 1:20 solid-to-liquid ratio was hence selected for RSM as a compromise.

Table 3. Effect of solid to liquid ratio on the extraction yield from LS and LoS.

Ratio (w/v)	LS (mg/100 g Sample)	LoS (mg/100 g Sample)
1:05	6.52 ± 0.04 [a]	8.76 ± 0.17 [a]
1:10	9.12 ± 0.13 [b]	11.04 ± 0.13 [b]
1:15	9.34 ± 0.13 [b]	11.94 ± 0.07 [c]
1:20	9.77 ± 0.06 [c]	12.39 ± 0.21 [d]
1:30	9.81 ± 0.06 [c]	13.30 ± 0.16 [e]
1:40	9.80 ± 0.16 [c]	13.87 ± 0.42 [f]

The values are all mean ± SD. Mean values in the same column with various superscripts differ significantly ($p < 0.05$).

3.2. Optimization of Extraction of Phenolic Compounds Using RSM

The RSM was used to optimize the extraction of phenolic antioxidants. The obtained results showed that 100 g of dry fruit seeds gave between 5.7 to 8.8 mg EY for LS. This study is supported by a little previous research. For instance, Daorueang [43] showed that EY of LS was 4.6 g/100 g DW. He also found that EPC in the LS was 89.5 mg GAE/g extract and LS inhibited the DPPH radical with an IC_{50} value of 0.65 mg/mL, demonstrating concentration-dependent antiradical action. However, Paliga et al. [27] found that the EY of LS was around 3.5 wt%, total phenolic content of about 126.4 mg GAE/100 g and AOA of up to 78.36%. Similarly, Prasad et al. [44] observed that maximum EY was found at 50% ethanol extract (26.8%), whereas 100% ethanol only produced 23.3%. The EPC of 50% ethanol extract was 239 GAE/g DW. However, there were no significant differences between 50% ethanol and 100% ethanol extract. The percent scavenging activity of 50% ethanol extracts was 48.9%. Therefore, the sequence of the DPPH radical-scavenging activity was 50% ethanol > ethanol > 50% methanol > methanol > BHT > distilled water. It was found that the LS extract exhibited antioxidant power of 134, 581, and 132 GAE/100 g dried seed for DPPH, ABTS and FRAP method, respectively. In general, the main PCs identified in the ethanolic extract of LS epicatechin, gallocatechin, epicatechin-3-gallate, gallic acid, and procyanidin B2 [6].

This study showed that 100 g of dry fruit seeds gave between 8.1 to 15.5 mg EY for LoS. The yield of PCs following traditional solid–liquid extraction was 46.86 mg/g utilizing an alkaline buffer as the extraction solvent; however, acid precipitation showed a phenol separation yield of 22.04 mg/g [25]. Chindaluang and Sriwattana [45] study was on the various extraction techniques for PCs in LoS. The extraction process using ethanol provided the lowest yield (27.7%), which requires the longest extraction time. The yield of the ultrasonic extraction method, which was 35%, was midway between the other two treatments but took the least amount of time, while the yield of the hot water extraction method, which generated the highest LoS extract yield, was 42.8%. While the antioxidant power of LoS extract was found to be 2442, 9173, and 3727 mg GAE/100 g of dried seed for DPPH, ABTS, and FRAP method, respectively. These values were similar with others reporting that LoS contain a phenolic content of 6300 mg GAE/100 g seeds [31].

However, Chindaluang and Sriwattana [45] extracted PCs by ethanol, ultrasonic, and hot water extraction techniques. They found that PCs concentrations were varied with the extraction techniques. For example, PCs for LoS were 11.717, 26.90, and 41.250 mg GAE/mL for ethanol, ultrasonic, and hot water extraction, respectively. In comparison to ascorbic acid (1.37 µg/mL), the AOA of crude extracts of LoS in 20% and 95% ethanol was in the range of 0.82 and 1.73 µg/mL [46]. The major PCs found in the LoS are gallic acid, corilagin, and ellagic acid with retention times of 3.6, 7.8, and 13.2 min, respectively, identified using HPLC by Hong-in et al. [47]. Analysis of variance (ANOVA) for the second order response surface model of each response is shown in Table 4 for both LS and LoS.

Table 4. Analysis of variance (ANOVA) of the second order response model of extraction yield, EPC, DPPH, ABTS, and FRAP values from LS and LoS.

Responses	Source	DF [1] (LS)	DF [1] (LoS)	SS [2] (LS)	SS [2] (LoS)	MS [3] (LS)	MS [3] (LoS)	F (LS)	F (LoS)	*p*-Value (LS)	*p*-Value (LoS)
Extraction yield	Lack-of-fit	8	3	0.5453	5.326	0.0685	1.7755				
	Pure error	2	8	0.0037	3.355	0.0018	0.4193	37.21	4.23	0.026	0.046
	Total	16	16	9.5443	56.048						
EPC	Lack-of-fit	8	9	44,518	2,049,453	5565	227,717				
	Pure error	2	2	45	229,404	22	114,702	247.66	1.99	0.004	0.380
	Total	16	16	489,630	17,224,657						
DPPH	Lack-of-fit	8	7	8222	147,221	1027.7	21,032				
	Pure error	2	2	109	7850	54.7	3925	18.78	5.36	0.052	0.166
	Total	16	16	112,382	2,352,154						
ABTS	Lack-of-fit	8	7	43,482	4,209,406	5435	601,344				
	Pure error	2	2	8511	155,653	4255	77,826	1.28	7.73	0.511	0.119
	Total	16	16	686,291	37,479,890						
FRAP	Lack-of-fit	8	3	8232	203,411	1029	67,804				
	Pure error	2	8	534	224,684	267	28,086	3.85	2.41	0.222	0.142
	Total	16	16	105,278	3,914,691						

[1] Degree of freedom, [2] Sum of squares, [3] Mean of square. LS = lychee seeds; LoS = Longan seeds. EPC = extractable phenolic content; DPPH = 2,2-Diphenyl-1-picryl-hydrazyl-hydrate; FRAP = ferric reducing antioxidant power; ABTS = 2,2′-azino-bis-(3-ethylbenzothiazoline-6-sulfonic) acid.

It was noted that the response of extraction yield and EPC for LS (EY for LoS) showed significant lack of fit ($p < 0.05$) presenting the model fits not well. However, there were significant ($p < 0.05$) effects on parameters on other responses, e.g., FRAP, DPPH, ABTS. The lack of fit tests, determined by dividing the residual error by the pure error from replicated design points to explain the sufficiency of the relationship between experimental factors and the responses. Table 5 shows the mathematical model for extraction yield; EPC, DPPH, ABTS, and FRAP.

Table 5. Regression coefficients of predicted models and independent effects of factors for the investigated responses of LS and LoS extracts.

Variables [a]	Yield CF (LS)	Yield CF (LoS)	EPC CF (LS)	EPC CF (LoS)	DPPH CF (LS)	DPPH CF (LoS)	ABTS CF (LS)	ABTS CF (LoS)	FRAP CF (LS)	FRAP CF (LoS)
β_0	7.586 ***,b	14.042 ***,b	939.26 ***	6015.86 ***	394.37 ***	2337.98 ***	1134.67 ***	8339.34 ***	387.25 ***	3567.43 ***
β_1	−0.718 ***	−1.062 ***	−82.34 **	−714.44 ***	−35.32 **	−213.78 ***	−90.88 ***	−496.76 *	−27.28 ***	−181.88 **
β_2	0.100 ns,c	0.676 *	60.85 *	42.34 ns,c	23.15 *	110.56 *	110.54 ***	−83.18 ns	50.29 ***	−67.09 ns
β_3	0.147 *		29.76 ns		3.91 ns	94.54 *	18.96 ns	−207.84 ns	11.63 ns	
β_{11}	−0.166 *	−1.270 ***	−129.05 ***	−771.02 ***	−66.64 ***	−254.23 ***	−135.22 ***	−1090.15 ***	−55.39 ***	−354.75 ***
β_{22}		−0.670 *	−83.64 **	−381.96 **	−47.88 ***	−207.99 ***	−110.73 ***	−1002.56 ***	−31.93 ***	−343.61 ***
β_{33}										
β_{12}		0.701 *				112.05 *		716.37 *		254.89 **
β_{13}	−0.333 **									
β_{23}	−0.189 *		−75.69 *		−36.59 *	107.24 *	−71.37 *	−598.95 *	−30.04 *	
R^2	0.94	0.85	0.91	0.86	0.93	0.93	0.92	0.88	0.92	0.89
Adj-R^2	0.91	0.77	0.85	0.82	0.88	0.88	0.88	0.79	0.87	0.84

[a] Polynomial model $Y = \beta_0 + \sum_{i=1}^{3} \beta_i X_i + \sum_{i=1}^{3} \beta_{ii} X_i^2 + \sum_{i-1}^{2} \sum_{j=2}^{3} \beta_{ij} X_i X_j$ adjusted by backward elimination at the level of 0.1% using the lack-of-fit test; β_0 = constant coefficient; β_i = linear coefficient (main effect); β_{ii} = quadratic coefficient; β_{ij} = two factors interaction coefficient; b,* significant at $p \leq 0.05$; ** significant at $p \leq 0.01$; *** significant at $p \leq 0.001$; c,ns = not significant ($p > 0.05$). CF = coefficients; LS = lychee seeds; LoS = longan seeds; EPC = extractable phenolics content; DPPH = 2,2-Diphenyl-1-picryl-hydrazyl-hydrate; FRAP = ferric reducing antioxidant power; ABTS = 2,2′-azino-bis-(3-ethylbenzothiazoline-6-sulfonic) acid.

The correlation coefficients (R^2) of each response were higher than 0.85 indicating a good relationship between extraction condition and phenolic antioxidant responses The data would properly fit with the statistical model if the R^2 values were higher than 0.80 [35,48]. The coefficients of the multiple regression equations that were generated as a result of the independent variables displayed in Table 5. Only the F value for the responses was significant for the model with a confidence level higher than 90 in order for the analysis of variance to fit the quadratic model of the extraction parameters. The adjusted polynomial second order models' regression coefficients and analysis of variance for antioxidant activity of LS and LoS extracts are summarized in Figures 2 and 3, respectively. According to the surface response analysis for antioxidant activity, temperatures and ethanol concentrations had the greatest impacts ($p \leq 0.01$ or $p \leq 0.001$), whereas time had no significant impact ($p > 0.05$). The model's suitability was confirmed by R^2, Adj-R^2, and a control of model parameters (Table 1).

The relationship between extraction yield and variables is described by the following second order polynomial equation of LS:

$$\text{Extraction yield (LS)} = 7.586 - 0.718X_1 + 0.100X_2 + 0.147X_3 - 0.166{X_1}^2 - 0.333X_1X_3 - 0.189X_2X_3. \quad (2)$$

For the response to DPPH, DPPH, the linear, and quadratic terms of X_1 and X_2 showed statistically significant in coefficients which are described by the following Equation (3):

$$\text{DPPH (LS)} = 394.37 - 35.32X_1 + 23.15X_2 + 3.91X_3 - 66.64{X_1}^2 - 47.88{X_2}^2 - 36.59X_2X_3. \quad (3)$$

Similarly, the linear and quadratic for ethanol proportion (X_1) and extraction temperature (X_2) for LoS were the statistically significant coefficients concerning the extraction yield and AAO. Then, the predicted model for extraction yield was calculated by the following Equation (4):

$$\text{Extraction yield (LoS)} = 14.04 - 1.06X_1 + 0.67X_2 - 1.27{X_1}^2 - 0.67{X_2}^2 - 0.70X_1X_2 \quad (4)$$

Therefore, the positive linear and quadratic effects of temperature (X_2) suggest that the increase in extraction temperature improved the phenolic compounds yield. Indeed, the solubility and diffusion coefficient of PCs was increased at high temperatures, permitting a higher extraction rate. A positive effect of time (X_3) was detected in LS extract but not significant in that of LoSs. Based on the predicted models, some responses depend on linear terms and quadratic terms, but some depend on interaction terms. It has been documented that ethanol concentrations and temperatures significantly affect the overall antioxidant activities of mango kernels [35], wheat [20], grape cans [21], apple pomaces [22], and wood apples [40].

$Y = 939.26 - 82.34X_1 + 60.85X_2 + 29.76X_3 - 129.05{X_1}^2 - 83.64{X_2}^2 - 75.69X_2X_3$ ($R^2 = 0.91$; Adj-$R^2 = 0.85$)

(A) **(E)**

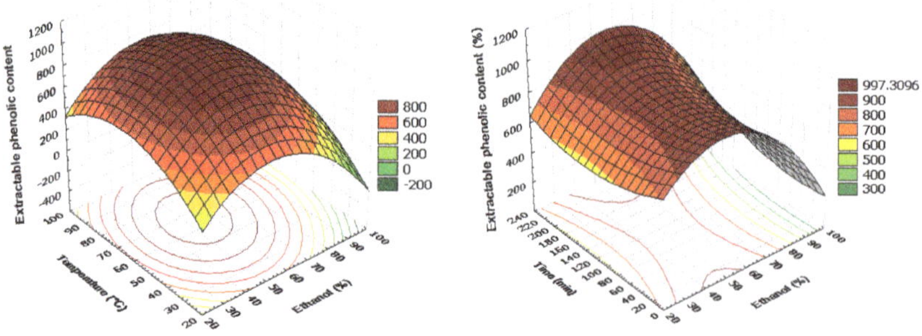

Figure 2. *Cont.*

$$Y = 394.37 - 35.32X_1 + 23.15X_2 + 3.91X_3 - 66.64X_1^2 - 47.88X_2^2 - 36.59X_2X_3\ (R^2 = 0.93;\ \text{Adj-}R^2 = 0.88)$$

(B) **(F)**

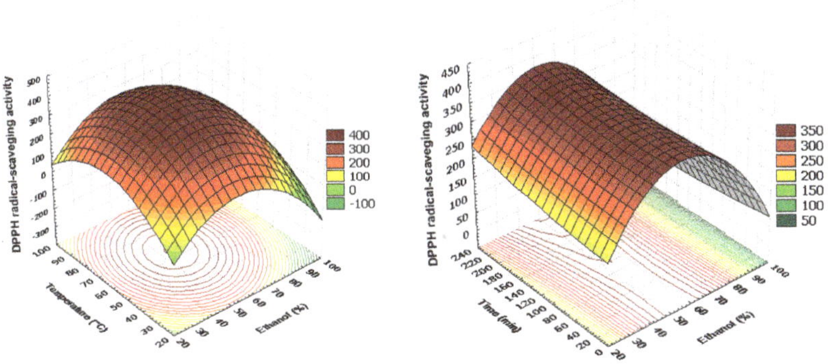

$$Y = 1134.67 - 90.88X_1 + 110.54X_2 + 18.96X_3 - 135.22X_1^2 - 110.73X_2^2 - 71.37X_2X_3\ (R^2 = 0.92;\ \text{Adj-}R^2 = 0.88)$$

(C) **(G)**

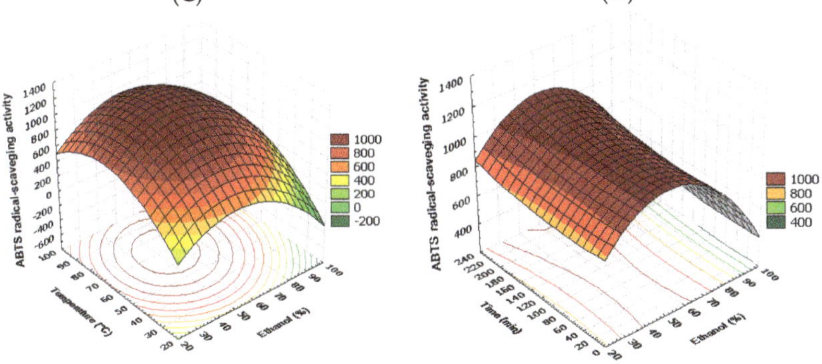

$$Y = 387.25 - 27.28X_1 + 50.29X_2 + 11.63X_3 - 55.39X_1^2 - 31.93X_2^2 - 30.04X_2X_3\ (R^2 = 0.92;\ \text{Adj-}R^2 = 0.87)$$

(D) **(H)**

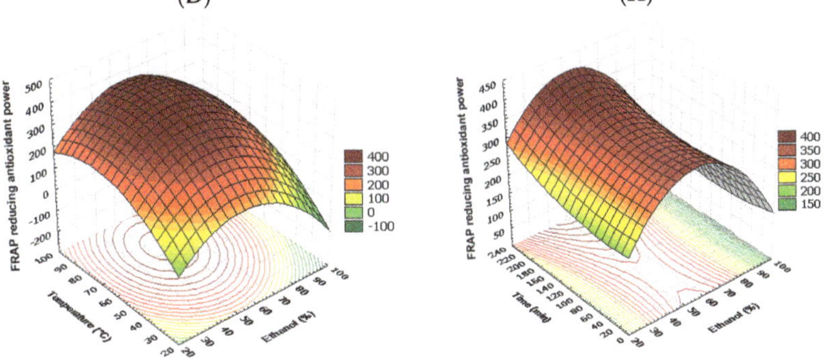

Figure 2. Response surface model plot and regression coefficients of predicted models demonstrating the effects of temperature and ethanol proportion (**A–D**) and time and ethanol proportion (**E–H**) on EPC, DPPH, ABTS, and FRAP (mg GAE/100 g dry sample) in LS extracts.

$$Y = 6015.86 - 714.44X_1 + 42.34X_2 - 771.02X_1^2 - 381.96X_2^2 \ (R^2 = 0.86;\ \text{Adj-}R^2 = 0.82)$$

(A)

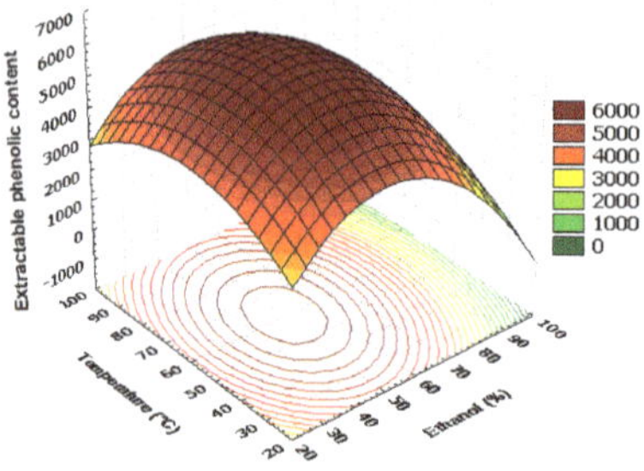

(E)

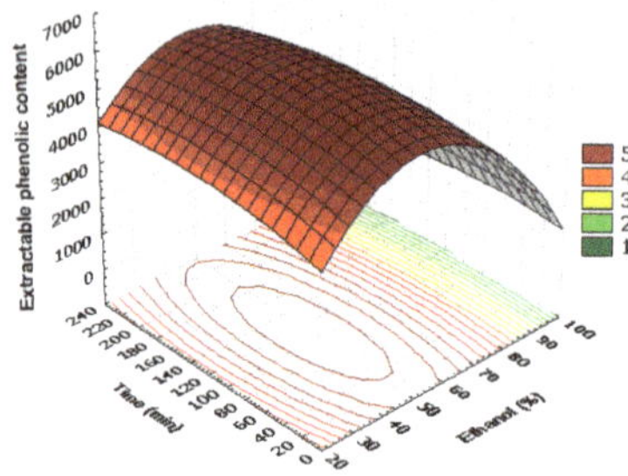

$$Y = 2337.98 - 213.78X_1 + 110.56X_2 + 94.54X_3 - 254.23X_1^2 - 207.99X_2^2 + 112.05X_1X_2 + 107.24\ X_2X_3 \ (R^2 = 0.93;\ \text{Adj-}R^2 = 0.88)$$

(B)

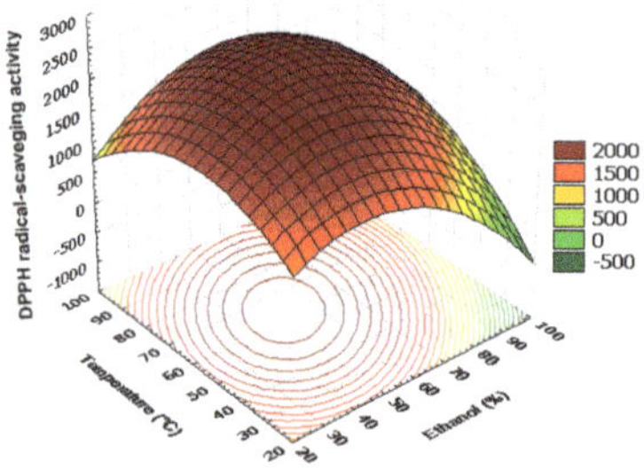

(F)

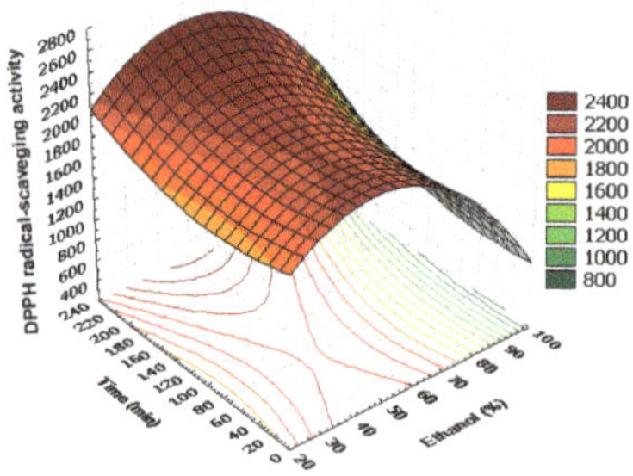

$$Y = 8339.34 - 496.76X_1 - 83.18X_2 - 207.84X_3 - 1090.15X_1^2 - 1002.56X_2^2 + 716.37X_1X_2 - 598.95X_2X_3 \ (R^2 = 0.88;\ \text{Adj-}R^2 = 0.79)$$

(C)

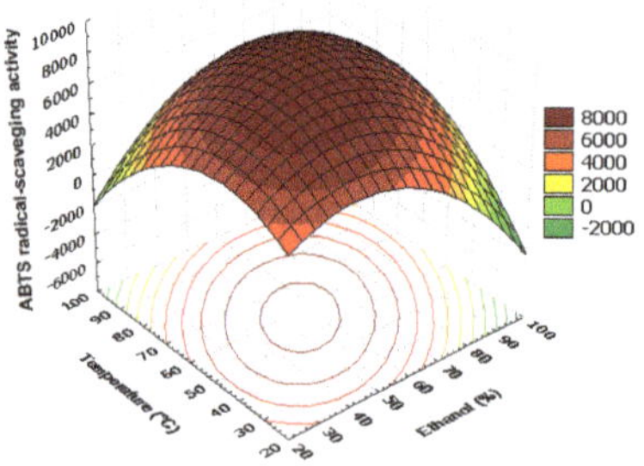

(G)

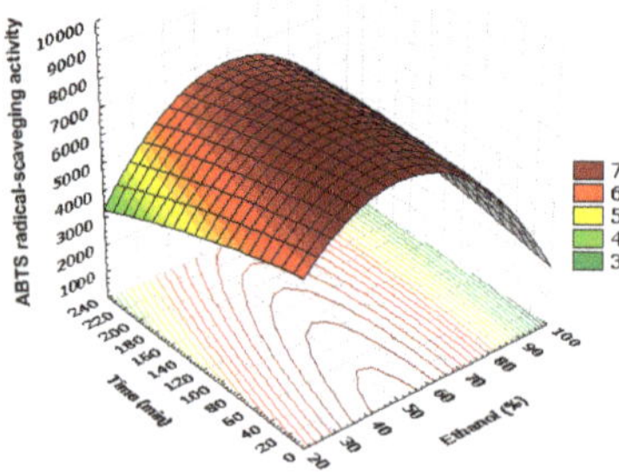

Figure 3. *Cont.*

$Y = 3567.43 - 181.88X_1 - 67.09X_2 - 354.75X_1^2 - 343.61X_2^2 + 254.89X_1X_2$ ($R^2 = 0.89$; Adj-$R^2 = 0.84$)

(D) (H)

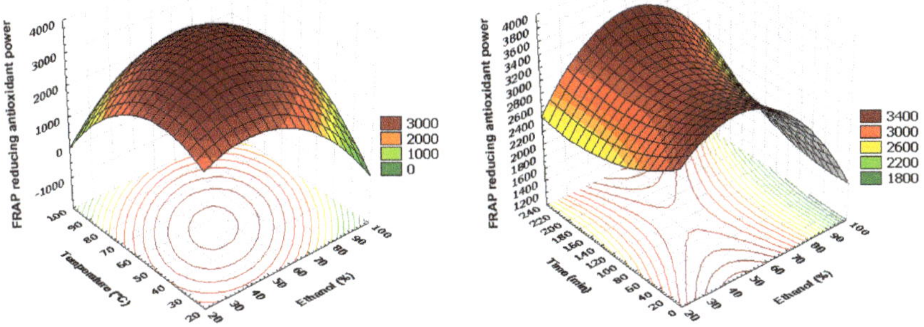

Figure 3. Response surface model plots and regression coefficients of predicted models demonstrating the effects of temperature and ethanol proportion (**A**–**D**) and time and ethanol proportion (**E**–**H**) on EPC, DPPH, ABTS, and FRAP (mg GAE/100 g dry sample) in LoS.

3.3. *Effect of Factors on the EPC and AAO*

The trend seen for EPC recovery from LS (Figure 2) and LoS (Figure 3) upon simultaneous variation in ethanol and temperature indicates that maximal yield can be achieved at the medium ethanol and temperature values. Most of the PCs available in the plant seeds are phenolic acids, flavonoids, tannins, and polysaccharides, which have a similarity in molecular structure for intermediate polar solvents due to dipole–dipole and dispersion forces, electrostatic, and hydrogen-bonding interactions [49]. At higher ethanol and temperature levels, EPC recovery showed a declining tendency. With regard to the extraction time, maximal EPC for LoS was found for a short duration, while for LS, medium and longer durations were proven favorable, respectively (Figure 3F). Increasing the temperature, increased the amount of EPC and consequently antioxidant activities ($p < 0.01$), although only up to 55 °C. Above this temperature, EPC and AAO reductions were observed. Both seed extracts were primarily impacted by the temperature and ethanol proportion interaction effects. The findings suggested that active chemicals from the solid matrix may be mobilized up to a specific threshold. This could be because of their high-temperature breakdown [20]. Generally, the high temperature enhances the mass transfer rate by decreasing the solvent viscosity. As a result, the solvent distribution to the plant tissues and cell membranes was greater, leading to an increase in phenolic content and antioxidant activities of bioactive compounds [50]. Cacace and-Mazza, 2002. J.E.Cacace and-G.Mazza, Extraction of anthocyanins and other phenolics from black currants with sulfured water, I Agr Food Chem 50 (2002), pp. 5939–5946. Full Text via CrossRef | View Record in Scopus | Cited By in Scopus (33). According to Cacace and Mazza [51], changing the temperature has an impact on a compound's solubility in the solvent and diffusion coefficient during extraction. As a result, a rise in temperature may enhance the diffusion coefficient and subsequently the rate of diffusion, which would shorten the extraction time [52].

The effects of ethanol proportion, extraction temperature and time on the amount of EPC are shown in Tables 4 and 5, and Figures 2 and 3. According to Equations (2) and (4), the ethanol concentration showed the highest impression on the extraction of EPC from LS and LoS, as its coefficient has the highest value. Linear and quadratic impacts of the ethanol concentration were seen for the apparent EPC content. The highest EPC was found at a certain ethanol concentration, according to the negative quadratic impact of X_1. In fact, the apparent EPC rose as the ethanol concentration rose, peaked at around 53% ethanol, and then started to fall (Figures 2 and 3). The coefficient of determination for EPC showed a good regression value ($R^2 \geq 0.91$ for LS, and 0.86 for LoS). Significant increases in EPC were observed at high ethanol concentrations and low temperatures. To obtain higher EPC, high ethanol concentrations increased the solubility and diffusion of the compounds, which enhanced the solvent ability to extract the compound into the matrix, leading to a decrease at higher temperatures [35]. On the other hand, higher temperatures generally degrade phenolic compounds, leading to a reduction in the EPC (Figures 2A and 3A). The relationships between the EPC and the ethanol content, extraction temperature, and extraction duration are shown in Figures 2 and 3 for LS and LoS, respectively. Strong regression values were shown by R^2 values of 0.91 for LS and 0.86 for LoS in the coefficient of determination for EPC. Low temperatures and high ethanol concentrations significantly increase the EPC. High ethanol concentrations enhance the solubility and diffusion of the chemicals. This enhanced ability of the solvent to extract the molecule into the matrix resulted in a decrease at higher temperatures [36]. However, higher temperatures frequently lead to the degradation of phenolic compounds, which reduced the EPC (Figures 2A and 3A). Furthermore, EPC decreased at extremely high temperatures, most likely as a result of the dwindling dielectric constant, which gave the EPC a less effective contribution [35,53]. Figures 2E and 3E illustrate the effects of time and ethanol concentration on EPC for LS and LoS, respectively. As opposed to this, EPC increases fast when the extraction time is extended from 60 to 139 min for LS and decreases after 140 min. However, it was 60 to 220 min for LoS, before starting to decline after 220 min. This is consistent with the contour plots in Figures 2E and 3E, as well as the findings in Table 5, which show that the interactions between the two variables are not statistically significant ($p > 0.05$). Numerous investigations have found that the amount of ethanol present in the extraction medium affects the yield of EPCs. Research revealed that piceatannol could be extracted most effectively from the seeds of passion fruit using an extraction method that used 80% aqueous ethanol [54]. The effect of the ethanol concentration results from its effect on the polarity of the extraction solvent and the consequent solubility of the phenolic compounds. According to the basic principle of like dissolves like, solvents only extract phytochemicals that are polarly similar to the solvent [55,56]. As with ethanol proportion, temperature also had linear and quadratic effects on the EPC extraction (Tables 4 and 5). Figures 2 and 3 demonstrate that increase in temperature increased the EPC extraction and peaked at around 51–58 °C, and then started to fall. The rate of diffusion of phenolic compounds may be improved by raising the extraction temperature and by raising the diffusion coefficient [52,57]. Beyond a certain point, however, the temperature may encourage the simultaneous breakdown of PCs that were previously mobilized at a lower temperature or even breakdown of PCs that are still present in the plant matrix [58]. This research shows that EPC only started to degrade when the temperature is above 65 °C (Figures 2 and 3), indicating that it is a somewhat thermo-resistant chemical that can withstand different food preparation procedures. Tables 4 and 5 show that extraction time had neither linear nor quadratic impact on the EPC extraction. The result suggested that a significant amount of EPC is extracted during the first few minutes of the extraction process. Thus, a study suggests that long extraction periods resulted in lower EPC content. Longer extraction durations may not always result in a complete extraction of phenolic chemicals [59]. Furthermore, phenolic oxidation or degradation brought on by exposure to light, oxygen, or high temperatures may result from lengthy extraction periods [21,58]. DPPH, ABTS, and FRAP were found to obey similar trends such as EPC for LS, but quite different for LoS (Figure 3). The RSM of DPPH (93% for both LS and LoS), ABTS (92% for LS and 88% for LoS), and FRAP (92% for LS and 89% for LoS) data indicated that the model was significant ($p < 0.01$). The two-dimensional contours and three-dimensional representation of the response surfaces of the model are shown in Figures 2 and 3. The DPPH and ABTS assays were almost similar as was shown by the shape of the contour plots. When the ethanol concentration and temperature were raised to 55% and 60 °C, respectively, the DPPH and ABTS assays increased as shown in Figures 2B,F and 3B,F. In general, chemicals may be easily removed from plant cells when the solvent's polarities are similar to those of the phenolic compounds [60]. Furthermore, a rise in the extractability of bioactive chemicals brought on by a high temperature may improve the antioxidants' yield [61]. Figures 2F,G and 3F,G illustrate the effects of ethanol concentration and extraction time on the DPPH and ABTS assays. The highest DPPH activities are reached when X_1 and X_3 are 55% and 139 min (LS) and 180 min (LoS), respectively, and subsequently the antioxidative activities swiftly increase at first and then decline with increasing ethanol concentration and extraction time. At X_1 = 65% and X_3 = 160 min (LS) and 220 min (LoS), the maximum ABTS activities are attained. It goes without saying that excessively high temperatures and ethanol concentrations always reduce antioxidative activity [35]. The response surface plot for the same variables on FRAP shows that the mutual interactions between ethanol concentration (X_1) and temperature (X_2) are significant, as seen by the elliptical contour plots in Figures 2D and 3D. Increasing the ethanol concentration and extraction temperature suggests a gradual increase of FRAP assays. As a result, high temperature promotes extraction by increasing the diffusion coefficient and solute's solubility [62]. Figures 2H and 3H illustrate the linkage between ethanol concentration and extraction time for FRAP assay. With increasing ethanol concentration, an increase in FRAP assay was seen, although the trend slowed after the ethanol concentration reached 65%. Additionally, longer extraction times were seen to result in a higher FRAP value. The effect of temperature and time was quite the opposite, as increases in temperature and time values afforded higher DPPH activity. In particular, DPPH activity from the two seeds was facilitated at medium ethanol proportion and temperature, whereas efficient extraction from LS required an ethanol proportion of around 55%. However, there has been a consistency in ethanol concentration and in all cases; medium ethanol (50%) was demonstrated as the most appropriate. Optimal extraction durations varied from 140 to 220 min (Figures 2 and 3). ABTS values were two to three times higher than DPPH and FRAP values. This might be due to the significant differences in their response to antioxidants, which ABTS and DPPH radicals show similar bi-phase kinetic reactions with many antioxidants, and the FRAP method is based on the reduction of a ferric analogue [63].

3.4. Experimental Validation of the Optimal Conditions

The predictive capacities of the model were verified by establishing the optimum condition using the simplex technique and the highest desirability for extraction yield, EPC, DPPH, ABTS, and FRAP antioxidant activity from LS and LoS. The extraction of PCs from LS and LoS has been greatly affected by ethanol concentration, extraction temperature, and extraction time. The optimum condition for ethanolic extraction for LS and LoS were pursued to maximize the EY, EPC, and AOA as determined by the DPPH, ABTS, and FRAP assays. Estimating economic circumstances that will decrease energy and solvent consumption while maintaining higher EY and AOA is plausible given the requirement to lower actual production expenses. The optimal conditions for this specific confluence of variables result in the optimal extraction condition within the target, as produced by the desire function approach. A 95% mean confidence range around the predicted value for those responses encompassed the measured values. Experimental data for the response extraction yield (mg/100 g DW), EPC (mg GAE/100 g DW), DPPH (mg GAE/100 g DW), ABTS (mg GAE/100 g DW) and FRAP (mg/100 g DW) of LS and LoS are shown in Tables 6 and 7, respectively, under different extraction conditions shown in Table 2. These findings support the capacity of the model to predict the extraction yield, antioxidant activity measured by EPC, DPPH, ABTS, and FRAP from LS and LoS under experimental conditions. The ethanol proportion of 41% and 53%, extraction temperature of 51 °C and 58 °C, and extraction time of 139 min and 220 min, were the optimal conditions that maximized the extraction yields, EPC and AAO from LS and LoS, respectively. For LS and LoS, the equivalent anticipated response values for yield, EPC, DPPH, ABTS, and FRAP under the optimal conditions were 8.9, 14.2; 967, 6144; 383, 2401; 1117, 8353, and 382, 3609, respectively. The consistency of the predicted and experimental results served to verify the RSM model for the extraction process. It was found that the EY found for fruit by-products, including grape seed (19.2%) [64]. At different solvent, temperature and time, total phenolic compounds of the jamun seed were 72 ± 2.5 mg GAE/g seed extract, guarana seed (119 mg GAE/g seed extract), date seed (55 mg GAE/g seed extract), lychee seed (17.9 mg GAE/g seed extract), grape seed (35–65 mg GAE/g seed extract), jackfruit seed (27.7 mg GAE/g seed extract), longan seed (62.6 mg GAE/g seed extract), and tamarind seed (94.5 mg GAE/g seed extract) [46]. The RSM was effectively used to optimize extraction yield and phenolic antioxidant chemicals from LS and LoS. The second order polynomial model well described the experimental data. In terms of effect on extraction performance, the three factors investigated in this study might be prioritized as follows: ethanol concentration > temperature >> extraction time. The EPC was shown to be significantly correlated to DPPH, ABTS, and FRAP.

Table 6. Experimental data for the response extraction yield, extractable phenolic compounds, DPPH, ABTS, and FRAP assay of LS under various extraction conditions as shown in Table 2.

St. Order	Response *									
	Extraction Yield, %		EPC		DPPH		ABTS		FRAP	
	Observed	Predicted	Observed	Predicted	Observed	Predicted	Observed	Predicted	Observed	Predicted
5	8.60 ± 0.1	8.62	841 ± 10	837	335 ± 0.7	333	971 ± 36	959	324 ± 2.5	327
10	5.71 ± 0.2	5.91	338 ± 29	434	134 ± 3.9	146	581 ± 33	599	132 ± 8.3	186
6	7.05 ± 0.4	6.69	765 ± 32	699	252 ± 0.9	262	743 ± 63	778	275 ± 3.6	254
3	7.83 ± 0.3	8.03	911 ± 9	923	361 ± 1.3	371	1138 ± 21	1143	387 ± 6.9	386
14	7.57 ± 0.1	7.89	987 ± 78	990	408 ± 3.4	401	1201 ± 35	1167	398 ± 3.0	403
8	6.38 ± 0.1	6.34	717 ± 31	695	218 ± 4.0	235	845 ± 59	856	330 ± 2.2	305
12	7.60 ± 0.1	7.69	819 ± 24	804	324 ± 2.8	298	1076 ± 11	1007	374 ± 2.7	383
15	7.62 ± 0.1	7.60	940 ± 18	937	406 ± 9.8	394	1097 ± 35	1135	400 ± 6.3	390
13	7.48 ± 0.3	7.39	829 ± 31	902	334 ± 3.8	388	995 ± 31	1103	340 ± 1.1	364
16	7.56 ± 0.2	7.60	938 ± 37	937	420 ± 3.7	394	1168 ± 98	1135	403 ± 7.6	390
2	6.59 ± 0.1	6.69	511 ± 16	463	214 ± 0.9	181	662 ± 93	597	217 ± 2.7	179
9	8.31 ± 0.2	8.33	743 ± 50	706	281 ± 4.6	265	942 ± 38	905	301 ± 20.5	278
4	7.26 ± 0.1	7.09	788 ± 18	737	314 ± 3.4	300	962 ± 38	961	373 ± 8.5	350
11	7.22 ± 0.1	7.36	519 ± 19	611	197 ± 0.7	220	585 ± 87	636	193 ± 10.9	213
1	7.39 ± 0.1	7.28	723 ± 51	690	262 ± 1.5	252	810 ± 100	779	231 ± 10.6	235
7	8.88 ± 0.2	8.62	749 ± 15	792	275 ± 0.5	306	926 ± 10	1038	341 ± 6.4	358
17	7.65 ± 0.4	7.60	947 ± 11	937	407 ± 0.7	394	1227 ± 5	1135	373 ± 17.4	390

* EPC, DPPH, ABTS, and FRAP (mg GAE/100 g dry sample).

Table 7. Experimental data for the response extraction yield, extractable phenolic compounds, DPPH, ABTS, and FRAP assay of LoS under various extraction conditions as shown in Table 2.

St. Order	Response *									
	Extraction Yield, %		EPC		DPPH		ABTS		FRAP	
	Observed	Predicted	Observed	Predicted	Observed	Predicted	Observed	Predicted	Observed	Predicted
5	12.21 ± 0.2	13.19	5309 ± 137	5472	2052 ± 171	2078	7569 ± 570	7934	3258 ± 88	3369
10	8.09 ± 0.3	8.67	2606 ± 133	2634	1160 ± 62	1259	3864 ± 657	4420	1977 ± 63	2271
6	9.99 ± 0.5	9.66	4192 ± 63	4043	1552 ± 60	1427	5613 ± 341	5508	2799 ± 15	2698
3	12.97 ± 0.3	13.14	5521 ± 118	5683	1925 ± 89	1886	7140 ± 470	6751	2876 ± 22	2868
14	13.75 ± 0.3	14.04	5979 ± 12	5910	2340 ± 9	2497	7511 ± 285	7990	3591 ± 158	3574
8	12.96 ± 0.3	12.42	3457 ± 38	4128	2177 ± 84	2086	6175 ± 924	5576	3032 ± 50	2921
12	13.28 ± 0.2	13.29	5301 ± 210	5007	1760 ± 31	1936	5324 ± 450	5364	2370 ± 39	2495
15	15.49 ± 0.4	14.04	5965 ± 155	6016	2287 ± 17	2338	9173 ± 62	8339	3563 ± 78	3603
13	13.32 ± 0.1	14.04	5286 ± 194	6122	2127 ± 19	2179	7616 ± 904	8689	3303 ± 222	3474
16	14.29 ± 0.2	14.04	6325 ± 165	6016	2382 ± 73	2338	8697 ± 532	8339	3546 ± 144	3603
2	9.23 ± 0.3	9.66	4124 ± 66	4169	1389 ± 27	1452	5070 ± 989	4726	2433 ± 29	2284
9	13.63 ± 0.2	12.24	5206 ± 167	5037	1898 ± 34	1978	6626 ± 141	6091	3021 ± 54	2882
4	12.47 ± 0.2	12.42	4788 ± 105	4254	1798 ± 6	1683	7090 ± 578	7190	3032 ± 41	2811
11	11.84 ± 0.1	11.01	4711 ± 23	4864	1560 ± 7	1564	5662 ± 209	5644	2692 ± 69	2721
1	12.41 ± 0.4	13.19	5835 ± 16	5598	2177 ± 42	2104	7048 ± 25	7152	3358 ± 35	3360
7	12.25 ± 0.5	13.14	5280 ± 159	5556	2442 ± 5	2290	4329 ± 280	5137	2533 ± 125	2573
17	14.05 ± 0.5	14.04	6642 ± 138	6016	2405 ± 11	2338	8682 ± 98	8339	3727 ± 58	3603

* EPC, DPPH, ABTS, and FRAP (mg GAE/100 g dry sample).

4. Conclusions

The experimental design method was successful in optimizing phenolic antioxidant extraction conditions from LS and LoS. The RSM has been shown to be useful in assessing the influence of three independent variables (ethanol content, extraction temperature, and time). In terms of effect on extraction performance, the three factors evaluated in this study might be prioritized as follows: ethanol concentration > extraction temperature >> extraction time. LS and LoS have high levels of phenolic compounds with antioxidant activity, which may be recovered using solid-to-water extraction. The solid-to-liquid ratio influenced the extraction of phenolic chemicals. A solid-to-liquid ratio of 1:20 (g/mL) had a substantial influence on EY, EPC, and AOA. This was selected as the optimum parameter. The findings of the experiments revealed that ethanol content and temperature had a substantial influence on the response values. The extraction yield of phenolic compounds, EPC, and FRAP rose as the solid-to-liquid ratio increased. The optimized conditions for maximum EY, EPC, and AOA of the LS and LoS were 41% and 53% ethanol concentration, temperatures of 51 °C and 58 °C, and times of 139 and 220 min, respectively. An increase in extraction temperature increased EPC quantity and, as a result, antioxidant activity ($p < 0.01$), but only up to 55 °C and reduced above this threshold. The interaction effects of temperature and ethanol percentage mostly impacted both seed extracts. It is anticipated that optimization techniques may help to extract phenolic antioxidants from LS and LoS more efficiently and hence more cost-effectively. The results suggested that active chemicals from the solid matrix might be mobilized to a certain extent. However, future research should focus on the utilization of novel and green extraction technologies such as supercritical fluid extraction, enzyme assisted extraction, ultrasonic assisted extraction, microwave assisted extraction, and as well as some new extraction solvents such as ionic liquid, low eutectic solvent and glycerol to optimize the extraction of phenolic antioxidant from food matrix.

Author Contributions: Conceptualization, S.R.; methodology, S.S.-U., P.K., M.A.R.M. and S.R.; formal analysis, S.S.-U., P.K., M.A.R.M. and S.R.; resources, S.R.; data curation, S.S.-U., P.K. and M.A.R.M.; writing—original draft preparation, S.S.-U., P.K. and M.A.R.M.; writing—review and editing, M.A.R.M. and S.R.; supervision, S.R.; project administration, S.R.; funding acquisition, S.R. All authors have read and agreed to the published version of the manuscript.

Funding: The authors gratefully acknowledge the financial support from Mae Fah Luang University via the postdoctoral fellowship by Mae Fah Luang University, Chiang Rai, Thailand (grant no.: 09/2023).

Data Availability Statement: Data are contained within the article.

Acknowledgments: The authors also warmly thank the staff of the Food Science and Technology Laboratory and Scientific and Technological Instruments Center (STIC), Mae Fah Luang University, for the facilities support and technical support regarding using any advanced instrument.

Conflicts of Interest: The authors declare no conflict of interest. The funders had no role in the design of the study; in the collection, analyses, or interpretation of data; in the writing of the manuscript; or in the decision to publish the result.

References

1. Production Volume of Lychee and Longan in Thailand in 2022, by Region. Available online: https://www.statista.com/statistics/1319829/thailand-lychee-production-by-region/#:~:text=In%202022%2C%20the%20production%20volume,to%20around%2094%20thousand%20rai (accessed on 26 June 2023).
2. He, N.; Wang, Z.; Yang, C.; Lu, Y.; Sun, D.; Wang, Y.; Shao, W.; Li, Q. Isolation and identification of polyphenolic compounds in longan pericarp. *Sep. Purif. Technol.* **2009**, *70*, 219–224. [CrossRef]
3. Soong, Y.Y.; Barlow, P.J. Isolation and structure elucidation of phenolic compounds from longan (*Dimocarpus lognan* Lour.) seed by high-performance liquid chromatography–electrospray ionization mass spectrometry. *J. Chromatogr. A* **2005**, *1085*, 270–277. [CrossRef] [PubMed]
4. Bhushan, B.; Kumar, S.; Mahawar, M.K.; Jalgaonkar, K.; Dukare, A.S.; Bibwe, B.; Meena, V.S.; Negi, N.; Narwal, R.K.; Pal, A. Nullifying phosphatidic acid effect and controlling phospholipase D associated browning in litchi pericarp through combinatorial application of hexanal and inositol. *Sci. Rep.* **2019**, *9*, 2402. [CrossRef] [PubMed]
5. Bhushan, B.; Pal, A.; Narwal, R.; Meena, V.S.; Sharma, P.C.; Singh, J. Combinatorial approaches for controlling pericarp browning in Litchi (*Litchi chinensis*) fruit. *J. Food Sci. Technol.* **2015**, *52*, 5418–5426. [CrossRef]
6. Punia, S.; Kumar, M. Litchi (*Litchi chinenis*) seed: Nutritional profile, bioactivities, and its industrial applications. *Trends Food Sci. Technol.* **2021**, *108*, 58–70. [CrossRef]

7. Guo, H.; Luo, H.; Yuan, H.; Xia, Y.; Shu, P.; Huang, X.; Lu, Y.; Liu, X.; Keller, E.T.; Sun, D.; et al. Litchi seed extracts diminish prostate cancer progression via induction of apoptosis and attenuation of EMT through Akt/GSK-3beta signaling. *Sci. Rep.* **2017**, *7*, 41656. [CrossRef]
8. Zhu, X.; Wang, H.; Sun, J.; Yang, B.; Duan, X.; Jiang, Y. Pericarp and seed of litchi and longan fruits: Constituent, extraction, bioactive activity, and potential utilization. *J. Zhejiang Uni. Sci. B* **2019**, *20*, 503–512. [CrossRef]
9. Wisitsak, P.; Nimkamnerd, J.; Thitipramote, N.; Saewan, N.; Chaiwut, P.; Pintathong, P. Comparison the bioactive compounds and their activities between longan and litchi seeds extracts. In Proceedings of the 1st Mae Fah Luang University International Conference, Chiang Rai, Thailand, 29 November–1 December 2012.
10. Tandee, K.; Kittiwachana, S.; Mahatheeranont, S. Antioxidant activities and volatile compounds in longan (*Dimocarpus longan* Lour.) wine produced by incorporating longan seeds. *Food Chem.* **2021**, *348*, 128921. [CrossRef]
11. Chen, J.-Y.; Xu, Y.-J.; Ge, Z.-Z.; Zhu, W.; Xu, Z.; Li, C.-M. Structural elucidation and antioxidant activity evaluation of key phenolic compounds isolated from longan (*Dimocarpus longan* Lour.) seeds. *J. Funct. Food* **2015**, *17*, 872–880. [CrossRef]
12. Rangkadilok, N.; Worasuttayangkurn, L.; Bennett, R.N.; Satayavivad, J. Identification and quantification of polyphenolic compounds in longan (*Euphoria longana* Lam.) fruit. *J. Agric. Food Chem.* **2005**, *53*, 1387–1392. [CrossRef]
13. Zheng, G.; Xu, L.; Wu, P.; Xie, H.; Jiang, Y.; Chen, F.; Wei, X. Polyphenols from longan seeds and their radical-scavenging activity. *Food Chem.* **2009**, *116*, 433–436. [CrossRef]
14. Anwar, F.; Mahmood, T.; Mehmood, T.; Aladedunye, F. Composition of fatty acids and tocopherols in cherry and lychee seed oils. *J. Adv. Biol.* **2014**, *1*, 586–593.
15. Koul, B.; Singh, J. Lychee biology and biotechnology. In *The Lychee Biotechnology*; Springer: Singapore, 2017; pp. 137–192.
16. Matthaus, B.; Vosmann, K.; Pham, L.Q.; Aitzetmüller, K. FA and tocopherol composition of Vietnamese oilseeds. *J. Am. Oil Chem. Soc.* **2003**, *80*, 1013–1020. [CrossRef]
17. Ding, G. The research progress of *Litchi chinensis*. *Lishizhen Med. Mater. Med. Res.* **1999**, *10*, 145–146.
18. Prasad, K.N.; Yang, B.A.O.; Zhao, M.; Ruenroengklin, N.; Jiang, Y.U.E.M.I.N.G. Application of ultrasonication or high-pressure extraction of flavonoids from litchi fruit pericarp. *J. Food Proc. Eng.* **2009**, *32*, 828–843. [CrossRef]
19. Yang, B.; Wang, H.; Prasad, N.; Pan, Y.; Jiang, Y. Use of litchi (*Litchi sinensis* sonn.) seeds in health. In *Nuts and Seeds in Health and Disease Prevention*; Academic Press: Cambridge, MA, USA, 2011; pp. 699–703.
20. Liyana-Pathirana, C.; Shahidi, F. Optimization of extraction of phenolic compounds from wheat using response surface methodology. *Food Chem.* **2005**, *93*, 47–56. [CrossRef]
21. Karacabey, E.; Mazza, G. Optimisation of antioxidant activity of grape cane extracts using response surface methodology. *Food Chem.* **2010**, *119*, 343–348. [CrossRef]
22. Wijngaard, H.H.; Brunton, N. The optimisation of solid–liquid extraction of antioxidants from apple pomace by response surface methodology. *J. Food Process Eng.* **2010**, *96*, 134–140. [CrossRef]
23. Yang, B.; Jiang, Y.; Shi, J.; Chen, F.; Ashraf, M. Extraction and pharmacological properties of bioactive compounds from longan (*Dimocarpus longan* Lour.) fruit—A review. *Food Res. Int.* **2011**, *44*, 1837–1842. [CrossRef]
24. Sudjaroen, Y.; Hull, W.E.; Erben, G.; Wurtele, G.; Changbumrung, S.; Ulrich, C.M.; Owen, R.W. Isolation and characterization of ellagitannins as the major polyphenolic components of Longan (*Dimocarpus longan* Lour) seeds. *Phytochemistry* **2012**, *77*, 226–237. [CrossRef]
25. Li, X.; Huang, J.; Wang, Z.; Jiang, X.; Yu, W.; Zheng, Y.; Li, Q.; He, N. Alkaline extraction and acid precipitation of phenolic compounds from longan (*Dimocarpus longan* L.) seeds. *Sep. Purif. Technol.* **2014**, *124*, 201–206. [CrossRef]
26. Potisate, Y.; Pintha, K. Optimum extraction condition and dehydration of lychee seeds extracted. *Naresuan Phayao J.* **2021**, *14*, 105–112.
27. Paliga, M.; Novello, Z.; Dallago, R.M.; Scapinello, J.; Magro, J.D.; Di Luccio, M.; Tres, M.V.; Oliveira, J.V. Extraction, chemical characterization and antioxidant activity of *Litchi chinensis* Sonn. and *Avena sativa* L. seeds extracts obtained from pressurized *n*-butane. *J. Food Sci. Technol.* **2017**, *54*, 846–851. [CrossRef] [PubMed]
28. Hossain, M.A.; Hossain, M.S. Optimization of antioxidative phenolic compound extraction from freeze-dried pulp, peel, and seed of Burmese grape (*Baccaurea ramifora* Lour.) by response surface methodology. *Biomass Convers. Biorefinery* **2023**, *13*, 8123–8137. [CrossRef]
29. Qu, W.; Pan, Z.; Ma, H. Extraction modeling and activities of antioxidants from pomegranate marc. *J. Food Eng.* **2010**, *99*, 1623. [CrossRef]
30. Mashkor, I.M.A.A. Total phenol, total favonoids and antioxidant activity of pomegranate peel. *Int. J. Chem. Technol. Res.* **2014**, *6*, 4656–4661.
31. Soong, Y.Y.; Barlow, P.J. Antioxidant activity and phenolic content of selected fruit seeds. *Food Chem.* **2004**, *88*, 411–417. [CrossRef]
32. Eberhardt, M.V.; Lee, C.Y.; Liu, R.H. Antioxidant activity of fresh apples. *Nature* **2000**, *405*, 903–904. [CrossRef]
33. Box, G.E.P.; Wilson, K.B. Experimental attainment of optimum conditions. *J. R. Stat. Soc.* **1951**, *13*, 267. [CrossRef]
34. Mazumder, M.A.R.; Tolaema, A.; Chaikhemarat, P.; Rawdkuen, S. Antioxidant and anti-cytotoxicity effect of phenolic extracts from *Psidium guajava* Linn. leaves by novel assisted extraction techniques. *Foods* **2023**, *12*, 2336. [CrossRef]
35. Rawdkuen, S.; Sai-Ut, S.; Benjakul, S. Optimizing the tyrosinase inhibitory and antioxidant activity of mango seed kernels with a response surface methodology. *Food Anal. Methods* **2016**, *9*, 3032–3043. [CrossRef]

36. Swain, T.; Hillis, W.E. The phenolic constituents of *Prunus domestica* 1. The quantitative analysis of phenolic constituents. *J. Sci. Food Agric.* **1959**, *10*, 63–68. [CrossRef]
37. Brand-Williams, W.; Cuvelier, M.E.; Berset, C. Use of a free radical method to evaluate antioxidant activity. *Food Sci. Technol.* **1995**, *28*, 25–30. [CrossRef]
38. Benzie, I.F.F.; Strain, J.J. The ferric reducing ability of plasma (FRAP) as a measure of antioxidant power: The FRAP assay. *Anal. Biochem.* **1996**, *239*, 70–76. [CrossRef]
39. Arnao, M.B.; Cano, A.; Acosta, M. Hydrophilic and lipophilic antioxidant activity contribution to total antioxidant activity. *Food Chem.* **2001**, *73*, 239–244. [CrossRef]
40. Ilaiyaraja, N.; Likhith, K.R.; Sharath Babu, G.R.; Khanum, F. Optimisation of extraction of bioactive compounds from *Feronia limonia* (wood apple) fruit using response surface methodology (RSM). *Food Chem.* **2015**, *173*, 348–354. [CrossRef]
41. Feng, S.; Luo, Z.; Tao, B.; Chen, C. Ultrasonic-assisted extraction and purification of phenolic compounds from sugarcane (*Saccharum officinarum* L.) rinds. *LWT-Food Sci. Technol.* **2015**, *60*, 970–976. [CrossRef]
42. Ondrejovic, M.; Benkovicva, H.; Silhar, S. Optimization of rosmarinic acid extraction from lemon balm (*Melissa officinalis*) Nova. *Biotechnologica* **2009**, *9*, 175.
43. Daorueang, D. Phenolic content and antioxidant activity of lychee seed extract. *Naresuan Phayao J.* **2019**, *12*, 25–27.
44. Prasad, K.N.; Yang, B.; Yang, S.; Chen, Y.; Zhao, M.; Ashraf, M.; Jiang, Y. Identification of phenolic compounds and appraisal of antioxidant and antityrosinase activities from litchi (*Litchi sinensis* Sonn.) seeds. *Food Chem.* **2009**, *116*, 1–7. [CrossRef]
45. Chindaluang, Y.; Sriwattana, S. Comparison of Ultrasonic Extraction with Conventional Extraction Methods of Phenolic Compounds in Longan (*Euphoria longana* Lamk.) Seed. *Food Appl. Biosci.* **2014**, *13*, 439–448. [CrossRef]
46. Rerk-am, U.; Runglert, K.; Eiamwat, J.; Kongsombat, B.; Sematong, T.; Tangsatirapakdee, S.; Thubthimthed, S. Anti-oxidant activities and polyphenolic compounds of Longan (*Dimocarpus longan* Lour) peel and seed extracts. *Thai J. Pharm. Sci.* **2016**, *40*, 120–122.
47. Hong-in, P.; Neimkhum, W.; Punyoyai, C.; Sriyab, S.; Chaiyana, W. Enhancement of phenolics content and biological activities of longan (*Dimocarpus longan* Lour.) treated with thermal and ageing process. *Sci. Rep.* **2021**, *11*, 15977. [CrossRef] [PubMed]
48. Joglekar, A.M.; May, A.T. Product excellence through design of experiments. *Cereal Food World* **1987**, *32*, 857–868.
49. Jung, E.P.; Conrado Thomaz, G.F.; de Brito, M.O.; de Figueiredo, N.G.; Kunigami, C.N.; de Oliveira Ribeiro, L.; Alves Moreira, R.F. Thermal-assisted recovery of antioxidant compounds from Bauhinia forficata leaves: Effect of operational conditions. *J. Appl. Res. Med. Aromat. Plants* **2021**, *22*, 100303. [CrossRef]
50. Chamali, S.; Bendaoud, H.; Bouajila, J.; Camy, S.; Saadaoui, E.; Condoret, J.S.; Romdhane, M. Optimization of accelerated solvent extraction of bioactive compounds from *Eucalyptus intertexta* using response surface methodology and evaluation of its phenolic composition and biological activities. *J. Appl. Res. Med. Aromat. Plants* **2023**, *35*, 100464. [CrossRef]
51. Cacace, J.E.; Mazza, G. Mass transfer process during extraction of phenolic compounds from milled berries. *J. Food Eng.* **2003**, *59*, 379–389. [CrossRef]
52. Pompeu, D.R.; Silva, E.M.; Rogez, H. Optimisation of the solvent extraction of phenolic antioxidants from fruits of *Euterpe oleracea* using Response Surface Methodology. *Bioresour. Technol.* **2009**, *100*, 6076–6082. [CrossRef]
53. Chen, M.; Zhao, Y.; Yu, S. Optimisation of ultrasonic-assisted extraction of phenolic compounds, antioxidants, and anthocyanins from sugar beet molasses. *Food Chem.* **2015**, *172*, 543–550. [CrossRef]
54. Ha Lai, T.N.; André, C.M.; Chirinos, R.C.; Thuy Nguyen, T.B.; Larondelle, Y.; Rogez, H. Optimisation of extraction of piceatannol from Rhodomyrtus tomentosa seeds using response surface methodology. *Sep. Purif. Technol.* **2014**, *134*, 139–146. [CrossRef]
55. Chew, K.K.; Ng, S.Y.; Thoo, Y.Y.; Khoo, M.Z.; Wan Aida, W.M.; Ho, C.W. Effect of ethanol concentration, extraction time and extraction temperature on the recovery of phenolic compounds and antioxidant capacity of *Centella asiatica* extracts. *Int. Food Res. J.* **2011**, *18*, 571–578.
56. Kossah, R.; Nsabimana, C.; Zhang, H.; Chen, W. Optimization of extraction of polyphenols from Syrian sumac (*Rhus coriaria* L.) and Chinese sumac (*Rhus typhina* L.) fruits. *Res. J. Phytochem.* **2010**, *4*, 146–153. [CrossRef]
57. Hismath, I.; Wan Aida, W.M.; Ho, C.W. Optimization of extraction conditions for phenolic compounds from neem (*Azadirachta indica*) leaves. *Int. Food Res. J.* **2011**, *18*, 931–939.
58. Chan, S.W.; Lee, C.Y.; Yap, C.F.; Wan Aida, W.M.; Ho, C.W. Optimisation of extraction conditions for phenolic compounds from limau purut (*Citrus hystrix*) peels. *Int. Food Res. J.* **2009**, *16*, 203–213.
59. Silva, E.M.; Rogez, H.; Larondelle, Y. Optimization of extraction of phenolics from Inga edulis leaves using response surface methodology. *Sep. Purif. Technol.* **2007**, *55*, 381–387. [CrossRef]
60. Gribova, N.Y.; Filippenko, T.A.; Nikolaevskii, A.N.; Belaya, N.I.; Tsybulenko, A.A. Optimization of conditions for the extraction of antioxidants from solid parts of medicinal plants. *J. Anal. Chem.* **2008**, *63*, 1034–1037. [CrossRef]
61. Dorta, E.; Lobo, M.G.; Gonzalez, M. Reutilization of mango byproducts: Study of the effect of extraction solvent and temperature on their antioxidant properties. *J. Food Sci.* **2012**, *77*, C80–C88. [CrossRef]
62. Şahin, S.; Aybastıer, Ö.; Işık, E. Optimisation of ultrasonic-assisted extraction of antioxidant compounds from *Artemisia absinthium* using response surface methodology. *Food Chem.* **2013**, *141*, 1361–1368. [CrossRef]

63. Amadou, I.; Yong-Hui, S.; Sun, J.; Guo-Wei, L. Fermented soybean products: Some methods, antioxidants compound extraction and their scavenging activity. *Asian J. Biochem.* **2009**, *4*, 68–76. [CrossRef]
64. Oktaya, M.; Gulcin, I.; Kufrevioglu, O.I. Determination of in Vitro Antioxidant Activity of Fennel (*Foeniculum vulgare*) Seed Extracts. *LWT Food Sci. Technol.* **2003**, *36*, 263–271. [CrossRef]

foods

MDPI

Article

Comparative Untargeted Metabolic Profiling of Different Parts of *Citrus sinensis* Fruits via Liquid Chromatography–Mass Spectrometry Coupled with Multivariate Data Analyses to Unravel Authenticity

Sherif M. Afifi [1,2,*], Eman M. Kabbash [3], Ralf G. Berger [4], Ulrich Krings [4] and Tuba Esatbeyoglu [2,*]

1 Pharmacognosy Department, Faculty of Pharmacy, University of Sadat City, Sadat City 32897, Egypt
2 Department of Food Development and Food Quality, Institute of Food Science and Human Nutrition, Gottfried Wilhelm Leibniz University Hannover, Am Kleinen Felde 30, 30167 Hannover, Germany
3 Phytochemistry Department, National Organization for Drug Control and Research, Giza 12622, Egypt
4 Institute of Food Chemistry, Gottfried Wilhelm Leibniz University Hannover, Callinstraße 5, 30167 Hannover, Germany
* Correspondence: sherif.afifi@fop.usc.edu.eg (S.M.A.); esatbeyoglu@lw.uni-hannover.de (T.E.)

Abstract: Differences between seven authentic samples of *Citrus sinensis* var. Valencia peel (albedo and flavedo) and juices from Spain and Uruguay, in addition to a concentrate obtained from Brazil, were investigated by untargeted metabolic profiling. Sixty-six metabolites were detected by nano-liquid chromatography coupled to a high-resolution electrospray-ionization quadrupole time-of-flight mass spectrometer (nLC-ESI-qTOF-MS) belonging to phenolic acids, coumarins, flavonoid glycosides, limonoids, terpenes, and fatty acids. Eleven metabolites were detected for the first time in *Citrus sinensis* and identified as citroside A, sinapic acid pentoside, apigenin-*C*-hexosyl-*O*-pentoside, chrysoeriol-*C*-hexoside, di-hexosyl-diosmetin, perilloside A, gingerol, ionone epoxide hydroxy-sphingenine, xanthomicrol, and coumaryl alcohol-*O*-hexoside. Some flavonoids were completely absent from the juice, while present most prominently in the *Citrus* peel, conveying more industrial and economic prospects to the latter. Multivariate data analyses clarified that the differences among orange parts overweighed the geographical source. PCA analysis of ESI-(−)-mode data revealed for hydroxylinoleic acid abundance in flavedo peel from Uruguay the most distant cluster from all others. The PCA analysis of ESI-(+)-mode data provided a clear segregation of the different *Citrus sinensis* parts primarily due to the large diversity of flavonoids and coumarins among the studied samples.

Keywords: albedo; flavedo; concentrate; juice; orange; metabolomics; flavonoids; coumarins

Citation: Afifi, S.M.; Kabbash, E.M.; Berger, R.G.; Krings, U.; Esatbeyoglu, T. Comparative Untargeted Metabolic Profiling of Different Parts of *Citrus sinensis* Fruits via Liquid Chromatography–Mass Spectrometry Coupled with Multivariate Data Analyses to Unravel Authenticity. *Foods* **2023**, *12*, 579. https://doi.org/10.3390/foods12030579

Academic Editors: Noelia Castillejo Montoya and Lorena Martínez-Zamora

Received: 22 December 2022
Revised: 18 January 2023
Accepted: 18 January 2023
Published: 29 January 2023

1. Introduction

Citrus trees bear some of the most popular fruits and are grown globally for food, medicinal and other industrial applications, with a total annual production of nearly 85 million tons [1]. Various species of *Citrus* genus are valuable, such as *C. medica* (citron), *C. limon* (lemon), *C. aurantium* (sour orange), *C. reticulata* (mandarin, tangerine), *C. paradisi* (grapefruit), *C. clementina* (clementine) and *C. sinensis* (sweet orange) [2]. They are either consumed as fresh fruits or after processing to juices, beverage products, and jams, with the peel being the main by-product of processing. Anatomically, the fruit consists of two parts—the outer peel and the pulp with juice sac glands [2]. The peel main parts include the outer pigmented flavedo with parenchymatous cells and cuticle, and the white albedo part lying beneath the flavedo [3]. Different plant parts find wide application in many countries in recipes to treat stomach disorders, skin inflammation, cough, muscular pain, and nausea, as well as being used as a slimming agent [4].

Phytochemical investigations have revealed the presence of bioactive coumarins (i.e., bergamottin, scopoletin, and umbelliferone), flavonoids (i.e., rutin, quercetin, and kaempferol), limonoids (i.e., limonin and nomilin), and acridone alkaloids (i.e., citruscridone and citrusinine-I) in *Citrus* plants [5]. Dietary fiber, minerals, carotenoids, and phenolic phytoconstituents, i.e., phenolic acids, flavanones, and polymethoxylated flavones, were all detected in orange peel [6]. *Citrus sinensis* var. Valencia fruits revealed superiority as an excellent source of phenolics, flavonoids and ascorbic acid compared to var. Mandora and some grapefruits grown in Cyprus [7]. Another study on var. Valencia pointed out that the flavedo part had highest vitamin C, flavones and carotenoid content, while the albedo part was the main source of flavanones and phenolics [8].

Moreover, previous clinical and animal studies have demonstrated that certain *Citrus* metabolites have antioxidant, anti-inflammatory, cytotoxic, antimicrobial, antiallergic and antiplatelet aggregation activities [9]. The metabolites in *Citrus* varieties influence their unique qualities, such as taste, appearance, and the supposed health benefits. Quantitative and qualitative differences in these metabolites are commonly affected by the cultivation environment and *Citrus* variety [10] and also vary significantly from one part to another within the *Citrus* fruit.

Although there is a wide variety of *Citrus* fruits that were intensely investigated for their metabolic composition [11], comparative studies of metabolite profiles within different fruit parts are limited. Because metabolite variability and modification are affected by the plant's adaptation to physiological, pathological, and diverse chemical stimuli, metabolomics enables quantitative investigation of these dynamic changes [12]. In-depth food analyses help to uncover discriminatory biomarkers, identify various pathways, find therapeutic targets, and discover new potential drug leads by evaluating and verifying the major variations in metabolite profiling [13].

Metabolite profiling by mass spectrometry is regarded as the most powerful analytical tool. Among the most recent detection techniques are: high-performance liquid chromatography–diode array detection (HPLC-DAD), ultra-high-performance liquid chromatography coupled to triple quadrupole mass spectrometry (UHPLC-QqQ-MS/MS), and ultra-performance liquid chromatography/quadrupole time-of-flight mass spectrometry (UPLC-QTOF-MS/MS) [14]. The analysis of the orange peel matrix has also been performed via HPLC-UV-ESI-MS/MS (high-performance liquid chromatography coupled with ultraviolet and electrospray ionization mass spectrometry) [6]. The current technique involves the use of nano-high-performance liquid chromatography (n-LC) coupled to high-resolution MS to provide fast metabolite analysis of orange peel, to the best of our knowledge, for the first time on a high sensitivity level. Previous studies have been performed via state-of-the-art nano-liquid chromatography–ESI–MS/MS technology to only separate chiral naringenin in *Citrus* pulp and peel [15].

By adding multivariate data analysis (MVA) techniques, a large-scale metabolomics dataset is presented in this study to provide a detailed insight into *C. sinensis* secondary metabolites. The use of MVA enables accurate specimen to specimen comparison and highlights distinctive traits [16].

The aim of this work was to investigate the differences in secondary metabolite composition in different parts of *C. sinensis* presenting the first comparative insights into metabolite profiles derived from various parts of Valencia orange fruits collected from different countries. A detailed identification of metabolites in different *C. sinensis* parts was considered together with an untargeted metabolic profile for the edible part and the peel as the main by-product during fruit processing.

2. Materials and Methods

2.1. Orange Samples and Chemicals

Valencia oranges (*C. sinensis*) from Spain and Uruguay, provided by Symrise (Holzminden, Germany) as depicted in Table 1, were cleaned and squeezed to obtain the juice, then flavedo and albedo were manually stripped by a fruit peeler. After that, all orange samples, including a Brazilian orange concentrate acquired from Symrise, were lyophilized and separately ground to fine powder by a mill rotor cyclone (Tecnal, Piracicaba, Brazil,

TE-651/2). Comminuted orange samples were stored at −80 °C until subsequent analysis. All chemicals and solvents were purchased from Sigma Aldrich (Steinheim, Germany).

Table 1. Source of the different *Citrus sinensis* (var. Valencia) specimens used in the analysis.

Code		Orange Part	Country
AS		Albedo peel	Spain
AU		Albedo peel	Uruguay
CB		Juice concentrate	Brazil
FS		Flavedo peel	Spain
FU		Flavedo peel	Uruguay
JS		Juice	Spain
JU		Juice	Uruguay

2.2. Metabolite Mass Fingerprinting

Orange specimens (2 mg), extracted by 1 mL methanol, were spiked with 4 μg mL $^{-1}$ hesperidin followed by sonication for 30 min, then centrifugation for 20 min at 15,000× *g* to eliminate any leftover debris. Solid-phase extraction was applied to each extract using a C_{18} cartridge (JT Baker, Phillipsburg, NJ, USA) as previously reported [17]. The resulting extracts were injected into a nano-LC system EASY-nLC II (Bruker, Bremen, Germany) equipped with a reversed phase column (150 × 0.1 mm, particle size 3 μm; Michrom Bioresources, Auburn, CA, USA) coupled to maXis impact quadrupole-time of-flight (qTOF) MS (Bruker, Bremen, Germany). A captive nano-spray ionization was operated in the negative and positive ion modes under conditions as previously reported [18]. Identification of metabolite mass fingerprints was carried out using exact parent ion masses as well as retention data, reference literature, fragmentation patterns, and the Phytochemical Dictionary of Natural Products Database (https://dnp.chemnetbase.com/ accessed on (16 March 2022)). Semi-quantification was based on the integrated peak areas of each compound after normalization to internal standard. Aiming at comparing the relative abundance of a given compound in the seven different samples, the determination of absolute concentrations by an external calibration of every compound was not required. Three independent replicates of each orange specimen were analyzed in parallel to evaluate the biological variance.

2.3. Multivariate Data Analyses (MVA)

Modeling viz. principal component analysis (PCA) and orthogonal projection to latent structures-discriminant analysis (OPLS-DA) was applied to a metabolite dataset of MS abundances produced by nLC-MS either in the negative or positive ion mode via the SIMCA-P+ 13.0 software package (Umetrics, Umeå, Sweden) to pinpoint various markers characterizing each group declared with correlation (pcor) and covariance (p). The iterative permutation testing and diagnostic indices, viz. R2 and Q2 values, were used to assess the validity of models, while all variables were mean centered and Pareto scaled.

3. Results and Discussion

3.1. Metabolite Identification via nLC–ESI-MS/MS Analysis

nLC-ESI-MS/MS analyses (Figure 1) of *Citrus* samples resulted in the identification of 66 metabolites, categorized into seven classes: phenolic acids, terpenes, limonoids, coumarins, flavonoids, fatty acids and nitrogenous compounds. The elution order of metabolites followed a sequence of decreasing polarity, whereby phenolic acids eluted first, followed by coumarins, flavonoid glycosides, limonoids, free aglyca and fatty acids. Samples were analyzed in both the negative and positive ionization modes to provide a greater coverage of the metabolome. Fatty acids and flavonoids were preferentially ionized under negative ionization conditions, while coumarins, limonoids and nitrogenous compounds showed better ionization in the positive mode. The list of identified compounds along with their retention time, characteristic molecular and fragment ions and occurrence is presented in Table 2.

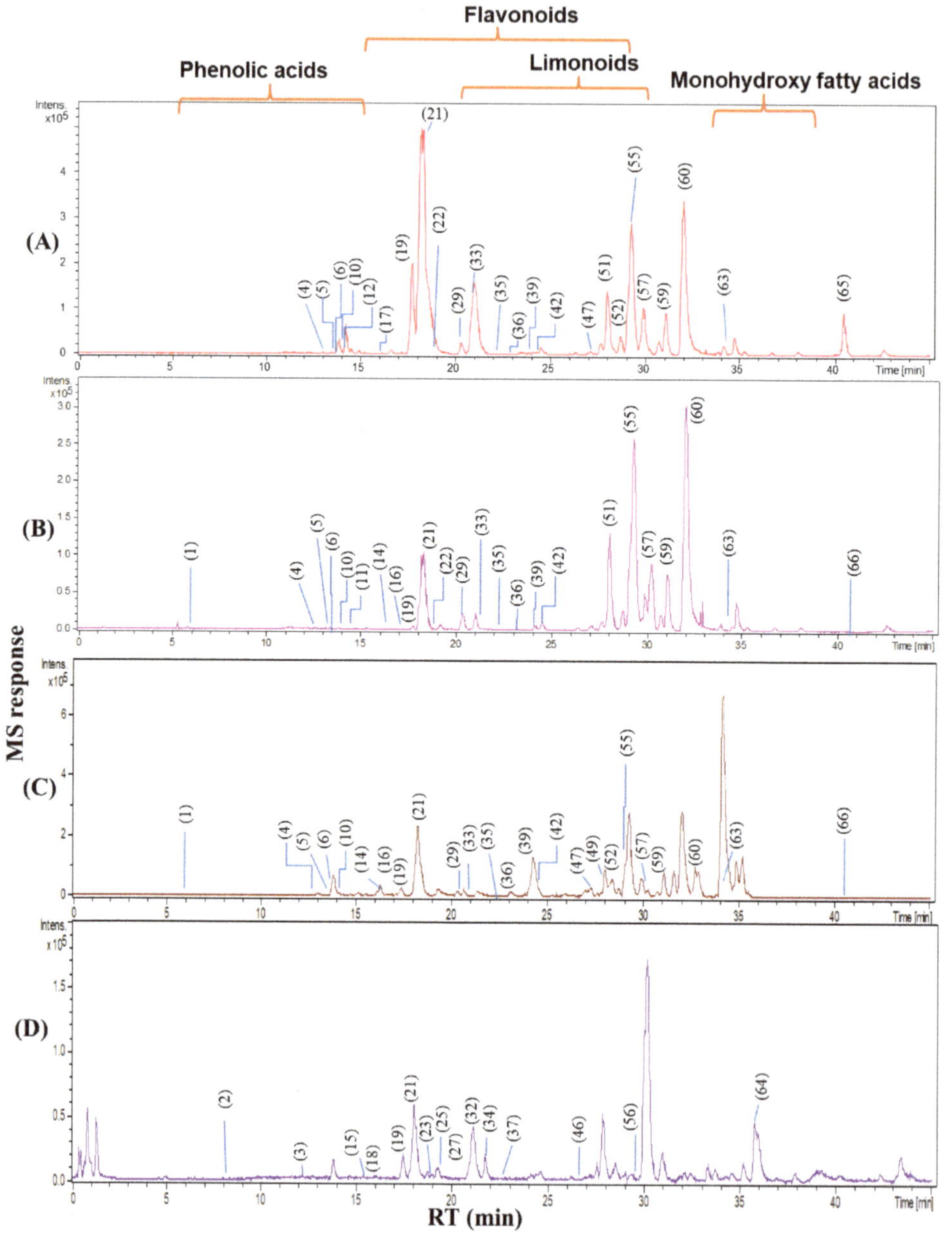

Figure 1. Representative nLC-MS base peak chromatogram of orange parts extracted by methanol on negative ionization (**A**) albedo from Uruguay, (**B**) orange concentrate from Brazil, (**C**) flavedo from Uruguay and on positive ionization, and (**D**) juice from Uruguay. For peak numbers, refer to Table 2.

3.1.1. Identification of Organic Acids and Phenolics

A number of organic acids were eluted in the first part of the nLC-ESI-MS/MS-chromatogram, as revealed from their MS spectral data. In the analyzed *Citrus* samples, the most representative phenolic and organic acids were citric, ferulic, sinapic acids and their glycosides. Peak 1—for peak numbers refer to Table 2—[(M−H)$^-$ *m/z* 191.0189 ($C_6H_7O_7$)$^-$] was identified as citric acid previously reported in *Citrus limon* ethanolic extract [19]. Peaks **6** and **8** (Suppl. Figure S1), [(M−H)$^-$ *m/z* 355.103 ($C_{16}H_{19}O_9$)$^-$] were annotated as ferulic acid hexoside and sinapic acid pentoside, respectively. Having the same molecular formula; peak **6** yielded a main fragment ion at *m/z* 193 [M−H−162 (hexose)]$^-$relative to ferulic acid, while peak **8** showed a fragment ion at *m/z* 223 [M−H−132 (pentose)]$^-$ relative to sinapic acid. Sinapic acid pentoside was detected for the first time in all of the examined *Citrus* samples except *Citrus sinensis* juice obtained from Uruguay (JU). In addition, citric acid was absent in the albedo of both suppliers (AS: albedo from Spain, and AU: albedo from Uruguay), while sinapoyl hexoside was absent in the juice (JS: juice from Spain, and JU: juice from Uruguay). Peak **17** [(M−H)$^-$ *m/z* 311.1139 ($C_{15}H_{19}O_7$)$^-$] with product ion at *m/z* 161 corresponding to negatively ionized hexose, was annotated as coumaryl alcohol-*O*-hexoside, first time to be reported in all *Citrus* samples. Peak **55** [(M−H)$^-$ *m/z* 293.1763 ($C_{17}H_{25}O_4$)$^-$] was a major constituent detected in all analyzed samples and identified as gingerol, a pungent principle previously detected in the peel of *Citrus reticulata* [20], but detected for the first time in *Citrus sinensis* and most prominent in albedo from Uruguay and juice from Spain (AU and JS). Ferulic and *p*-coumaric acid were major hydroxycinnamic acids detected in both free and bound form in the methanolic extracts of bitter orange (*Citrus microcarpa*) peel [6].

3.1.2. Identification of Coumarins

Peak **11** [(M−H)$^-$ *m/z* 205.0504 ($C_{11}H_9O_4$)$^-$] yielded fragment ions at *m/z* 175 [M−H−30 (C_2H_6)]$^-$ and *m/z* 101 corresponding to the loss of CO_2 from the pyrone ring, and was identified as citropten, the most abundant coumarin in *Citrus* (Suppl. Figure S2) [19]. Peak **9** [(M−H)$^-$ *m/z* 175.0392 ($C_{10}H_7O_3$)$^-$] was identified as methoxycoumarin and detected in all *Citrus* samples. Peak **26** [(M−H)$^-$ *m/z* 367.0668 ($C_{16}H_{15}O_{10}$)$^-$ showed a characteristic fragment ion at *m/z* 191 [M−H−176 (glucuronyl)]$^-$ was identified as scopoletin-*O*-glucuronide. The flavedo part from Uruguay showed the highest level of the identified coumarins. This finding was in accordance with results previously reported by Liu et al. (2017) showing that the highest coumarin content was found in the flavedo part among different studied *Citrus* cultivars [21], conveying a particular interest regarding the presumed anticancer and antidiabetic effects of this part.

Table 2. Metabolites annotated in various orange samples via nLC -ESI-MS/MS using both the positive and negative ionization modes.

No.	Rt (min)	Compound Name	Chemical Class	$[M-H]^-$/ $[M+H]^+$	Molecular Formula	Mass Error	MS/MS frag.	Ref.	AS	AU	CB	FS	FU	JS	JU
1.	5.7	Citric acid	Organic acid	191.0189	$C_6H_7O_7^-$	4.3	111, 87	[19]	-	-	++	+	+	++	++
2.	11.1	Feruloylagmatine	Cinnamic acid amide	307.177	$C_{15}H_{23}N_4O_3^+$	1.7	264, 177		++	++	+	+	-	+	+
3.	13.0	Hydroxy-phenyl-valeric acid-*O*-sulphate	Sulphate ester	275.0607	$C_{11}H_{15}O_6S^+$	8.4	195		+	-	+	+	++	+	+
4.	13.2	Dihydroxymegastigmadienone -*O*-hexoside (Citroside A)	Terpene	385.1860	$C_{19}H_{29}O_8^-$	0.8	223, 153	[22]	++	+	+	+	++	-	-
5.	13.9	Sinapoyl hexoside	Phenolic acid	385.1141	$C_{17}H_{21}O_{10}^-$	−1.8	223, 190	[19]	++	+	+	+	++	-	-
6.	13.9	Ferulic acid hexoside	Phenolic acid	355.1030	$C_{16}H_{19}O_9^-$	0.0	193, 175, 160	[19]	++	++	++	++	++	+	++
7.	13.7	Citropten	Coumarin	205.0504	$C_{11}H_9O_4^-$	1.1	175, 101	[19]	+	+	+	+	++	+	+
8.	13.9	Sinapic acid pentoside	Phenolic acid	355.1037	$C_{16}H_{19}O_{10}^-$	−0.7	223, 205	[23]	++	+	+	++	+	+	-
9.	13.9	Methoxycoumarin	Coumarin	175.0392	$C_{10}H_7O_3^-$	0.2	-	[24]	+	+	+	+	+	+	+
10.	14.0	Sinapic acid	Phenolic acid	223.0607	$C_{11}H_{11}O_5^-$	0.2	-	[19]	++	++	++	+	++	+	+
11.	14.2	Di-hexosyl-diosmetin	Flavone	625.1757	$C_{28}H_{33}O_{16}^+$	−1	-	[25]	+	+	+	++	++	-	-
12.	14.3	Naringenin-*O*-hexoside	Flavanone	433.1145	$C_{21}H_{21}O_{10}^-$	1.2	271	[26]	++	++	-	+	-	-	-
13.	14.7	Limononic acid	Limonoid degradation	183.1024	$C_{10}H_{15}O_3^-$	1.6	-		++	+	+	+	++	-	+
14.	13.2	Methyl epijasmonate	Fatty acid degradation	223.1337	$C_{13}H_{19}O_3^-$	1.3	-	[27]	++	++	+	+	++	+	+
15.	16.4	Apigenin-di-*C*-hexoside	Flavone	595.1654	$C_{27}H_{31}O_{15}^+$	1.5	385	[26]	+	+	+	++	++	+	+
16.	16.5	Apigenin-di-*O*-hexoside	Flavone	593.1502	$C_{27}H_{29}O_{15}^-$	8.6	431, 269	[26]	+	-	+	+	++	-	-
17.	16.6	Coumaryl alcohol-*O*-hexoside	Cinnamyl alcohol glycoside	311.1139	$C_{15}H_{19}O_7^-$	1.1	269, 161		+	+	+	+	++	+	+
18.	16.8	Apigenin-*C*-hexosyl-*O*-pentoside	Flavone	563.1408 565.1550	$C_{26}H_{27}O_{14}^-$ $C_{26}H_{29}O_{14}^+$	0.3 0.3	311	[28]	-	-	+	+	++	+	+
19.	17.5	Naringenin-*O*-hexosyldeoxyhexoside	Flavanone	579.1720 581.1867	$C_{27}H_{31}O_{14}^-$ $C_{27}H_{33}O_{14}^+$	1.0 −0.5	271 419, 273	[29]	++	++	++	+	+	+	+
20.	17.6	Chrysoeriol-*C*-hexoside	Flavone	463.1238	$C_{22}H_{23}O_{11}^+$	−0.1	343	[26]	+	+	+	++	++	+	+
21.	18.2	Hesperidin (Hesperetin-*O*-hexosyldeoxyhexoside)	Flavanone	609.1833 611.1971	$C_{28}H_{33}O_{15}^-$ $C_{28}H_{35}O_{15}^+$	−2.2 0.1	301 -	[29]	++	++	++	++	++	++	+
22.	18.3	Limonin-hexoside	Limonoid	649.2507	$C_{32}H_{41}O_{14}^-$	0.7	-	[30]	+	+	+	+	+	+	+
23.	18.7	Hesperetin-*O*-deoxyhexoside	Flavanone	449.1625	$C_{22}H_{25}O_{10}^+$	−0.4	303	[19]	++	+	-	+	+	-	-
24.	19.1	Azelaic acid	Dicarboxylic acid	187.0969	$C_9H_{15}O_4^-$	2.1	169, 125		+	+	+	+	+	+	+
25.	19.3	Deacetylnomilin	Limonoid	473.2167	$C_{26}H_{33}O_8^+$	−1.7	455, 411, 161		++	+	+	+	+	+	+
26.	19.4	Scopoletin-*O*-glucuronide	Coumarin	367.0668	$C_{16}H_{15}O_{10}^-$	−0.1	191		+	+	+	+	++	+	+
27.	20.2	Pyridoxamine phosphate	Vitamin B6 phosphate	249.0614	$C_8H_{14}N_2O_5P^+$	−8.4	169, 81	[31]	+	+	++	+	+	+	+
28.	20.3	Unknown	Alkaloid	368.1921	$C_{15}H_{30}NO_9^+$	−13	-		+	+	-	++	++	+	-
29.	20.4	Decatrienoic acid	Fatty acid	165.0913	$C_{10}H_{13}O_2^-$	−5.0	-		+	+	+	+	+	+	+
30.	21.0	Sakuranetin-*O*-hexosyl-*O*-deoxyhexoside	Flavanone	595.2020	$C_{28}H_{35}O_{14}^+$	−1.2	433, 287	[28]	++	++	-	+	+	-	-

Table 2. *Cont.*

No.	Rt (min)	Compound Name	Chemical Class	[M−H]⁻/ [M + H]⁺	Molecular Formula	Mass Error	MS/MS frag.	Ref.	AS	AU	CB	FS	FU	JS	JU
31.	20.7	Hydroxyhexadecanedioic acid	Fatty acid	303.2171	$C_{16}H_{31}O_5^-$	1.1	259, 214	[32]	+	-	-	+	++	-	-
32.	20.9	Sakuranetin-*O*-hexosyl deoxyhexoside	Flavanone	593.1835 595.2018	$C_{28}H_{33}O_{14}^-$ $C_{28}H_{35}O_{14}^+$	1.7 −1.5	285		+	+	++	++	++	++	+
33.	21.1	Nomilin-hexoside	Limonoid	693.2723	$C_{34}H_{45}O_{15}^-$	5.9	159	[19]	++	++	+	+	+	++	+
34.	21.2	Sakuranetin	Flavanone	285.0770 287.1042	$C_{16}H_{13}O_5^-$ $C_{16}H_{15}O_5^+$	2.5 0.7	255 237	[29]	++	++	+	+	+	+	+
35.	22.5	Perilloside A	Terpene	313.1651	$C_{16}H_{25}O_6^-$	−0.3	-	[33]	++	+	+	++	+	+	-
36.	23.0	Trihydroxy-octadecadienoic acid (trihydroxy-linoleic acid)	Fatty acid	327.2175	$C_{18}H_{31}O_5^-$	0.9	-	[28]	+	+	+	+	++	+	+
37.	23.7	Hydroxy-sphingenine	Ceramide	316.2827	$C_{18}H_{38}NO_3^+$	−6.1	-	[34]	+	+	++	+	+	+	-
38.	23.9	Linoleamide	Fatty acid amide	280.2677	$C_{18}H_{34}NO^+$	1.4	-	[35]	+	-	-	+	-	++	-
39.	24.1	Trihydroxy-octadecenoic acid	Fatty acid	329.2333	$C_{18}H_{33}O_5^-$	1.5	-		+	+	+	+	++	+	+
40.	24.2	Demethylnobiletin	Methoxy-flavone	389.1234	$C_{20}H_{21}O_8^+$	−0.8	-	[36]	-	-	-	+	++	-	-
41.	24.3	Methyl dihydrojasmonate	Organic acid	225.1496	$C_{13}H_{21}O_3^-$	−0.4	-		+	+	+	+	+	+	+
42.	17.7	Naringenin	Flavanone	271.0606	$C_{15}H_{11}O_5^-$	0.0	-	[37]	+	+	++	++	+	++	++
43.	26.1	Nobiletin	Methoxy-flavone	403.1385	$C_{21}H_{23}O_8^+$	−1.2	388, 373, 342	[38]	+	+	+	++	++	-	-
44.	18.5	Hesperetin	Flavanone	301.0717	$C_{16}H_{13}O_6^-$	0.9	-	[29]	++	++	+	+	+	+	+
45.	26.5	Unknown	Methoxy-flavone	403.1568	$C_{25}H_{23}O_5^+$	7.4	373		-	-	+	+	++	+	+
46.	27.2	Tangeretin	Methoxy-flavone	373.1312	$C_{20}H_{21}O_7^+$	6.7	343, 312	[39]	+	+	+	+	++++		+
47.	27.7	Methyl-dodecadienoate	Fatty acid ester	209.1548	$C_{13}H_{21}O_2^-$	2.4	-		+	+	+	+	+	++	+
48.	27.8	Unknown	Alkaloid	272.1862	$C_{14}H_{26}NO_4^+$	−1.8	255, 237		++	+	+	+	+	+	+
49.	27.9	Dimethylkaempferol	Methoxy-flavone	313.0718	$C_{17}H_{13}O_6^-$	1.9	163, 117	[40]	-	+	+	++	++	-	-
50.	28.0	γ-Lactone hydroxy-dodecenedioic acid methyl ester	Fatty acid ester	253.1443	$C_{18}H_{15}O_7^-$	0.0	209		+	+	+	+	+	+	+
51.	28.1	Limonin	Limonoid	469.1869 471.2018	$C_{26}H_{29}O_8^-$ $C_{26}H_{31}O_8^+$	1.5 0.2	- 453, 161	[30]	++	+	+	+	+	+	+
52.	28.3	Methyllimonexic acid	Limonoid	515.1925	$C_{27}H_{31}O_{10}^-$	1.6	469, 229	[41]	++	+	-	+	+	+	+
53.	28.4	Dihydroxytrimethoxyflavone (xanthomicrol)	Methoxy-flavone	343.0826	$C_{18}H_{15}O_7^-$	0.2	298, 270, 242		+	-	+	++	++	-	-
54.	29.0	Tetramethoxyflavone (Tetra-*O*-methylscutellarein)	Methoxy-flavone	343.1201	$C_{19}H_{19}O_6^+$	7.2	313, 282		+	+	+	++	++	-	-
55.	29.2	Gingerol	Phenylpropen-oid polyketide	293.1763	$C_{17}H_{25}O_4^-$	3.8	193	[20]	+	++	+	+	+	++	+
56.	29.5	Heptamethoxyflavone	Methoxy-flavone	455.1313 433.1511	$C_{22}H_{24}O_9Na^+$ $C_{22}H_{25}O_9^+$	−1.1 −4.1	433 403	[42]	-	-	+	+++++		-	-
57.	29.9	Ionone epoxide	Terpene	207.1392	$C_{13}H_{19}O_2^-$	−3.4	-	[43]	+	+	+	+	+	+	+
58.	28.7	Hydroxy-tetradecatrienoyl glycerol	Acylglycerol	311.1857	$C_{17}H_{27}O_5^-$	−2.2	-		+	+	+	+	++	+	+
59.	31.6	Dihydroxyoctadecadienoic acid	Fatty acid	311.2225	$C_{18}H_{31}O_4^-$	1.2	211, 171, 129		+	+	+	++	++	+	+
60.	32.0	*N*-Phenylacetylglycine	Nitrogenous compound	194.0821	$C_{10}H_{12}NO_3^-$	0.8	-		+	+	++	+	+	++	+

Table 2. *Cont.*

No.	Rt (min)	Compound Name	Chemical Class	[M−H]−/ [M + H]+	Molecular Formula	Mass Error	MS/MS frag.	Ref.	AS	AU	CB	FS	FU	JS	JU
61.	32.5	Hydroxy-oxohexadecanoic acid	Fatty acid	285.2069	$C_{16}H_{29}O_4^-$	0.7	-	[44]	++	++	+	+	+	+	+
62.	32.5	Hydroxylinolenic acid	Fatty acid	293.2115	$C_{18}H_{29}O_3^-$	−2.4	275, 235	[45]	+	+	-	+	++	-	-
63.	34.1	Hydroxyoctadecadienoic acid (Hydroxylinoleic acid)	Fatty acid	295.2276	$C_{18}H_{31}O_3^-$	1.6	-		+	+	+	+	++	+	+
64.	35.7	Heptyl caffeate	Cinnamic acid ester	301.1421 279.1600	$C_{16}H_{22}O_4Na^+$ $C_{16}H_{23}O_4^+$	1.7 −3.3	279		+	+	-	++	++	+	++
65.	39.1	Erucamide (Docosenamide)	Fatty acid amide	338.3428	$C_{22}H_{44}NO^+$	−3.1	321	[35]	-	-	+	+	-	++	-
66.	40.5	Linoleic acid	Fatty acid	279.2326	$C_{18}H_{31}O_2^-$	1.3	-	[45]	+	++	+	+	+	+	+

AS: Albedo from Spain; AU: Albedo from Uruguay; CB: concentrate from Brazil; FS: Flavedo from Spain; FU: Flavedo from Uruguay; JS: Juice from Spain; JU: Juice from Uruguay. Signs (-, +, ++ and +++) indicate relative semi-quantification of each compound based on the integrated peak areas of the total ion; - = absent, + = present, ++ = present in higher amount, and +++ = present in highest amount.

3.1.3. Identification of Flavonoids

A total of 22 flavonoid derivatives were identified with better ionization in the negative ion mode as a result of their rapid deprotonation ability. Peak **20** [(M + H)$^+$ *m/z* 463.1238 ($C_{22}H_{23}O_{11}$)$^+$], with a fragment ion at *m/z* 343 [M + H−120]$^+$ was identified as chrysoeriol-C-hexoside. Peak **15** [(M + H)$^+$ *m/z* 595.1654 ($C_{27}H_{31}O_5$)$^+$] showed a fragment ion at *m/z* 385 [M + H−120−90]$^+$, indicating di-C-hexoside, and was identified as apigenin-di-C-hexoside (vicenin-2). Vicenin-2 was previously identified in the peels and edible pulp of *C. aurantiifolia,* and *C. unshiu* [46]. Both flavones, chrysoeriol-C-hexoside and vicenin-2, were detected in all the analyzed samples with relatively high concentration in flavedo peel of both suppliers.

Peaks **12** and **19** showed the same deprotonated aglycone fragment ion at *m/z* 271, and were identified as naringenin derivatives. Peak **12** [(M−H)$^-$ *m/z* 433.1145 ($C_{21}H_{21}O_{10}$)$^-$] with MS^2 fragment at *m/z* 271 [M−H−162 (hexose)]$^-$ was identified as naringenin-*O*-hexoside. Peak **19** [(M + H)$^+$ *m/z* 581.1867 ($C_{27}H_{33}O_{14}$)$^+$] showed fragment ions (Suppl. Figure S3) at *m/z* 419 [M + H−162 (hexose)]$^+$ followed by successive loss of a deoxyhexose at *m/z* 273 [M + H−162 (hexose)−146 (deoxyhexose)]$^+$, while peak **31** showed a base signal at *m/z* 273.0756 [M + H−308 (rutinose)]$^+$ indicating that both carbohydrate moieties were linked through an -*O*-glycosidic bond [47]. Thus, peak **19** was identified as naringenin-*O*-hexosyldeoxyhexoside.

Likewise, Peaks **16** and **18** showed the same deprotonated fragment ion at *m/z* 269 indicating apigenin derivatives. Peak **16** [(M−H)$^-$ *m/z* 593.1502 ($C_{27}H_{29}O_{15}$)$^-$] with MS^2 fragments at *m/z* 431 [M−H−162 (hexose)]$^-$ and *m/z* 269 [M−H−162 (hexose)−162 (hexose)]$^-$, indicating successive losses of two hexose moieties, was identified as apigenin-di-*O*-hexoside previously detected in different *Citrus* species [26]. Peak **18** [(M−H)$^-$ *m/z* 563.1408 ($C_{26}H_{27}O_{14}$)$^-$]/[(M + H)$^+$ *m/z* 565.1550 ($C_{26}H_{29}O_{14}$)$^+$] was a mixed *O*-*C* flavone, with characteristic fragment ion at *m/z* 311 [M−H−120 (cross ring cleavage of C-hexosyl)−132 (pentose)]$^-$. It was identified as apigenin-*C*-hexosyl-*O*-pentoside, first reported in *Citrus sinensis* (Suppl. Figure S4). In the studied samples, flavedo peel from Uruguay was found more enriched in apigenin derivatives, while naringenin derivatives were found most abundant in the albedo part of both suppliers.

Peaks **30** showed deprotonated fragment ion at *m/z* 285, indicating sakuranetin derivatives. Peak **30** showed fragment ions at *m/z* 433 [M + H−162 (hexose)]$^+$ followed by successive loss of a deoxyhexose at *m/z* 287 [M + H−162 (hexose)−146 (deoxyhexose)]$^+$. Thus, peak **30** was identified as sakuranetin-*O*-hexosyl-*O*-deoxyhexoside.

Peak **21** [(M−H)$^-$ *m/z* 609.1833 ($C_{28}H_{33}O_{15}$)$^-$]/[(M + H)$^+$ *m/z* 611.1971 ($C_{28}H_{35}O_{15}$)$^+$] showed a MS^2 fragment ion in the negative ionization mode at *m/z* 301 [M−H−308 (rutinose)]$^-$ and was the main flavanone glycoside identified in all samples (Suppl. Figure S5). It was annotated as hesperidin, previously reported as the main flavanone in the peel of *Citrus sinensis* L. varieties, which suffers dramatic losses in filtered peel juice due to its relatively low water solubility [48]. Hesperidin is well known for its supposed effects on health including antimicrobial, anticancer, antihypertensive and antiulcer effects, thus attracting medicinal interest to orange peel [49].

A number of polymethoxy flavones (PMF) were identified in the studied samples being more readily ionized in the positive ion mode than in the negative mode. In general, the PMF MS^2 spectra had characteristic fragments at [M + H−nCH$_3$]$^+$, [M + H−2CH$_3$−CO]$^+$ and [M + H−2CH$_3$−H$_2$O]$^+$ [50]. Peak **43** [(M + H)$^+$ *m/z* 403.1385 ($C_{21}H_{23}O_8$)$^+$], peak **46** [(M + H)$^+$ *m/z* 373.1312 ($C_{20}H_{21}O_7$)$^+$] and peak **53** [(M−H)$^-$ *m/z* 343.0826 ($C_{18}H_{15}O_7$)$^-$] showed typical fragmentation patterns for PMF and were identified as nobiletin, tangeretin and dihydroxytrimethoxy-flavone, respectively. Xanthomicrol (dihydroxytrimethoxy-flavone) was previously isolated from *Citrus sudachi* [51] but reported herein for first time in *C. sinensis*. Likewise, peak **54** [(M + H)$^+$ *m/z* 343.1201 ($C_{19}H_{19}O_6$)$^+$] and peak **56** [(M + H)$^+$ *m/z* 433.1511 ($C_{22}H_{25}O_9$)$^+$] were annotated as tetra-*O*-methylscutellarein and heptamethoxyflavone, respectively. In addition to xanthomicrol, other hydroxylated PMFs were detected as demethylnobiletin (peak **40**), and dimethylkaempferol (peak **49**). Generally, PMFs exhibited extensive range of biological actions, i.e., antiatherogenic, and anti-inflammation activities [52]. However, hydroxylated PMFs revealed potent antineoplastic effects against various cancer types [53]. Flavedo peel from Uruguay was the richest part in methoxy flavones in contrast to the juice from both suppliers. The flavedo extracts of various *Citrus* fruits encompassed flavanone glycosides (poncirin, hesperidin, neohesperidin, dydimin, naringin, narirutin, and neoeriocitrin), flavonol glycosides (rutin), and flavone glycosides (diosmin, rhoifolin, and isorhoifolin) [6].

3.1.4. Identification of Limonoids and Terpenes

Limonoids are tetranortriterpenoids, found extensively in Rutaceae and Meliaceae [54]. They are widely distributed in different *Citrus* fruits, such as grapefruit (*Citrus paradisi*), sweet orange (*Citrus sinensis*), sour orange (*Citrus aurantium*), lemon (*Citrus limon*) and lime (*Citrus aurantiifolia*) [55]. Water-insoluble limonoid aglycones are mainly distributed within seeds and peels. In contrast, the water-soluble limonoid glycosides are more abundant within juices and pulps [56]. In the studied *Citrus* samples, limonoids occur in significant amounts as aglycone and glycoside forms. Peak **25** [$(M + H)^+$ *m/z* 473.2167 $(C_{26}H_{33}O_8)^+$] showed MS^2 fragments at *m/z* 455 [M + H−18 (H_2O)]$^+$, 411 [M + H−62 (CH_2O_3)]$^+$ and the characteristic ion for *Citrus* limonoids at *m/z* 161 attributed to the furan ring bound to a lactone ring moiety [57] and was identified as deacetylnomilin (Suppl. Figure S6). It was found more abundant in the peel than in juice or concentrate; this may be related to its relatively low water solubility. Peak **51** with protonated and deprotonated molecular ions [$(M + H)^+$ *m/z* 471.2018 $(C_{26}H_{31}O_8)^+$]/[(M−H) *m/z* 469.1869 $(C_{26}H_{29}O_8)^-$] showed a fragment ion *m/z* 453 [M + H−18 (H_2O)]$^+$ and the characteristic fragment ion for *Citrus* limonoids at *m/z* 161. It was identified as limonin, a well-known limonoid that possesses various biological, mainly anti-inflammatory activities [58]. Limonin was detected in more abundant content in albedo from Spain (AS), suggestive to have a slightly bitter taste compared to albedo from Uruguay (AU) upon providing a likely sustainable source of food additive [59].

Terpenes are important for plant aroma and flavor playing key roles in fruit quality, plant defense and pollinator attraction. A total of three terpenes was detected ionized most preferentially in the negative ionization mode, from which a gluco-conjugated megastigmadienone was identified as peak **4** [$(M-H)^-$ *m/z* 385.1860 $(C_{19}H_{29}O_8)^-$] known as citroside A. Its MS^2 fragments showed the characteristic fragment ion at *m/z* 223 [M−H−162 (hexose)]$^-$ relative to the loss of a hexose moiety. Citroside A was reported previously to be a precursor of damascenone and 3-hydroxydamascone, two important industrial flavor compounds. It has been reported in *Citrus unshiu* leaves, *Solanum quitoense* leaves [60] and *Gynostemma pentaphyllum* [61]. Herein it was detected for the first time in the peel of *Citrus sinensis* being more prominent in albedo peel from Spain and flavedo peel from Uruguay (AS and FU). A number of monoterpenes belonging to different classes were identified: Peak **35** [(M−H) *m/z* 313.1651 $(C_{16}H_{25}O_6)^-$] was a monoterpene glycoside identified as perilloside A, detected for the first time in *Citrus* and found most prominent in albedo and flavedo peel from Spain (AS and FS), suggestive to be a marker to distinguish between peels from Spain and Uruguay. Peak **57** [$(M-H)^-$ *m/z* 207.1392 $(C_{13}H_{19}O_2)^-$] was an apo-carotenoid monoterpene identified as ionone epoxide, a flavoring substance with a fruity and woody flavor previously identified in many foods, such as apricot, raspberry, tea and lemon balm [43].

3.1.5. Identification of Fatty Acids and Fatty Acid Amides

Saturated and unsaturated fatty acids in addition to low molecular mass amide compounds were detected in several orange peel compartments. For details, refer to Supplementary Materials [28,35,44].

3.1.6. Identification of Nitrogenous Compounds

Other metabolites containing nitrogen were detected at trace levels. For details, refer to Supplementary Materials [34].

3.2. Multivariate Data Analyses of Orange Samples

The datasets encompassed a total of 42 nLC-ESI-MS/MS runs (seven different orange samples with three biological replicates each in both the negative and positive ionization modes). Thus, unsupervised and supervised multivariate data analyses were adopted to simplify interpretation of such complex datasets allowing better biomarkers characterization and sample classification [62]. Principal component analysis (PCA), as an unsupervised approach, was reported to evaluate the variance within various samples involving no prior knowledge of the datasets [63]. Table 1 shows the color-coded source of the *Citrus sinensis* specimens compared.

PCA on negative ionization runs as depicted in Figure 2A–C for all orange samples was illustrated by PC1 and PC2 accounting for 87% of the total variance (Figure 2A). Along PC1, there was a clear separation between flavedo specimens from Spain (FS) and Uruguay (FU), while the latter was distinguished from the other samples and located at upper right quadrant with positive score values, suggesting the impact of geographical source based on their metabolites. The albedo samples were positioned at the lower right quadrant, whereas juice samples were located at upper left quadrant with negative PC1 values. Conversely, orange concentrate from Brazil (CB) appeared near the origin (Figure 2A). The respective loading plot demonstrated that mono-methoxy flavonoids, i.e., hesperetin and sakuranetin along with hydroxy-oxohexadecanoic acid were found more abundant in albedo specimens (Figure 2B). On the other hand, hydroxylinoleic acid was major contributor to flavedo from Uruguay (FU) being the most distant data point. Naringenin was responsible for the clustering of orange juice of both suppliers and flavedo from Spain (FS) in the upper left quadrant, while *N*-phenylacetylglycine was more abundant in juice from Spain (JS) and orange concentrate (CB). The dendrogram of HCA (Hierarchical clustering analysis), as depicted in Figure 2C, revealed that flavedo from Uruguay (FU) represented one cluster, whereas the other cluster was divided into albedo specimens at one sub-group and the other sub-group encompassed the remaining samples.

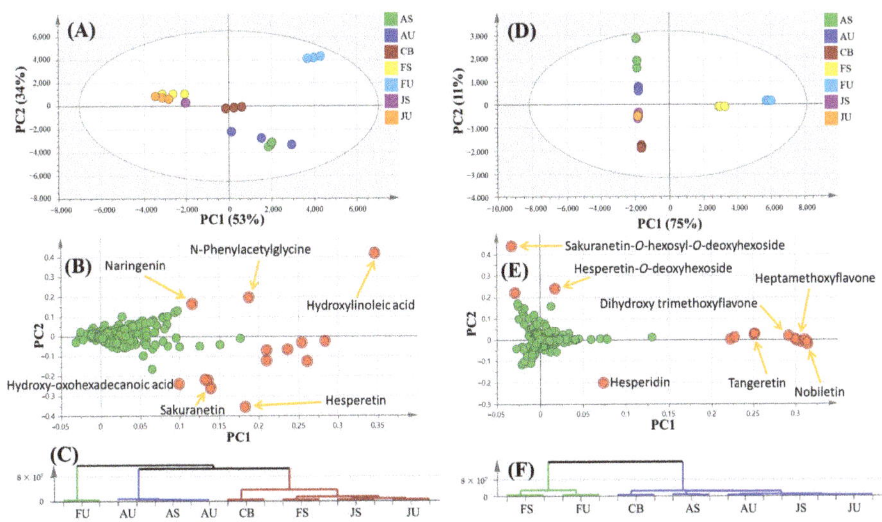

Figure 2. MS-based PCA of all orange samples ($n = 3$) negative ionization (**A**) score plot, (**B**) relevant loading plot and (**C**) HCA; positive ionization (**D**) score plot, (**E**) relevant loading plot and (**F**) HCA. AS: Albedo from Spain; AU: Albedo from Uruguay; CB: concentrate from Brazil; FS: Flavedo from Spain; FU: Flavedo from Uruguay; JS: Juice from Spain; JU: Juice from Uruguay.

Likewise, PCA on positive ionization as illustrated in Figure 2D–F revealed 86% of the total variance, albeit with better segregation (Figure 2D) compared to its negative ionization counterpart. Particularly, all replicates from each orange sample were tightly clustered indicating the superb reproducibility of the experimental analysis in both the negative and positive ionization modes. In agreement with the previous PCA model, flavedo from Uruguay (FU) was also the most distant group. Flavedo specimens showed positive score values separable from other orange parts, while orange juice and concentrate had negative score values. Albedo from both countries (AU and AS) was separated in the upper left quadrant. The respective loading plot (Figure 2E) demonstrated high levels of nobiletin, tangeretin, dihydroxytrimethoxyflavone and heptamethoxyflavone in flavedo samples. Albedo specimens were enriched in sakuranetin-*O*-hexosyl-*O*-deoxyhexoside and hesperetin-*O*-deoxyhexoside, whereas high levels of hesperidin were observed in orange juice and concentrate. The HCA dendrogram, as depicted in Figure 2F, revealed that the flavedo from both suppliers represented one cluster, while the other cluster included the remaining orange parts.

OPLS-DA (orthogonal projection to latent structures-discriminant analysis) as a supervised approach was reported to possess a great potentiality maximizing the segregation of overlapping sample groups by identifying chemical determinants [64]. Being the most distant cluster, flavedo from Uruguay (FU) was further subjected to OPLS-DA against all other specimens (Figure 3) to assess sample discrimination with p value less than 0.001. The first OPLS model on negative ionization (Figure 3A) exhibited 0.96 model predictability (Q2) and 97% total variance (R2). The relevant loading S-plot showed that FU included abundant levels of hydroxylated fatty acids, i.e., hydroxylinoleic, trihydroxylinoleic, and dihydroxyoctadecadienoic acids (Figure 3B). Conversely, the second OPLS model on positive ionization (Figure 3C) exhibited 0.62 model predictability (Q2) and 72% total variance (R2). The relevant loading S-plot revealed that nobiletin, dihydroxytrimethoxyflavone, demethylnobiletin and tangeretin were detected in higher levels in FU (flavedo from Uruguay) (Figure 3D). Then, several OPLS models were constructed (Suppl. Figures S7–S10) with p value less than 0.001 to assess chemical determinants in orange parts derived from the various countries. Hence, a model of flavedo from Uruguay (FU) against its counterpart from Spain (FS) was performed (Suppl. Figure S7). On negative mode, OPLS model (Suppl. Figure S7A) revealed the enrichment of flavedo from Uruguay (FU) in hydroxyl-linoleic acid, dimethylkaempferol, apigenin-di-*O*-hexoside and citropten (Suppl. Figure S7B). Notably, hydroxyl fatty acids other than hydroxylinoleic acid did not appear in this model, which was attributed to their similar levels present in both samples. Conversely, OPLS model in positive mode (Suppl. Figure S7C) exhibited the enrichment of flavedo from Uruguay (FU) in nobiletin, demethylnobiletin, and tangeretin, whereas heptamethoxyflavone was abundant in FS (Suppl. Figure S7D). Another OPLS model was performed with albedo from Uruguay (AU) versus its counterpart from Spain (AS) (Suppl. Figure S8). OPLS model in negative ionization (Suppl. Figure S8A) demonstrated the particular abundance of naringenin-*O*-hexoside and linoleic acid in albedo from Uruguay (AU), while hesperetin, sakuranetin and hydroxy-oxohexadecanoic acid were found more abundant in albedo from Spain (Suppl. Figure S8B). In positive ionization, the OPLS model (Suppl. Figure S8C) showed significantly higher levels of glycosyl flavonoids, i.e., sakuranetin-*O*-hexosyl-*O*-deoxyhexoside, limonin and hesperetin-*O*-deoxyhexoside in AS (albedo from Spain) (Suppl. Figure S8D). Further and for better discrimination assessment of orange juices derived from various sources, the OPLS model in negative mode (Suppl. Figure S9A) revealed abundant levels of *N*-phenylacetylglycine, nomilin-hexoside and gingerol in orange juice from Spain (JS) (Suppl. Figure S9B). Lastly, the OPLS model including orange juice from both suppliers and orange concentrate from Brazil (CB) was implemented in positive mode (Suppl. Figure S10A). The relevant loading plot (Suppl. Figure S10B) showed that hesperidin, hydroxy-sphingenine and pyridoxamine phosphate amounted for the major metabolites in orange concentrate (CB). The obvious segregation between orange juice from Uruguay (JU) and Spain (JS) was ascribed to the abundance of heptyl caffeate in the former, while the later was enriched in nomilinhexoside and sakuranetin-*O*-hexosyl-deoxyhexoside.

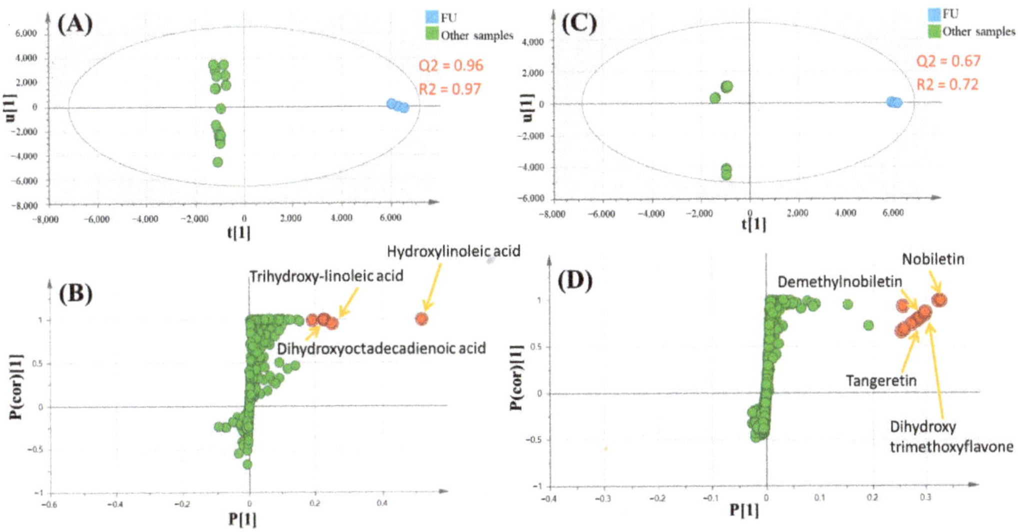

Figure 3. MS-based OPLS of flavedo from Uruguay (FU) against all other samples negative ionization (**A**) score plot and (**B**) relevant loading S-plot; positive ionization (**C**) score plot and (**D**) relevant loading S-plot.

4. Conclusions

In this study, the metabolites in the orange peels (albedo and flavedo parts) together with the juice and concentrate from different suppliers were systematically analyzed and identified using state-of-the-art nLC-ESI-MS/MS-based, widely non-targeted metabolome analysis. A total of 66 metabolites were annotated, 37 of which were compounds shared by all samples. Furthermore, 29 differential metabolites were detected, 15 of which were mainly flavonoids and completely absent from the juice. A total of eleven metabolites were detected for the first time in *Citrus sinensis:* citroside A, sinapic acid pentoside, di-hexosyl-diosmetin, apigenin-C-hexosyl-*O*-pentoside, chrysoeriol-C-hexoside, perilloside A, hydroxy-sphingenine, xanthomicrol, coumaryl alcohol-*O*-hexoside, gingerol and ionone epoxide. The annotation of the novel metabolites warrants in-depth functional genomic mining to identify their various biosynthetic pathways. Comparing different fruit parts, a number of flavonoids with proposed preventive therapies against some diseases, i.e., naringenin-*O*-hexoside, hesperetin-*O*-deoxyhexoside, sakuranetin-*O*-hexosyl-*O*-deoxyhexoside, and demethylnobiletin, were completely absent from the juices, but present most prominent in the peel. *Citrus* peel is thus considered a renewable bio-resource of functional foods. In addition, hesperetin-*O*-deoxyhexoside, sakuranetin-*O*-hexosyl-*O*-deoxyhexoside and hydroxylinolenic acid were detected only in albedo and flavedo of both suppliers, albeit absent in orange juices and concentrate, suggestive to be a distinctive marker for orange peel and to increase the potential transform of *Citrus* by-products into valuable food ingredients, nutraceuticals, and perhaps even pharmaceuticals.

The comprehensive nLC-ESI-MS/MS metabolic profiling followed by multivariate data analyses suggested that the differences between orange parts were much more obvious than the geographical source. Generally, albedo was richer in mono-methoxylated flavonoids, while flavedo was richer in poly-methoxylated flavonoids and hydroxylated fatty acids. Moreover, further research is required to more accurately evaluate the effectiveness, the toxicity, and the mechanism of action of many PMFs as well as to increase their bioavailability by using readily accessible and appropriate drug delivery vehicles.

Further analysis of more orange samples from other sources has yet to be evaluated to obtain a bigger picture of the entire respective population of orange samples. In addition, other factors, i.e., seasonal variation, storage conditions and agricultural practices may be assessed using the same platform.

Supplementary Materials: The following supporting information can be downloaded at: https://www.mdpi.com/article/10.3390/foods12030579/s1, Figure S1: MS^2 Spectra of A: ferulic acid hexoside $[M-H]^-$ 355.1024, $C_{16}H_{19}O_9^-$, B: sinapic acid pentoside $[M-H]^-$ 355.1025, $C_{16}H_{19}O_9^-$ Figure S2: MS^2 spectra of citropten $[M-H]^-$ 205.0495, $C_{11}H_9O_4^-$. Figure S3: MS^2 spectra of naringenin-*O*-hexosyldeoxyhexoside $[M-H]^-$ 579. 1687 $C_{27}H_{31}O_{14}^-$. Figure S4: MS^2 spectra of apigenin-C-hexoside-*O*-pentoside $[M-H]^-$ 563.1416, $C_{26}H_{27}O_{14}^-$. Figure S5: MS^2 spectra of hesperidin $[M-H]^-$ 609.1832, $C_{28}H_{33}O_{15}^-$. Figure S6: MS^2 spectra of deacetylnomilin $[M + H]^+$ 473.222, $C_{26}H_{33}O_8^+$. Figure S7: MS-based OPLS of flavedo from Uruguay (FU) against that from Spain (FS) (n = 3) negative ionization (**A**) score plot and (**B**) relevant loading S-plot; positive ionization (**C**) score plot and (D) relevant loading S-plot. Figure S8: MS-based OPLS of albedo from Uruguay (AU) against that from Spain (AS) (n = 3) negative ionization (**A**) score plot and (**B**) relevant loading S-plot; positive ionization (**C**) score plot and (**D**) relevant loading S-plot. Figure S9: MS-based OPLS of orange juice from Uruguay (JU) against that from Spain (JS) (n = 3) negative ionization (**A**) score plot and (**B**) relevant loading S-plot. Figure S10: MS-based OPLS of orange juice from Uruguay (JU) and Spain (JS) and orange concentrate (CB) (n = 3) positive ionization (**A**) score plot and (**B**) relevant loading plot. References [63,64] are cited in the supplementary materials.

Author Contributions: Conceptualization, S.M.A. and T.E.; methodology, S.M.A., E.M.K., U.K., R.G.B. and T.E.; validation, S.M.A. and E.M.K.; formal analysis, S.M.A., U.K. and R.G.B.; investigation, S.M.A., E.M.K., U.K., R.G.B. and T.E.; writing—original draft preparation, S.M.A. and E.M.K.; writing—review and editing, S.M.A., E.M.K., U.K., R.G.B. and T.E.; funding acquisition, R.G.B. and T.E. All authors have read and agreed to the published version of the manuscript.

Funding: This research project was partially supported by the German academic exchange program (DAAD; German Egyptian Research Short-Term Scholarship—GERSS program). The publication of this article was funded by the Open Access Fund of Leibniz Universität Hannover.

Data Availability Statement: Not applicable.

Acknowledgments: We thank Silke Hillebrand from Symrise (Holzminden, Germany) for providing the *Citrus* samples.

Conflicts of Interest: The authors declare no conflict of interest.

References

1. Paw, M.; Begum, T.; Gogoi, R.; Pandey, S.K.; Lal, M. Chemical Composition of *Citrus limon* L. Burmf Peel Essential Oil from North East India. *J. Essent. Oil Bear. Plants* **2020**, *23*, 337–344. [CrossRef]
2. Favela-Hernández, J.M.J.; González-Santiago, O.; Ramírez-Cabrera, M.A.; Esquivel-Ferriño, P.C.; Camacho-Corona, M.D.R. Chemistry and pharmacology of *Citrus sinensis*. *Molecules* **2016**, *21*, 247. [CrossRef] [PubMed]
3. Suri, S.; Singh, A.; Nema, P.K. Current applications of citrus fruit processing waste: A scientific outlook. *Appl. Food Res.* **2022**, *2*, 100050. [CrossRef]
4. Olaiya, C.; Adigun, A. Chemical manipulation of tomato growth and associated biochemical implications on flavonoid, lycopene and mineral contents. *Afr. J. Plant Sci.* **2010**, *4*, 167–171.
5. Peixoto, J.S.; Comar, J.F.; Moreira, C.T.; Soares, A.A.; De Oliveira, A.L.; Bracht, A.; Peralta, R.M. Effects of *Citrus aurantium* (bitter orange) fruit extracts and p-synephrine on metabolic fluxes in the rat liver. *Molecules* **2012**, *17*, 5854–5869. [CrossRef]
6. Singh, B.; Singh, J.P.; Kaur, A.; Singh, N. Phenolic composition, antioxidant potential and health benefits of citrus peel. *Food Res. Int.* **2020**, *132*, 109114. [CrossRef]
7. Goulas, V.; Manganaris, G.A. Exploring the phytochemical content and the antioxidant potential of *Citrus* fruits grown in Cyprus. *Food Chem.* **2012**, *131*, 39–47. [CrossRef]
8. Escobedo-Avellaneda, Z.; Gutiérrez-Uribe, J.; Valdez-Fragoso, A.; Torres, J.A.; Welti-Chanes, J. Phytochemicals and antioxidant activity of juice, flavedo, albedo and comminuted orange. *J. Funct. Foods* **2014**, *6*, 470–481. [CrossRef]
9. Fattahi, J.; Hamidoghli, Y.; Fotouhi, R.; Ghasemnejad, M.; Bakhshi, D. Assessment of fruit quality and antioxidant activity of three citrus species during ripening. *South-West. J. Hortic. Biol. Environ.* **2011**, *2*, 113–128.
10. Zhao, X.J.; Xing, T.T.; Li, Y.F.; Jiao, B.N. Analysis of phytochemical contributors to antioxidant capacity of the peel of Chinese mandarin and orange varieties. *Int. J. Food Sci. Nutr.* **2019**, *70*, 825–833. [CrossRef] [PubMed]

11. Salonia, F.; Ciacciulli, A.; Poles, L.; Pappalardo, H.D.; La Malfa, S.; Licciardello, C. New plant breeding techniques in citrus for the improvement of important agronomic traits. A Review. *Front. Plant Sci.* **2020**, *11*, 1234. [CrossRef] [PubMed]
12. Farag, M.A.; Fathi, D.; Shamma, S.; Shawkat, M.S.A.; Shalabi, S.M.; El Seedi, H.R.; Afifi, S.M. Comparative metabolome classification of desert truffles *Terfezia claveryi* and *Terfezia boudieri* via its aroma and nutrients profile. *LWT* **2021**, *142*, 111046. [CrossRef]
13. Ammar, N.M.; Hassan, H.A.; Ahmed, R.F.; AE, G.E.-G.; Abd-ElGawad, A.M.; Farrag, A.R.H.; Farag, M.A.; Elshamy, A.I.; Afifi, S.M. Gastro-protective effect of *Artemisia Sieberi* essential oil against ethanol-induced ulcer in rats as revealed via biochemical, histopathological and metabolomics analysis. *Biomark. Biochem. Indic. Expo. Response Susceptibility Chem.* **2022**, *27*, 247–257. [CrossRef] [PubMed]
14. Kim, D.-S.; Lee, S.; Park, S.M.; Yun, S.H.; Gab, H.-S.; Kim, S.S.; Kim, H.-J. Comparative metabolomics analysis of Citrus varieties. *Foods* **2021**, *10*, 2826. [CrossRef] [PubMed]
15. Yao, Z.; Wu, S.; Zhang, H.; Feng, X.; Wang, Z.; Lin, M. Chiral determination of naringenin by ultra-performance liquid chromatography-tandem mass spectrometry and application in Citrus peel and pulp. *Front. Nutr.* **2022**, *9*, 906859. [CrossRef]
16. Afifi, S.M.; El-Mahis, A.; Heiss, A.G.; Farag, M.A. Gas chromatography-mass spectrometry-based classification of 12 fennel (*Foeniculum vulgare* Miller) varieties based on their aroma profiles and estragole levels as analyzed using chemometric tools. *ACS Omega* **2021**, *6*, 5775–5785. [CrossRef]
17. El-Newary, S.A.; Afifi, S.M.; Aly, M.S.; Ahmed, R.F.; El Gendy, A.E.-N.G.; Abd-ElGawad, A.M.; Farag, M.A.; Elgamal, A.M.; Elshamy, A.I. Chemical profile of *Launaea nudicaulis* ethanolic extract and its antidiabetic effect in streptozotocin-induced rats. *Molecules* **2021**, *26*, 1000. [CrossRef]
18. Farag, M.A.; Kabbash, E.M.; Mediani, A.; Döll, S.; Esatbeyoglu, T.; Afifi, S.M. Comparative metabolite fingerprinting of four different cinnamon species analyzed via UPLC-MS and GC-MS and chemometric tools. *Molecules* **2022**, *27*, 2935. [CrossRef]
19. Ledesma-Escobar, C.; Priego-Capote, F.; Luque de Castro, M. Characterization of lemon (*Citrus limon*) polar extract by liquid chromatography–tandem mass spectrometry in high resolution mode. *J. Mass Spectrom.* **2015**, *50*, 1196–1205. [CrossRef]
20. Wang, F.; Huang, Y.; Wu, W.; Zhu, C.; Zhang, R.; Chen, J.; Zeng, J. Metabolomics analysis of the peels of different colored citrus fruits (*Citrus reticulata* cv.'Shatangju') during the maturation period based on UHPLC-QQQ-MS. *Molecules* **2020**, *25*, 396. [CrossRef]
21. Liu, Y.; Ren, C.; Cao, Y.; Wang, Y.; Duan, W.; Xie, L.; Sun, C.; Li, X. Characterization and purification of bergamottin from *Citrus grandis* (L.) Osbeck cv. Yongjiazaoxiangyou and its antiproliferative activity and effect on glucose consumption in HepG2 cells. *Molecules* **2017**, *22*, 1227. [CrossRef] [PubMed]
22. Jo, M.S.; Lee, S.; Yu, J.S.; Baek, S.C.; Cho, Y.-C.; Kim, K.H. Megastigmane derivatives from the cladodes of *Opuntia humifusa* and their nitric oxide inhibitory activities in macrophages. *J. Nat. Prod.* **2020**, *83*, 684–692. [CrossRef] [PubMed]
23. Oszmiański, J.; Kolniak-Ostek, J.; Wojdyło, A. Application of ultra performance liquid chromatography-photodiode detector-quadrupole/time of flight-mass spectrometry (UPLC-PDA-Q/TOF-MS) method for the characterization of phenolic compounds of *Lepidium sativum* L. sprouts. *Eur. Food Res. Technol.* **2013**, *236*, 699–706. [CrossRef]
24. Tine, Y.; Renucci, F.; Costa, J.; Wélé, A.; Paolini, J. A method for LC-MS/MS profiling of coumarins in *Zanthoxylum zanthoxyloides* (Lam.) B. Zepernich and Timler extracts and essential oils. *Molecules* **2017**, *22*, 174. [CrossRef] [PubMed]
25. Zagurskaya, Y.V.; Vasil'ev, V.; Bogatyrev, A.; Bayandina, I.; Kukina, T. Flavonoids and hydroxycinnamic acids from *Leonurus quinquelobatus*. *Chem. Nat. Compd.* **2015**, *51*, 156–157. [CrossRef]
26. Brito, A.; Ramirez, J.E.; Areche, C.; Sepúlveda, B.; Simirgiotis, M.J. HPLC-UV-MS profiles of phenolic compounds and antioxidant activity of fruits from three citrus species consumed in Northern Chile. *Molecules* **2014**, *19*, 17400–17421. [CrossRef] [PubMed]
27. Katsuno, T.; Kasuga, H.; Kusano, Y.; Yaguchi, Y.; Tomomura, M.; Cui, J.; Yang, Z.; Baldermann, S.; Nakamura, Y.; Ohnishi, T.; et al. Characterisation of odorant compounds and their biochemical formation in green tea with a low temperature storage process. *Food Chem.* **2014**, *148*, 388–395. [CrossRef] [PubMed]
28. Fayek, N.M.; Farag, M.A.; Abdel Monem, A.R.; Moussa, M.Y.; Abd-Elwahab, S.M.; El-Tanbouly, N.D. Comparative metabolite profiling of four citrus peel cultivars via ultra-performance liquid chromatography coupled with quadrupole-time-of-flight-mass spectrometry and multivariate data analyses. *J. Chromatogr. Sci.* **2019**, *57*, 349–360. [CrossRef] [PubMed]
29. Cho, H.E.; Ahn, S.Y.; Kim, S.C.; Woo, M.H.; Hong, J.-T.; Moon, D.C. Determination of flavonoid glycosides, polymethoxyflavones, and coumarins in herbal drugs of citrus and poncirus fruits by high performance liquid chromatography–electrospray ionization/tandem mass spectrometry. *Anal. Lett.* **2014**, *47*, 1299–1323. [CrossRef]
30. Gualdani, R.; Cavalluzzi, M.M.; Lentini, G.; Habtemariam, S. The chemistry and pharmacology of citrus limonoids. *Molecules* **2016**, *21*, 1530. [CrossRef]
31. Li, K.; Wang, H. Simultaneous determination of matrine, sophoridine and oxymatrine in *Sophora flavescens* Ait. by high performance liquid chromatography. *Biomed. Chromatogr.* **2004**, *18*, 178–182. [CrossRef]
32. Ding, S.; Zhang, J.; Yang, L.; Wang, X.; Fu, F.; Wang, R.; Zhang, Q.; Shan, Y. Changes in cuticle components and morphology of 'Satsuma'mandarin (*Citrus unshiu*) during ambient storage and their potential role on Penicillium digitatum infection. *Molecules* **2020**, *25*, 412. [CrossRef] [PubMed]
33. Yang, H.; Tang, Y.; Wang, L.; Guo, Y.; Luo, D.; Wu, L.; Deng, X.; Li, X.; Bai, L. The potential of *Myosoton aquaticum* extracts and compounds to control barnyardgrass (*Echinochloa crus-galli*). *Weed Res.* **2021**, *61*, 317–326. [CrossRef]

34. Mukai, K.; Takeuchi, M.; Ohnishi, M.; Kudoh, M.; Imai, H. Characterization of ceramides and glucosylceramides of the Satsuma mandarin (*Citrus unshiu*) fruit. *J. Oleo Sci.* **2022**, *71*, 535–540. [CrossRef] [PubMed]
35. Ou, M.-C.; Liu, Y.-H.; Sun, Y.-W.; Chan, C.-F. The composition, antioxidant and antibacterial activities of cold-pressed and distilled essential oils of *Citrus paradisi* and *Citrus grandis* (L.) Osbeck. *Evid. -Based Complement. Altern. Med.* **2015**, *2015*, 804091. [CrossRef] [PubMed]
36. Dugo, P.; Bonaccorsi, I.; Ragonese, C.; Russo, M.; Donato, P.; Santi, L.; Mondello, L. Analytical characterization of mandarin (*Citrus deliciosa* Ten.) essential oil. *Flavour Fragr. J.* **2011**, *26*, 34–46. [CrossRef]
37. Lee, S.Y.; Shaari, K. LC–MS metabolomics analysis of *Stevia rebaudiana* Bertoni leaves cultivated in Malaysia in relation to different developmental stages. *Phytochem. Anal.* **2022**, *33*, 249–261. [CrossRef]
38. Dadwal, V.; Joshi, R.; Gupta, M. A multidimensional UHPLC-DAD-QTOF-IMS gradient approach for qualitative and quantitative investigation of Citrus and Malus fruit phenolic extracts and edibles. *ACS Food Sci. Technol.* **2021**, *1*, 2006–2018. [CrossRef]
39. Liu, E.H.; Zhao, P.; Duan, L.; Zheng, G.-D.; Guo, L.; Yang, H.; Li, P. Simultaneous determination of six bioactive flavonoids in Citri Reticulatae Pericarpium by rapid resolution liquid chromatography coupled with triple quadrupole electrospray tandem mass spectrometry. *Food Chem.* **2013**, *141*, 3977–3983. [CrossRef]
40. Falcão, S.I.; Vale, N.; Gomes, P.; Domingues, M.R.; Freire, C.; Cardoso, S.M.; Vilas-Boas, M. Phenolic profiling of Portuguese propolis by LC–MS spectrometry: Uncommon propolis rich in flavonoid glycosides. *Phytochem. Anal.* **2013**, *24*, 309–318. [CrossRef]
41. Shi, Y.-S.; Zhang, Y.; Li, H.-T.; Wu, C.-H.; El-Seedi, H.R.; Ye, W.-K.; Wang, Z.-W.; Li, C.-B.; Zhang, X.-F.; Kai, G.-Y. Limonoids from Citrus: Chemistry, anti-tumor potential, and other bioactivities. *J. Funct. Foods* **2020**, *75*, 104213. [CrossRef]
42. Tsukayama, M.; Ichikawa, R.; Yamamoto, K.; Sasaki, T.; Kawamura, Y. Microwave-assisted rapid extraction of polymethoxyflavones from dried peels of *Citrus yuko* Hort. ex Tanaka. *Nippon. Shokuhin Kagaku Kogaku Kaishi = J. Jpn. Soc. Food Sci. Technol.* **2009**, *56*, 359–362. [CrossRef]
43. Lu, V.; Bastaki, M.; Api, A.M.; Aubanel, M.; Bauter, M.; Cachet, T.; Demyttenaere, J.; Diop, M.M.; Harman, C.L.; Hayashi, S.-m.; et al. Dietary administration of β-ionone epoxide to Sprague-Dawley rats for 90 days. *Curr. Res. Toxicol.* **2021**, *2*, 192–201. [CrossRef]
44. Li, S.; Jiang, Z.; Thamm, L.; Zhou, G. 10-Hydroxy-2-decenoic acid as an antimicrobial agent in draft keg-conditioned wheat beer. *J. Am. Soc. Brew. Chem.* **2010**, *68*, 114–118. [CrossRef]
45. Yang, M.; Jiang, Z.; Wen, M.; Wu, Z.; Zha, M.; Xu, W.; Zhang, L. Chemical variation of Chenpi (Citrus peels) and corresponding correlated bioactive compounds by LC-MS metabolomics and multibioassay analysis. *Front. Nutr.* **2022**, *9*, 825381. [CrossRef] [PubMed]
46. Takiyama, M.; Matsumoto, T.; Watanabe, J. LC-MS/MS detection of citrus unshiu peel-derived flavonoids in the plasma and brain after oral administration of yokukansankachimpihange in rats. *Xenobiotica* **2019**, *49*, 1494–1503. [CrossRef] [PubMed]
47. Vukics, V.; Guttman, A. Structural characterization of flavonoid glycosides by multi-stage mass spectrometry. *Mass Spectrom. Rev.* **2010**, *29*, 1–16. [CrossRef]
48. Lim, T.K. (Ed.) Citrus × *aurantium* sweet orange group. In *Edible Medicinal and Non-Medicinal Plants: Volume 4, Fruits*; Springer: Dordrecht, The Netherlands, 2012; pp. 806–831.
49. Ganeshpurkar, A.; Saluja, A. The pharmacological potential of hesperidin. *Indian J. Biochem. Biophys. (IJBB)* **2019**, *56*, 287–300.
50. Zhang, J.-Y.; Li, N.; Che, Y.-Y.; Zhang, Y.; Liang, S.-X.; Zhao, M.-B.; Jiang, Y.; Tu, P.-F. Characterization of seventy polymethoxylated flavonoids (PMFs) in the leaves of *Murraya paniculata* by on-line high-performance liquid chromatography coupled to photodiode array detection and electrospray tandem mass spectrometry. *J. Pharm. Biomed. Anal.* **2011**, *56*, 950–961. [CrossRef]
51. Fattahi, M.; Cusido, R.M.; Khojasteh, A.; Bonfill, M.; Palazon, J. Xanthomicrol: A comprehensive review of its chemistry, distribution, biosynthesis and pharmacological activity. *Mini Rev. Med. Chem.* **2014**, *14*, 725–733. [CrossRef]
52. Gao, Z.; Gao, W.; Zeng, S.-L.; Li, P.; Liu, E.H. Chemical structures, bioactivities and molecular mechanisms of citrus polymethoxyflavones. *J. Funct. Foods* **2018**, *40*, 498–509. [CrossRef]
53. Lai, C.-S.; Wu, J.-C.; Ho, C.-T.; Pan, M.-H. Disease chemopreventive effects and molecular mechanisms of hydroxylated polymethoxyflavones. *BioFactors* **2015**, *41*, 301–313. [CrossRef] [PubMed]
54. Tundis, R.; Loizzo, M.R.; Menichini, F. An overview on chemical aspects and potential health benefits of limonoids and their derivatives. *Crit. Rev. Food Sci. Nutr.* **2014**, *54*, 225–250. [CrossRef]
55. Russo, M.; Arigò, A.; Calabrò, M.L.; Farnetti, S.; Mondello, L.; Dugo, P. Bergamot (*Citrus bergamia* Risso) as a source of nutraceuticals: Limonoids and flavonoids. *J. Funct. Foods* **2016**, *20*, 10–19. [CrossRef]
56. Breksa III, A.P.; Hidalgo, M.B.; Yuen, M.L. Liquid chromatography–electrospray ionisation mass spectrometry method for the rapid identification of citrus limonoid glucosides in citrus juices and extracts. *Food Chem.* **2009**, *117*, 739–744. [CrossRef]
57. Pessoa, L.G.A.; Pessoa, L.A.; da Silva Santos, É.; Pilau, E.J.; Gonçalves, J.E.; Gonçalves, R.A.C.; de Oliveira, A.J.B. Limonoid detection and profile in callus culture of sweet orange. *Acta Scientiarum. Biol. Sci.* **2021**, *43*, e53075. [CrossRef]
58. Vu, T.O.; Seo, W.; Lee, J.H.; Min, B.S.; Kim, J.A. Terpenoids from *Citrus unshiu* peels and their effects on NO production. *Nat. Prod. Sci.* **2020**, *26*, 176–181.
59. Wedamulla, N.E.; Fan, M.; Choi, Y.-J.; Kim, E.-K. Citrus peel as a renewable bioresource: Transforming waste to food additives. *J. Funct. Foods* **2022**, *95*, 105163. [CrossRef]
60. Dembitsky, V.M.; Maoka, T. Allenic and cumulenic lipids. *Prog. Lipid Res.* **2007**, *46*, 328–375. [CrossRef] [PubMed]

51. Takita, M.A.; Berger, I.J.; Basílio-Palmieri, A.C.; Borges, K.M.; Souza, J.M.d.; Targon, M.L. Terpene production in the peel of sweet orange fruits. *Genet. Mol. Biol.* **2007**, *30*, 841–847. [CrossRef]
52. Farag, M.A.; Afifi, S.M.; Rasheed, D.M.; Khattab, A.R. Revealing compositional attributes of *Glossostemon bruguieri* Desf. root geographic origin and roasting impact via chemometric modeling of SPME-GC-MS and NMR metabolite profiles. *J. Food Compos. Anal.* **2021**, *102*, 104073. [CrossRef]
53. Ammar, N.M.; Hassan, H.A.; Abdallah, H.M.I.; Afifi, S.M.; Elgamal, A.M.; Farrag, A.R.H.; El-Gendy, A.E.-N.G.; Farag, M.A.; Elshamy, A.I. Protective effects of naringenin from *Citrus sinensis* (var. Valencia) peels against CCl_4-induced hepatic and renal injuries in rats assessed by metabolomics, histological and biochemical analyses. *Nutrients* **2022**, *14*, 841. [CrossRef] [PubMed]
54. Farag, M.A.; Khattab, A.R.; Shamma, S.; Afifi, S.M. Profiling of primary metabolites and volatile determinants in mahlab cherry (*Prunus mahaleb* L.) seeds in the context of its different varieties and roasting as analyzed using chemometric tools. *Foods* **2021**, *10*, 728. [CrossRef] [PubMed]

foods

MDPI

Article

Revalorization of Melon By-Product to Obtain a Novel Sparkling Fruity-Based Wine

José Ángel Salas-Millán [1,2,3], Encarna Aguayo [1,2,*], Andrés Conesa-Bueno [3] and Arantxa Aznar [4,*]

1 Postharvest and Refrigeration Group, Universidad Politécnica de Cartagena (UPCT), 30202 Cartagena, Spain
2 Food Quality and Health Group, Institute of Plant Biotechnology (UPCT), Campus Muralla del Mar, 30202 Cartagena, Spain
3 JimboFresh International SLL, C/Mina Buena Suerte, 1, La Unión, 30360 Murcia, Spain
4 Department of Agronomical Engineering, Institute of Plant Biotechnology, UPCT, Paseo Alfonso XIII, 48, 30203 Cartagena, Spain
* Correspondence: encarna.aguayo@upct.es (E.A.); arantxa.aznar@upct.es (A.A.)

Abstract: Fresh melons not meeting cosmetic standards were revaluated into sparkling melon-based wine. Firstly, still melon wine was elaborated and bottled into 750 mL bottles, closed with a crown seal, and stored for 10-weeks at 14 °C. The oenological parameters and polar compounds in must, still wine, and during the sparkling process were evaluated during the experiment. The volatile profile was qualified by GC-MS, and the odor activity value (OAV) and relative odor contribution (ROC) were measured for aroma characterization. Results show that sparkling wine resulted in 12% *v*/*v* ethanol. Certain amino acids contributed to the transformation and increase of volatile compounds via Ehrlich's pathway: leucine to isoamyl alcohol; valine to iso-butyl alcohol; and phenylalanine to phenethyl alcohol. The volatile compounds also increased after the first fermentation, principally in acetate and ethyl esters, and higher alcohols. Isoamyl acetate, ethyl decanoate, 3,6-nonadienyl acetate, and (E,Z)-nonadien-1-ol had the highest OAV and ROC values among the volatiles; this contributed to the sweet, fruity, banana, tropical, nutty and melon aroma in this sparkling wine. Sensory evaluation (100 to 40) was evaluated according to International Organisation of Vine and Wine compendium, the final product (10-week) scored 92 points, with great visual, nose, and taste values. This study demonstrates how by-products revalorization can provide new products such as this novel sparkling wine with a characteristic and distinctive aroma, good sensory acceptance and market potential.

Keywords: fermentative process; melon; beverage; aroma; sparkling; fruit wine

Citation: Salas-Millán, J.Á.; Aguayo, E.; Conesa-Bueno, A.; Aznar, A. Revalorization of Melon By-Product to Obtain a Novel Sparkling Fruity-Based Wine. *Foods* **2023**, *12*, 491. https://doi.org/10.3390/foods12030491

Academic Editor: María Jose Esteve

Received: 7 December 2022
Revised: 5 January 2023
Accepted: 16 January 2023
Published: 20 January 2023

1. Introduction

Food loss refers to any food that is discarded, incinerated, or otherwise disposed of along the food supply chain from harvest/catch/slaughter up to but excluding retail level, and which is not used for any other productive use, such as animal feed or seed [1]. In this context, fresh products can be rejected owing to superficial cosmetic imperfections, color, shape, and size after preparation and packaging. These imposed cosmetic quality standards may lead to food losses with significant environmental impacts (land use, water consumption, greenhouse gas emissions, etc.) and financial implications. However, reducing and preventing food loss and waste can increase food security, foster productivity and economic efficiency, promote resource and energy conservation, and address climate change, which in turn, could also decrease climate change-related shocks to the supply chain [2]. In this context, the agriculture sector must tackle the issue of stimulating a circular economy model which enables superficially cosmetic imperfect fruits and vegetables to be utilized in novel food products such as fruit beverages and fruit-based wines by alcoholic fermentation [3]. This research presents an example of revalorizing melon by-products—melons rejected for failing to meet cosmetic standards—to obtain a novel sparkling melon-based wine.

Grapes are traditionally the most used fruit in wine production, but other examples of fermented beverages can be found, made from rice, honey, and fruits including persimmon and kiwi [4–6]. Various fruits are produced in large amounts around the world to produce alcoholic beverages through the fermentation process. The technique is related to traditional winemaking in that it involves alcoholic fermentation using yeast—usually *Saccharomyces cerevisiae*—to produce ethanol, carbon dioxide (CO_2), and other secondary metabolites that enhance the volatile aroma profile, such as esters and higher alcohols [7]. Some wines are made from fruits other than grapes, such as cider which is made from fermented apples and is one of the most popular types of fruit-based wine [8], and it is commonly used in Europe. Fruits such as strawberries, plums, and peaches are used in the USA and Canada for fruit-based wine, whilst mango and pineapple are used in Asia [9]. Depending on the CO_2 content in the wine, they are classified into "still" or "sparkling" wines [10]. In sparkling wine production, CO_2 is generated which produces effervescence due to the use of yeast in a second step after the first alcoholic fermentation. Sparkling wines are frequently classified according to the method of production, the three main approaches are: traditional (*champenoise*), transfer, and bulk (*charmat*). With the traditional method, sparkling wines are produced by two steps of fermentation. After the first fermentation is prepared, the *cuvée*, where a rich saccharose solution and nutrients (*liqueur de tirage*) and yeast is added to start the second fermentation to produce CO_2 inside sealed bottles [4]. These wines are considered for special occasions due to their additional value, their positive mouthfeel increment, their perception of volatile compounds, and their sweetness in consumers [11]. During this second fermentation, the yeast metabolism affects the aroma composition and the chemical composition which may improve the organoleptic perception [12].

Melon (*Cucumis melo* L.) is cultivated in several parts of the world thanks to its adaptability to many types of soils and temperatures. According to FAOSTAT [13], China was the world's foremost producer of melon (49% of the total, 14 million tons), followed by Turkey and India, although Spain was the world's main exporter, producing 600,000 t of which 440,000 t were exported, which accounts for around 20% of total global exportation. There are different *C. melo* botanical varieties with morphological, physiological, and organoleptic profiles, such us *cantalupensis, reticulatus, inodorus, ameri, flexuosus, chate, dudaim, tibish, acidulus, momordica, conomon, makuwa* and *chinensis* [14]. *C. melo* var. *reticulatus* is the most accepted for its sweetness, pulp texture, and aroma, with highly volatile contents such as esters (butyl acetate, ethyl isobutyrate, hexyl acetate, propyl 2-methyl butanoate, 3-methyl-2-butenyl acetate) that improve its aroma attributes [15]. Previous authors have reported melon-based wines [3,16] and have described the main volatiles in melon distillates [17] but, for the moment, there has been no scientific research focused on the development of a sparkling melon-based wine, from *C. melo* var. *reticulatus*, using fresh melon rejected for the cosmetic reasons mentioned above. These melons cultivar have a huge potential aroma profile for alcoholic fermentation. Thus, the aim of this research is to obtain a sparkling wine using fresh melons with cosmetic imperfections, evaluating the physico-chemical, polar compounds, and aroma changes during the sparkling process and then performing a sensory evaluation at the end of the process.

2. Materials and Methods

2.1. Pilot-Scale Melon-Based Wine

Fresh melon (*Cucumis melo* var. *reticulatus*) Okashi® cultivar was obtained from JimboFresh International Coop. (La Unión, Murcia, Spain). The melons received had failed to meet cosmetic standards, mainly because of their tiny caliber and a few sunspots on the skin. Following previous melon-based wine reports [3], we scaled up the process in a winery pilot plant Figure 1, located in the University (Universidad Politécnica de Cartagena, Spain). Fresh melons were hand peeled and cut into four pieces before being blended in a commercial crusher-destemmer (Enoitalia, Florence, Italy) prior to being pressed with a pneumatic press (Bucher Vaslin, Chalonnes-sur-Loire, France) to make the initial must. The must was obtained after decanting for 24 h, at 5 °C adding 0.04

g/L of sodium metabisulfite ($Na_2S_2O_5$) and enriched with 5 g/L of tartaric and malic acids (1:1, *w/w*), 0.2 g/L of commercial yeast (Zymaflore®, Laffort, Bordeaux, France), 0.2 g/L of nutrients yeast (Superstart® Blanc & Rosé, Laffort, Bordeaux, France) and added commercial saccharose until 21.3 °Brix was obtained in the must Figure 1.

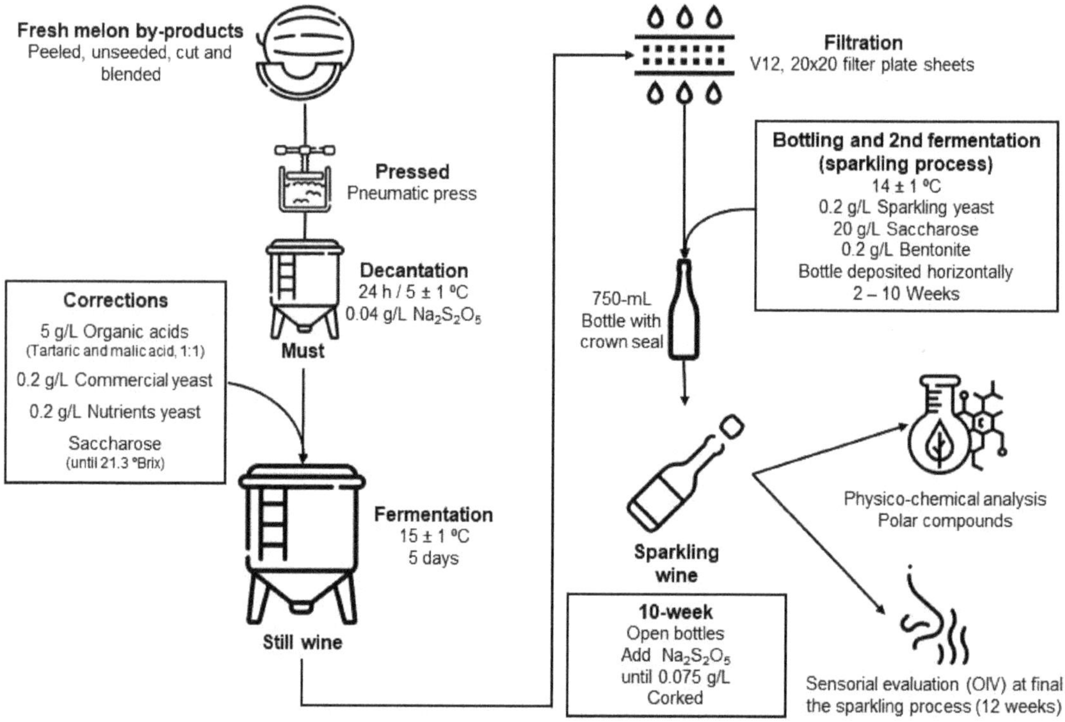

Figure 1. Scheme of the production of melon wine and sparkling wine.

After five days at 15 °C, the first fermentation finished obtaining a melon wine named "Still Wine". Subsequently, that wine was filtered through filter plate sheets (V12, 20 × 20, Gruppo Cordenons SpA, Milano, Italy) with a filter plate (Filtro Jolly 20, MORI, Tavarnelle Val di Pesa, Italy). The sparkling process was based on Martínez-García et al. [12] with slight modifications: adding 0.2 g/L of commercial sparkling yeast (Zymaflore® Spark, Laffort, Bordeaux, France) used for the second fermentation in this base melon-based wine and 0.2 g/L of yeast nutrients, and 20 g/L of saccharose, and 0.2 g/L of bentonite were also added. The wine was then bottled in 750 mL bottles closed with a crown seal and stored horizontally in a chamber at 14 °C. At the end of the process (10 weeks), the bottles were placed in a desk and the lees were mixed (*remuage*) and gradually turned and inclined into a vertical position, so the sediment from the lees was deposited in the neck of the bottle and then withdrawn at the end of the process. Sodium metabisulfite was added to reach 0.075 g/L of total SO_2 and the bottles were corked. The physico-chemical changes and aroma obtained during this second fermentation, sparkling wine, were monitored at six time points of the process (0, 2, 4, 6, 8, and 10 weeks).

2.2. Physico-Chemical Analysis

Physico-chemical parameters such as alcohol strength (% *v/v*), pressure, total soluble solids (TSS), residual sugar, sugar-free extracts, and total and free sulfide, were measured using the standardized method by OIV [18] in the melon must, the still wine, and during the sparkling process. The pH, total acidity (TA), volatile acidity (VA) and color were determined following the methods from Salas-Millán et al. [3].

The total polyphenol content (TPC) and antioxidant capacities (FRAP and TEAC) were ascertained with a multiscan plate spectrophotometer (Tecan infinite M200, Männedorf, Switzerland after diluting the melon must, still wine, and sparkling wine samples in water (1:5), as per Salas-Millán et al. [19].

2.3. *Analysis of Polar Compounds*

2.3.1. Individual Sugars and Organic Acids

The analysis was performed as per Ortiz-Duarte et al. [20], using ultra-high-performance liquid chromatography (UHPLC) instrument (Shimadzu, Kyoto, Japan) which included a DGU-20 A degasser, LC-170 30AD quaternary pump, SIL-30AC autosampler, CTO-10AS column heater, refractive index detector (RID), and SPDM-20 A diode array detector (DAD). Chromatographic separation was performed at 65 °C with a mobile phase of 2.5 mM H_2SO_4 at 0.6 mL/min for 30 min using a Rezex RAM column (300 × 7.8 mm, 8 μm particle size; Phenomenex, Macclesfield, UK). Authentic standards were used to identify and quantify sugars and organic acids (Sigma, St Louis, MO, USA). Calibration curves were created for each standard using at least six data points. A 5 mL must, still wine or sparkling wine aliquot was centrifuged at 14,000× *g* for 15 min at 4 °C, and the supernatant was further purified by solid phase extraction Phenomenex C18-SPE (Torrance, CA, USA) conditioned columns (5 mL of MeOH + 5 mL of water + 5 mL of air). The purified extract was filtered with a polyamide 0.20 μm syringe filter and diluted tenfold prior to analyses. Individual sugars and organic acids were expressed in g/L, except for fumaric acid which was presented in mg/L.

2.3.2. Individual Amino Acids

The separation and analysis of samples was performed with an HPLC/MS system consisting of an Agilent 1290 Infinity II Series HPLC (Agilent Technologies, Santa Clara, CA, USA) equipped with an Automated Multisampler module and a High-Speed Binary Pump and connected to an Agilent 6550 Q-TOF Mass Spectrometer (Agilent Technologies, Santa Clara, CA, USA) using an Agilent Jet Stream Dual electrospray (AJS-Dual ESI) interface. Experimental parameters for HPLC and Q-TOF were set in MassHunter Workstation Data Acquisition software (Agilent Technologies, Santa Clara, CA, USA, Rev. B.10.1.48). Standards with known concentrations of amino acids were prepared in water. Both standards and samples were passed through 0.22 μm filters. Then 20 μL of each standard or sample was injected onto a Zorbax Eclipse Plus C18 HPLC column (100 × 2.1 mm, 1.8 μm,), thermostatted at 40 °C, and eluted at a flow rate of 0.4 mL/min. Chromatographic conditions were run in accordance with Giordano et al. [21] with a slight modification: mobile phase A, consisting of 0.1% TDFHA (tridecafluoroheptanoic acid) (*w/v*) in MilliQ water and mobile phase B, consisting of 0.1% TDFHA (*w/v*) in acetonitrile, were used for the chromatographic separation. The initial HPLC running conditions were solvent A:B 90:10 (*v/v*). The gradient elution program was 10% solvent B for 3 min; a linear gradient from 10 to 40% solvent B in 5 min; another linear gradient from 40 to 100% solvent B in 5 min; 2 min at constant 100% solvent B. The column was equilibrated with the starting composition of the mobile phase for 5 min before each analytical run.

The mass spectrometer was operated in the positive mode. The nebulizer gas pressure was set to 40 psi, whilst the drying gas flow was set to 14 L/min at a temperature of 275 °C, and the sheath gas flow was set to 12 L/min at a temperature of 300 °C. The capillary spray, nozzle, fragmentor, and octopole RF Vpp voltages were 3500 V, 100 V, 360 V, and 750 V respectively. Profile data in the 50–400 m/z range were acquired for MS scans in 2 GHz extended dynamic range mode. A reference mass of 121.0509 was used. Data analysis was performed with MassHunter Qualitative Analysis Navigator software (Agilent Technologies, Santa Clara, CA, USA, Rev. B.80.00). Extracted ion chromatograms of different amino acids, were obtained from their molecular formula (Table 1) and were measured in µmol per liter of must, still wine, or sparkling wine (µM).

Table 1. Amino acids and molecular formula, molecular mass (MS), and qualifier fragment (MS/MS).

Amino Acid	Abbreviature	Formula	MS	MS/MS
Glycine	Gly	$C_2H_5NO_2$	76.0393	
Alanine	Ala	$C_3H_7NO_2$	90.055	
Serine	Ser	$C_3H_7NO_3$	106.0499	88.0389
Proline	Pro	$C_5H_9NO_2$	116.0706	70.0645
Valine	Val	$C_5H_{11}NO_2$	118.0863	72.0801
Threonine	Thr	$C_4H_9NO_3$	120.0655	74.0601
Cysteine	Cys	$C_3H_7NO_2S$	122.027	76.0208
Isoleucine	Ile	$C_6H_{13}NO_2$	132.1019	86.0959
Leucine	Leu	$C_6H_{13}NO_2$	132.1019	86.0959
Asparagine	Asn	$C_4H_8N_2O_3$	133.0608	87.0551
Aspartate	Asp	$C_4H_7NO_4$	134.0448	88.0387
Glutamine	Gln	$C_5H_{10}N_2O_3$	147.0764	84.0442
Lysine	Lys	$C_6H_{14}N_2O_2$	147.1128	84.0802
Glutamate	Glu	$C_5H_9NO_4$	148.0604	84.0442
Methionine	Met	$C_5H_{11}NO_2S$	150.0583	104.0527
Histidine	His	$C_6H_9N_3O_2$	156.0768	110.0711
Phenylalanine	Phe	$C_9H_{11}NO_2$	166.0863	120.0807
Arginine	Arg	$C_6H_{14}N_4O_2$	175.119	116.0703
Tyrosine	Tyr	$C_9H_{11}NO_3$	182.0812	165.0540
Tryptophane	Trp	$C_{11}H_{12}N_2O_2$	205.0972	188.0699
Cystine	Cys-Cys	$C_6H_{12}N_2O_4S_2$	241.0311	122.0273

2.4. Analysis of Volatile Compounds by GC-MS

The volatile profiles were extracted from samples using headspace solid-phase micro-extraction (HS-SPME), and identified utilizing gas chromatography (Agilent 7890B) linked to a mass spectrometer (Agilent MSD 5977A) with an autosampler (Gerstel MPS 2XL Twister) according to Salas-Millán et al. [3]. The NIST database provided the mass spectrum and retention index (RI) via the Kovats Index (KI) for comparison in the identification of the volatile substances. RI values were computed employing the same GC-MS settings and an n-alkane external standard solution C8-C20 (Sigma-Aldrich, St. Louis, MO, U.S.A.). As a semi-quantification, the GC peak area ratio of each volatile in the total ion chromatogram to the internal standard peak area was utilized [22], and measured in mg/L of the must, still wine, or sparkling wine.

2.5. Odor Activity Value (OAV) and Relative Odor Contribution (ROC)

The traditional indicators of odor activity values (OAV) and relative odor contributions (ROC) were employed to quantify the sensory contribution of aromatic compounds to wine flavor [23]. A compound's concentration divided by its odor threshold value gave the OAV, as stated by other authors [24–30]. The ratio of the compound's OAV to overall OAV for each wine is used when calculating the ROC for each aroma component.

2.6. Sensory Evaluation

Sensory evaluation was conducted in a normalized tasting room (22 °C) using standardized wine glasses containing 15 mL of melon sparkling wine. The sensory panel was mainly composed of 15 research group judges (eight women and seven men between the ages of 30 and 55), and the sensory evaluation assessment was performed following OIV 332A/2009 resolution [31], in which judges rated several aspects (Table 2). The scores for each sensory attribute were written down, with the overall score being produced by adding individual attribute values. This trained sensory evaluation does not require an ethical statement. Before starting this sensory evaluation, the research team explained the scope and details of the project to the participants, including the purpose of the research, the identity of the researchers, information on data protection, privacy, and data retention, the right not to take part (participation was voluntary), the right to withdraw, and contact details for any questions. Finally, participants signed a written consent form, confirming that they had read it and questions had been answered.

Table 2. Organoleptical characteristics and definitions used in sensory evaluation.

Organoleptical Characteristic	Definition	Range (Excellent to Inadequate)
Visual	**Discrimination of differences in outside world with sensory impressions from visible light rays**	**(15–3)**
Limpidity	Measure of cloudiness.	(5–1)
Aspect other than limpidity	Determine the full spectrum of visible properties of a product	(10–2)
Nose	**Sensations perceived by the olfactory organ when stimulated by certain volatile substances**	**(30–12)**
Genuineness	Measure degree of sensation perceived (magnitude) by the nose, of a viticulture, oenological defect of product	(6–2)
Positive intensity	Degree (magnitude) of full spectrum of qualitative odors perceived by nose.	(8–2)
Quality	Spectrum of properties and characteristics of a wine that gives an aptitude to satisfy nose, implicit or expressed needs	(16–8)
Taste	**Full spectrum of sensations perceived with wine mouthfeel.**	**(44–18)**
Genuineness	Measure degree of sensation perceived (magnitude) by the taste, of a viticulture, oenological defect of product	(6–2)
Positive intensity	Degree (magnitude) of full spectrum of qualitative odors perceived by taste.	(8–2)
Harmonious persistence	To measure the length of residual olfactory-gustatory sensation, corresponding to the sensation perceived when the product is in mouth and length of time is measured.	(8–4)
Quality	Degree (magnitude) of full spectrum of qualitative odors perceived by taste	(22–10)
Harmony-Overall judgement	**Corresponds to overall appraisal of a product.**	**(11–7)**

2.7. Statistical Analysis

A completely randomized design was performed with three replicates in the must and still wine, and per week during the sparkling process. A one-way ANOVA ($p < 0.01$) was carried out to determine the effect of fermentation and time of storage for sparkling wine. Mean values were compared by LSD multiple range test to identify significant differences among samples.

3. Results and Discussion

3.1. Physico-Chemical Characterization for First Alcoholic Fermentation (Still Wine) and During the Second Fermentation Process (Sparkling Wine)

Table 3 shows the principal physico-chemical parameters of the must, still wine, and sparkling melon wine. After melon must optimization (Section 2.1), once commercial saccharose had been added, the TSS reached 21.3 °Brix and 90.55 g/L of residual sugar (as the sum of total individual sugar, Table 4). The enrichment with tartaric acid and malic acid (5 g/L of 1:1 tartaric and malic acid) provided 6.44 g/L for TA, 0.10 g/L for volatile acidity, and a pH of 4.4. The initial sodium metabisulfite added in the must provided 11.5 mg/L and 25.6 mg/L, free and total SO_2, respectively, and it avoided prior contamination before alcoholic fermentation. The color of the must was defined as a pale greenness (109 °h) resembling the typical green color of the melon pulp.

Table 3. Physico-chemical parameters of enriched must, still wine, and during the sparkling process of melon-based wine (10 weeks).

Parameter	Must	Still Wine	Sparkling Wine (Weeks)				
			2	4	6	8	10
Alcohol (%; v/v)		12.3 ± 0.1 ns	12.3 ± 0.0	12.5 ± 0.1	12.4 ± 0.0	12.3 ± 0.1	12.4 ± 0.1
TSS (°Brix)	21.3 Z ± 0.7 a	8.7 ± 0.0 b	8.6 ± 0.0 b	8.5 ± 0.2 b	8.5 ± 0.0 b	8.5 ± 0.0 b	8.5 ± 0.0 b
Residual sugar (g/L)	90.55 ± 3.49 a	5.72 ± 0.73 b	10.23 ± 1.25 b	6.03 ± 2.12 b	8.09 ± 0.04 b	7.20 ± 0.52 b	5.38 ± 0.02 b
Sugar-free extract (g/L)	142.0 ± 4.4 a	25.1 ± 0.9 b	20.6 ± 1.2 b	23.2 ± 1.5 b	22.8 ± 0.1 b	22.4 ± 0.7 b	24.6 ± 0.0 b
pH	4.4 ± 0.1 a	3.9 ± 0.0 b	4.0 ± 0.0 b	4.0 ± 0.0 b	4.0 ± 0.0 b	4.1 ± 0.0 b	4.1 ± 0.0 b
TA (g/L)	6.44 ± 0.24 ns	5.62 ± 0.13	6.40 ± 0.06	6.63 ± 0.08	6.53 ± 0.01	6.51 ± 0.48	6.76 ± 0.17
Volatile acidity (g/L)	0.10 ± 0.01 c	0.27 ± 0.02 ab	0.28 ± 0.00 ab	0.35 ± 0.02 a	0.32 ± 0.00 ab	0.23 ± 0.01 b	0.34 ± 0.05 a
SO_2 free (mg/L)	11.5 ± 0.0 b	0.0 ± 0.0 c	0.0 ± 0.0 c	0.0 ± 0.0 c	0.0 ± 0.0 c	0.0 ± 0.00 c	12.8 ± 0.7 a
SO_2 total (mg/L)	25.6 ± 0.0 b	26.2 ± 0.9 b	26.9 ± 0.4 b	25.0 ± 2.1 b	25.1 ± 0.2 b	22.7 ± 3.4 b	75.7 ± 2.3 a
L*	39.2 ± 0.2 a	33.5 ± 0.2 bc	33.8 ± 0.0 bc	33.4 ± 0 c	33.6 ± 0.1 bc	33.6 ± 0.0 bc	33.9 ± 0.1 b
°Hue	109.0 ± 0,5 b	178.5 ± 0.0 a	178.5 ± 0.0 a	178.6 ± 0.1 a	178.5 ± 0.0 a	178.5 ± 0.0 a	178.6 ± 0.0 a
Chroma	20.7 ± 0.6 a	0.9 ± 0.0 b	0.8 ± 0.0 b	1.0 ± 0.1 b	0.8 ± 0.1 b	1.2 ± 0.1 b	1.1 ± 0.1 b
TPC	173.6 ± 4.6 ab	153.9 ± 2.4 c	164.6 ± 3.4 bc	157.6 ± 2.0 c	158.4 ± 0.1 c	156.2 ± 1.3 c	182.4 ± 3.0 a
FRAP	0.94 ± 0.01 b	0.73 ± 0.02 c	0.76 ± 0.01 c	0.72 ± 0.03 c	0.66 ± 0.01 c	0.63 ± 0.02 c	1.15 ± 0.08 a
TEAC	0.47 ± 0.01 b	0.42 ± 0.02 b	0.46 ± 0.01 b	0.45 ± 0.01 b	0.42 ± 0.00 b	0.43 ± 0.01 b	0.78 ± 0.05 a

Z Means ($n = 3 \pm$ SE). TSS: Total soluble solids. TA (g TE/L): g tartaric acid equivalent per liter. Volatile acidity (g/L): g acetic acid equivalent per liter. TPC: Total phenolic compounds, mg gallic acid equivalent/L (mg GAE/L). FRAP: Ferric reducing antioxidant capacity, mmol Fe^{+2} equivalent/L (mmol Fe^{+2}/L). TEAC: Trolox equivalent antioxidant capacity, mmol Trolox equivalent/L (mmol TE/L). Different letters in the same row indicate significant differences ($p < 0.01$) among must, still wine, and sparkling process. ns: statistically non-significant differences.

Table 4. Evolution of individual sugars and organic acids of enriched must, still wine, and during the sparkling process of melon-based wine (10 weeks).

Polar Compounds	Must	Still Wine	Sparkling Wine (Weeks)				
			2	4	6	8	10
Individual sugar (g/L)							
Saccharose	53.63 Z ± 2.58 a	n.d.	n.d.	n.d.	n.d.	n.d.	n.d.
Fructose	10.90 ± 0.55 a	2.95 ± 0.40 c	5.01 ± 0.73 b	3.27 ± 0.03 c	4.29 ± 0.02 bc	4.05 ± 0.66 bc	3.10 ± 0.02 c
Glucose	11.34 ± 0.64 a	1.87 ± 0.36 c	4.09 ± 0.83 b	1.74 ± 0.01 c	2.70 ± 0.02 bc	2.27 ± 0.19 c	1.41 ± 0.03 c
∑ Total	**75.87 ± 3.77 a**	**4.82 ± 0.75 b**	**9.10 ± 1.55 b**	**5.01 ± 0.04 b**	**6.98 ± 0.04 b**	**6.32 ± 0.59 b**	**4.51 ± 0.04 b**
Organic acids (g/L)							
Fumaric acid (mg/L)	11.49 ± 1.59 a	7.66 ± 1.46 bc	10.67 ± 1.53 ab	9.80 ± 0.63 abc	11.72 ± 0.05 a	6.84 ± 0.13 b	9.66 ± 1.45 abc
Succinic acid	1.25 ± 0.20 a	0.15 ± 0.02 c	0.28 ± 0.03 b	0.20 ± 0.00 c	0.25 ± 0.00 bc	0.16 ± 0.02 c	0.19 ± 0.01 c
Malic acid	3.50 ± 0.43 a	2.37 ± 0.03 b	2.52 ± 0.27 b	2.39 ± 0.09 b	1.80 ± 0.30 b	1.88 ± 0.09 b	1.80 ± 0.08 b
Citric acid	2.07 ± 0.24 b	2.92 ± 0.22 ab	3.19 ± 0.50 a	2.83 ± 0.04 ab	3.12 ± 0.11 a	2.93 ± 0.40 ab	2.59 ± 0.39 ab
Tartaric acid	1.56 ± 0.22 b	1.62 ± 0.37 ab	1.31 ± 0.23 b	1.57 ± 0.07 b	1.12 ± 0.05 b	1.97 ± 0.25 a	1.59 ± 0.05 b
∑ Total	**8.40 ± 1.08 ns**	**7.07 ± 0.32**	**7.31 ± 0.02**	**7.00 ± 0.15**	**6.31 ± 0.46**	**6.95 ± 0.51**	**6.18 ± 0.43**

Mean ± SE ($n = 3$). Different letters in the same row indicate significant differences ($p < 0.05$) among must, still wine, and sparkling process. n.d.: not detected. ns: statistically non-significant differences.

After the first alcoholic fermentation, the still wine was obtained and TSS dropped to <9 °Brix and remained stable during the sparkling process without significant differences. The alcoholic grade reached 12.3° and was stable in the sparkling melon process (12.3° to 12.5°). Likewise, residual sugar decreased with fermentation, from 90.55 g/L in the must to 5.72 g/L in the still melon wine, and after the incorporation of 20 g/L of commercial saccharose before the second fermentation, the residual sugar decreased to 5.38–10.23 g/L in the sparkling wine; a decrease in the sugar concentration with the storage time was detected.

Regarding the sugar-free extract, the high value obtained in the must (142 g/L) was due to saccharose added and the presence of suspension pulp, dropping with the fermentation process and the filtration process in the still melon wine (25 g/L) and remaining quite stable during the sparkling wine process (20.6 to 25.1 g/L). Sugar-free extract is an important qualitative parameter for evaluating fullness and harmony in wine, this parameter should usually be below 25 g/L [23], as obtained in the sparkling wine. The pressure reached 2 atm in the second week after second fermentation and remained stable throughout the sparkling process.

TA was maintained in the range of 5.62 to 6.76 g TA/L, without significant differences during the time studied for the sparkling process. As expected, the volatile acidity increased with the first alcoholic fermentation, 0.27 g/L was obtained in the still melon wine, and it ranged from 0.23 to 0.35 g/L during the sparkling process. The pH slightly decreased to 3.9 in the still wine with similar and stable values for the sparkling wines (4.0 to 4.1).

Regarding the color parameters, the first fermentation, induced an increase in the °h (178°) and a slight reduction in L*, with a green color being obtained that was maintained during the studied sparkling process. Chrome decreased from 20.7 (melon must) toward 0.8 to 1.2, which means a decline in the green color intensity, due to the filtration process and its clarification [32].

The TPC and antioxidant capacities (FRAP and TEAC) in enriched melon must were 173.6 mg GAE/L, 0.94 mmol Fe^{+2}/L, and 0.47 mmol TE/L, respectively. Fermentation led to a decrease in those values, with 153 mg GAE/L being obtained in the still wine, and the antioxidant capacity was 0.73 mg Fe^{+2}/L and 0.43 mmol TE/L. These values were similar for the sparkling melon wine during the eight weeks of storage, excepting the end of the storage time when an increase in TPC (182 mg GAE/L) and antioxidant activity (1.15 mg Fe^{+2}/L and 0.78 mmol TE/L) was found, probably due to the interference of metabisulfite as an antioxidant agent [33], which was added in week 10.

3.2. Polar Compounds Quantification

3.2.1. Individual Sugar and Organic Acids Content

The total sugar content for the melon must was 75.87 g/L; this high concentration was achieved after adding saccharose. In the first fermentation, the still melon wine obtained 4.82 g/L and the sparkling melon wine ranged from 4.51 to 9.10 g/L, with the lowest data being found after 10 weeks Table 4. This decrease in total sugars was the effect of the fermentation processes in which *S. cerevisiae*, which has the ability to ferment sugars, produced ethanol, CO_2, and other secondary metabolites as the product of the alcoholic fermentation carried out [3]. The reduction in sugars, specifically glucose, in the sparkling melon wine after 10 weeks of storage shows that the fermentation process continued slowly inside the bottles [34]. Saccharose was only measured in the melon must and was completely consumed in the first alcoholic fermentation. A different trend was observed in the fructose and glucose concentrations, which decreased in the still wine with the first fermentation (2.95 and 1.87 g/L, respectively), but rose in the sparkling wine, due to the saccharose added (20 g/L) to initiate the second fermentation. After two weeks in the sparkling process, the fructose and glucose contents were 5.01 and 4.09 g/L, due to the *S. cerevisiae* Suc2 invertase enzyme which hydrolyzed extracellularly saccharose into glucose and fructose and increased the content of those monosaccharides [35]. After that, the fructose and glucose concentrations were remained stable throughout the sparkling process for 10 weeks Table 4. Similar trends were reported by Berbegal et al. [36] for sparkling wine with a dry yeast preparation.

Regarding total organic acids, the melon must presented a total of 8.40 g/L, decreasing slightly with the fermentations, 7.07 g/L in the still wine and 6.18 g/L in the sparkling wine (10 weeks). As mentioned in Section 2.1, the melon must was enriched with tartaric and malic acids (5 g/L, 1:1), providing a higher content for these organic acids. There were slight differences with a stabilization trend for tartaric acid throughout both fermentations and the sparkling melon wine storage (1.12–1.62 g/L). A similar trend in tartaric acid has been reported in other sparkling wines from grapes, in the 3rd and 12th months [37]. The malic acid concentration in the must (3.50 g/L) decreased after the first fermentation (2.37 g/L), due to malolactic fermentation, in which malic acid is transformed into lactic acid, normally by *Oenococcus*, *Lactobacillus* and *Pediococcus* species in typical grape wines, such production reduces the perceived acidity [4]. Malic acid continued to decrease slightly during the sparkling process (from 2.52 to 1.80 g/L). Succinic acid also decreased with the first fermentation, from 1.25 g/L in the must to 0.15 g/L in the still wine and remained stable during the sparkling process (0.16–0.28 g/L).

3.2.2. Individual Amino Acids Content

Amino acids were quantified into four groups in accordance with Liang et al. [38] bitter, sweet, umami, and other amino acids Table 5. The bitter amino acids included histidine, arginine, leucine, lysine, valine, phenylalanine, and isoleucine; sweet amino acids consisted of glycine, alanine, proline, serine, threonine, methionine, and cysteine, the umami amino acids were aspartic acid and glutamate; and other amino acids such as tyrosine, asparagine, glutamine, tryptophan, and cistin as the dimerization of two cysteines (Cys-Cys). The concentration of total amino acids in the melon must was 33.06 μM, although this dropped with the first alcoholic fermentation (18.31 μM) and was stable during the sparkling wine process (15.10 to 16.29 μM). This reduction was caused by the yeasts, whose population increased or retained a high viability during that period, and the use of amino acids for protein, RNA, and DNA synthesis, and storage as a reserve in vacuoles [34]. This trend was mainly detected in the total bitter amino acids, with arginine, phenylalanine, and histidine being the most abundant amino acids in that group. Arginine and phenylalanine decreased by 60 to 70% in this first fermentation and remained stable during the sparkling process. This behavior has been previously reported in wine during alcoholic fermentation by Wang et al. [39], who noted that these amino acids were consumed by the yeasts during the fermentation process as a nitrogen source. By contrast, histidine did not suffer a drop after fermentation and ranged between 5.54 and 6.45 μM during the experiment.

Table 5. Evolution of amino acids (μM) of enriched must, still wine, and during the sparkling process of melon-based wine (10 weeks).

Polar Compounds		Must	Still Wine	Sparkling Wine (Weeks)				
				2	4	6	8	10
TBAA	Histidine	6.19 ^{z} ± 0.14 ^{ab}	6.45 ± 0.21 ^{a}	5.54 ± 0.04 ^{c}	5.97 ± 0.18 ^{bc}	6.02 ± 0.10 ^{ab}	6.19 ± 0.19 ^{ab}	6.22 ± 0.22 ^{ab}
	Arginine	13.67 ± 0.51 ^{a}	3.97 ± 0.08 ^{b}	3.61 ± 0.00 ^{b}	3.79 ± 0.08 ^{b}	3.91 ± 0.02 ^{b}	3.86 ± 0.05 ^{b}	3.94 ± 0.01 ^{b}
	Leucine	0.26 ± 0.08 ^{a}	0.05 ± 0.01 ^{b}	0.05 ± 0.00 ^{b}	0.04 ± 0.00 ^{b}	0.04 ± 0.00 ^{b}	0.03 ± 0.00 ^{b}	0.03 ± 0.00 ^{b}
	Lysine	0.13 ± 0.01 ^{a}	0.12 ± 0.00 ^{a}	0.09 ± 0.00 ^{b}	0.12 ± 0.01 ^{a}	0.06 ± 0.01 ^{c}	0.08 ± 0.01 ^{bc}	0.06 ± 0.01 ^{c}
	Valine	0.04 ± 0.00 ^{ns}	0.05 ± 0.00	0.05 ± 0.00	0.04 ± 0.01	0.04 ± 0.00	0.04 ± 0.00	0.04 ± 0.00
	Phenylalanine	10.44 ± 1.45 ^{a}	3.95 ± 0.05 ^{b}	3.27 ± 0.03 ^{bc}	2.84 ± 0.15 ^{bc}	2.50 ± 0.05 ^{bc}	2.15 ± 0.05 ^{c}	2.12 ± 0.01 ^{c}
	Isoleucine	0.13 ± 0.05 ^{a}	0.01 ± 0.00 ^{b}	0.02 ± 0.00 ^{b}	0.02 ± 0.00 ^{b}	0.02 ± 0.00 ^{b}	0.02 ± 0.00 ^{b}	0.02 ± 0.00 ^{b}
	∑ TBAA	**30.87 ± 2.0** ^{a}	**14.60 ± 0.33** ^{b}	**12.63 ± 0.08** ^{b}	**12.83 ± 0.28** ^{b}	**12.59 ± 0.14** ^{b}	**12.38 ± 0.10** ^{b}	**12.45 ± 0.21** ^{b}
TSAA	Glycine	0.07 ± 0.01 ^{c}	0.32 ± 0.03 ^{a}	0.26 ± 0.00 ^{b}	0.28 ± 0.01 ^{ab}	0.29 ± 0.03 ^{ab}	0.27 ± 0.01 ^{ab}	0.09 ± 0.01 ^{c}
	Alanine	0.67 ± 0.07 ^{ns}	0.80 ± 0.06	0.86 ± 0.01	0.73 ± 0.03	0.73 ± 0.02	0.74 ± 0.03	0.67 ± 0.03
	Proline	0.07 ± 0.01 ^{ns}	0.16 ± 0.09	0.22 ± 0.09	0.05 ± 0.06	0.05 ± 0.00	0.04 ± 0.00	0.05 ± 0.00
	Serine	0.01 ± 0.00 ^{d}	0.27 ± 0.01 ^{a}	0.25 ± 0.00 ^{b}	0.22 ± 0.01 ^{c}	0.22 ± 0.00 ^{c}	0.21 ± 0.00 ^{c}	0.21 ± 0.00 ^{c}
	Threonine	0.03 ± 0.00 ^{bc}	0.03 ± 0.00 ^{ab}	0.04 ± 0.00 ^{a}	0.03 ± 0.00 ^{bc}	0.03 ± 0.00 ^{bc}	0.03 ± 0.00 ^{bc}	0.03 ± 0.00 ^{c}
	Metionine	0.14 ± 0.04 ^{ns}	0.12 ± 0.00	0.13 ± 0.00	0.10 ± 0.01	0.09 ± 0.01	0.10 ± 0.01	0.08 ± 0.01
	Cysteine	0.01 ± 0.01 ^{c}	0.34 ± 0.03 ^{b}	0.41 ± 0.01 ^{a}	0.31 ± 0.03 ^{b}	0.32 ± 0.01 b	0.32 ± 0.00 ^{b}	0.32 ± 0.03 ^{b}
	∑ TSAA	**0.92 ± 0.11** ^{c}	**1.73 ± 0.12** ^{a}	**1.90 ± 0.09** ^{a}	**1.44 ± 0.13** ^{b}	**1.43 ± 0.02** ^{b}	**1.44 ± 0.04** ^{b}	**1.37 ± 0.03** ^{b}
TUAA	Aspartate	0.07 ± 0.06 ^{c}	1.01 ± 0.04 ^{a}	0.87 ± 0.01 ^{abc}	0.87 ± 0.05 ^{bc}	0.82 ± 0.04 ^{b}	0.99 ± 0.05 ^{ab}	0.84 ± 0.04 ^{bc}
	Glutamate	0.00 ± 0.00 ^{d}	0.02 ± 0.00 ^{a}	0.01 ± 0.00 ^{ab}	0.00 ± 0.00 ^{d}	0.01 ± 0.00 ^{b}	0.01 ± 0.00 ^{bc}	0.01 ± 0.00 ^{c}
	∑ TUAA	**0.07 ± 0.06** ^{d}	**1.03 ± 0.04** ^{a}	**0.89 ± 0.01** ^{bc}	**0.87 ± 0.05** ^{c}	**0.83 ± 0.04** ^{c}	**1.01 ± 0.06** ^{ab}	**0.85 ± 0.05** ^{c}
Others	Tyrosine	0.01 ± 0.00 ^{b}	0.02 ± 0.00 ^{a}	0.01 ± 0.00 ^{a}	0.02 ± 0.00 ^{a}	0.01 ± 0.00 ^{a}	0.01 ± 0.00 ^{a}	0.02 ± 0.00 ^{a}
	Asparagine	0.00 ± 0.00 ^{c}	0.08 ± 0.01 ^{a}	0.08 ± 0.00 ^{ab}	0.07 ± 0.01 ^{ab}	0.07 ± 0.01 ^{b}	0.08 ± 0.00 ^{ab}	0.08 ± 0.00 ^{ab}
	Glutamine	0.15 ± 0.02 ^{a}	0.02 ± 0.00 ^{b}	0.02 ± 0.01 ^{b}	0.01 ± 0.00 ^{b}	0.01 ± 0.00 ^{b}	0.01 ± 0.00 ^{b}	0.01 ± 0.00 ^{b}
	Tryptophan	0.88 ± 0.08 ^{a}	0.25 ± 0.01 ^{b}	0.22 ± 0.00 ^{bc}	0.20 ± 0.02 ^{bc}	0.16 ± 0.01 ^{bc}	0.16 ± 0.01 ^{bc}	0.14 ± 0.00 ^{c}
	Cistin	0.10 ± 0.00 ^{d}	0.26 ± 0.02 ^{a}	0.27 ± 0.01 ^{a}	0.20 ± 0.02 ^{b}	0.19 ± 0.00 ^{b}	0.15 ± 0.01 ^{c}	0.10 ± 0.01 ^{d}
	∑ Total AA	**33.06 ± 2.18** ^{a}	**18.31 ± 0.49** ^{b}	**16.29 ± 0.03** ^{bc}	**15.92 ± 0.23** ^{bc}	**15.59 ± 0.09** ^{c}	**15.50 ± 0.17** ^{c}	**15.10 ± 0.15** ^{c}

Mean ± SE (n = 3). TBAA: Total bitter amino acids. TSAA: Total sweet amino acids. TUAA: Total umami amino acids. TAA: Total amino acids. Different letters in the same row indicate significant differences ($p < 0.05$) among must, still wine, and sparkling process. ns: statistically non-significant differences.

The total sweet amino acids rose from 0.92 μM in the must to 1.73 μM in the still wine, after the first fermentation, and slightly increased to 1.90 μM after two weeks of sparkling wine but decreased in the following weeks. The main sweet amino acid was alanine whose concentration changed following the same trend as mentioned for the total sweet amino acids. Some authors have suggested that the yeast's own autolysis contributes to increase the amino acids content [39].

In the total umami amino acids, aspartate showed the highest concentration, rising from 0.07 μM to 1.01 μM in the first fermentation wine and decreasing to 0.84 μM at the end of the sparkling process. Other amino acids such as tyrosine, asparagine, and glutamine presented very low concentrations in the still wine and during the sparkling process (<0.08 μM). Tryptophan decreased from 0.88 μM in the must to 0.25 μM in the still wine and ranged from 0.14 to 0.22 μM during the sparkling process. By contrast, cistin (dimerization of cisteine) increased from the must to the still wine (0.10 to 0.26 μM, respectively) and decreased during the sparkling process (0.26 to 0.10 μM).

3.3. Volatile Compounds in Melon Wines and Sparkling Wine

The volatile compounds identified in these melon wines were classified as: ethyl acetate, acetic acid, acetate higher alcohols (AHA), short-chain (C3-C5) fatty acid ethyl ester (SCFAEE), medium-chain (C6-C12) fatty acid ethyl ester (MCFAEE), long-chain (C13-C22) fatty acid ethyl ester (LCFAEE), other esters, ethanol, higher alcohols (HA), and other compounds. The aroma evolution during the first alcoholic fermentation and the sparkling process is shown in Figure 2, except for other ester and other compound groups which can be consulted in Table 6.

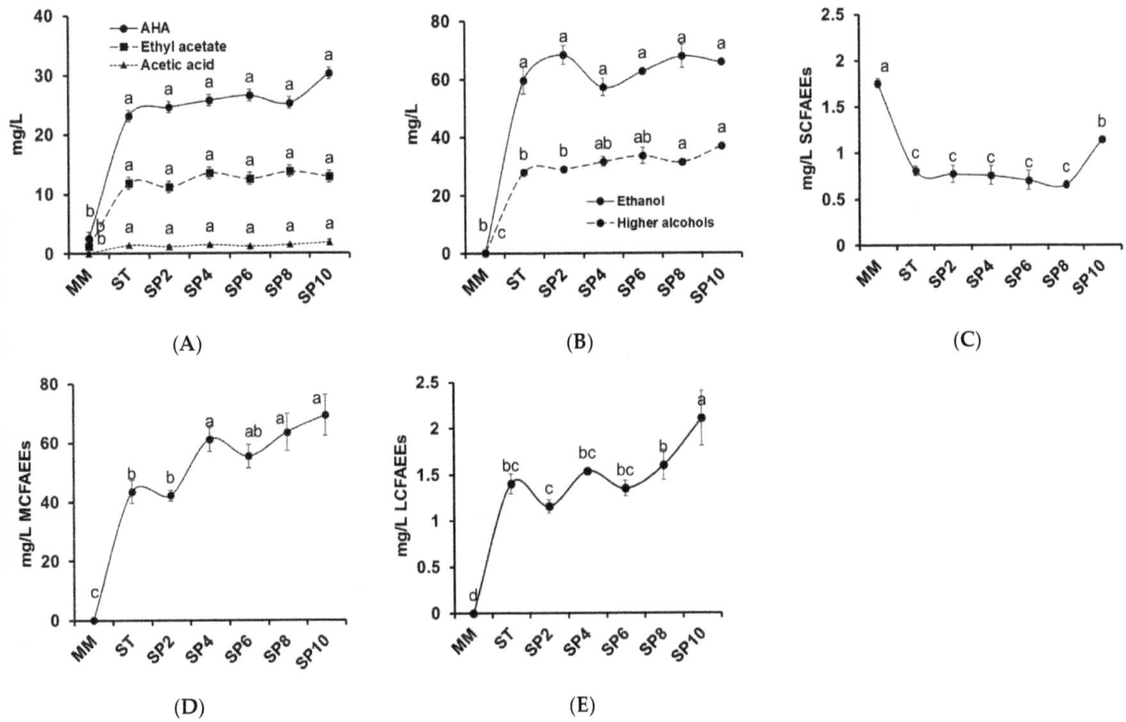

Figure 2. Evolution of aroma volatiles for enriched must, still wine, and during the sparkling process of melon-based wine (10 weeks). Must (MM); Still wine (ST); Sparkling wine (SP) in weeks (2, 4, 6, 8, and 10). (**A**) Acetate higher alcohol (AHA), ethyl acetate and acetic acid. (**B**) Ethanol and higher alcohols. (**C**) Short-chain fatty acid ethyl ester (SCFAEEs). (**D**) Medium-chain fatty acid ethyl ester (MCFAEEs). (**E**) Long-chain fatty acid ethyl ester (LCFAEEs). Different letters in the same image indicate significant differences ($p < 0.05$).

Table 6. Evolution of aroma volatiles (mg/L) in enriched must, still wine, and during the sparkling process of melon-based wine (10 weeks).

Group	RT	Compounds	KI	Must	Still Wine	Sparkling Wine (Weeks)				
						2	4	6	8	10
	2.465	Ethyl acetate	<1100	1.30 ± 0.13 [b]	11.88 ± 1.88 [a]	11.22 ± 1.01 [a]	13.57 ± 1.96 [a]	12.60 ± 1.62 [a]	13.79 ± 2.51 [a]	12.92 ± 0.93 [a]
	21.181	Acetic acid	1436	n.d.	1.46 ± 0.04 [ab]	1.23 ± 0.24 [b]	1.56 ± 0.08 [ab]	1.27 ± 0.06 [b]	1.53 ± 0.09 [ab]	1.95 ± 0.40 [a]
AHA	**3.785**	Isobutyl acetate	<1100	1.11 ± 0.11 [c]	3.50 ± 0.17 [b]	4.76 ± 0.71 [ab]	4.71 ± 0.64 [ab]	6.75 ± 1.44 [a]	4.19 ± 0.29 [b]	3.91 ± 0.41 [b]
	4.816	Butyl acetate	<1100	0.24 ± 0.02 [a]	0.12 ± 0.01 [b]	n.d.	n.d.	n.d.	n.d.	n.d.
	5.910	Isoamyl acetate	<1100	0.53 ± 0.03 [c]	14.43 ± 0.36 [b]	14.88 ± 0.06 [b]	15.39 ± 1.03 [ab]	14.51 ± 1.19 [b]	15.10 ± 0.96 [b]	19.40 ± 2.80 [a]
	7.439	Pentyl acetate	<1100	0.02 ± 0.00	n.d.	n.d.	n.d.	n.d.	n.d.	n.d.
	11.531	Hexyl acetate	1237	0.25 ± 0.03	n.d.	n.d.	n.d.	n.d.	n.d.	n.d.
	13.563	3-Hexenyl acetate	1286	0.10 ± 0.01	n.d.	n.d.	n.d.	n.d.	n.d.	n.d.
	23.062	2,3-Butanediol, diacetate	1470	0.07 ± 0.01	n.d.	n.d.	n.d.	n.d.	n.d.	n.d.
	25.216	2,3-Butanediol, diacetate	1506	0.05 ± 0.00	n.d.	n.d.	n.d.	n.d.	n.d.	n.d.
	32.218	3,6-Nonadienyl-acetate	1598	0.01 ± 0.00 [c]	1.25 ± 0.01 [ab]	0.89 ± 0.39 [bc]	1.62 ± 0.26 [ab]	1.09 ± 0.34 [b]	1.98 ± 0.46 [a]	1.83 ± 0.27 [ab]
	33.439	Phenylmethyl acetate	1612	0.16 ± 0.02	n.d.	n.d.	n.d.	n.d.	n.d.	n.d.
	35.311	2-Phenylethyl acetate	1843	0.03 ± 0.00 [c]	3.84 ± 0.33 [b]	4.10 ± 0.24 [ab]	4.07 ± 0.31 [ab]	4.24 ± 0.45 [ab]	4.04 ± 0.16 [b]	5.13 ± 0.60 [a]
		∑AHA		2.57 ± 0.19 [c]	23.14 ± 0.59 [b]	24.63 ± 0.14 [ab]	25.79 ± 1.04 [ab]	23.30 ± 0.45 [b]	25.31 ± 1.54 [ab]	27.81 ± 2.83 [a]
SCFAEE	**3.063**	Ethyl propanoate	<1100	0.08 ± 0.01	n.d.	n.d.	n.d.	n.d.	n.d.	n.d.
	3.160	Ethyl isobutyrate	<1100	0.39 ± 0.04	n.d.	n.d.	n.d.	n.d.	n.d.	n.d.
	4.160	Ethyl butanoate	<1100	0.68 ± 0.07 [a]	0.64 ± 0.04 [b]	0.57 ± 0.12 [b]	0.55 ± 0.08 [b]	0.51 ± 0.09 [b]	0.46 ± 0.03 [b]	0.82 ± 0.09 [b]
	4.451	Ethyl 2-methylbutyrate	<1100	0.77 ± 0.07 [a]	0.17 ± 0.01 [c]	0.20 ± 0.03 [c]	0.21 ± 0.02 [c]	0.19 ± 0.01 [c]	0.20 ± 0.01 [c]	0.32 ± 0.04 [b]
	6.204	Ethyl pentanoate	<1100	0.02 ± 0.00	n.d.	n.d.	n.d.	n.d.	n.d.	n.d.
		∑ SCFAEE		1.76 ± 0.05 [a]	0.81 ± 0.05 [c]	0.77 ± 0.09 [c]	0.76 ± 0.10 [c]	0.70 ± 0.10 [c]	0.66 ± 0.04 [c]	1.13 ± 0.13 [b]
MCFAEEs	**9.750**	Ethyl hexanoate	1185	0.19 ± 0.02 [b]	2.36 ± 0.13 [a]	2.41 ± 0.24 [a]	2.24 ± 0.05 [a]	2.07 ± 0.01 [a]	2.18 ± 0.21 [a]	2.34 ± 0.24 [a]
	19.923	Ethyl octanoate	1409	n.d.	23.78 ± 2.03 [b]	20.46 ± 0.94 [b]	33.42 ± 3.23 [ab]	33.32 ± 2.83 [ab]	35.44 ± 3.22 [a]	44.55 ± 7.79 [a]
	31.144	Ethyl decanoate	1591	n.d.	15.78 ± 1.60 [d]	16.74 ± 0.25 [cd]	22.50 ± 0.65 [abc]	18.44 ± 1.03 [bcd]	24.23 ± 2.63 [ab]	28.39 ± 3.67 [a]
	32.928	Ethyl 9-decenoate	1606	n.d.	1.48 ± 0.33 [n.s.]	1.67 ± 0.07	1.42 ± 0.19	1.12 ± 0.04	1.35 ± 0.14	1.89 ± 0.35
	36.046	Ethyl laurate	1790	n.d.	0.13 ± 0.02 [de]	0.98 ± 0.23 [b]	1.62 ± 0.12 [a]	0.52 ± 0.10 [c]	0.36 ± 0.03 [cd]	0.35 ± 0.05 [cd]
		∑ MCFAEEs		0.19 ± 0.02 [c]	43.52 ± 4.03 [b]	43.25 ± 1.65 [b]	61.86 ± 4.20 [a]	55.70 ± 4.01 [ab]	64.51 ± 6.19 [a]	69.23 ± 6.87 [a]

Table 6. *Cont.*

Group	RT	Compounds	KI	Must	Still Wine	Sparkling Wine (Weeks)				
						2	4	6	8	10
LCFAEEs	39.044	Ethyl tetradecanoate	2007	n.d.	0.73 ± 0.01 $^{n.s.}$	0.57 ± 0.10	0.69 ± 0.04	0.70 ± 0.09	0.75 ± 0.08	0.72 ± 0.09
	40.829	Ethyl hexadecanoate	2136	n.d.	0.68 ± 0.10 ^{b}	0.59 ± 0.02 ^{b}	0.85 ± 0.03 ^{b}	0.65 ± 0.01 ^{b}	0.85 ± 0.08 ^{b}	1.39 ± 0.24 ^{a}
		∑ LCFAEEs		n.d.	1.40 ± 0.11 ^{b}	1.16 ± 0.07 ^{b}	1.54 ± 0.02 ^{b}	1.35 ± 0.08 ^{b}	1.60 ± 0.16 ^{b}	2.11 ± 0.30 ^{a}
Other Ester	**5.193**	Isobutyl butyrate	<1100	0.02 ± 0.00	n.d.	n.d.	n.d.	n.d.	n.d.	n.d.
	6.296	Isobutyl butyrate	<1100	0.03 ± 0.00	n.d.	n.d.	n.d.	n.d.	n.d.	n.d.
	28.261	Ethyl 3-acetoxybutyrate	1552	0.03 ± 0.00	n.d.	n.d.	n.d.	n.d.	n.d.	n.d.
	37.298	Hydrocinnamyl isobutyrate	1833	0.02 ± 0.00	n.d.	n.d.	n.d.	n.d.	n.d.	n.d.
		∑ Other Ester		0.09 ± 0.01	n.d.	n.d.	n.d.	n.d.	n.d.	n.d.
	2.825	Ethanol	<1100	0.06 ± 0.03 ^{b}	59.73 ± 4.65 ^{a}	68.43 ± 3.21 ^{a}	57.21 ± 2.93 ^{a}	62.70 ± 0.63 ^{a}	67.91 ± 4.07 ^{a}	65.83 ± 12.30 ^{a}
HA	**5.658**	Isobutyl alcohol	<1100	n.d.	1.56 ± 0.08 ^{ab}	1.61 ± 0.17 ^{ab}	1.40 ± 0.02 ^{b}	1.93 ± 0.18 ^{a}	1.56 ± 0.09 ^{ab}	1.72 ± 0.29 ^{ab}
	9.202	2-Methyl-1-butanol	1168	0.04 ± 0.01	n.d.	n.d.	n.d.	n.d.	n.d.	n.d.
	9.733	Isoamyl alcohol	1185	n.d.	18.70 ± 0.76 ^{b}	19.21 ± 0.29 ^{ab}	20.21 ± 0.84 ^{ab}	22.07 ± 1.62 ^{ab}	20.11 ± 0.65 ^{ab}	24.17 ± 3.45 ^{a}
	19.630	2-Octanol	1404	0.02 ± 0.00 ^{a}	n.d.	n.d.	n.d.	n.d.	n.d.	n.d.
	23.385	2-Ethyl-1-hexanol	1475	0.05 ± 0.00	n.d.	n.d.	n.d.	n.d.	n.d.	n.d.
	26.011	2,3-Butanediol	1518	n.d.	1.08 ± 0.04 ^{ab}	0.62 ± 0.10 ^{bc}	1.38 ± 0.34 ^{a}	0.41 ± 0.11 ^{cd}	0.42 ± 0.22 ^{cd}	n.d.
	32.746	(Z)-3-Nonen-1-ol	1604	0.02 ± 0.00	n.d.	n.d.	n.d.	n.d.	n.d.	n.d.
	33.213	Methionol	1609	n.d.	n.d.	n.d.	n.d.	n.d.	0.32 ± 0.04 ^{a}	0.25 ± 0.00 ^{b}
	33.598	(6 Z)-Nonen-1-ol	1613	0.03 ± 0.01	n.d.	n.d.	n.d.	n.d.	n.d.	n.d.
	34.218	(E,Z)-3,6-Nonadien-1-ol	1620	0.06 ± 0.01 ^{c}	0.12 ± 0.01 ^{ab}	0.12 ± 0.00 ^{ab}	0.12 ± 0.02 ^{ab}	0.11 ± 0.00 ^{bc}	0.10 ± 0.02 ^{bc}	0.17 ± 0.04 ^{a}
	36.150	Benzyl alcohol	1794	0.01 ± 0.00 ^{c}	n.d.	n.d.	n.d.	0.98 ± 0.14 ^{a}	0.18 ± 0.01 ^{b}	0.16 ± 0.02 ^{bc}
	36.696	2-Phenylethanol	1813	0.02 ± 0.00 ^{d}	4.71 ± 0.24 ^{c}	5.46 ± 0.44 ^{bc}	6.45 ± 0.48 ^{b}	5.75 ± 0.53 ^{bc}	6.80 ± 0.30 ^{ab}	8.16 ± 0.96 ^{a}
	41.030	2,4-Di-tert-butylphenol	2213	0.01 ± 0.00 ^{c}	1.75 ± 0.04 ^{b}	1.99 ± 0.08 ^{ab}	1.94 ± 0.02 ^{ab}	2.35 ± 0.27 ^{a}	1.88 ± 0.12 ^{ab}	2.28 ± 0.29 ^{a}
		∑ HA		0.27 ± 0.03 ^{c}	27.92 ± 0.98 ^{b}	29.01 ± 0.57 ^{ab}	31.51 ± 1.66 ^{ab}	33.60 ± 2.73 ^{ab}	31.38 ± 0.44 ^{ab}	36.92 ± 4.88 ^{a}
Other volatiles	**10.207**	Styrene	1199	n.d.	n.d.	n.d.	0.27 ± 0.03 $^{n.s.}$	0.27 ± 0.03	0.22 ± 0.01	0.29 ± 0.06
	19.291	1,3-Di-tert-butylbenzene	1392	0.02 ± 0.01 ^{c}	0.67 ± 0.06 ^{b}	0.64 ± 0.03 ^{b}	0.88 ± 0.08 ^{ab}	1.04 ± 0.24 ^{a}	0.85 ± 0.05 ^{ab}	0.86 ± 0.07 ^{ab}
	34.810	Oxime-methoxy phenyl	1803	0.02 ± 0.00 ^{c}	0.47 ± 0.04 ^{b}	0.56 ± 0.04 ^{ab}	0.60 ± 0.04 ^{ab}	0.69 ± 0.06 ^{a}	0.69 ± 0.07 ^{a}	0.76 ± 0.13 ^{a}
	38.577	Octanoic acid	1876	n.d.	1.87 ± 0.36 ^{b}	1.97 ± 0.28 ^{b}	2.06 ± 0.18 ^{b}	1.54 ± 0.12 ^{b}	2.01 ± 0.25 ^{b}	3.02 ± 0.53 ^{a}

Mean ($n = 3 \pm$ SE). RT: Retention time. KI: Kovats index. n.d.: not detected. Different letters in the same row indicate significant differences ($p < 0.05$) among must, still wine, and sparkling process. AHA: Acetate higher alcohol. SCFAEEs: Short-chain fatty acid ethyl ester. MCFAEEs: Medium-chain fatty acid ethyl ester. LCFAEEs: Long-chain fatty acid ethyl ester. HA: Higher alcohols. ns: statistically non-significant differences.

Acetate esters and acetic acid are presented in Figure 2A. Initially, the content of ethyl acetate in the must was low (1.30 mg/L) compared to its increment in the first alcoholic fermentation (11.88 mg/L) and stabilization during the sparkling process (11.22 to 13.79 mg/L). These acetate esters, common in wine, are responsible for the negative effect on flavor and spoilage character when they exceed 150 to 200 mg/L [40], far higher than our values. However, acetate esters under 80 mg/L concentration may improve the aroma of wine [41]. Acetic acid was not detected in must but appeared with the fermentations, 1.23 to 1.95 mg/L.

AHA increased from 2.57 mg/L in the must to 23.14 in the still wine and to 27.81 mg/L in the sparkling wine (10 weeks). Isoamyl acetate was the principal AHA Table 6, representing around 60% of all AHA compound. In wines, this compound is responsible for providing a fruity character in the wine aroma [42]. Zhang et al. [43] found that *S. cerevisiae* catabolites the leucine into 3-methyl-butyraldehyde and isoamyl alcohol, via the Ehrlich pathway Figure 3, and the final esterification step with active acetic acid or acetyl-CoA to isoamyl acetate; its yields are regulated by the initial concentration of leucine. Our results follow that trend, with an increased concentration of isoamyl acetate, showing a negative correlation with the leucine concentration.

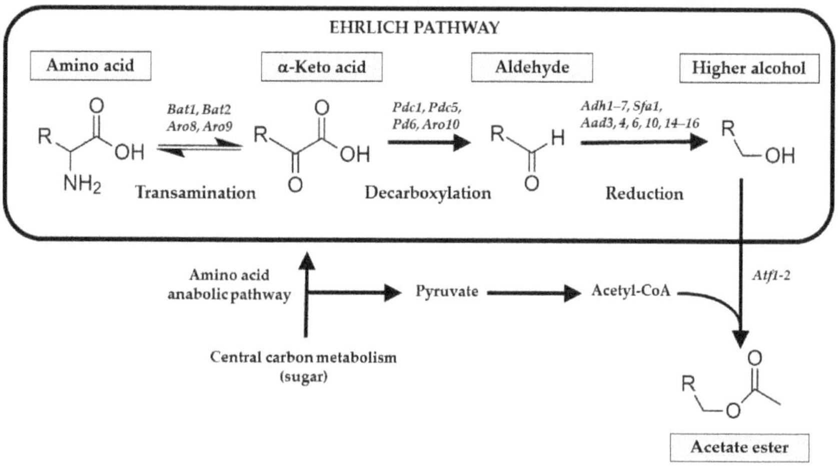

Figure 3. Ehrlich pathway (square) in Saccharomyces cerevisiae, biochemistry and main genes encoding enzymes (italics) involved in this catabolic pathway. It consists of three stages in which higher alcohols are produced from assimilated amino acids. In the first step, amino acids are deaminated in a reversible transamination reaction to the corresponding α-keto acids, followed by a decarboxylation in the second step of α-keto acids to aldehydes. In the third step, the reduction of aldehydes to the corresponding alcohol by alcohol dehydrogenases catalyzes this last step of the Ehrlich pathway. The higher alcohols can be esterified to produce the equivalent acetate esters once they have been synthesized. The biosynthesis of amino acids from a carbon source, or the anabolic pathway, may also provide the α-keto acid.

The next principal AHA compounds were 2-phenylethyl acetate and isobutyl acetate, which increased during the first fermentation (3.84 mg/L) and the sparkling process (5.13 mg/L). As with the isoamyl acetate and leucine correlation, a similar scenario occurred with 2-phenylethyl acetate and phenylalanine, which *S. cerevisiae* catabolized into 2-phenylethanol and 2-phenylethyl acetate [30], and this metabolic trend is shown in the decrement of phenylalanine and the increment in its alcoholic and acetate metabolites. Another AHA that increased during the first alcoholic fermentation was 3,6-nonadien-1-yl acetate, probably derived from 3,6-nonadien-1-ol due to the activity of alcohol acetyltransferases, which catalyzed alcohols and acetyl-coenzyme A for the formation of acetate ester [44].

Ethanol and higher alcohols are presented in Figure 2B. The melon wine fermentations converted sugar into alcohol; an ethanol concentration of 59.73 mg/L was obtained in the still wine, and it reached 65.83 mg/L in the sparkling wine (10 weeks).

Higher alcohols also increased with a slight rise during the sparkling process, principally isoamyl alcohol, derived from the deamination, decarboxylation, and reduction of leucine as explained above, via the Ehrlich pathway. Isoamyl alcohol was not detected in the must Table 6 and developed with the first fermentation and remained stable throughout the sparkling processes (18.70 to 24.17 mg/L). 2-Phenylethanol was the third most abundant higher alcohol, derived from phenylalanine, and ranged from 4.71 mg/L in the still wine to 8.16 mg/L in the sparkling processes (10 weeks). Isobutyl alcohol and 2,3-butanediol were identified with alcoholic fermentation. As reported by Wess et al. [45], *S. cerevisiae* metabolizes valine amino acid by the Ehrlich pathway to produce isobutyl alcohol in the cytosol. Another alcohol that competes with the ethanol pathway is 2,3-butanediol which, from 2-acetolactate after a spontaneous decarboxylation and followed by oxidation by butanediol dehydrogenases, is reduced to 2,3-butanediol [45]; this was detected in very low concentrations in both fermentation processes. Another higher alcohol identified was (E,Z)-3,6-nonadien-1-ol, which was found in the must, as reported in other melon cultivars [46] and doubled its concentration with fermentations.

An aromatic higher alcohol, 2,4-di-tert-butylphenol increased from 1.75 to 2.35 mg/L after the first alcoholic fermentation, probably due to the oxidation of 1,3-di-tertbutylbenzene, which was also identified Table 6.

Fatty acid ethyl esters Figure 2C–E were produced by *S. cerevisiae* during the first alcoholic fermentation and sparkling process and determined the complex flavor and fruity aroma, obtained from the condensation of acyl-CoA with ethanol, catalyzed by ethanol-O-acyl transferase [47].

The most abundant group of ethyl esters was MCFAEEs, followed by LCFAEEs. Both groups increased during the first fermentation and sparkling process, ranging from 43.25 to 69.23 mg/L and 1.16 to 2.11 mg/L, respectively. In contrast, SCFAEEs decreased with the first fermentation (0.81 mg/L) and showed a slight increase at the end of the studied sparkling process (1.13 mg/L). Five SCFAEEs were identified in the must Table 6, principally ethyl 2-methyl butyrate and ethyl butanoate, followed by ethyl isobutyrate, ethyl propanoate, and ethyl pentanoate. These last three compounds decreased to a non-detected limit after the first alcoholic fermentation. However, ethyl 2-methyl butyrate decreased from 0.77 mg/L in the must to 0.17–0.32 mg/L after the first fermentation and sparkling process. Ethyl butanoate ranged from 0.46 to 0.68 mg/L during the experiment, increasing to 0.82 mg/L after 10 weeks. Generally, ethyl 2-methyl butyrate and ethyl butanoate are commonly identified in a large range of white and rosé wines fermented with *S. cerevisiae* [48].

Regarding MCFAEEs, five ethyl esters were identified throughout the experiment, with a gradual increase during the first alcoholic fermentation and sparkling process, with the maximum concentration being obtained at the end of the experiment (69.23 mg/L) Figure 2D. Ethyl hexanoate was the only compound identified in the must. Instead, the first alcoholic fermentation provided the formation of new aromas such as ethyl octanoate, ethyl decanoate, ethyl hexanoate, ethyl 9-decenoate and ethyl laurate, with octanoate and decanoate being the compounds found in the highest concentrations. All these MCFAEE volatile compounds have been identified in other alcoholic fermentative beverages such as wines from grapes [48,49] and kiwi wines [6].

Concerning LCFAEEs, ethyl tetradecanoate and ethyl hexadecanoate were not identified in the must but developed with alcoholic fermentation. The concentration of these compounds was low, with a stable trend throughout the sparkling process (0.57 to 0.75 mg/L, and 0.59–1.39 mg/L, respectively) Table 6. Both compounds have been reported in other sparkling grape wines [50] and fruit-based wines such as pomegranate and kiwi [6,51].

3.4. Odor Activity Value (OAV) and Relative Odor Contribution (ROC)

Table 7 shows the 15 individual volatile compounds which had at least OAV > 1 and contributed to the aroma of the melon-based wine. Figure 4 illustrates the principal aromas with a ROC value above 0.1. In the AHA groups, only four compounds exceeded the odor threshold: isobutyl acetate, isoamyl acetate, 3,6-nonadienyl acetate, and 2-phenylethyl acetate. As mentioned before, these compounds increased with the first alcoholic fermentation and sparkling process. Isoamyl acetate, followed by 3,6-nonadienyl acetate, were the AHA compounds with the highest OAV values in the must, with an increase in the alcoholic fermentations providing sweet banana and fruity odors in the still and sparkling wines Table 7. Isoamyl acetate, with an initial (must) OAV value of 17.7, increased to 481 in the still wine and 647 at the end of the sparkling process and its ROC increased from 0.2 (must) to 0.55–0.61, during the first alcoholic fermentation and the sparkling process, meaning that the aroma was a major contribution to the melon-based wine.

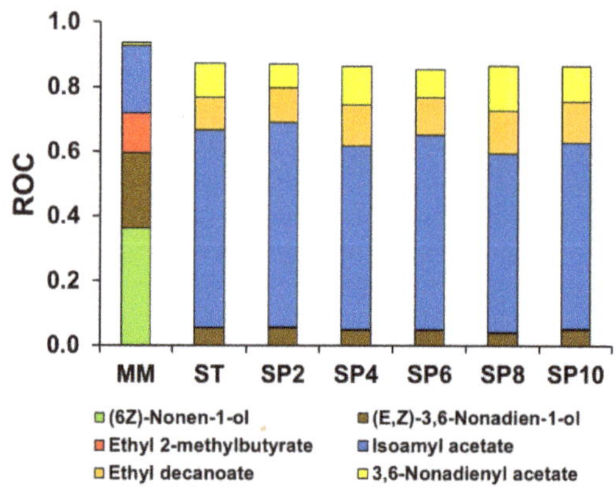

Figure 4. Evolution of principal relative odor contribution value (ROC > 0.1) in enriched must, still wine, and during the sparkling process of melon-based wine (10 weeks). MM: Must. ST: Still wine. SP: Sparkling wine in weeks (2, 4, 6, 8, and 10 weeks).

Regarding the SCFAEEs, two volatile compounds contributed to the aroma of the melon-based wines: ethyl butanoate and ethyl 2-methyl butyrate. The first aroma was detected in the must (1.1), the still wine (1.1) and at the end of the sparkling process (1.4). Ethyl 2-methyl butyrate decreased its OAV value from 10.5 in the must to 2.3 in the still wine and increased to 4.3, at the end of the sparkling period studied. Both aromas provide a sweet and fruity connotation; however, only ethyl 2-methylbutyrate reached a ROC > 0.1 value in the must, after which other aromas became more relevant Figure 4.

Table 7. Evolution of principal odor activity value (OAV > 1) in enriched must, still wine, and during the sparkling process of melon-based wine (10 weeks).

Compounds	OTH	Must	Still Wine	Sparkling Wine (Weeks)					Odor Descriptor
				2	4	6	8	10	
AHA									
Isobutyl acetate	1.605	<1	2.2 ± 0.1 ^{b}	3.0 ± 0.4 ^{b}	2.9 ± 0.4 ^{b}	4.2 ± 0.9 ^{a}	2.6 ± 0.2 ^{b}	2.4 ± 0.3 ^{b}	Sweet, fruity, banana, tropical
Isoamyl acetate	0.03	17.7 ± 1.1 ^{c}	480.9 ± 11.8 ^{b}	495.9 ± 2.1 ^{b}	512.9 ± 34.2 ^{b}	483.7 ± 39.7 ^{b}	503.4 ± 32.1 ^{b}	646.8 ± 93.2 ^{a}	Sweet banana, and fruity
3,6-Nonadienyl acetate	0.015	<1	83.1 ± 0.5 ^{bc}	59.2 ± 26.2 ^{c}	108.2 ± 17.4 ^{abc}	72.3 ± 22.4 ^{c}	132.1 ± 30.6 ^{ab}	121.7 ± 18.3 ^{a}	Fruity
2-Phenylethyl acetate	0.25	<1	15.4 ± 1.3 ^{b}	16.4 ± 1.0 ^{b}	16.3 ± 1.3 ^{b}	17.0 ± 1.8 ^{b}	16.1 ± 0.6 ^{b}	20.5 ± 2.4 ^{a}	Roses and honey aroma
SCFAEE									
Ethyl butanoate	0.6	1.1 ± 0.1 ^{ab}	1.1 ± 0.1 ^{b}	<1	<1	<1	<1	1.4 ± 0.2 ^{a}	Fruity, sweet, tutti frutti, apple
Ethyl 2-methylbutyrate	0.074	10.5 ± 0.9 ^{a}	2.3 ± 0.2 ^{c}	2.7 ± 0.4 ^{bc}	2.8 ± 0.3 ^{bc}	2.6 ± 0.2 ^{c}	2.7 ± 0.1 ^{bc}	4.3 ± 0.6 ^{b}	Sharp, sweet, green apple, fruity
MCFAEE									
Ethyl hexanoate	0.1	1.9 ± 0.2 ^{b}	23.6 ± 1.3 ^{a}	24.1 ± 2.4 ^{a}	22.4 ± 0.5 ^{a}	20.7 ± 0.1 ^{a}	21.8 ± 2.1 ^{a}	23.4 ± 2.4 ^{a}	Apple peel fruit
Ethyl octanoate	0.6	<1	39.6 ± 3.4 ^{cd}	34.1 ± 1.6 ^{d}	55.7 ± 5.4 ^{bc}	55.5 ± 4.7 ^{bc}	59.1 ± 5.4 ^{b}	74.3 ± 13.0 ^{a}	Pineapple, pear, soapy
Ethyl decanoate	0.2	<1	78.9 ± 8.0 ^{d}	83.7 ± 1.3 ^{d}	112.5 ± 3.2 ^{bc}	92.2 ± 5.2 ^{cd}	121.1 ± 13.2 ^{b}	141.9 ± 18.3 ^{a}	Sweet, fruity, nuts, dried fruit
Ethyl 9-decenoate	0.1	<1	14.8 ± 3.3 ^{ab}	16.7 ± 0.7 ^{ab}	14.2 ± 1.9 ^{b}	11.2 ± 0.4 ^{b}	13.5 ± 1.4 ^{b}	18.9 ± 3.5 ^{a}	Floral
Ethyl laurate	0.5	<1	<1	2.0 ± 0.5 ^{b}	3.2 ± 0.2 ^{a}	1.0 ± 0.2 ^{c}	<1	<1	Sweet
LCFAEE									
Ethyl tetradecanoate	0.5	<1	1.5 ± 0.0 ^{b}	1.1 ± 0.2 ^{bc}	1.4 ± 0.1 ^{bc}	1.4 ± 0.2 ^{bc}	1.5 ± 0.2 ^{a}	1.4 ± 0.2 ^{bc}	Sweet fruit, butter, fatty odor
Ethyl hexadecanoate	1	<1	<1	<1	<1	<1	<1	1.4 ± 0.2	Fruity, concord grape
HA									
(6Z)-Nonen-1-ol	0.001	32.2 ± 5.3	<1	<1	<1	<1	<1	<1	Sweet, fatty, melon-like
(E,Z)-3,6-Nonadien-1-ol	0.003	20.2 ± 2.0 ^{c}	41.6 ± 2.9 ^{b}	41.3 ± 0.2 ^{b}	41.6 ± 5.2 ^{b}	36.9 ± 0.7 ^{bc}	33.6 ± 6.9 ^{bc}	56.4 ± 14.5 ^{a}	Melon-like, leafy, nutty odor

Mean ($n = 3 \pm$ SE). Different letters in the same row indicate significant differences ($p < 0.05$) among must, still wine, and sparkling process. OAV value < 1. OTH: Odor threshold.

For the MCFAEEs, the trend of these fatty acid ethyl esters was, after AHA, the group with the largest increment in OAV; it increased with the fermentation process, and its evolution followed a similar pattern to other grape and fruit wines [5,52]. Ethyl hexanoate was the only MCFAEEs, with OAV > 1, present in the must (1.9), which increased with the fermentations and remained stable, ranging from 20.7 to 24.1. However, the compounds with high OAV were ethyl decanoate (78.9 to 141.9), whose odor description corresponds to sweet, fruity, nuts, and dried fruit, and ethyl octanoate (34.1 to 74.3) which provides pineapple, pear, and a soapy odor. The highest levels were obtained at the end of the sparkling process. However, only ethyl decanoate managed to contribute a ROC > 0.1, between 0.10 to 0.13, in the still and sparkling wines. Ethyl 9-decenoate had a value between 11.2 to 18.9 after alcoholic fermentations, providing a floral odor with a ROC > 0.1 Figure 4 For LCFAEEs group, only ethyl tetradecanoate had an OAV > 1 after first fermentation (1.5) and ranged 1.10—1.50 during the sparkling process. Regarding ethyl hexadecanoate, it only reached an OAV > 1 with 1.39 at 10 weeks of the sparkling process. Both volatile aromas contributed a fruity and fatty odor but with low contribution to the aroma of the still and sparkling melon wine (ROC > 0.1).

(6Z)-Nonen-1-ol and (E,Z)-3,6-nonadien-1-ol are common aromas in some melon cultivars [46], and they were the only HA that reached their odor threshold with a melon-like and sweet odor description in the initial must (32.2 and 20.2, respectively), and with the highest ROC values of 0.4 and 0.23, respectively. After the first alcoholic fermentation, the contribution of (6Z)-nonen-1-ol disappeared but that of (E,Z)-3,6-nonadien-1-ol increased, with an OAV value of 33.6–56.44 during the experiment, with an ROC value of around 0.05.

3.5. Sensory Evaluation of Final Sparkling Wine

The sensory evaluation of the sparkling melon, after 10 weeks in the bottle, is shown in Table 8. In accordance with the categories described by OIV (2009), this sparkling melon-based wine, obtained a total mark of 92.11 points, after the sum of all markers (visual, nose, taste, and harmony), achieving the "Grand Gold" category (>92 points). Considering visual aspects (14.44), the judges rated the sparkling wine as limpid to excellent limpidity, which means a good visual impression without cloudiness. Regarding the sensation perceived in the nose, melon-based wine had a very total absence of defects in genuineness (5.78), very strong qualitative intensity (7.78), and a very-to-excellent impression of quality (15.11). These results refer to the intensity of OAV Table 7, such as isoamyl acetate, ethyl decanoate, 3,6-nonadienyl acetate and (E,Z)-3,6-nonadien-1-ol, which are the most relevant volatile aromas providing the highest ROC values that contributed to the complexity and fruity-aroma character in this sparkling melon wine. Concerning taste values (39.11), sparkling melon wine was characterized as having a total absence of defects in genuineness, strong intensity (7.11), a very good persistence of residual olfactory-gustatory sensation of flavors (7.22), and a very good impression for the taste quality (19.33). These results suggest that this melon wine is absent in oxidation flavor, volatile acidity, and a negative flavor that affects the taste and mouthfeel. Finally, the judges marked a very good general impression (9.89), which along with the total marks received, defined our sparkling melon-based wine as a potential marketable and novel product.

Table 8. Sensory evaluation of the melon sparkling wine at 10 weeks of storage.

Organoleptical Characteristics	Sparkling Melon-Wine	Range (Excellent to Inadequate)
Visual	**14.44 ^{z} ± 0.24**	**(15–3)**
Limpidity	4.67 ± 0.17	(5–1)
Aspect other than limpidity	9.78 ± 0.22	(10–2)
Nose	**28.67 ± 0.69**	**(30–12)**
Genuineness	5.78 ± 0.15	(6–2)
Positive intensity	7.78 ± 0.15	(8–2)
Quality	15.11 ± 0.48	(16–8)
Taste	**39.11 ± 0.99**	**(44–18)**
Genuineness	5.44 ± 0.18	(6–2)
Positive intensity	7.11 ± 0.11	(8–2)
Harmonious persistence	7.22 ± 0.22	(8–4)
Quality	19.33 ± 0.60	(22–10)
Harmony-Overall judgement	**9.89 ± 0.20**	**(11–7)**
TOTAL	**92.11 ± 1.77**	**(100–40)**

^{z} Mean (n = 14 ± SE).

4. Conclusions

In this study, a novel melon-based sparkling wine, using melons that failed to meet cosmetic standards, was developed and characterized. In this sparkling melon wine, a second fermentation takes place in the bottle, from a still wine (first fermentation). In our case, during the second fermentation, *S. cerevisiae* provided significant physico-chemical and aroma changes. Certain amino acids contributed to the transformation and increase in some volatile compounds via the Ehrlich pathway, in this case, leucine to isoamyl alcohol, valine to isobutyl alcohol and phenylalanine to phenethyl alcohol. Furthermore, during the second fermentation principally medium- and long-chain fatty acid ethyl ester increased and reached its odor threshold and contributed to the sweet, fruity, banana, tropical, nutty and melon aroma character, mainly by isoamyl acetate, ethyl decanoate, 3,6-nonadienyl acetate, and (E,Z)-3,6-nonadien-1-ol, with the highest ROC values during the sparkling process. In summary, this study is an example of how the agriculture sector should support a circular economy model with by-products revalorization such as fresh fruit that does not reach aesthetic standards, providing a novel fruit-based wine with a characteristic and distinctive aroma, good sensory acceptance and with market potential.

Author Contributions: Conceptualization, J.Á.S.-M., A.A. and E.A.; methodology, J.Á.S.-M. and A.A.; formal analysis, J.Á.S.-M.; investigation, J.Á.S.-M., A.A. and E.A.; Writing—Original Draft, J.Á.S.-M.; supervision, A.A., A.C.-B. and E.A.; funding acquisition, A.A. and E.A., project administration, A.A. and E.A.; resources, A.C.-B.; writing—review & editing, E.A. All authors have read and agreed to the published version of the manuscript.

Funding: This research was funded by RTI2018-099139-B-C21 project from the Ministry of Science and Innovation (Spain)—National Research Agency (MCIN/AEI/10.13039/501100011033) and by "ERDF A way of making Europe", of the European Union and by 21645/PDC/21 project from "Fundación Séneca" of Murcia Region (Spain). José-Angel Salas-Millán acknowledges financial support for "Industrial PhD" grant (DIN2019-010837) from the Ministry of Science and Innovation.

Institutional Review Board Statement: Ethical review and approval were waived for this study since it is not required for the sensory analysis performed in this experiment.

Informed Consent Statement: Written informed consent was obtained from all subjects in the sensory panel involved in the study.

Data Availability Statement: The datasets supporting the findings of this article are available upon reasonable request.

Conflicts of Interest: The authors declare that they have no known competing financial interests or personal relationships that could have appeared to influence the work reported in this paper Author Andrés Conesa is employed by the JimboFresh company, and he contributed to the paper as a researcher for supervision tasks and resources (providing fresh melon). However, the JimboFresh company did not contribute financially, nor in the optimization, analysis of the results, or writing of the paper. Therefore, there is no conflict of interest in relation with JimboFresh company. The remaining authors declare that the research was conducted in the absence of any commercial or financial relationships that could be construed as a potential conflict of interest.

References

1. FAO. Food Loss Measurement | Technical Platform on the Measurement and Reduction of Food Loss and Waste | Food and Agriculture Organization of the United Nations. Available online: https://www.fao.org/platform-food-loss-waste/food-waste/introduction/en/ (accessed on 13 April 2022).
2. *EPA From Farm to Kitchen: The Environmental Impacts of US Food Waste*; U.S. Environmental Protection Agency: Washington, DC, USA, 2021; pp. 1–106.
3. Salas-Millán, J.-Á.; Aznar, A.; Conesa, E.; Conesa-Bueno, A.; Aguayo, E. Fruit Wine Obtained from Melon By-Products: Physico-Chemical and Sensory Analysis, and Characterization of Key Aromas by GC-MS. *Foods* **2022**, *11*, 3619. [CrossRef]
4. Jackson, R.S. *Wine Science: Principles and Application*; Academic Press: Cambridge, MA, USA, 2008; ISBN 9780333227794.
5. Lu, Y.; Liu, Y.; Lv, J.; Ma, Y.; Guan, X. Changes in the Physicochemical Components, Polyphenol Profile, and Flavor of Persimmon Wine during Spontaneous and Inoculated Fermentation. *Food Sci. Nutr.* **2020**, *8*, 2728–2738. [CrossRef] [PubMed]
6. Sun, N.; Gao, Z.; Li, S.; Chen, X.; Guo, J. Assessment of Chemical Constitution and Aroma Properties of Kiwi Wines Obtained from Pure and Mixed Fermentation with Wickerhamomyces Anomalus and Saccharomyces Cerevisiae. *J. Sci. Food Agric.* **2022**, *102*, 175–184. [CrossRef] [PubMed]
7. Eder, M.; Sanchez, I.; Brice, C.; Camarasa, C.; Legras, J.L.; Dequin, S. QTL Mapping of Volatile Compound Production in Saccharomyces Cerevisiae during Alcoholic Fermentation. *BMC Genom.* **2018**, *19*, 1–19. [CrossRef]
8. Joshi, V.K.; Panesar, P.S.; Rana, V.S.; Kaur, S. Science and Technology of Fruit Wines: An Overview. In *Science and Technology of Fruit Wine Production*; Elsevier Inc.: Amsterdam, The Netherlands, 2017; pp. 1–72. ISBN 9780128010341.
9. Matei, F. Technical Guide for Fruit Wine Production. In *Science and Technology of Fruit Wine Production*; Elsevier Inc.: Amsterdam, The Netherlands, 2017; pp. 663–703. ISBN 9780128010341.
10. European Commision. List and Description of the Files of the OIV Code of Oenological Practices Regulation (EU) 2019/934, PUB. 2022 Eur-Lex.Europa.Eu. 2019. Available online: https://eur-lex.europa.eu/legal-content/EN/TXT/PDF/?uri=CELEX:52022XC0506(04)&from=EN (accessed on 6 January 2023).
11. Wendrick, N.A.; Sims, C.A.; Macintosh, A.J.; Durán-Guerrero, E. The Effect of Carbonation Level on the Acceptability and Purchase Intent of Muscadine and Fruit Wines. *Beverages* **2021**, *7*, 66. [CrossRef]
12. Martínez-García, R.; García-Martínez, T.; Puig-Pujol, A.; Mauricio, J.C.; Moreno, J. Changes in Sparkling Wine Aroma during the Second Fermentation under CO2 Pressure in Sealed Bottle. *Food Chem.* **2017**, *237*, 1030–1040. [CrossRef]
13. FAOSTAT. Available online: https://www.fao.org/faostat/es/#data/TCL (accessed on 12 April 2022).
14. Esteras, C.; Rambla, J.L.; Sánchez, G.; López-Gresa, M.P.; González-Mas, M.C.; Fernández-Trujillo, J.P.; Bellés, J.M.; Granell, A.; Picó, M.B. Fruit Flesh Volatile and Carotenoid Profile Analysis within the Cucumis Melo, L. Species Reveals Unexploited Variability for Future Genetic Breeding. *J. Sci. Food Agric.* **2018**, *98*, 3915–3925. [CrossRef] [PubMed]
15. Farcuh, M.; Copes, B.; Le-Navenec, G.; Marroquin, J.; Cantu, D.; Bradford, K.J.; Guinard, J.X.; Van Deynze, A. Sensory, Physicochemical and Volatile Compound Analysis of Short and Long Shelf-Life Melon (*Cucumis Melo* L.) Genotypes at Harvest and after Postharvest Storage. *Food Chem. X* **2020**, *8*, 100107. [CrossRef]
16. Minkova, S.; Vlaeva, I.; Nikolova, K.; Petkova, N.; Gentscheva, G.; Tzvetkova, C. Assessment of the Elemental Composition, Antioxidant Activity, and Optical Properties of Non-Traditional Bulgarian Fruit Wines. *Bulg. Chem. Commun.* **2022**, *54*, 26–30. [CrossRef]
17. Gómez, L.F.H.; Úbeda, J.; Briones, A. Characterisation of Wines and Distilled Spirits from Melon (*Cucumis Melo* L.). *Int. J. Food Sci. Technol.* **2008**, *43*, 644–650. [CrossRef]
18. OIV International Organisation of Vine and Wine. Compendium of International Methods of Analysis -OIV Chromatic Characteristics Method OIV-MA-AS2-11. 2009. Determination of Chromatic Characteristics According to CIELab (Resolution Oeno 1/2006). Available online: https://www.oiv.int/public/medias/7907/oiv-vol1-compendium-of-international-methods-of-analysis.pdf (accessed on 2 June 2022).
19. Salas-Millán, J.-Á.; Aznar, A.; Conesa, E.; Conesa-Bueno, A.; Aguayo, E. Functional Food Obtained from Fermentation of Broccoli By-Products (Stalk): Metagenomics Profile and Glucosinolate and Phenolic Compounds Characterization by LC-ESI-QqQ-MS/MS. *LWT* **2022**, *169*, 113915. [CrossRef]
20. Ortiz-Duarte, G.; Pérez-Cabrera, L.E.; Artés-Hernández, F.; Martínez-Hernández, G.B. Ag-Chitosan Nanocomposites in Edible Coatings Affect the Quality of Fresh-Cut Melon. *Postharvest Biol. Technol.* **2019**, *147*, 174–184. [CrossRef]

21. Giordano, G.; Gucciardi, A.; Pirillo, P.; Naturale, M. Quantification of Underivatized Amino Acids on Dry Blood Spot, Plasma, and Urine by HPLC-ESI-MS/MS. In *Methods in Molecular Biology*; Alterman, M.A., Ed.; Humana Press Inc.: Totowa, NJ, USA, 2019; Volume 2030, pp. 153–172.
22. Sun, Y.; Peng, W.; Zeng, L.; Xue, Y.; Lin, W.; Ye, X.; Guan, R.; Sun, P. Using Power Ultrasound to Release Glycosidically Bound Volatiles from Orange Juice: A New Method. *Food Chem.* **2021**, *344*, 128580. [CrossRef] [PubMed]
23. Korenika, A.M.J.; Biloš, J.; Kozina, B.; Tomaz, I.; Preiner, D.; Jeromel, A. Effect of Different Reducing Agents on Aromatic Compounds, Antioxidant and Chromatic Properties of Sauvignon Blanc Wine. *Foods* **2020**, *9*, 996. [CrossRef]
24. Acree, T.; Arn, H. Flavornet and Human Odor Space. Available online: http://flavornet.org/flavornet.html (accessed on 6 January 2023).
25. Christlbauer, M.; Schieberle, P. Evaluation of the Key Aroma Compounds in Beef and Pork Vegetable Gravies a La Chef by Stable Isotope Dilution Assays and Aroma Recombination Experiments. *J. Agric. Food Chem.* **2011**, *59*, 13122–13130. [CrossRef]
26. Fan, W.; Xu, Y.; Jiang, W.; Li, J. Identification and Quantification of Impact Aroma Compounds in 4 Nonfloral Vitis Vinifera Varieties Grapes. *J. Food Sci.* **2010**, *75*, S81–S88. [CrossRef]
27. Feng, Y.; Su, G.; Zhao, H.; Cai, Y.; Cui, C.; Sun-Waterhouse, D.; Zhao, M. Characterisation of Aroma Profiles of Commercial Soy Sauce by Odour Activity Value and Omission Test. *Food Chem.* **2015**, *167*, 220–228. [CrossRef]
28. Rahayu, Y.Y.S.; Yoshizaki, Y.; Yamaguchi, K.; Okutsu, K.; Futagami, T.; Tamaki, H.; Sameshima, Y.; Takamine, K. Key Volatile Compounds in Red Koji-Shochu, a Monascus-Fermented Product, and Their Formation Steps during Fermentation. *Food Chem.* **2017**, *224*, 398–406. [CrossRef]
29. Welke, J.E.; Zanus, M.; Lazzarotto, M.; Alcaraz Zini, C. Quantitative Analysis of Headspace Volatile Compounds Using Comprehensive Two-Dimensional Gas Chromatography and Their Contribution to the Aroma of Chardonnay Wine. *Food Res. Int.* **2014**, *59*, 85–99. [CrossRef]
30. Fairbairn, S.; McKinnon, A.; Musarurwa, H.T.; Ferreira, A.C.; Bauer, F.F. The Impact of Single Amino Acids on Growth and Volatile Aroma Production by Saccharomyces Cerevisiae Strains. *Front. Microbiol.* **2017**, *8*, 2554. [CrossRef]
31. OIV Standard for International Wine Competitions and Spiritous Beverages of Vitivinicultural Origin (OIV-Concours 332A-2009); Zagreb, Croatia. 2009. Available online: https://www.oiv.int/public/medias/4661/oiv-concours-332a-2009-en.pdf (accessed on 17 January 2023).
32. Ennouri, M.; Ben Hassan, I.; Ben Hassen, H.; Lafforgue, C.; Schmitz, P.; Ayadi, A. Clarification of Purple Carrot Juice: Analysis of the Fouling Mechanisms and Evaluation of the Juice Quality. *J. Food Sci. Technol.* **2015**, *52*, 2806–2814. [CrossRef] [PubMed]
33. De Albuquerque Gil, D.M.; Rebelo, M.J.F. Metabisulfite Interference in Biosensing and Folin-Ciocalteu Analysis of Polyphenols. *Microchim. Acta* **2009**, *167*, 253–258. [CrossRef]
34. Martínez-Rodríguez, A.; Carrascosa, A.; Martín-Álvarez, P.J.; Moreno-Arribas, V.; Polo, M. Influence of the Yeast Strain on the Changes of the Amino Acids, Peptides and Proteins during Sparkling Wine Production by the Traditional Method. *J. Ind. Microbiol. Biotechnol.* **2002**, *29*, 314–322. [CrossRef] [PubMed]
35. Ostergaard, S.; Olsson, L.; Nielsen, J. Metabolic Engineering of Saccharomyces Cerevisiae. *Microbiol. Mol. Biol. Rev.* **2000**, *64*, 34–50. [CrossRef]
36. Berbegal, C.; Polo, L.; García-Esparza, M.J.; Álvarez, I.; Zamora, F.; Ferrer, S.; Pardo, I. Influence of the Dry Yeast Preparation Method on Final Sparkling Wine Characteristics. *Fermentation* **2022**, *8*, 313. [CrossRef]
37. Sartor, S.; Toaldo, I.M.; Panceri, C.P.; Caliari, V.; Luna, A.S.; de Gois, J.S.; Bordignon-Luiz, M.T. Changes in Organic Acids, Polyphenolic and Elemental Composition of Rosé Sparkling Wines Treated with Mannoproteins during over-Lees Aging. *Food Res. Int.* **2019**, *124*, 34–42. [CrossRef] [PubMed]
38. Liang, Z.; Lin, X.; He, Z.; Su, H.; Li, W.; Ren, X. Amino Acid and Microbial Community Dynamics during the Fermentation of Hong Qu Glutinous Rice Wine. *Food Microbiol.* **2020**, *90*, 103467. [CrossRef]
39. Wang, Y.Q.; Ye, D.Q.; Zhu, B.Q.; Wu, G.F.; Duan, C.Q. Rapid HPLC Analysis of Amino Acids and Biogenic Amines in Wines during Fermentation and Evaluation of Matrix Effect. *Food Chem.* **2014**, *163*, 6–15. [CrossRef]
40. Rojas, V.; Gil, J.V.; Piñaga, F.; Manzanares, P. Acetate Ester Formation in Wine by Mixed Cultures in Laboratory Fermentations. *Int. J. Food Microbiol.* **2003**, *86*, 181–188. [CrossRef]
41. Plata, C.; Millán, C.; Mauricio, J.C.; Ortega, J.M. Formation of Ethyl Acetate and Isoamyl Acetate by Various Species of Wine Yeasts. *Food Microbiol.* **2003**, *20*, 217–224. [CrossRef]
42. Cameleyre, M.; Lytra, G.; Tempere, S.; Barbe, J.C. 2-Methylbutyl Acetate in Wines: Enantiomeric Distribution and Sensory Impact on Red Wine Fruity Aroma. *Food Chem.* **2017**, *237*, 364–371. [CrossRef] [PubMed]
43. Zhang, L.; Zhang, Y.; Hu, Z. The Effects of Catabolism Relationships of Leucine and Isoleucine with BAT2 Gene of Saccharomyces Cerevisiae on High Alcohols and Esters. *Genes* **2022**, *13*, 1178. [CrossRef] [PubMed]
44. Li, W.; Cui, D.Y.; Wang, J.H.; Liu, X.E.; Xu, J.; Zhou, Z.; Zhang, C.Y.; Chen, Y.F.; Xiao, D.G. Overexpression of Different Alcohol Acetyltransferase Genes with BAT2 Deletion in Saccharomyces Cerevisiae Affects Acetate Esters and Higher Alcohols. *Eur. Food Res. Technol.* **2018**, *244*, 555–564. [CrossRef]
45. Wess, J.; Brinek, M.; Boles, E. Improving Isobutanol Production with the Yeast Saccharomyces Cerevisiae by Successively Blocking Competing Metabolic Pathways as Well as Ethanol and Glycerol Formation. *Biotechnol. Biofuels* **2019**, *12*, 1–15. [CrossRef]
46. Kourkoutas, D.; Elmore, J.S.; Mottram, D.S. Comparison of the Volatile Compositions and Flavour Properties of Cantaloupe, Galia and Honeydew Muskmelons. *Food Chem.* **2006**, *97*, 95–102. [CrossRef]

47. Saerens, S.M.G.; Delvaux, F.; Verstrepen, K.J.; Van Dijck, P.; Thevelein, J.M.; Delvaux, F.R. Parameters Affecting Ethyl Ester Production by Saccharomyces Cerevisiae during Fermentation. *Appl. Environ. Microbiol.* **2008**, *74*, 454–461. [CrossRef]
48. Del Barrio-Galán, R.; del Valle-Herrero, H.; Bueno-Herrera, M.; López-de-la-Cuesta, P.; Pérez-Magariño, S. Volatile and Non-Volatile Characterization of White and Rosé Wines from Different Spanish Protected Designations of Origin. *Beverages* **2021**, *7*, 49 [CrossRef]
49. Stój, A.; Czernecki, T.; Domagała, D.; Targoński, Z. Comparative Characterization of Volatile Profiles of French, Italian, Spanish, and Polish Red Wines Using Headspace Solid-Phase Microextraction/Gas Chromatography-Mass Spectrometry. *Int. J. Food Prop.* **2017**, *20* (Suppl. S1), S830–S845. [CrossRef]
50. Del González-Jiménez, M.C.; Moreno-García, J.; García-Martínez, T.; Moreno, J.J.; Puig-Pujol, A.; Capdevilla, F.; Mauricio, J.C. Differential Analysis of Proteins Involved in Ester Metabolism in Two Saccharomyces Cerevisiae Strains during the Second Fermentation in Sparkling Wine Elaboration. *Microorganisms* **2020**, *8*, 403. [CrossRef]
51. Wang, X.; Ren, X.; Shao, Q.; Peng, X.; Zou, W.; Sun, Z.; Zhang, L.; Li, H. Transformation of Microbial Negative Correlations into Positive Correlations by Saccharomyces Cerevisiae Inoculation during Pomegranate Wine Fermentation. *Appl. Environ. Microbiol.* **2020**, *86*, e01847-20. [CrossRef]
52. Izquierdo-Cañas, P.M.; González Viñas, M.A.; Mena-Morales, A.; Navarro, J.P.; García-Romero, E.; Marchante-Cuevas, L.; Gómez-Alonso, S.; Sánchez-Palomo, E. Effect of Fermentation Temperature on Volatile Compounds of Petit Verdot Red Wines from the Spanish Region of La Mancha (Central-Southeastern Spain). *Eur. Food Res. Technol.* **2020**, *246*, 1153–1165. [CrossRef]

foods

MDPI

Article

Microencapsulation of Carotenoid-Rich Extract from Guaraná Peels and Study of Microparticle Functionality through Incorporation into an Oatmeal Paste

Lorena Silva Pinho [1,2], Bhavesh K. Patel [2], Osvaldo H. Campanella [2], Christianne Elisabete da Costa Rodrigues [1] and Carmen Sílvia Favaro-Trindade [1,*]

1 Departamento de Engenharia de Alimentos (ZEA), Faculdade de Zootecnia e Engenharia de Alimentos (FZEA), Universidade de São Paulo (USP), Pirassununga 13635-900, São Paulo, Brazil; spinholorena@gmail.com (L.S.P.)

2 Department of Food Science and Technology, College of Food, Agricultural, and Environmental Sciences, Ohio State University (OSU), Columbus, OH 43210, USA

* Correspondence: carmenft@usp.br

Abstract: The peels of guaraná (*Paullinia cupana*) fruit contain abundant carotenoid content, which has demonstrated health benefits. However, these compounds are unstable in certain conditions, and their application into food products can be changed considering the processing parameters. This study aimed to encapsulate the carotenoid-rich extract from guaraná peels by spray drying (SD), characterize the microparticles, investigate their influence on the pasting properties of oatmeal paste, and evaluate the effects of temperature and shear on carotenoid stability during the preparation of this product. A rheometer with a pasting cell was used to simulate the extrusion conditions. Temperatures of 70, 80, and 90 °C and shear rates of 50 and 100 1/s were the parameters evaluated. Microparticles with a total carotenoid content between 40 and 96 μg/g were obtained. Over the storage period, carotenoid stability, particle size, color, moisture, and water activity varied according to the core:carrier material proportion used. Afterward, the formulation SD1:2 was selected to be incorporated in oatmeal, and the paste viscosity was influenced by the addition of this powder. β-carotene retention was higher than that of lutein following the treatment. The less severe treatment involving a temperature of 70 °C and a shear rate of 50 1/s exhibited better retention of total carotenoids, regardless of whether the carotenoid-rich extract was encapsulated or non-encapsulated. In the other treatments, the thermomechanical stress significantly influenced the stability of the total carotenoid. These results suggest that the addition of encapsulated carotenoids to foods prepared at higher temperatures has the potential for the development of functional and stable products.

Keywords: mechanical stress; thermal stress; β-carotene; lutein; stability; *Paullinia cupana*

Citation: Pinho, L.S.; Patel, B.K.; Campanella, O.H.; Rodrigues, C.E.d.C.; Favaro-Trindade, C.S. Microencapsulation of Carotenoid-Rich Extract from Guaraná Peels and Study of Microparticle Functionality through Incorporation into an Oatmeal Paste. *Foods* **2023**, *12*, 1170. https://doi.org/10.3390/foods12061170

Academic Editors: Noelia Castillejo Montoya and Lorena Martínez-Zamora

Received: 15 January 2023
Revised: 2 March 2023
Accepted: 3 March 2023
Published: 10 March 2023

1. Introduction

Carotenoids are among the major classes of pigments found in tree leaves, fruits, and vegetables. Consumption of these compounds has been considered to bring health benefits. According to the literature, carotenoids may act as reducing agents for cancer [1–3], cardiovascular diseases [4,5] and macular degeneration [6], as well as antioxidants [7] and provitamin A [8]. The nutritive and coloring properties of carotenoids make them be an ideal food additive to develop functional products with desirable appearance.

The interest in recovering carotenoids from agro-industrial waste has increased, considering their high potential to enhance the valorization of by-products. Several studies [9–12] have reported the extraction of carotenoids from different vegetable wastes. However, guaraná residues, with potential uses in the production of soft drinks and food ingredient industries, have been rarely explored as a possible source of carotenoids in foods.

Carotenoids are highly susceptible to harsh external conditions, such as high temperature and shear, presence of oxygen, light, acidity, and pro-oxidant agents [13–15] Encapsulation has emerged as a key approach for making the incorporation of these compounds in processed foods feasible. In the food industry, spray drying is the most widely used technology for the entrapment of bioactive molecules in the form of microparticles The technique consists of atomizing a dispersion or emulsion, containing the component of interest and drying adjuvants, followed by its dehydrating forming microparticles. This encapsulation facilitates transport, prolongs shelf-life, and reduces the risk of carotenoid degradation during food processing [16,17].

Several additives are used as carrier materials to facilitate the drying, handling and application of these bioactive compounds. In addition, the property of these carrier materials can guarantee the stability of carotenoids against oxidation, storage conditions and processing in the food industry. Tuyen et al. [18] reported the effect of different concentrations of maltodextrin as carrier material and different inlet temperatures on color preservation, total carotenoid content and antioxidant activity of spray-dried gac powder. In addition, Hojjati et al. [19] investigated the influence of different concentrations of soluble soy polysaccharides on the properties of microcapsules loaded with canthaxanthin obtained by spray drying. Indeed, the encapsulation of canthaxanthin in this study resulted in a more significant storage stability of the samples. Highlighting the variety of carrier materials, Etzbach et al. [20] evaluated the effect of using different carriers such as maltodextrin, modified starch, inulin, alginate, gum arabic and cellobiose for spray drying golden blackberry juice rich in carotenoids. In this study, cellobiose's proposed alternative carrier showed a high capacity to protect carotenoids from degradation processes by exposure to light, high temperature and oxygen, possibly due to a more compact particle wall and larger particle sizes.

The development of starchy products includes the investigation of new formulations and new processing techniques. However, typical technologies used to favor gelatinization and texturization of starchy products apply a combination of heat and shear over these products, which can be harmful to carotenoids. During gelatinization, starch swells, destabilizing its crystalline structure and loses birefringence. The main starch fractions, amylose and amylopectin, disperse and lead to the production of a paste [21,22]. After disruption of their granular structure, during processing, starch is subjected to mechanical stresses, mainly shear provoked by processing.

The rheological properties of starch solutions can vary according to factors such as amylose and amylopectin content, the presence of functional groups, and granularity. In the case of the current research, these factors can be influenced by adding microparticles loaded with carotenoids into the oat paste, which may change the paste's rheological behavior. On the other hand, the processing of oat paste, which involves heat and shear conditions, can lead to the degradation of thermosensitive compounds such as carotenoids. From that, the proposal focused on understanding the effects of these conditions on microparticles rich in carotenoids, also evaluating the effectiveness of microencapsulation by spray drying. Oats have wide applications, and the carotenoids from the guarana by-product are still atypical but with great potential for exploration and use by the food industry.

To simulate a pasting environment where temperature and shear are applied, a rheometer equipped with a pasting cell was used to produce an oatmeal paste enriched with encapsulated carotenoids. Thus, this study was aimed to prepare a carotenoid-rich extract from guaraná peels, encapsulate this extract, investigate the influence of microparticles on the pasting properties of oatmeal, and evaluate the effect of thermo- and mechanical stress on carotenoid stability under different conditions.

2. Material and Methods

2.1. Materials

The Executive Commission for Cocoa Cultivation Planning, CEPLAC (Taperoá, Bahia, Brazil), provided the guaraná fruits. The fruit peels were removed from seeds and pulp followed by being washed with water. The reddish peels were dried in a convection oven (Marconi, MA035/1152) at 50 °C for 18 h [23]. They were then milled and stored under dark at −20 °C until analysis.

β-carotene (CAS 7235-40-7) and lutein (CAS 127-40-2) standards were purchased from Sigma-Aldrich (Saint Louis, USA). Analytical-grade ethanol (CAS 64-17-5), hexane (CAS 110-54-3), and acetone (CAS 67-64-1) were obtained from Fisher Scientific (Waltham, MA, USA).

The sunflower oil used during the guaraná peel extract preparation was from Cargill (brand Liza). The microparticles rich in carotenoids produced by spray drying (SD) were prepared using gum arabic as carrier material, obtained from Nexira, Brazil. Oat flakes (the Oat Quaker Company, Chicago, IL, USA) were purchased from a local supermarket at Columbus, OH, USA.

2.2. Carotenoid-Rich Extract Preparation

Carotenoid-rich extract preparation from guaraná peels was performed following [24], using ethanol as solvent, at a ratio of 1:10 (peel: solvent, *w/v*), during 4 h at 50 °C. The concentration of 3% of sunflower oil was added to the extract to reduce the carotenoid degradation detected during preliminary experiments. The final concentration was determined considering the liquid–liquid equilibrium for the mixture composed of oil and ethanol [25]. As follows, the material was concentrated using a rotary evaporator (TE-211 Tecnal, Piracicaba, Brazil) at 48 ± 2 °C to 20% of the initial volume. The concentrated extract was named guaraná peel extract (GPE).

2.3. Microencapsulation by Spray Drying

Formulations were prepared with a ratio of 1:2, 1:3, and 1:4 of concentrated ethanolic extract:gum arabic in aqueous solution (20% *w/v*), *v/v*, using Ultra-Turrax ® IKA T25 (Labotechnic, Staufen, Germany) at 11,200× *g* for 3 min. The mixture was atomized according to Rocha et al. [26], with modifications. The spray dryer (Model MSD 1.0, Labmaq do Brasil, Ribeirão Preto, Brazil) was used coupled with a spray nozzle of 1.2 mm, with an inlet air temperature of 100 °C, an air drying speed of 2.5 m/s, a feed flow of 10 mL/min, and air pressure of 8.4 kgf/cm^2.

2.4. Total Carotenoid Content

The microparticles were blended with hexane for 1 min and ultrasonicated for 20 min using an ultrasound bath, Branson 1800 (Branson Ultrasonics Corporation, Danbury, CT, USA) to extract carotenoids. Absorbances of extracts were measured at 450 nm wavelength and recorded using a UV–visible spectrophotometer (Thermo Scientific, Waltham, MA, USA, Genesys 10S). The β-carotene standard was used for quantification, and the results were expressed as μg β-carotene/g sample [27]. The test was conducted in triplicate.

The retention after encapsulation was calculated as a ratio of the total carotenoid amount in the microparticles to that in the feed materials before atomization.

2.5. Stability Study of Microparticles

The samples were placed in glass vials covered with aluminum foil and stored in desiccators containing saturated solutions of magnesium chloride $MgCl_2$ (relative humidity, RH of 32.8%). The desiccators were kept at a temperature of 25 °C, and the storage period lasted for 90 days under the specified conditions [28]. The samples were evaluated every 15 days in terms of carotenoid content and color. Particle size distribution, mean diameter, moisture content, and water activity were analyzed at the initial time and after 90 days of storage, in triplicate.

2.5.1. Carotenoid Degradation Kinetics

Carotenoid stability was determined by comparing the total carotenoid concentration at the initial time and over the storage of encapsulated and non-encapsulated carotenoid-rich GPE. Previous studies [29,30] have hypothesized that a first-order kinetics describes adequately the reaction of carotenoids degradation. Thereby, to investigate the stability of our samples, the degradation constant (k) and the half-life ($t_{1/2}$) were determined following the first-order kinetic model, according to Equations (1) and (2).

$$\ln C(t)/C_0 = -kt \tag{1}$$

$$t_{1/2} = \ln(2)/k \tag{2}$$

where C = carotenoid concentration at time t (μg/g); C_0 = initial carotenoid concentration (μg/g); t = time (days).

2.5.2. Color

The microparticles color defined by the parameters L* (Brightness), a* (red–green), and b*(yellow–blue) were determined using a HunterLab Mini Scan XE colorimeter (Reston, VA, USA). *Chroma* (color saturation) was calculated according to the Minolta procedure [31]:

$$Chroma = \sqrt{(a^*)^2 + (b^*)^2} \tag{3}$$

The total color difference (ΔE) was calculated according to Equation (4).

$$\Delta E = \sqrt{(L_f^* - L_0^*)^2 + (a_f^* - a_0^*)^2 + (b_f^* - b_0^*)^2} \tag{4}$$

where ΔE = total color difference; L_f^* = final L*; L_0^* = initial L*; a_f^* = final a*; a_0^* = initial a*; b_f^* = final b*; b_0^* = initial b*.

2.5.3. Mean Diameter

Mean diameter of particles were measured using the SALD-201V laser diffraction particle analyzer, Shimadzu (Kyoto, Japan). Ethanol was used as a dispersing liquid. The measurements were conducted at 25 °C.

2.5.4. Moisture Content and Water Activity

The moisture content of the microparticles was determined in a moisture analyzer model MB 35 from Ohaus (Parsippany, OH, USA), in triplicate. The results are expressed in percentage. The determination of water activity (a_w) was achieved by direct reading, on an Aqualab hygrometer, model CX-2T, from Decagon Devices Inc., Pullman, WA, USA. The readings were performed at 25 °C.

2.6. Dynamic Vapor Sorption (DVS) of Microparticles and Oat Flakes

Water vapor sorption isotherms of the microparticles and oat flakes were determined at 25 °C using a Dynamic Vapor Sorption instrument (Surface Measurement Systems Ltd., Allentown, PA, USA). Under a continuous airflow (200 mL/min), the system was pre-equilibrated at 5% relative humidity (RH). The samples were exposed sequentially to different relative humidities (RH) from 30 to 95%. The RH transitions and mass variations of the sample were monitored continuously. The moisture sorption isotherms were determined using the DVS Analysis Macro V6.1 software, in duplicate.

2.7. Thermal Properties of Oat Flakes and Oat Flakes Containing Microparticles

The thermal behavior of oat flakes and oat flakes containing microparticles was described by the parameters 'onset of gelatinization' (To), 'peak gelatinization temperature' (Tp), and 'gelatinization enthalpy' (J/g) and measured by a Multi-Cell Differential Scanning Calorimeter (MC-DSC, TA Instruments, New Castle, DE, USA) equipped with the TRIOS software (TA Instruments, New Castle, DE, USA). The samples dispersed in water (with a moisture content of 80%) were weighed into ampoules and sealed. An empty pan was used as the reference. The samples were equilibrated at 5 °C and then heated to 140 °C at a rate of 1 °C/min. Each sample was run in triplicate, and the average results are shown.

2.8. Pasting Properties

The oatmeal paste was prepared using the Discovery Hybrid Rheometer 3 (DHR-3, TA Instruments Ltd., New Castle, DE, USA) and a pasting cell geometry to simulate operating conditions where temperature and shear are applied during the process. The experiment was conducted by mixing (i) oat flakes and distilled water (80% *w*/*w*); (ii) oat flakes blended with microparticles (6% *w*/*w*), and distilled water (80% *w*/*w*); (iii) oat flakes, GPE (6% *w*/*w*), and distilled water (80% *w*/*w*).

The samples were prepared using the following procedure: (1) conditioning the sample at 25 °C for 2 min; (2) heating ramp to selected temperatures of 70, 80 or 90 °C at 5 °C/min with a shear rate of 50 or 100 1/s; (3) flow peak hold at the selected temperature and shear conditions for 120 s; (4) cooling ramp to 25 °C at 5 °C/min; (5) oscillation frequency at 25 °C from 0.01 to 10 Hz with a 0.5% strain. The testing parameters and treatments are shown in Table 1. Each sample was run in triplicate.

Table 1. Conditions used for the preparation of oatmeal pastes in DHR-3.

Treatments	Parameters	
	Temperature (°C)	**Shear (1/s)**
70/50	70	50
80/50	80	50
90/50	90	50
70/100	70	100
80/100	80	100
90/100	90	100

2.9. Retention of β-Carotene, Lutein, and Total Carotenoid Content in the Oatmeal Paste

Total carotenoid content of oatmeal paste containing encapsulated and free GPE was analyzed using spectrophotometry, as described in Section 2.4. The main carotenoids extracted from guaraná peels were β-carotene and lutein [24]. The contents of the incorporated carotenoids were quantified in the oatmeal paste and the raw material by HPLC to evaluate the effect of the process on the retention of these compounds, in duplicate. Carotenoid extraction was carried out following Kopec et al. [32], with some modifications according to the solvents used. Sequential extractions were performed using methanol and a mixture of ethanol: acetone: hexane (1:1:1, *v*/*v*/*v*). The extract was then injected into the HPLC instrument Agilent 1260 ultra-high-performance liquid chromatograph with a diode array detector (UHPLC-DAD), using a C30 column (YMC Inc., Meridian, ID, USA, 4.6 × 250 mm, 3 μm) [33]. The contents of β-carotene and lutein were calculated from their peak areas in comparison to standards with known concentration, using a calibration curve. Carotenoid retention was determined by comparing their content before and after the pasting process and expressed as a percentage.

2.10. Statistics

The data were analyzed using ANOVA and Tukey's test in SAS statistical software (version 8.02, Statistic Analysis System). Significant differences were defined at $p < 0.05$. All data were expressed as the means ± standard deviation (SD). The pasting process was performed in duplicate.

3. Results and Discussion

3.1. Total Carotenoid Content of Microparticles

Microparticles rich in carotenoids were produced by spray drying using three formulations to assess the best treatment for further applications. Total carotenoid content and retention after encapsulation are presented in Table 2. The SD1:2 treatment showed a significantly higher carotenoid content compared to the others. This difference was consistent since the microparticles containing more carotenoid-rich extract in the proportion core:carrier material would present higher content of carotenoid.

Table 2. Total carotenoid content and retention in microparticles after the spray drying (SD) process.

Formulations	Carotenoid Content (μg/g)	Carotenoid Retention (%)
SD1:2	96.0 ± 1.0 [a]	100.0 ± 1.0 [a]
SD1:3	57.6 ± 0.7 [b]	99.4 ± 0.2 [a]
SD1:4	40.7 ± 0.9 [c]	96.7 ± 0.3 [b]

In formulations, the name is the proportion of the core:carrier material. The results are expressed as the mean ± standard deviation (n = 3). Different letters in the columns represent a significant difference ($p \leq 0.05$).

Regarding the protective effect of carrier material on carotenoid stability after encapsulation, the SD1:4 treatment had the lowest retention. Although this sample had the highest carrier material proportion, the atomization temperature was not sufficiently high to evaporate the water and dry the microparticles effectively. Consequently, this caused the adherence of semi-moist powder to the wall of the drying chamber, which may lead to the additional exposition to the temperatures of the process, leading to degradative reactions of the carotenoids.

Carotenoids are sensitive compounds and well documented, and the temperature applied for the atomization can influence their stability. The high retention observed in SD1:2 and SD1:3 was probably associated with the relationship between the suitable degree of heat treatment and feed material of these formulations during the process. This result formalizes our hypothesis that among the main factors that affect carotenoid retention during encapsulation by spray drying is the core:carrier material proportion associated with the operational condition of atomization.

3.2. *Stability Study of Microparticles during Storage*

3.2.1. Carotenoid Retention

Figure 1 shows the retention of total carotenoids in the microparticles and free GPE during storage under dark at 25 °C. Following 90 days, a carotenoid loss of 45% was found in the free GPE, whereas only ~30% was observed in the microparticles. It is essential to highlight that free GPE, considered a control, contains sunflower oil, as mentioned in Section 2.2. Sunflower oil was added to alleviate significant losses observed in obtaining the extract, which certainly protected the carotenoids during storage.

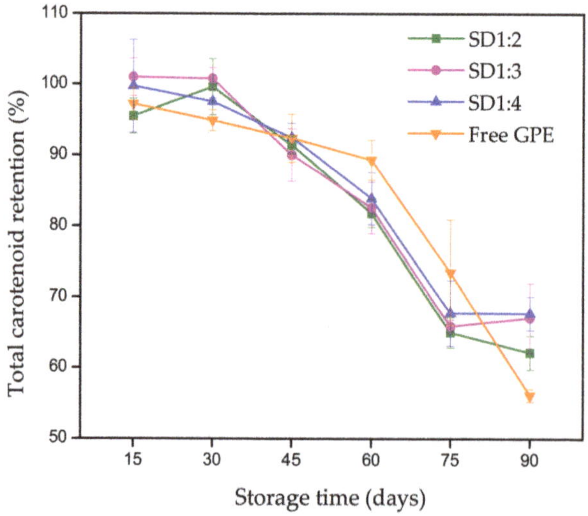

Figure 1. Total carotenoid retention in free and encapsulated guaraná peel extract (GPE) produced by spray drying at a temperature of 100 °C. The formulations represent the proportion of the core:carrier material.

Regarding the microparticles, the carrier material acts as a physical barrier against environmental conditions, and the core was efficiently entrapped by the gum arabic. This reduced the pronounced degradation of the active component observed in free GPE. However, isomerization and/or oxidation of these compounds were observed in the encapsulated samples. In this system, the high degree of carotenoid unsaturation concomitant with the presence of oxygen was crucial to their degradation over time.

First-order kinetics was followed to assess the stability of carotenoids in encapsulated and non-encapsulated GPE, and the results are reported in Table 3. In a first-order mechanism, the reaction rate depends on variations in the amount of only one reactant; in this case, the reactant was the carotenoids. All microparticles showed low values for the first-order rate constant (k), indicating a reduced degradation, as evidenced by the half-life ($t_{1/2}$). It can be noted that the loss of carotenoids occurred faster in the non-encapsulated extract than those in the encapsulated. The findings denote that the spray drying technique improved the stability of GPE due to the protective potential of the carrier material under storage conditions, promoting a longer shelf life. Similar results were reported when carotenoid stability was evaluated in spray-dried samples [20,34].

Table 3. Kinetic parameters for carotenoid degradation in non-encapsulated and encapsulated extract during storage.

Sample	k (s^{-1})	$t_{1/2}$ (Days)	R^2
Free extract	0.006	108.082	0.811
SD1:2	0.005	141.924	0.888
SD1:3	0.005	152.286	0.897
SD1:4	0.005	144.676	0.874

Formulation SD (core:carrier material ratio).

3.2.2. Color Parameters

The powder's color, measured by the parameters L*, a*, b*, *Chroma*, and ΔE (total color difference), may indirectly indicate the degradation of carotenoids during storage. Overall, the color characteristics of the samples were influenced by core:carrier material concentration in the feed material formulations (Figure 2). Comparing the treatments, an increase in the L* values was observed by increasing the ratio of gum arabic solution in the samples following the order: SD14 > SD1:3 > SD1:2. Further, the formulations with the highest concentration of carotenoid-rich extract showed higher values of a* and b*, as expected. Chroma represents color saturation, which was in the first quadrant of the CIELab chart, corresponding to the vivid red-yellow color. A slight reduction in Chroma was observed in all formulations during storage, indicating loss of color intensity and degradation of the pigment, which corroborates with the results found in the stability study.

Regarding the trend for the total color difference, the expression of ΔE values for SD1:2 (7.31), SD1:3 (6.19), and SD1:4 (5.19) exhibited the overall variation between samples at initial time and samples stored 90 days in the absence of light. This result is related to the degradation of carotenoids during the period, considering stability to the component of interest conferred by the carrier agent (gum arabic). Indeed, the color variation over storage had some influence on the powder quality, considering the color property as well as the bioactivity capacity of these pigments.

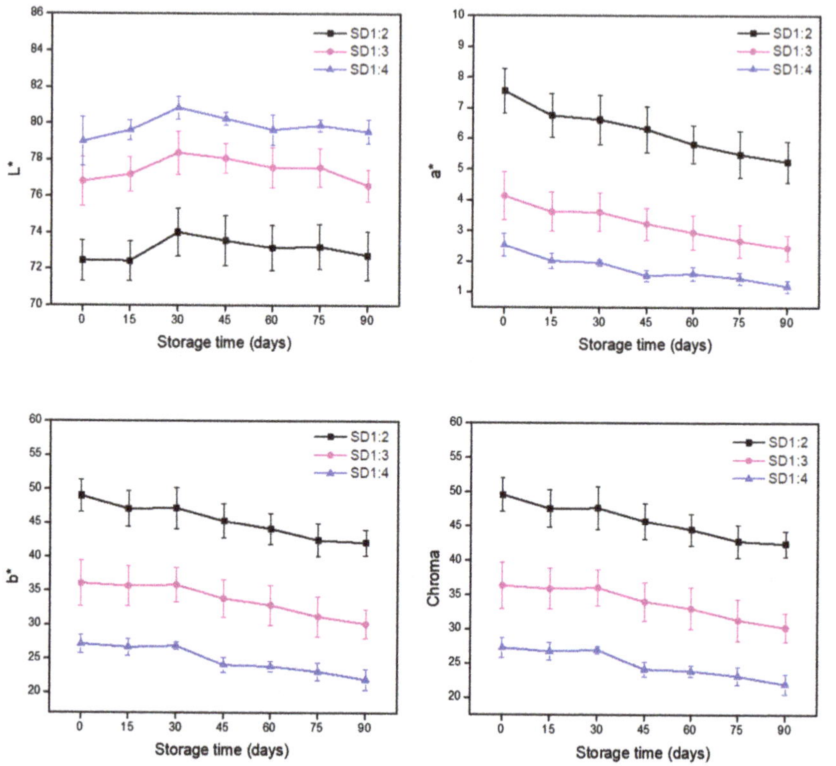

Figure 2. L*, a*, b*, and Chroma parameters obtained for each formulation produced by spray drying in instrumental color analysis during storage at 25 °C. The formulations denote the proportion of the core:carrier material.

3.2.3. Mean Particle Diameter

Particle size is an important property for application as an ingredient in food products. Particle size distribution can be influenced by external and/or internal factors. The conditions used in the spray drying procedure, such as temperature, pressure nozzle, airspeed, and feed flow, are some of the external factors. However, the feed material formulation and its preparation process may act as the internal factors [35].

Unlike formulations SD1:2 and SD1:3, SD1:4 microparticles exhibited a remarkable size variation, in which the median diameter ranged from 9 to 18 µm, over storage (Table 4). This phenomenon was attributed to agglomeration. The hydrophilic active sites of gum arabic (carrier material) can absorb water-favoring adherence properties of the samples. Due to the electrostatic effects and covalent bonds, adhesion between the wetted microparticles occurs, and the contact or collision among them can generate new agglomerate structures [36,37]. Considering the higher proportion of gum arabic in the SD1:4 formulation, the effect of this phenomenon was maximized and reflected in the size variation after 90 days.

The particle size was within the typical range for atomization, which varies from 5 to 150 µm. Their dimensions were below the value (<100 µm) of particles which have been found to cause little or no interference sensory when being added to food. Moreover, this parameter may be associated with bioavailability and solubility of the active components entrapped, considering that the more surface area of the particle the more the bioactivity of the compounds [38].

Table 4. Mean and standard deviation of median diameter sizes (μm) expressed in volume for each formulation of microparticles obtained by spray drying at 0 and 90 days of storage at 25 °C.

Time (days)	Formulations	Median Diameter (μm)
0	SD1:2	11.9 ± 1.5 [a]
0	SD1:3	9.1 ± 1.8 [ab]
0	SD1:4	8.9 ± 0.9 [b]
90	SD1:2	15.1 ± 2.8 [a]
90	SD1:3	13.4 ± 2.4 [a]
90	SD1:4	18.6 ± 5.5 [a]

Formulation SD (core:carrier material ratio). Results are expressed as the mean ± standard deviation (n = 3). Different letters in the columns represent a significant difference ($p \leq 0.05$).

3.2.4. Moisture Content and a_w

Moisture and a_w are indicators of drying efficiency and are the main factors that affect powder stability. The mean values for these parameters before and after storage are shown in Table 5.

Table 5. Moisture and water activity (a_w) of encapsulated carotenoid-rich guaraná peel extract.

Time (Days)	Formulations	Parameters	
		Moisture (%)	a_w
0	SD1:2	3.9 ± 0.4 [d]	0.21 ± 0.02 [c]
0	SD1:3	4.8 ± 0.4 [c]	0.24 ± 0.03 [c]
0	SD1:4	4.1 ± 0.7 [cd]	0.16 ± 0.02 [d]
90	SD1:2	6.9 ± 0.3 [b]	0.46 ± 0.01 [b]
90	SD1:3	8.1 ± 0.3 [a]	0.48 ± 0.01 [a]
90	SD1:4	8.2 ± 0.1 [a]	0.50 ± 0.02 [a]

Formulation SD (core:carrier material ratio). The results are expressed as the mean ± standard deviation (n = 3). Different letters in the columns represent a significant difference ($p \leq 0.05$).

After 90 days, the moisture and water activity of the powders increased significantly ($p < 0.05$). This was due to the water uptake of the powder during storage under a relative humidity of 33% and 25 °C. However, even after this period, the samples showed low values of water activity. The moisture content and water activity can influence microbial growth, in addition to enabling biochemical reactions. According to the literature, to avoid microbial growth, the water activity of products must be less than 0.6 [18,39]; therefore, the powders may be considered microbiologically safe. Further, a_w values near 0.3 imply that the food products are less sensitive against non-enzymatic browning and enzymatic activities during storage [40,41].

The relevance of powdery ingredients lies in their large application in the food industry. The incorporation of microparticles enriched with carotenoids in a model food requires investigating the characteristics of the powder in general. In this way, it would be possible to achieve a suitable application considering the convenient features for processing. The findings were expected to suggest a reasonable formulation, considering the carotenoid content and particle behavior. Thus, the formulation SD1:2 was selected to be incorporated in oatmeal paste, to better understand the effect of conditions prevalent on processing and/or preparation on carotenoid stability and paste properties.

3.3. *Dynamic Vapor Sorption (DVS) of Oat Flakes and Microparticles*

Most materials are sensitive to the presence of water vapor or moisture content in a system [42], also shown in the storage study performed in this work and described in previous section. Investigation of interactions between water and oat flakes, as well as between water and microparticles is essential to improve the conception of the effect of the encapsulated GPE incorporation on the characteristics of oatmeal pastes. As shown in Figure 3, the equilibrium moisture content of the microparticles was 5-fold higher than that of oat flakes at RH of 60%. The DVS curves for the samples were type III (non-sigmoidal).

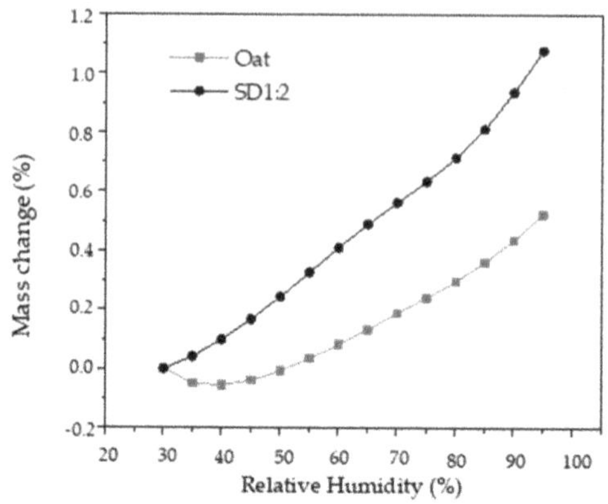

Figure 3. Dynamic vapor sorption isotherms of particles produced by spray drying (SD1:2) and oat flakes.

A moisture sorption isotherm is determined by subjecting a material to different increasing relative humidity and monitoring the change in mass due to water absorption The absorption of water depends on the number of available sites in the material capable of binding water molecules [43,44].

The structural difference between oat flakes and encapsulated GPE may be related with their distinct absorption rates. Gum arabic is a complex polysaccharide with a highly branched structure, which facilitates polar interactions with water by hydrogen bonds at room temperature [45], whereas oat flake comprises mainly starch and fiber. Starch contains amylose, a linear molecule, and amylopectin, a non-linear and highly branched molecule [46,47], both with a hydrophilic nature. In the crystalline regions of the granule, the intermolecular interactions among the chains are very strong, and the diffusion of a plasticizer, such as water, is slow at room temperature.

3.4. Thermal Properties

The onset temperature and the peak temperature for oat flakes and oat flakes incorporated with encapsulated GPE with a moisture content of 80% are listed in Table 6. The thermal properties were significantly affected ($p \leq 0.05$) by the addition of microparticles, showing a decrease in the enthalpies and a slight increase in gelatinization temperatures. This indicates that in samples containing oat flakes and microparticles, the starch was less gelatinized which is likely attributed to the decreased availability of water [48,49]. In the system studied, microparticles and starch competitively bind water to form hydrogen bonds. As part of the water is bound to the microparticles with larger water absorption capacity shown in Figure 3, the available water for starch is reduced, which could elevate the starch gelatinization temperature due to insufficient hydration and swelling.

Table 6. DSC gelatinization properties of oat flakes and oat flakes enriched with encapsulated GPE.

	Onset Temp. (°C)	Peak Temp. (°C)	Enthalpy (J/g)
Oat	53.05 ± 0.01 [b]	60.1 ± 0.3 [b]	3.1 ± 0.1 [a]
Oat + Encapsulated GPE	54.3 ± 0.6 [a]	61.0 ± 0.3 [a]	2.6 ± 0.1 [b]

The results are expressed as the mean ± standard deviation ($n = 3$). Different letters in the columns represent a significant difference ($p \leq 0.05$).

3.5. Pasting Properties

The viscosity of the oat flakes starch with and without GPE was investigated during the heating–cooling procedure, as described in Section 2.8, to evaluate their influence on the oatmeal paste properties. In Figure 4, gelatinization of the samples was observed at the temperature of 70 °C or above, showing a rapid increase in viscosity.

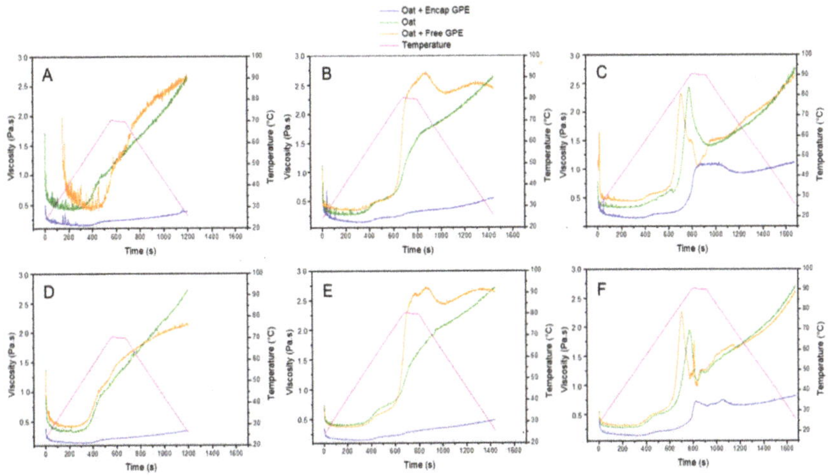

Figure 4. DHR-3 pasting curves of oatmeal paste enriched with encapsulated GPE and free GPE. Testing conditions: (**A**) temperature 70 °C and shear rate 50 1/s; (**B**) temperature 80 °C and shear rate 50 1/s, (**C**) temperature 90 °C and shear rate 50 1/s; (**D**) temperature 70 °C and shear rate 100 1/s; (**E**) temperature 80 °C and shear rate 100 1/s; (**F**) temperature 90 °C and shear rate 100 1/s.

Temperature significantly affected the pasting behavior of the samples. At 70 °C and 80 °C, the viscosity of oatmeal paste with encapsulated GPE was lower without a noticeable peak. The increase in starch viscosity indicates the swelling of the starch molecules. Further heating resulted in molecular disorganization and the leaching of amylose and amylopectin into the solvent. Shear facilitated the formation of the paste by promoting non-covalent interactions between starch molecules, which can be affected by temperature and sample composition [50,51].

Microparticles produced using gum arabic as the carrier material hindered the swelling/gelatinization of the grains. Singh, Geveke, and Yadav [52] suggested that gum arabic might cover the surface of the starch during processing, leading to a reduced interaction among neighboring starch granules, which may control the swelling power of starch and restrict the increment of viscosity. In this case, the system would require higher temperatures to completely gelatinize the starch. Shahzad et al. [53] reported similar findings.

At 80 °C, the oatmeal paste with free GPE showed significantly higher peak viscosity than the control (only oat flakes and water). This may be attributed to the presence of residual ethanol from GPE in the starch granules, which favors maximum swelling and leads to the gelatinization of most of the starch in the system [54]. Previous researchers [55] have reported that an aqueous ethanol medium promoted the formation of hydrogen bonds with starch molecules, resulting in stronger gels than those prepared with pure water. Alternatively, GPE contains a high amount of carotenoids. These compounds are hydrophobic, which may interact with double helices structures via hydrophobic interactions, and contributes to higher viscosity.

Singh et al. [54] suggested that gum arabic might cover the surface of the starch during processing, leading to the reduced interaction among neighboring starch granules, which may control the swelling power of starch and restricted the increment of viscosity.

At 80 °C, the oatmeal paste with free GPE showed significantly higher peak viscosity than the control. As before, this is attributed to the presence of residual ethanol in the extracted GPE, which favors maximum swelling and leads to the gelatinization of most of the starch in the system [56]. Previous researchers [57] have reported that an aqueous ethanol medium promoted the formation of hydrogen bonds with starch molecules, resulting in stronger gels compared to those prepared with pure water.

3.6. Carotenoid Retention

The formulation of foods enriched with active compounds is a matter of concern for the food industry, considering the impact of processing on its bioactivity. For this reason, the retention of β-carotene and lutein under the thermomechanical treatment applied during the production of oatmeal paste was investigated. Results are illustrated in Figure 5.

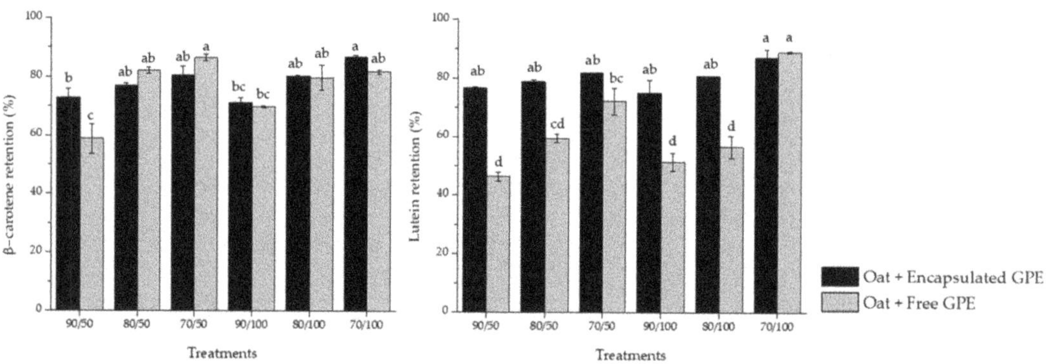

Figure 5. β-carotene and lutein retention in oatmeal paste incorporated with encapsulated and free guaraná peel extract (GPE) after the process at different conditions (temperature/shear rate). Different letters in the columns represent a significant difference ($p < 0.05$).

The behavior of β-carotene revealed higher stability under temperature and shear conditions applied during the paste preparation, regardless of whether it was in an encapsulated or non-encapsulated form, with retention ranging from 59 to 87%. The lutein stability was significantly higher in samples enriched with encapsulated GPE compared to the sample with free GPE, varying from 51 to 89%. Regarding the testing conditions, 90 °C and the shear rate of 100 1/s exhibited a larger detrimental effect on the carotenoids, either free or in their encapsulated forms.

Carotenoids are classified into carotenes and xanthophylls, and the latter are more polar. The first group, comprising pro-vitamin A carotenoids (β-carotene), demonstrated greater stability. Higher retention of β-carotene in an aqueous medium may be due to its physical stability towards environmental conditions. In contrast, the reactivity of xanthophylls, such as lutein, is highly affected during food processing due to its structure, characterized by the presence of oxygen in the molecule chain. Trapped oxygen within the food matrix composed of water and oat flakes may contribute to the oxidation of the carotenoids in the free guaraná peel extract, and as a result, this system led to the substantial loss of lutein. Therefore, the encapsulation technique can reduce the degradation of this component. Dhuique-Mayer et al. [56] assessed the stability of carotenes and xanthophylls from citrus juice upon thermal treatment, and they suggested that the former may react less in a polar solvent. Interestingly, on the opposite, when the stability of β-carotene and lutein in oil (i.e., a non-polar solvent) was investigated during heat treatment, they found that β-carotene had the highest reactivity [57].

Carotenoids have highly unsaturated chains and are easily oxidized. In addition, the isomerization of these compounds also affects their stability through food processing [58,59]. During the preparation of the oatmeal paste, extrinsic and intrinsic factors can be considered the promoters of oxidation/isomerization reactions. Thermomechanical stress, the incidence of light, and the presence of oxygen may be the main external reasons contributing to the loss of carotenoids. The composition of the food matrix, the physical status of the carotenoid, and the air incorporation into the sample may be considered the internal factors affecting the active retention [27,60].

Indeed, degrading reactions are promoted by the structure of the carotenoid as well as the factors mentioned previously. Similar results have been reported by other researchers [61–63].

4. Conclusions

Microencapsulation of carotenoid-rich extract from guaraná peels enhanced its stability over storage. The sample SD1:2 exhibited promising features among other formulations, including high carotenoid content, suitable particle size, and intense color. The addition of microparticles rich in bioactive compounds into oatmeal paste increased the onset and peak temperature of starch but decreased its enthalpy. Further, it reduced the viscosity of the system, whereas the opposite trend was observed for the samples with free GPE. Overall, the gelatinization of oat flakes starch was affected by the decrease in accessible water, due to the presence of the microparticles.

Testing conditions, such as temperature and shear, favored heat transfer and oxygen incorporation in the system, reducing the total carotenoid content of the samples. In addition, the treatment significantly altered the lutein content, when compared to β-carotene. However, pigment loss decreased by encapsulation.

The recovery of bioactive components from by-products, such as β-carotene (provitamin A) and lutein (recognized as an agent that prevents macular degeneration), reveals the importance of the present research on adding value to a waste of a food process encouraging sustainable development. In addition, the study of applications of these materials, as suggested in the current work, represents a suitable trend for obtaining functional foods. The range of oat products and derivatives is wide, including oatcake, oatmeal, and porridge, which can be enriched with bioactive compounds to add health benefits. In addition, the use of oats as an ingredient in newer sectors, such as plant proteins, has grown and has opened up possibilities for applications and, consequently, new processes.

Author Contributions: L.S.P.: conceptualization, methodology, validation, formal analysis, investigation, data curation, writing—original draft, writing—review and editing, visualization, and project administration. B.K.P.: investigation and methodology. O.H.C.: conceptualization, formal analysis, and writing—review and editing. C.E.d.C.R.: visualization and writing—review and editing. C.S.F.-T.: conceptualization, supervision, and writing—review and editing. All authors have read and agreed to the published version of the manuscript.

Funding: This research was funded by Fundação de Amparo à Pesquisa do Estado de São Paulo (FAPESP) grants numbers 2016/24916-2 and 2019/11113-7; Coordenação de Aperfeiçoamento de Pessoal de Nível Superior—Brasil (CAPES) Finance Code 001; and Conselho Nacional de Desenvolvimento Científico e Tecnológico (CNPq) grant number 310686/2022-9.

Data Availability Statement: The data presented in this study are available upon request.

Acknowledgments: The authors thank the Fundação de Amparo à Pesquisa do Estado de São Paulo (FAPESP) for fellowship granted to L.S. Pinho (#2016/24916-2 and #2019/11113-7) and Coordenação de Aperfeiçoamento de Pessoal de Nível Superior—Brasil (CAPES) (Finance Code 001). The authors also thank Conselho Nacional de Desenvolvimento Científico e Tecnológico (CNPq) for the research fellowship granted to C.S. Favaro-Trindade (#305115/2018-9) and partial financial support of the Ohio State University.

Conflicts of Interest: The authors declare no conflict of interest.

References

1. Micozzi, M.S.; Beecher, G.R.; Taylor, P.R.; Khachik, F. Carotenoid analyses of selected raw and cooked foods associated with a lower risk for cancer. *JNCI J. Natl. Cancer Inst.* **1990**, *82*, 282–285. [CrossRef]
2. Toma, S.; Losardo, P.L.; Vincent, M.; Palumbo, R. Effectiveness of beta-carotene in cancer chemoprevention. *Eur. J. Cancer Prev.* **1995**, *4*, 213–224. [CrossRef]

3. Kumar, S.R.; Hosokawa, M.; Miyashita, K. Fucoxanthin: A marine carotenoid exerting anti-cancer effects by affecting multiple mechanisms. *Mar. Drugs* **2013**, *11*, 5130–5147. [CrossRef]
4. Riccioni, G. Carotenoids and cardiovascular disease. *Curr. Atheroscler. Rep.* **2009**, *11*, 434–439. [CrossRef]
5. Monroy-Ruiz, J.; Sevilla, M.Á.; Carrón, R.; Montero, M.J. Astaxanthin-enriched-diet reduces blood pressure and improves cardiovascular parameters in spontaneously hypertensive rats. *Pharmacol. Res.* **2011**, *63*, 44–50. [CrossRef]
6. Gale, C.R.; Hall, N.F.; Phillips, D.I.; Martyn, C.N. Lutein and zeaxanthin status and risk of age-related macular degeneration. *Investig. Ophthalmol. Vis. Sci.* **2003**, *44*, 2461–2465. [CrossRef]
7. Stahl, W.; Sies, H. Antioxidant activity of carotenoids. *Mol. Asp. Med.* **2003**, *24*, 345–351. [CrossRef]
8. Olson, J.A. Provitamin A function of carotenoids: The conversion of β-carotene into vitamin A. *J. Nutr.* **1989**, *119*, 105–108. [CrossRef]
9. Chen, B.H.; Tang, Y.C. Processing and stability of carotenoid powder from carrot pulp waste. *J. Agric. Food Chem.* **1998**, *46*, 2312–2318. [CrossRef]
10. Ajila, C.M.; Bhat, S.G.; Rao, U.P. Valuable components of raw and ripe peels from two Indian mango varieties. *Food Chem.* **2007**, *102*, 1006–1011. [CrossRef]
11. De Andrade Lima, M.; Kestekoglou, I.; Charalampopoulos, D.; Chatzifragkou, A. Supercritical fluid extraction of carotenoids from vegetable waste matrices. *Molecules* **2019**, *24*, 466. [CrossRef]
12. De Souza Mesquita, L.M.; Martins, M.; Maricato, É.; Nunes, C.; Quinteiro, P.S.; Dias, A.C.; Coutinho, J.A.; Pisani, L.P.; de Rosso, V.V.; Ventura, S.P. Ionic liquid-mediated recovery of carotenoids from the Bactris gasipaes fruit waste and their application in food-packaging chitosan films. *ACS Sustain. Chem. Eng.* **2020**, *8*, 4085–4095. [CrossRef]
13. Caris-Veyrat, C.; Schmid, A.; Carail, M.; Böhm, V. Cleavage products of lycopene produced by in vitro oxidations: Characterization and mechanisms of formation. *J. Agric. Food Chem.* **2003**, *51*, 7318–7325. [CrossRef]
14. Boon, C.S.; McClements, D.J.; Weiss, J.; Decker, E.A. Factors influencing the chemical stability of carotenoids in foods. *Crit. Rev. Food Sci. Nutr.* **2010**, *50*, 515–532. [CrossRef]
15. Xiao, Y.D.; Huang, W.Y.; Li, D.J.; Song, J.F.; Liu, C.Q.; Wei, Q.Y.; Zhang, M.; Yang, Q.M. Thermal degradation kinetics of all-trans and cis-carotenoids in a light-induced model system. *Food Chem.* **2018**, *239*, 360–368. [CrossRef]
16. Sagar, V.R.; Kumar, P.S. Recent advances in drying and dehydration of fruits and vegetables: A review. *J. Food Sci. Technol.* **2010**, *47*, 15–26. [CrossRef]
17. Subtil, S.F.; Rocha-Selmi, G.A.; Thomazini, M.; Trindade, M.A.; Netto, F.M.; Favaro-Trindade, C.S. Effect of spray drying on the sensory and physical properties of hydrolysed casein using gum arabic as the carrier. *J. Food Sci. Technol.* **2014**, *51*, 2014–2021. [CrossRef]
18. Tuyen, C.K.; Nguyen, M.H.; Roach, P.D. Effects of spray drying conditions on the physicochemical and antioxidant properties of the Gac (*Momordica cochinchinensis*) fruit aril powder. *J. Food Eng.* **2010**, *98*, 385–392. [CrossRef]
19. Hojjati, M.; Razavi, S.H.; Rezaei, K.; Gilani, K. Spray drying microencapsulation of natural canthaxantin using soluble soybean polysaccharide as a carrier. *Food Sci. Biotechnol.* **2011**, *20*, 63–69. [CrossRef]
20. Etzbach, L.; Meinert, M.; Faber, T.; Klein, C.; Schieber, A.; Weber, F. Effects of carrier agents on powder properties, stability of carotenoids, and encapsulation efficiency of goldenberry (*Physalis peruviana* L.) powder produced by co-current spray drying. *Curr. Res. Food Sci.* **2020**, *3*, 73–81. [CrossRef]
21. Ai, Y.; Jane, J.L. Gelatinization and rheological properties of starch. *Starch-Stärke* **2015**, *67*, 213–224. [CrossRef]
22. Wang, J.; Xia, S.; Wang, B.; Ali, F.; Li, X. Effect of twin-screw extrusion on gelatinization characteristics of oat powder. *J. Food Process Eng.* **2019**, *42*, e13014. [CrossRef]
23. Silva, M.P.; Thomazini, M.; Holkem, A.T.; Pinho, L.S.; Genovese, M.I.; Fávaro-Trindade, C.S. Production and characterization of solid lipid microparticles loaded with guaraná (*Paullinia cupana*) seed extract. *Food Res. Int.* **2019**, *123*, 144–152. [CrossRef]
24. Pinho, L.P.; Silva, M.P.; Thomazini, M.; Cooperstone, J.L.; Campanella, O.; Rodrigues, C.E.C.; Fávaro-Trindade, C.S. Guaraná (*Paullinia cupana*) by-product as a source of bioactive compounds and as a natural antioxidant for food applications. *Food Process. Preserv.* **2021**, *45*, e15854. [CrossRef]
25. Cuevas, M.S.; Rodrigues, C.E.; Gomes, G.B.; Meirelles, A.J. Vegetable Oils Deacidification by Solvent Extraction: Liquid-Liquid Equilibrium Data for Systems Containing Sunflower Seed Oil at 298.2 K. *J. Chem. Eng. Data* **2010**, *55*, 3859–3862. [CrossRef]
26. Rocha, G.A.; Fávaro-Trindade, C.S.; Grosso, C.R.F. Microencapsulation of lycopene by spray drying: Characterization, stability and application of microcapsules. *Food Bioprod. Process.* **2012**, *90*, 37–42. [CrossRef]
27. Rodriguez-Amaya, D.B. *A Guide to Carotenoid Analysis in Foods*; ILSI Press: Washington, DC, USA, 2001.
28. Tonon, R.V.; Brabet, C.; Hubinger, M.D. Anthocyanin stability and antioxidant activity of spray-dried açai (*Euterpe oleracea* Mart.) juice produced with different carrier agents. *Food Res. Int.* **2010**, *43*, 907–914. [CrossRef]
29. Hidalgo, A.; Brandolini, A. Kinetics of carotenoids degradation during the storage of einkorn (*Triticum monococcum* L. ssp. *monococcum*) and bread wheat (*Triticum aestivum* L. ssp. *aestivum*) flours. *J. Agric. Food Chem.* **2008**, *56*, 11300–11305. [CrossRef]
30. Song, J.; Wang, X.; Li, D.; Liu, C. Degradation kinetics of carotenoids and visual colour in pumpkin (*Cucurbita maxima* L.) slices during microwave-vacuum drying. *Int. J. Food Prop.* **2017**, *20* (Suppl. S1), S632–S643. [CrossRef]
31. Minolta Corporation. *Precise Color Communication: Color Control from Feeling to Instrumentation*; Minolta Corporation: Ramsey, NJ, USA, 2008; pp. 1–49.

32. Kopec, R.E.; Cooperstone, J.L.; Schweiggert, R.M.; Young, G.S.; Harrison, E.H.; Francis, D.M.; Clinton, S.K.; Schwartz, S.J. Avocado consumption enhances human postprandial provitamin a absorption and conversion from a novel high-β-carotene tomato sauce and from carrots. *J. Nutr.* **2014**, *144*, 1158–1166. [CrossRef]
33. Cooperstone, J.L.; Ralston, R.A.; Riedl, K.M.; Haufe, T.C.; Schweiggert, R.M.; King, S.A.; Timmers, C.D.; Francis, D.M.; Lesinski, G.B.; Clinton, S.K.; et al. Enhanced bioavailability of lycopene when consumed as cis-isomers from tangerine compared to red tomato juice, a randomized, cross-over clinical trial. *Mol. Nutr. Food Res.* **2015**, *59*, 658–669. [CrossRef]
34. Rascón, M.P.; Beristain, C.I.; García, H.S.; Salgado, M.A. Carotenoid retention and storage stability of spray-dried encapsulated paprika oleoresin using gum Arabic and soy protein isolate as wall materials. *LWT-Food Sci. Technol.* **2011**, *44*, 549–557. [CrossRef]
35. Favaro-Trindade, C.S.; Okuro, P.K.; de Matos, F.E., Jr. Encapsulation via spray. In *Handbook of Encapsulation and Controlled Release*; CRC Press: Boca Raton, FL, USA, 2015; pp. 71–88.
36. Walton, D.E.; Mumford, C.J. Spray dried products—Characterization of particle morphology. *Chem. Eng. Res. Des.* **1999**, *77*, 21–38. [CrossRef]
37. Ghosal, S.; Indira, T.N.; Bhattacharya, S. Agglomeration of a model food powder: Effect of maltodextrin and gum Arabic dispersions on flow behavior and compacted mass. *J. Food Eng.* **2010**, *96*, 222–228. [CrossRef]
38. Ezhilarasi, P.N.; Karthik, P.; Chhanwal, N.; Anandharamakrishnan, C. Nanoencapsulation techniques for food bioactive components: A review. *Food Bioprocess Technol.* **2013**, *6*, 628–647. [CrossRef]
39. Chiou, D.; Langrish, T.A.G. Development and characterisation of novel nutraceuticals with spray drying technology. *J. Food Eng.* **2007**, *82*, 84–91. [CrossRef]
40. Labuza, T.P.; Dugan, L.R., Jr. Kinetics of lipid oxidation in foods. *Crit. Rev. Food Sci. Nutr.* **1971**, *2*, 355–405. [CrossRef]
41. Álvarez-Henao, M.V.; Saavedra, N.; Medina, S.; Cartagena, C.J.; Alzate, L.M.; Londoño-Londoño, J. Microencapsulation of lutein by spray-drying: Characterization and stability analyses to promote its use as a functional ingredient. *Food Chem.* **2018**, *256*, 181–187. [CrossRef]
42. Hettrich, K.; Fanter, C. Novel xylan gels prepared from oat spelts. In *Macromolecular Symposia*; WILEY-VCH Verlag: Weinheim, Germany, 2010; Volume 294, pp. 141–150. [CrossRef]
43. Al-Muhtaseb, A.H.; McMinn, W.A.M.; Magee, T.R.A. Moisture sorption isotherm characteristics of food products: A review. *Food Bioprod. Process.* **2002**, *80*, 118–128. [CrossRef]
44. Steenberg, E.; Harris, P.J. Adsorption of carboxymethylcellulose, guar gum and starch onto talc, sulphides, oxides and salt type minerals. *South Afr. J. Chem.* **1984**, *37*, 85–90. [CrossRef]
45. Gabas, A.L.; Telis, V.R.N.; Sobral, P.J.A.; Telis-Romero, J. Effect of maltodextrin and arabic gum in water vapor sorption thermodynamic properties of vacuum dried pineapple pulp powder. *J. Food Eng.* **2007**, *82*, 246–252. [CrossRef]
46. Damodaran, S.; Parkin, K.L.; Fennema, O.R. (Eds.) *Fennema's Food Chemistry*; CRC Press: Boca Raton, FL, USA, 2007.
47. Brett, B.; Figueroa, M.; Sandoval, A.J.; Barreiro, J.A.; Müller, A.J. Moisture sorption characteristics of starchy products: Oat flour and rice flour. *Food Biophys.* **2009**, *4*, 151–157. [CrossRef]
48. Díaz-Calderón, P.; MacNaughtan, B.; Hill, S.; Foster, T.; Enrione, J.; Mitchell, J. Changes in gelatinisation and pasting properties of various starches (wheat, maize and waxy maize) by the addition of bacterial cellulose fibrils. *Food Hydrocoll.* **2018**, *80*, 274–280. [CrossRef]
49. Eliasson, A.C. Differential scanning calorimetry studies on wheat starch—Gluten mixtures: I. Effect of gluten on the gelatinization of wheat starch. *J. Cereal Sci.* **1983**, *1*, 199–205. [CrossRef]
50. Punia, S.; Sandhu, K.S.; Dhull, S.B.; Siroha, A.K.; Purewal, S.S.; Kaur, M.; Kidwai, M.K. Oat starch: Physico-chemical, morphological, rheological characteristics and its applications-A review. *Int. J. Biol. Macromol.* **2020**, *154*, 493–498. [CrossRef]
51. Taghizadeh, A.; Favis, B.D. Effect of high molecular weight plasticizers on the gelatinization of starch under static and shear conditions. *Carbohydr. Polym.* **2013**, *92*, 1799–1808. [CrossRef]
52. Singh, A.; Geveke, D.J.; Yadav, M.P. Improvement of rheological, thermal and functional properties of tapioca starch by using gum arabic. *LWT* **2017**, *80*, 155–162. [CrossRef]
53. Shahzad, S.A.; Hussain, S.; Mohamed, A.A.; Alamri, M.S.; Ibraheem, M.A.; Qasem, A.A.A. Effect of hydrocolloid gums on the pasting, thermal, rheological and textural properties of chickpea starch. *Foods* **2019**, *8*, 687. [CrossRef]
54. Farrag, Y.; Sabando, C.; Rodríguez-Llamazares, S.; Bouza, R.; Rojas, C.; Barral, L. Preparation of donut-shaped starch microparticles by aqueous-alcoholic treatment. *Food Chem.* **2018**, *246*, 1–5. [CrossRef]
55. Sun, Y.; Li, F.; Luan, Y.; Li, P.; Dong, X.; Chen, M.; Dai, L.; Sun, Q. Gelatinization, pasting, and rheological properties of pea starch in alcohol solution. *Food Hydrocoll.* **2021**, *112*, 106331. [CrossRef]
56. Dhuique-Mayer, C.; Tbatou, M.; Carail, M.; Caris-Veyrat, C.; Dornier, M.; Amiot, M.J. Thermal degradation of antioxidant micronutrients in citrus juice: Kinetics and newly formed compounds. *J. Agric. Food Chem.* **2007**, *55*, 4209–4216. [CrossRef]
57. Achir, N.; Randrianatoandro, V.A.; Bohuon, P.; Laffargue, A.; Avallone, S. Kinetic study of β-carotene and lutein degradation in oils during heat treatment. *Eur. J. Lipid Sci. Technol.* **2010**, *112*, 349–361. [CrossRef]
58. Aman, R.; Schieber, A.; Carle, R. Effects of heating and illumination on trans− cis isomerization and degradation of β-carotene and lutein in isolated spinach chloroplasts. *J. Agric. Food Chem.* **2005**, *53*, 9512–9518. [CrossRef]
59. Waramboi, J.G.; Gidley, M.J.; Sopade, P.A. Carotenoid contents of extruded and non-extruded sweetpotato flours from Papua New Guinea and Australia. *Food Chem.* **2013**, *141*, 1740–1746. [CrossRef]

60. Vásquez-Caicedo, A.L.; Schilling, S.; Carle, R.; Neidhart, S. Impact of packaging and storage conditions on colour and β-carotene retention of pasteurised mango purée. *Eur. Food Res. Technol.* **2007**, *224*, 581–590. [CrossRef]
61. Gama, J.J.T.; de Sylos, C.M. Effect of thermal pasteurization and concentration on carotenoid composition of Brazilian Valencia orange juice. *Food Chem.* **2007**, *100*, 1686–1690. [CrossRef]
62. Provesi, J.G.; Dias, C.O.; Amante, E.R. Changes in carotenoids during processing and storage of pumpkin puree. *Food Chem.* **2011**, *128*, 195–202. [CrossRef]
63. Pinho, L.S.; de Lima, P.M.; de Sá, S.H.G.; Chen, D.; Campanella, O.H.; da Costa Rodrigues, C.E. Favaro-Trindade, C.S. Encapsulation of Rich-Carotenoids Extract from Guaraná (*Paullinia cupana*) Byproduct by a Combination of Spray Drying and Spray Chilling. *Foods* **2022**, *11*, 2557. [CrossRef]

foods

MDPI

Article

Effect of Ultrasound and High Hydrostatic Pressure Processing on Quality and Bioactive Compounds during the Shelf Life of a Broccoli and Carrot By-Products Beverage

Pablo Pérez [1,2], Seyedehzeinab Hashemi [1,3], Marina Cano-Lamadrid [1,3], Lorena Martínez-Zamora [1,3,4,*], Perla A. Gómez [3] and Francisco Artés-Hernández [1,3,*]

1 Postharvest and Refrigeration Group, Department of Agricultural Engineering, Universidad Politécnica de Cartagena, 30203 Cartagena, Region of Murcia, Spain; pabloofed@hotmail.com (P.P.); seyedehzeinab.hashemi@edu.upct.es (S.H.); marina.cano@upct.es (M.C.-L.)

2 Laboratorio de Investigación en Tecnología de Alimentos, Instituto de Tecnologías y Ciencias de la Ingeniería (INTECIN), Facultad de Ingeniería, Departamento de Ingeniería Química, Consejo Nacional de Investigaciones Científica y Técnicas (CONICET), Universidad de Buenos Aires, C.A.B.A, Buenos Aires C1428EGA, Argentina

3 Institute of Plant Biotechnology, Universidad Politécnica de Cartagena, 30202 Cartagena, Region of Murcia, Spain; perla.gomez@upct.es

4 Department of Food Technology, Nutrition, and Food Science, Faculty of Veterinary Sciences, University of Murcia, 30071 Espinardo, Region of Murcia, Spain

* Correspondence: lorena.martinez23@um.es (L.M.-Z.); fr.artes-hdez@upct.es (F.A.-H.); Tel.: +34-968325509 (F.A.-H.)

Abstract: Vegetable beverages are a convenient strategy to enhance the consumption of horticultural commodities, with the possibility of being fortified with plant by-products to increase functional quality. The main objective was to develop a new veggie beverage from broccoli stalks and carrot by-products seasoned with natural antioxidants and antimicrobial ingredients. Pasteurization, Ultrasound (US), and High Hydrostatic Pressure (HHP) and their combinations were used as processing treatments, while no treatment was used as a control (CTRL). A shelf-life study of 28 days at 4 °C was assayed. Microbial load, antioxidant capacity, and bioactive compounds were periodically measured. Non-thermal treatments have successfully preserved antioxidants (~6 mg/L ΣCarotenoids) and sulfur compounds (~1.25 g/L ΣGlucosinolates and ~5.5 mg/L sulforaphane) throughout the refrigerated storage, with a longer shelf life compared to a pasteurized beverage. Total vial count was reduced by 1.5–2 log CFU/mL at day 0 and by 6 log CFU/mL at the end of the storage in HHP treatments. Thus, the product developed in this study could help increase the daily intake of glucosinolates and carotenoids. These beverages can be a good strategy to revitalize broccoli and carrot by-products with high nutritional potential while maintaining a pleasant sensory perception for the final consumer.

Keywords: broccoli stalks; non-thermal processing; juices; revalorization; food loss; cold pressed; phytochemicals

Citation: Pérez, P.; Hashemi, S.; Cano-Lamadrid, M.; Martínez-Zamora, L.; Gómez, P.A.; Artés-Hernández, F. Effect of Ultrasound and High Hydrostatic Pressure Processing on Quality and Bioactive Compounds during the Shelf Life of a Broccoli and Carrot By-Products Beverage. *Foods* **2023**, *12*, 3808. https://doi.org/10.3390/foods12203808

Academic Editor: Isabel Hernando

Received: 12 September 2023
Revised: 10 October 2023
Accepted: 12 October 2023
Published: 17 October 2023

1. Introduction

During the last few years, consumers' interest in new ready-to-eat plant products has increased due to their fresh characteristics and high nutritional value [1–3]. Predominantly, the consumption of cruciferous vegetables, such as broccoli, is linked to health benefits for consumers associated with the high content of nutrients and phytochemicals, particularly glucosinolates and phenolic compounds [4]. However, it is estimated that approximately 35–40% of the world's broccoli production is lost or wasted [5]. Several studies have demonstrated that the nutritional value of broccoli by-products is comparable to that of its florets [6–8]. As a result, these by-products have the potential to serve as ingredients for developing and/or fortifying new products, thereby reducing agricultural losses [4].

In this way, for the present study, broccoli and carrot by-products have been chosen due to their valuable nutritional and complementary compositions. In fact, broccoli is rich in glucosinolates and their derived isothiocyanates, which have been demonstrated to be potential anticarcinogenic and anti-inflammatory compounds [9–11]. Carrot has a high carotene content, especially β-carotene, which is responsible for its orange color and is widely known for its antioxidant and anti-inflammatory effects due to its prevention against oxidative stress [12].

Beverages containing fruit and vegetable by-products as ingredients could have a high content of bioactive compounds, making them attractive to consumers [1]. However these plant products have a short shelf life limited by the growth of certain microorganisms such as mesophilic bacteria (>10^7), molds and yeast (>10^4), and the apparition of some pathogens such as *Listeria* or *Salmonella* [13,14]. Traditionally, the industry commonly applies physical thermal treatments, such as pasteurization, to destroy microorganisms and inactivate enzymes that shorten the commercial shelf life. Nevertheless, such physical treatments may have a negative impact on the sensory and nutritional quality of these fresh products [15].

Consequently, there is growing interest in emerging non-thermal processing treatments such as ultrasounds (US) and high hydrostatic pressures (HHP). HHP is a non-thermal physical preservation technology where the packaged product is exposed to high pressures, typically around 500 MPa [16–18]. These conditions guarantee microbiological safety without an increase in temperature during the process (around 35 °C), while maintaining the nutritional value and fresh characteristics of the product [19]. On the other hand, the US can increase the shelf-life of vegetable products through cell disruption without negatively impacting their bioactive compounds [20,21]. The effectiveness of these emerging technologies has been demonstrated in different foods. It has been reported that the combined HHP and US have a positive effect on the physicochemical and nutritional quality of cold-brew tea, guaranteeing its microbiological safety [22].

Other researchers have previously studied the development of vegetable beverages enriched with broccoli by-products rich in sulfur compounds [23]. However, to the best of our knowledge, a broccoli stalk beverage as its main ingredient was not earlier reported in the scientific literature, which is the main novelty of the present work. The mixture with other fruit and vegetable commodities and derivates rich in other compounds, such as carotenoids, can be convenient. For instance, a blend of broccoli stalk juice with carrot juice, as a main source of sugars to improve sensory perception and carotenoids as antioxidants, could be a good combination for beverage fortification.

The objective of this research was to explore the revalorization of vegetable by-products by developing a new veggie beverage obtained from broccoli and carrot discards, mixed with other natural antioxidant and antimicrobial ingredients. Then, conventional pasteurization and alternative HHP and US processing treatments, and their combinations, were studied to preserve the microbiological and functional quality during a refrigerated shelf-life of 28 days.

2. Materials and Methods

2.1. Beverage Formulation and Preparation

Broccoli stalks (Naxos F1 Hybrid Broccoli, Sakata) were provided by Grupo Lucas (Murcia, Spain), making up 15% (*w*/*w*) of the harvested broccoli. On the other hand, carrots not meeting the quality parameters of size and shape to be fresh commercialized were used as by-products (Hortalizas Requena S.L., Murcia, Spain). These vegetables were transported to the laboratory under refrigerated conditions, washed for disinfection with chlorinated water (100 mg L^{-1} of free chlorine; pH 6.5; 5 °C, for 2 min), and rinsed with cold (5 °C) tap water for 1 min. Then, carrots were cut into transverse slices, while broccoli stalks were cut into longitudinal portions, taking care to remove browned areas. After that, the cut vegetables were blanched in a microwave oven to inactivate the browning enzymes [2,21,24]. For that, samples of 300 g of each product were exposed twice to 700 W during 2 min, with 1 min between the first exposition and the second. After blanching, the ingredients were immediately cooled with ice, and then they were placed in trays covered by a plastic film and kept under refrigeration (5 °C) for 16 h. After this time, they were separately squeezed for juice extraction with a Robot Coupe J80 Ultra (Vincennes, Île-de-France, France). Broccoli stalks yielded 37.5% of juice, and carrots 39.4%. Then, the vegetable juices were elaborated according to the formulations shown in Table 1, previously chosen based on their sensory perception by a trained panel of experts. The ingredients were mixed and emulsified using a hand blender, which avoided the separation of fat from the surface of the beverage. Samples of 125 mL were packaged into aseptic polypropylene pouches (Infantino, San Diego, CA, USA) using a sterilized funnel. The presence of air bubbles that could adversely interfere during processing treatments was avoided. Finally, the pouches were hermetically closed with caps and stored at 4 °C, which was monitored with a Tinytag ULTRA 2 Data logger (TGU-4500; Chichester, UK).

Table 1. Recipe composition of vegetable juices elaborated.

	Recipe Composition (%)
Broccoli stalk juice	82.5
Carrot by-product juice	17.5
EVOO	1
Lemon juice	0.25
Black pepper	0.1
Salt	0.1
Garlic powder	0.1

EVOO: Extra Virgin Olive Oil.

Physicochemical analyses, including pH, titratable acidity (TA), total soluble solid content (SST), and color, were monitored in quintuplicate during the shelf-life study following the methods previously described in our research work [25]. Briefly, the standard parameters L*, a*, and b* were recorded according to the CIE Lab system for color determination using a colorimeter (Konica Minolta CR-400). TA, determined by titration (automatic titrator T50, Metter Toledo; Milan, Lombardia, Italy), and SST, measured with a handheld refractometer (Atago N1; Tokyo, Kanto, Japan), were expressed as g citric acid 100 mL^{-1} and °Brix, respectively. The initial values of the blended juice before processing treatments were: 45.7, 3.0, 19.0, 6.2, 6.2, and 0.1 for L*, a*, b*, pH, SST, and TA, respectively (specified in Section 2.2, Table S1, and Figure 1).

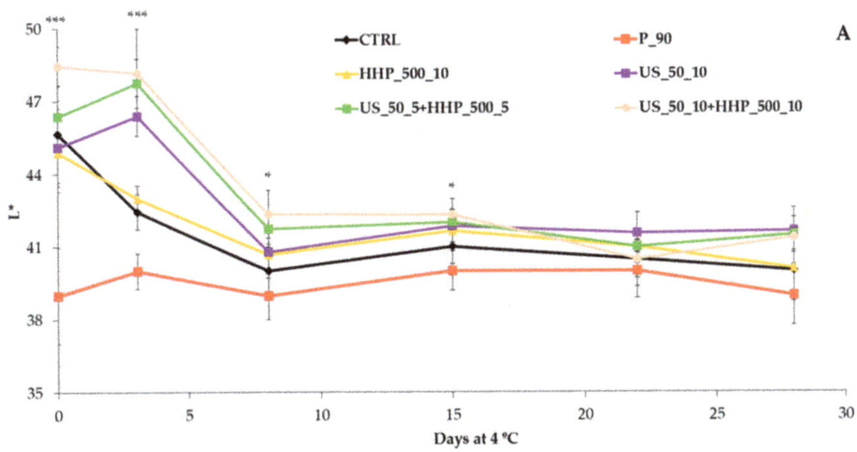

Figure 1. *Cont.*

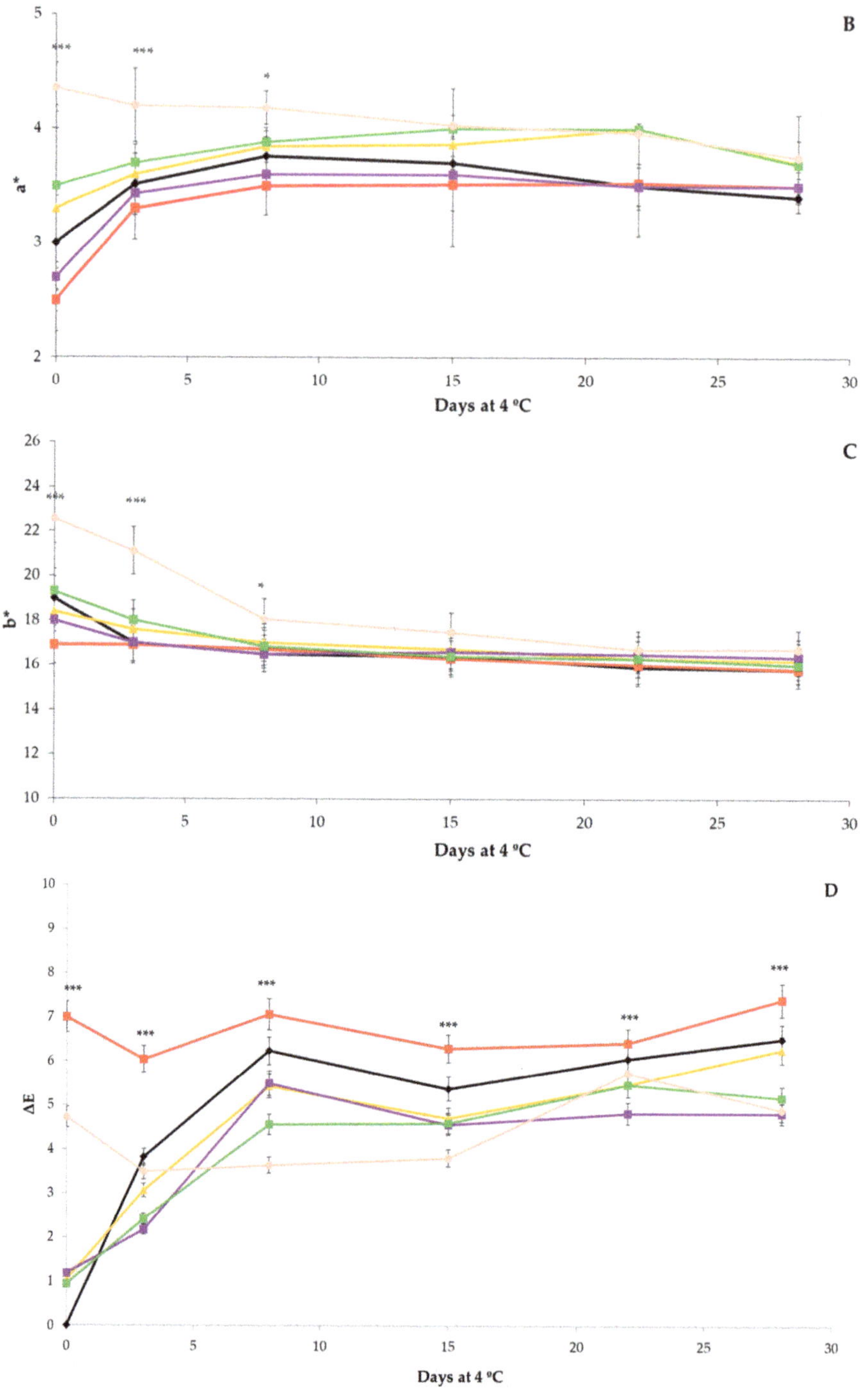

Figure 1. *Cont.*

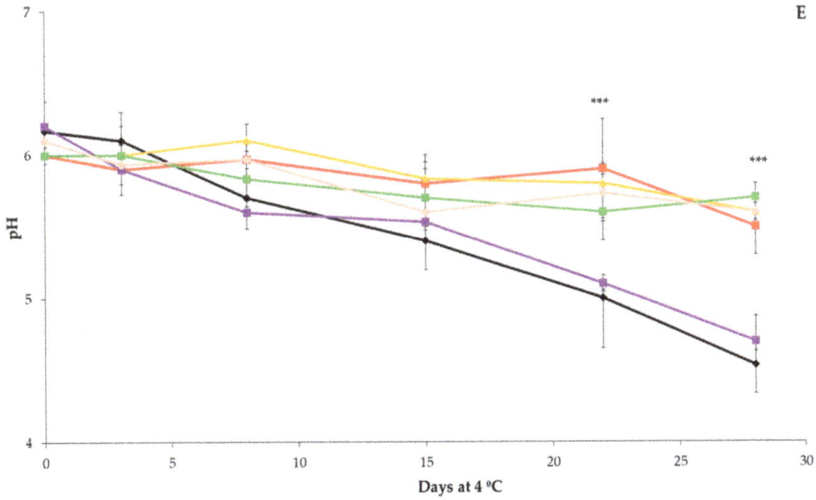

Figure 1. L* (**A**), a* (**B**), b* (**C**), ΔE (**D**), and pH (**E**) change during 28 days at 4 °C in a broccoli and carrot by-product beverage according to several processing techniques (n = 5). The ANOVA information (*: $p < 0.05$; ***: $p < 0.001$) refers to the differences among treatments at the same sampling time.

2.2. Processing Treatments and Storage Conditions

The applied processing treatments were:

- **CTRL**: Fresh blended beverage without any processing treatment. This treatment was used as a control.
- **P_90**: Pasteurization (90 °C; 10 min) in an agitated water bath to ensure the temperature distribution (J.P. Selecta, Barcelona, Spain). This temperature and time were chosen according to our previous experiments and previous works on similar beverages, since this kind of product with a pH >4.6 needs stronger thermal treatments to avoid microbial spoilage [26–29].
- **HHP_500_10**: HHP (500 MPa, 10 min) using a high-pressure Iso-Lab system (Stansted Fluid Power Ltd., Harlow, UK), as recommended by previous authors [2,30].
- **US_50_10**: US (720 W, 35 kHz, 50 °C, 10 min) using an ultrasound bath (Sonorex Digiplus, Helsinki, Finland), based on previous works [26,31,32].
- **US_50_5+HHP_500_5**: US (720 W, 35 kHz, 50 °C, 5 min) combined with HHP (500 MPa, 5 min).
- **US_50_10+HHP_500_10**: US (720 W, 35 kHz, 50 °C, 10 min) combined with HHP (500 MPa, 10 min).

Then, the samples were stored for 28 days at 4 °C (4 weeks), the minimum shelf-life required for this kind of product in Spain, until they were analyzed. Microbiological quality and bioactive compounds were periodically measured throughout storage as indicators to assess the effectiveness of the preservation methods. The analysis was carried out on five sampling days (0, 7, 14, 21, and 28). For each treatment, three samples were taken on each sampling day, and physicochemical and microbial analyses were performed. The samples intended for the determination of bioactive compounds and antioxidants were frozen and stored at −80 °C to avoid alterations to these parameters until analysis.

2.3. Microbial Analyses

Mesophilic and psychrophilic aerobic bacteria, *Enterobacteriaceae*, and mold and yeast were evaluated. For this, 10 mL of the sample were mixed with 90 mL of sterile peptone water. Decimal dilutions were made with sterile peptone water, and seeding was performed on plates containing agar for total count (for mesophilic and psychrophilic aerobic bacteria) McConckey agar (for *Enterobacteriaceae*), and dichlorane rose Bengal chloramphenicol medium (for mold and yeast). The plates were incubated (in the incubator Incubate (J.P. Selecta, Barcelona, Spain)) at 37 °C for 24–48 h (for bacteria) and 25 °C for 4 days (for molds and yeasts). Microbial counts were reported as log CFU per mL with a detection limit of 2 log (CFU/mL). The analysis was carried out in triplicate, and three dilutions per treatment were analyzed on each sampling day.

2.4. *Bioactive Compounds*

2.4.1. Free Polyphenol Content (FPC) and Antioxidant Capacity (AC)

Extracts for FPC and AC were obtained from 250 µL of frozen samples mixed in plastic tubes with 750 µL of methanol: water (80:20, *v*/*v*). The extraction was carried out in triplicate with an orbital shaker (Stuart, Stone, UK) for 1 h at 200 rpm in darkness at 4 °C. Finally, the extracts were centrifuged at 3220× *g* for 10 min at 4 °C, with the supernatant being used for analysis. FPC determination was carried out according to Singleton et al. [33] with some modifications. Briefly, 19 µL of the supernatant extract was dispensed into a 96-well plate, followed by the addition of 29 µL of 1 N Folin–Ciocalteu reagent. The plate was then incubated in darkness at room temperature (20 °C) for 3 min Then, 192 µL of Na_2CO_3 (0.4%) and NaOH (2%) were added, and the mix was incubated at room temperature (in darkness) for 1 h. Finally, each sample was spectrophotometrically measured at a wavelength of 750 nm in a microplate reader (Infinite PRO 2000, Tecan Trading AG, Männedorf, Switzerland). The FPC was calculated using a gallic acid standard and expressed as mg of gallic acid equivalent per L of sample.

The determination of AC was carried out following the DPPH method [4] and the iron reduction power assay (FRAP) [34]. For the DPPH assay, 194 µL of DPPH solution were added to 21 µL of extract in a 96-well plate. The mixture was incubated for 30 min at room temperature (20 °C) in darkness. The absorbance was measured by changes at 515 nm. For the FRAP method, 198 µL of daily FRAP solution were added to 6 µL of extract in a 96-well plate. The mixture was incubated for 30 min at room temperature (20 °C) in darkness, and the absorbance was measured by changes at 493 nm. For both methods, the antioxidant capacity was calculated using a Trolox standard and expressed as mg of Trolox equivalents per L of sample.

2.4.2. Carotenoid Content

The extraction and analysis of carotenoids were performed in triplicate, according to Martínez-Zamora et al. [35], based on the method developed by Gupta et al. [36]. Carotenoids were quantified with an Ultra High Performance Liquid Chromatography (UHPLC) instrument (Shimadzu, Kyoto, Japan) equipped with a C32 column, DGU-20A degasser, LC-30CE quaternary pump, SIL-30AC autosampler, CTO-10AS column heater, and SPDM-20A photodiode array detector. The carotenoid content (9-cis-β-carotene, 13-cis-β-carotene, all-trans-β-carotene, and lutein) was expressed as mg per L of sample. For that, β-carotene and lutein standards were used (Sigma-Aldrich, St. Louis, MO, USA).

2.4.3. Glucosinolate and Isothiocyanate Content

For the determination of glucosinolates, myrosinase inactivation, desulfated step, purification, identification, and quantification were carried out in triplicate according to Martínez-Zamora et al. [37] based on the method previously described by Kiddle et al. [38]. The glucosinolate content (ds-glucoraphanin, ds-4-methoxy-glucobrassicin, and ds-neoglucobrassicin) was expressed as mg per L of sample, and glucoraphanin (PhytoLab GmbH & Co. KG, Vestenbergsgreuth, Germany) was used as standard. Sulforaphane was quantified using the DL-sulforaphane standard (Sigma-Aldrich, St. Louis, MO, USA), and results were expressed as mg/L. Each sample was analyzed in triplicate. A UPLC instrument (Shimadzu, Kyoto, Japan) equipped with a C18 column, DGU-20A degasser, LC-

30AD quaternary pump, SIL-30AC autosampler, CTO-10AS column heater, and SPDM-20A photodiode array detector was used as previously described [37] for both analysis.

2.5. Sensory Analysis

A trained panel consisting of five experts on fruit-and-vegetable-based products, including brassica species, from the Institute of Plant Biotechnology (IBV), which belongs to the Universidad Politécnica de Cartagena (UPCT) (Murcia, Spain), performed the sensory study. Tests were conducted in a standard room [39] equipped with ten individual taste booths.

To select the base formulation for our product, an informal and preliminary study was carried out comparing the main sensory characteristics among the prototypes of the vegetable beverages. The panelists focused on optimizing the percentage of broccoli and carrot juice following a previous study in which millennial consumers overall liking of beverages was negatively correlated with the level of sweetness and earthy, carrot, beetroot, and pear flavors. penalty analysis showed that beverages based on vegetables (smoothies) need improvement, mainly dealing with sweetness, bitterness, and vegetable flavors [40]. Moreover, the decision was made due to the know-how of the research group about the description and development of products based on brassicas [24,41]. Samples were served into odor-free, disposable 50 mL covered plastic cups at room temperature and coded with 3-digit randomized numbers. Each judge tested ~10 g of each sample in a randomized order. Water and unsalted crackers were provided to the judges for palate cleansing.

On the other hand, a descriptive sensory study was carried out, focusing on the negative attributes "off-flavor", "cooked flavor", and "fermented flavor", following the descriptors included in previous studies [40,42]. These attributes were evaluated at each sampling time to establish and limit the shelf life of the beverage. The scale was from 1 to 5 (1: absence; 2: slight presence; 3: moderate presence, as the consumption limit was decided by the authors; 4: significant presence; and 5: extreme presence).

2.6. Statistical Analysis

Box plots and scatter graphs using XLSTAT Premium 2016 (Addingsoft, Barcelona, Spain) were conducted. Results were analyzed using the SSPS program (IBM SPSS Statist cs, Chicago, IL, USA). Data were submitted to analysis of variance (ANOVA), and significant differences were determined using the Tukey test ($p < 0.05$). The results after Tukey's test using sampling time as a factor for each of the treatments are shown in tables. Figures show the analysis of variance between treatments for each sampling time ($n = 3$).

3. Results and Discussion

3.1. Physicochemical Analysis

Figure 1 shows the statistics among treatments for CIELab color coordinates and pH at each sampling time, while Table S1 shows both statistics among sampling times for each treatment and among treatments at each sampling time. On processing day, a significant effect of treatments on L* values were observed, especially with thermal pasteurization showing the lowest luminosity (L* values). The same trend was detected for b* (Figure 1). Related to a*, the US_50_10+HHP_500_10 treatment showed differences in the first 8 days compared with the rest of the treatments, with the US_50_10 + HHP_500_10 treatment having the highest value. The ΔE of samples varied between 0.0 and 7.4, considering all the treatments and sampling days. Values ranging from 0.0 to 0.5 mean no color differences; from 0.5 to 1.0 means a difference only perceivable for experienced observers; from 1.0 to 2.0 means a minimal color difference; from 2.0 to 4.0 means a perceivable color difference; and more than 4.0 means a significant color difference [43]. In that sense, finding low ΔE values indicates high similarity to the ideal color of the fresh beverage and means that the treatment was good for preserving color and storability. At day 0, no significant color differences were detected in HHP_500_10, US_50_10, and US_50_5+HHP_500_5 (values < 4.0) compared to CTRL. However, a significant effect (ΔE higher than 4.0) was observed after P_90 and US_50_10+HHP_500_10 samples, respectively.

Storage time also affected color parameters. L* tended to decrease during shelf life, with a decrease after 8 days, except for the pasteurization treatment, which remained stable after a high initial decrease (Table S1). The a* values increased throughout storage at 4 °C for the treatments CTRL, P_90, and US_50_10. For the remaining treatments, there were no differences during storage (Table S1). As expected, ΔE levels gradually increased throughout storage in all treatments except for pasteurization (P_90) and the combination of US and HHP (US_50_10+HHP_500_10). The samples CTRL, HHP_500_10, US_50_10, and US_50_5+HHP_500_5 after 8 d at 4 °C showed different ΔE values compared to the fresh beverage. As the differences in color (>4.0) were observed after treatments, there was no variation in ΔE during storage. The values found were lower than those obtained in previous studies in which a red smoothie was developed (based on broccoli and carrot, with tomato, red pepper, and spices), with the ΔE of untreated and heat-treated smoothies between 8.6 and 34.7 [25]. Untreated samples during shelf life reached around 20 units higher ΔE than the treated ones. The authors of this study proposed that thermal inactivation of browning enzymes, specifically polyphenoloxidase and peroxidase, will lead to a reduction in ΔE values [25]. In our research, we applied a pre-juicing step involving microwaving the plant material to effectively inactivate a great amount of such browning enzymes.

The initial pH of the untreated beverage (6.2) was higher than the values reported by other authors in a similar formulation [25]. The pH did not change after the thermal and non-thermal treatment (0 d 4 °C) and during 21 days of storage (Figure 1), at which point differences were observed. From that day onwards, the CTRL and US_50_10 samples showed a decrease in pH, probably related to the microbial growth, as shown in Figure 2 and Table S2.

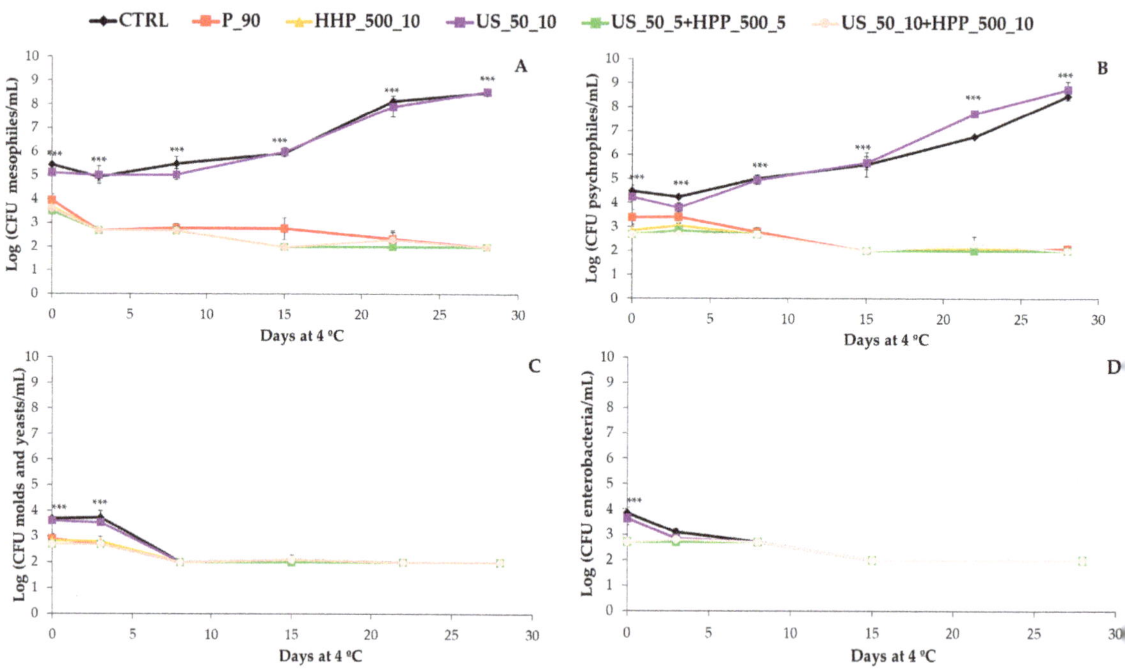

Figure 2. Aerobic mesophilic bacteria (**A**), psicrophylles (**B**), mold and yeast (**C**), and enterobacteria (**D**) (log CFU/mL) changes during 28 days at 4 °C in a broccoli and carrot by-products beverage according to several processing techniques (n = 3). The ANOVA information (***: $p < 0.001$) refers to the differences among treatments at the same sampling time.

3.2. Microbial Load

Figure 2 depicts the effects of treatments on the microbial load, while Table S2 shows the numerical data and the statistics at different sampling times and among the studied treatments. The P_90, HHP_500_10, US_50_5+HHP_500_5, and US_50_10+HHP_500_10 processing treatments reduced the mesophilic bacteria by 1.5–2.0 log CFU/mL at day 0, which in turn resulted in a reduction of even 6 log CFU/mL at the end of the shelf-life compared to CTRL treatment. In this sense, treatments allowed a reduction of the microbial load by 27–36% from the beginning of the experiments in comparison with CTRL samples, which ensured the microbial safety of these beverages. Nevertheless, US_50_10 treatment seems to reach a similar mesophilic bacteria load to CTRL-untreated samples; therefore, CTRL and US broccoli-carrot beverages remain microbiologically stable for up to 15 days at 4 °C. After that, an increase in microbial counts was observed, making the consumption of this product not advisable. Further microbial analysis of pathogenic bacteria must be performed in order to meet EU legislation for such products, although exceeding the limit of 10^7 in the aerobic mesophilic bacteria counts after 15 days of storage already indicates the loss of microbiological quality of the CTRL and US_50_10 beverages. The same trend was observed for psychrophilic bacteria, as can be seen from the data in Table S2 and Figure 2. Contrary to this, US has been demonstrated to be effective for the reduction and inhibition of microbial growth due to the cavitation force made by this technology [44], leading to the production of free radicals, the rupture of the cell membrane, and a localized temperature increase, which has been demonstrated to be effective in different fruit juices yet [45].

For enterobacteria and mold and yeast, all treatments had a reducing effect on the microbial load at day 0 compared to CTRL and US_50_10 treatments, respectively (Figure 2). At day 7, the enterobacteria, mold, and yeast loads of all treatments reached a steady state. This can be attributed to differences in the competitive abilities among groups of microorganisms, with mesophilic and psychrophilic bacteria showing higher survival rates compared to enterobacteria, molds, and yeasts under the specific conditions observed in this study [46].

US has been considered a good tool to preserve fruit juices with a low pH, and probably that is the reason why this technology was not effective in the present work, with a nearly neutral pH of the broccoli-carrot beverage that favored the microbiological growth of competing mesophilic bacteria.

3.3. Free Polyphenolic Content (FPC) and Antioxidant Capacity (AC)

Figure 3 and Table S3 show FPC and AC indicating differences among treatments at each sampling time and among sampling times. FPC by the Folin–Ciocalteau method measures the reducing capacity of compounds with the Mo^{6+} complex [33]. The method is not specific to polyphenols because other reducing compounds, such as ascorbic acid, sugars, and proteins, are also included in the quantification. Most of those compounds are heat-sensitive and are also degraded during cold storage. Pasteurized samples showed an important reduction of FPC compared to CTRL just after treatment on 0d. Then, significant differences among treatments at all sampling times were noticed, with P_90 presenting the lowest values. This behavior can be explained by the fact that above 60–70 °C, most antioxidant compounds, such as ascorbic acid and phenolics, can be degraded [47]. During cold storage, a reduction of FPC compared to the initial values for all the treatments was found (Table S3), being significant after 8 d at 4 °C for P_90 and US_50_10 and after 15 d at 4 °C for CTRL, HHP_500_10, US_50_5+HHP_500_5, and US_50_10+HHP_500_10. This reduction can be explained by the degradation of phenolic compounds during storage that are continuously acting as antioxidants during this period against the oxidative damage produced by the senescence of the product. This effect can be easily detected through spoilage or the reduction of quality during the shelf life that has been widely demonstrated in several food matrixes [48,49].

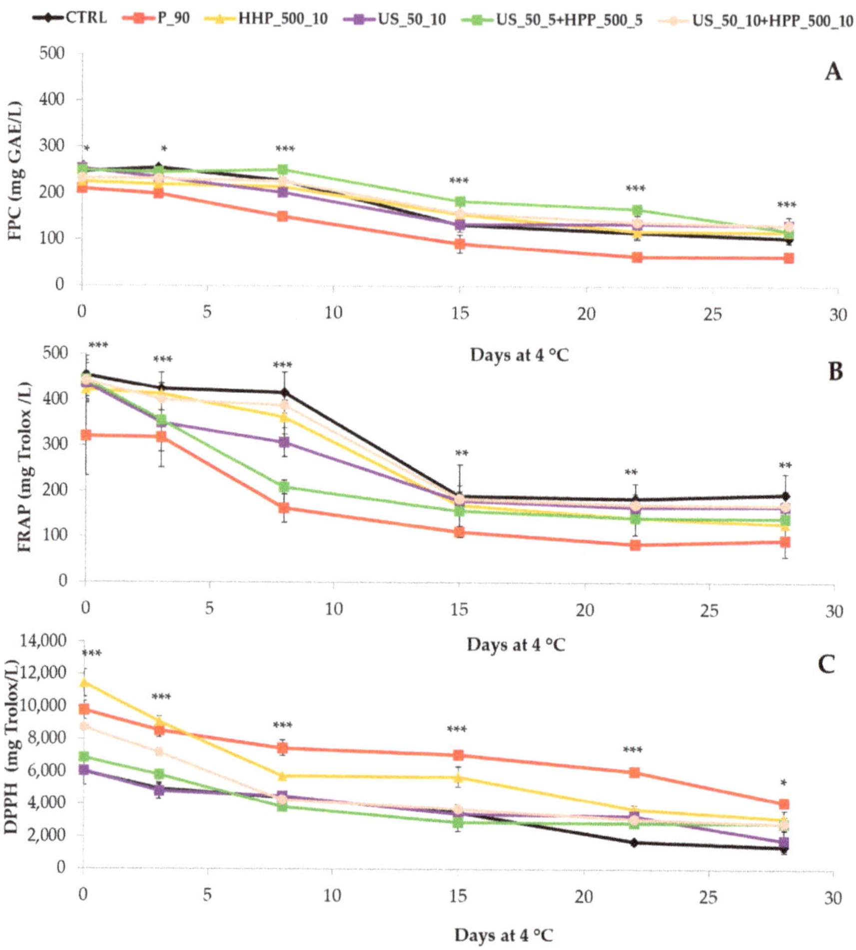

Figure 3. Free polyphenolic content (mg GAE/L) (**A**), FRAP (**B**), and DPPH (mg TE/L) (**C**) changes during 28 days at 4 °C in a broccoli and carrot by-products beverage according to several processing techniques (n = 3). The ANOVA information (*: $p < 0.05$; **: $p < 0.005$; ***: $p < 0.001$) refers to the differences among treatments at the same sampling time.

In the case of US-treated samples, it can be appreciated as these beverages have shown a higher content of FPC, which can be justified by the fact that cavitation phenomena form during sonication. This phenomenon can promote the extraction of phenolic compounds and carotenoids [29,50], which justifies the increase in FPC of studied beverages treated with US, single or combined with HHP.

The FRAP assay measures the reduction of the ferric ion (Fe^{3+})-ligand complex to the intensely blue-colored ferrous (Fe^{2+}) complex by antioxidant compounds in an acidic medium [34]. FRAP values presented the same trend observed for FPC. In our study, a significant Pearson correlation between FRAP and FPC was observed (0.893; p-value < 0.0001; R^2 = 0.797). The main reason is that both are based on a single electron transfer reaction: in the presence of an antioxidant, Fe^{3+} gets reduced to Fe^{2+} and Mo^{6+} to Mo^{5+}, respectively.

Finally, the DPPH assay measures the presence of antioxidant compounds thanks to a single electron and hydrogen atom transfer reaction, being the most powerful in scavenging DPPH radicals, the structural classes of catechins (flavan-3-ols), proanthocyanidins, and flavonols [51]. As shown in Table S3 and Figure 3C, the results obtained through this method showed a different trend compared to FRAP and FPC content. In this sense, a significant, although low, Pearson correlation between DPPH and FPC was observed (0.462; p-value = 0.005; R^2 = 0.213). After processing treatments, HHP_500_10 had the highest antioxidant capacity, followed by P_90, US_50_10+HHP_500_10, US_50_5+HHP_500_5, US_50_10, and CTRL. However, after four weeks of refrigerated storage, all the studied treatments tended to lose their DPPH scavenging activity by reducing between 2.3 and 4.3 fold compared to the initial values after processing.

The differences in results obtained for the different methods can be justified by the fact that this method measures the ability to scavenge DPPH free radicals in a methanolic medium, in which hydrophilic and hydrophobic antioxidants are included [51]. In this sense, carotenoids, and particularly β-carotene, are the main antioxidants found in carrots, and they are very stable even at high temperatures; hence, they are not destroyed by heat treatments, as P_90 [52]. As previously reported during the processing of Momordica charantia fruit or in fruit beverages, heat treatment (90–100 °C) for a short time could increase carotenoid availability as well as other antioxidant compounds that are highly non-heat-sensitive, such as flavan-3-ols or flavonols, while decreasing non-enzymatic browning during storage [28,53]. Furthermore, the low correlation, although significant, found between FPC and DPPH in the present study can also be explained by the fact that the Folin method is specific for phenolic compounds and carotenoids are not detected. This hypothesis is based on previous findings in orange, carrot, and tomato juices, which have shown increaments in the carotenoid and DPPH activity after a pasteurization process [54]. These authors showed increases by ~12% in the carotenoids and AC values in orange juices after 1 min at 90 °C and 30 s at 70 °C regarding a HHP treatment of 400 MPa per 1 min at 40 °C [54]. The same authors also reported a similar behavior in traditional pasteurized tomato juices [55]. More recently, Dallagi et al. have shown that processing conditions can also enhance the AC and carotenoids values of carrot juices by blanching at 95 °C for 9 min, which reported the highest yield of carotenoids and AC measured by DPPH [56].

3.4. Carotenoids and Sulfur Compounds Content

Non-significant differences were observed among treatments and sampling times regarding bioactive compound content (data shown in Table S4). A boxplot and a scatter graph were represented to better explain the obtained results. The boxplot includes each group of functional compounds (Figure 4). The red crosses correspond to the means; the central horizontal bars are the medians; the lower and upper limits of the box are the first and third quartiles, respectively; and the blue dots are the minimum and maximum values. The horizontal width of the box has no statistical significance. On the other hand, the scatter graph was included to show the individual carotenoids and glucosinolates. The red crosses correspond to the means; the central horizontal red bars are the medians; and the lower and upper black dots are the minimum and maximum values. Despite the treatment, the resulting beverage offers a potential means to enhance the recommended daily intake of carotenoids [57]. A single serving (250 mL) of the beverage contains approximately 1.5 mg of ∑carotenoids, covering 30%. This valuable contribution to carotenoid consumption makes the beverage a promising dietary option for promoting overall health and nutrition.

Given the low sensitivity of the carotenoids to heat treatments [57], it is coherent that in our research no differences were found between heated and non-heated samples. This can be due to the previous blanching with microwave technology, a treatment that was aimed at inhibiting browning and carotenoid-degrading enzymes [53]. The most common carotenoids in European diets are α-carotene, β-carotene, and lutein, among others. Among the food constituents that were the subject of the evaluation for obtaining the authorized claims by the EFSA panel, 'carotenoids from fruits and vegetable juices' are related to health claims effects described as 'antioxidant properties/protection of DNA', 'health during pregnancy/bioavailability', and 'skin protector' [58]. Different carotenoids may have different antioxidant and pro-vitamin A activities and a different capacity to absorb UV radiation in the tissues where they are accumulated [58]. The EFSA Panel concluded that a cause-and-effect relationship cannot be established between the consumption of carotenoids and the claimed advantages considered in this section; however, there is scientific evidence that reports the above-mentioned properties. Despite non-authorized claims for carotenoids, it is important to note that the condition of use 'carotenoids from fruits and vegetable juices' for '*carotenoids contained in this product ensure antioxidant action and protective effect on the organism*' was subject to the following conditions of use: 4 mg, and 30% of observed intakes per day (1.5 mg) [58].

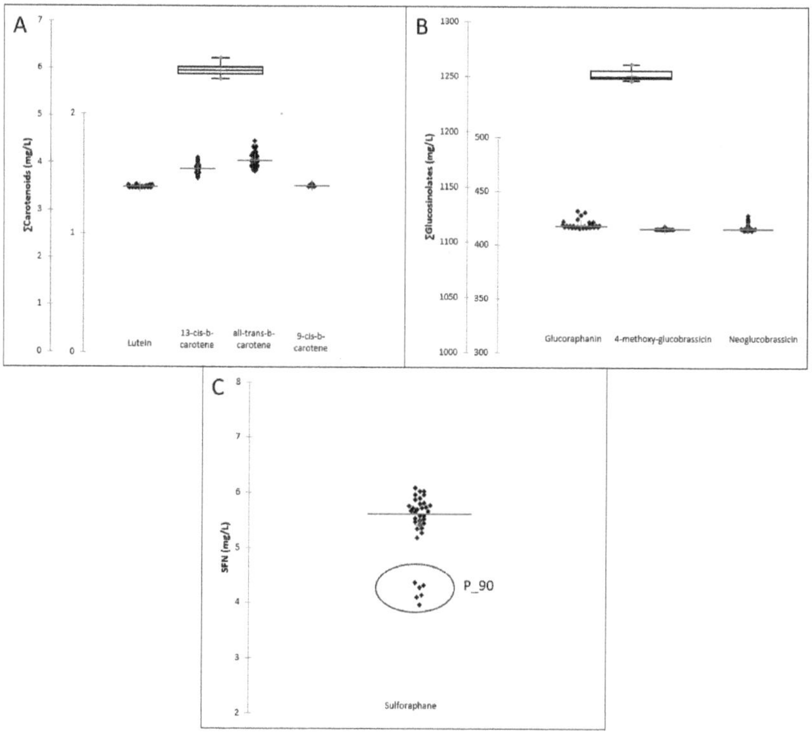

Figure 4. Box-plot and scatter-graphs of ∑Carotenoids and individuals carotenoids (**A**), ∑Glucosinolates and individuals glucosinolates (**B**), and sulforaphane (SFN) (**C**) of the studied treatments during the shelf life study.

Thermal processing conditions degrade sulfur and other non-sulfured compounds such as vitamins and phenolics. Previous reports indicated that temperature and exposure time should be kept as low as possible. Several studies concluded that steaming is the most efficient process to retain glucosinolates in cruciferous vegetables when compared with blanching, microwaving, and other processing technologies [59]. In our research, no differences were found between heated and non-heated samples because previous blanching with microwave technology was applied to the vegetal material; this treatment was aimed at inhibiting degrading enzymes [60–62]. It has been well documented in the scientific literature that isothiocyanates derived from glucosinolates have significant connections with immune health. Despite non-authorized claims for glucosinolates, it is convenient to highlight that the condition of using 'glucosinolates' for '*Foods containing glucosinolates help strengthen our body's defenses (glucosinolates as precursors to contribute to the proper functioning of the cells/support) a proper functioning immune system*' was subject to the following conditions: phytoconstituent's content in vegetables expressed in comparison with the daily needs and threshold for activity: up to 20 mg. Although the EFSA Panel considered that the claimed effects are general and non-specific (not sufficiently defined, and no further details were given in the proposed wording, the references, or the clarifications provided by Member States), the product developed in this study could help to increase the daily intake of glucosinolates. Around 20 mg of ∑glucosinolates should be within a 25 mL serving of this developed beverage [63].

On the other hand, despite non-authorized claims for sulforaphane (derived from glucosinolates), it is important to note that the unique condition of using 'sulforaphane' from 'broccoli sprouts can enhance antioxidant activity and boost the elimination of free radicals'. This was subject to the following conditions of use: *"Verify levels present in the Broccoli strain used by analysis. We need to confirm that the seed variant used to produce the product is one of the variants producing elevated levels of sulforaphane. Not all broccoli seeds do. To also refer to the consumption of broccoli sprouts in a balanced diet as part of the "Five a day" dietary recommendations"* [64].

The impact of pasteurization at 90 °C on the beverages resulted in a reduction of 35% in SFN compared to the CTRL and the rest of the treatments. This decrease could have been previously expected since myrosinase activity is inhibited at temperatures above 70 °C, leading to the inability to convert glucoraphanin into sulforaphane [65,66]. The SFN content in the beverages treated with green technologies remained the same as the untreated ones (CTRL). These observations are consistent with the findings of those who demonstrated that combining green technology with temperature (blanching at 60 °C for 4 min) resulted in the highest SFN content in broccoli florets, mainly due to the reduced myrosinase activity [67].

3.5. Sensory Analysis

As mentioned in Section 2.5, a value of 3 denotes the limit of consumption due to tolerable off-flavor and cooked flavor; this value is the midpoint of the scale selected for the descriptive evaluation. Undesirable sensory attributes (off-flavor, cooked flavor, and fermented flavor) were not found on processing day in the samples, except P_90 (1.5, 2.5, and 0, respectively). During the shelf life of CTRL and P_90 treatments, off-flavor made the rating values be above 3 from day 14 onwards. As to cooked flavor, P_90 reached 3 at 7 d at 4 °C, which was unacceptable on the following days. Fermented flavor at unacceptable values was found in CTRL samples at 14 d 4 °C. The remaining treatments did not have attribute values equal to or greater than three during their lifetime, which justified that their shelf life was greater than 28 days.

4. Conclusions

An innovative seasoned veggie beverage made of broccoli and carrot by-products has been developed within a circular economy scenario. Broccoli stalk by-products, as inedible parts usually wasted in the industry, have been revalorized and applied as sources of bioactive compounds to biofortify the beverage. In addition, due to the application of green processing treatments as HHP (single or combined with US), health-promoting compounds as carotenoids (~6 mg/L ΣCarotenoids) and sulfur compounds (~1.25 g/L ΣGlucosinolates and ~5.5 mg/L sulforaphane) have been preserved during a refrigerated shelf-life period of 4 weeks, in which the total microbial counts were reduced by ~1.5–2 log CFU/mL at day 0, and by ~6 log CFU/mL at the end of the storage thanks to these technologies applied in combination. Nevertheless, this technique must be optimized, finding the best conditions or combinations with other sustainable processing technologies to ensure food safety while preserving functional and sensory quality.

Supplementary Materials: The following supporting information can be downloaded at: https://www.mdpi.com/article/10.3390/foods12203808/s1, Table S1: L*, a*, b*, ΔE, and pH changes during 28 days at 4 °C in a broccoli and carrot by-products beverage according to several processing treaments; Table S2: Aerobic mesophilic bacteria, psychrophiles, mold & yeast, and enterobacteria (log CFU/mL) changes during 28 days at 4 °C in a broccoli and carrot by-products beverage according to several processing treaments; Table S3: Free polyphenolic content (mg GAE/L), FRAP, and DPPH (mg TE/L) changes during 28 days at 4 °C in a broccoli and carrot by-products beverage according to several processing treaments; Table S4: Range of individual identified carotenoids and glucosinolates content (mg/L) during 28 d refrigeration (from 0 to 28 d).

Author Contributions: Conceptualization, M.C.-L., L.M.-Z. and F.A.-H.; methodology, P.P. and S.H. software, P.P. and S.H.; validation, M.C.-L. and P.A.G.; formal analysis. P.P. and M.C.-L.; investigation P.P., M.C.-L., L.M.-Z. and P.A.G.; resources, F.A.-H. and P.A.G.; data curation, M.C.-L. and L.M.-Z.; writing—original draft preparation, M.C.-L., P.P. and L.M.-Z.; writing—review and editing P.A.G. and F.A.-H.; visualization, M.C.-L.; supervision, F.A.-H. and P.A.G., project administration F.A.-H.; funding acquisition, F.A.-H. All authors have read and agreed to the published version of the manuscript.

Funding: Project PID2021-123857OB-I00 is financed by the Spanish Ministry of Science and Innovation (MCIN), the Spanish State Research Agency/10.13039/501100011033, and FEDER. This work is also a result of the AGROALNEXT program and was supported by MCIN with funding from EU NextGeneration (PRTR-C17.I1) and by the Seneca Foundation with funding from the Autonomous Community of the Region of Murcia (CARM).

Data Availability Statement: The data used to support the findings of this study can be made available by the corresponding author upon request.

Acknowledgments: The L.M.-Z. contract has been financed by the Program for the Re-qualification of the Spanish University System, funded by the EU–NextGeneration, Margarita Salas modalityby the University of Murcia. The M.C.-L. contract has been co-financed by Juan de la Cierva-Formación (FJC2020-043764-I) from the Spanish Ministry of Education (MCIN/AEI/10.13039/501100011033) and EU–NextGeneration. P.P. mobility to perform this experiment at the Universidad Politécnica de Cartagena (Spain) was partially financed by the AUIP (Asociación Universitaria Iberoamericana de Postgrado) International Program of Academic Mobility.

Conflicts of Interest: The authors declare no conflict of interest.

References

1. Arjmandi, M.; Otón, M.; Artés, F.; Artés-Hernández, F.; Gómez, P.A.; Aguayo, E. Continuous Microwave Pasteurization of a Vegetable Smoothie Improves Its Physical Quality and Hinders Detrimental Enzyme Activity. *Food Sci. Technol. Int.* **2016**, *23*, 36–45. [CrossRef] [PubMed]
2. Klug, T.V.; Martínez-Sánchez, A.; Gómez, P.A.; Collado, E.; Aguayo, E.; Artés, F.; Artés-Hernández, F. Improving Quality of an Innovative Pea Puree by High Hydrostatic Pressure. *J. Sci. Food Agric.* **2017**, *97*, 4362–4369. [CrossRef] [PubMed]
3. McCrory, M.A.; Hamaker, B.R.; Lovejoy, J.C.; Eichelsdoerfer, P.E. Pulse Consumption, Satiety, and Weight Management. *Adv. Nutr.* **2010**, *1*, 17–30. [CrossRef] [PubMed]
4. Castillejo, N.; Martínez-Hernández, G.B.; Artés-Hernández, F. Revalorized Broccoli By-Products and Mustard Improved Quality during Shelf Life of a Kale Pesto Sauce. *Food Sci. Technol. Int.* **2021**, *27*, 734–745. [CrossRef] [PubMed]
5. Artés-Hernández, F.; Martínez-Zamora, L.; Cano-Lamadrid, M.; Hashemi, S.; Castillejo, N. Genus Brassica By-Products Revalorization with Green Technologies to Fortify Innovative Foods: A Scoping Review. *Foods* **2023**, *12*, 561. [CrossRef] [PubMed]
6. Campas-Baypoli, O.N.; Sánchez-Machado, D.I.; Bueno-Solano, C.; Núñez-Gastélum, J.A.; Reyes-Moreno, C.; López-Cervantes, J. Biochemical Composition and Physicochemical Properties of Broccoli Flours. *Int. J. Food Sci. Nutr.* **2009**, *60*, 163–173. [CrossRef] [PubMed]
7. Krupa-Kozak, U.; Drabińska, N.; Rosell, C.M.; Fadda, C.; Anders, A.; Jeliński, T.; Ostaszyk, A. Broccoli Leaf Powder as an Attractive By-Product Ingredient: Effect on Batter Behaviour, Technological Properties and Sensory Quality of Gluten-Free Mini Sponge Cake. *Int. J. Food Sci. Technol.* **2019**, *54*, 1121–1129. [CrossRef]
8. Domínguez-Perles, R.; Martínez-Ballesta, M.C.; Carvajal, M.; García-Viguera, C.; Moreno, D.A. Broccoli-Derived By-Products—A Promising Source of Bioactive Ingredients. *J. Food Sci.* **2010**, *75*, C383–C392. [CrossRef]
9. Vanduchova, A.; Anzenbacher, P.; Anzenbacherova, E. Isothiocyanate from Broccoli, Sulforaphane, and Its Properties. *J. Med. Food* **2019**, *22*, 121–126. [CrossRef]
10. Ho, E.; Clarke, J.D.; Dashwood, R.H. Dietary Sulforaphane, a Histone Deacetylase Inhibitor for Cancer Prevention. *J. Nutr.* **2009**, *139*, 2393–2396. [CrossRef]
11. Clarke, J.D.; Dashwood, R.H.; Ho, E. Multi-Targeted Prevention of Cancer by Sulforaphane. *Cancer Lett.* **2008**, *269*, 291–304. [CrossRef] [PubMed]
12. Arscott, S.A.; Tanumihardjo, S.A. Carrots of Many Colors Provide Basic Nutrition and Bioavailable Phytochemicals Acting as a Functional Food. *Compr. Rev. Food Sci. Food Saf.* **2010**, *9*, 223–239. [CrossRef]
13. Bevilacqua, A.; Petruzzi, L.; Perricone, M.; Speranza, B.; Campaniello, D.; Sinigaglia, M.; Corbo, M.R. Nonthermal Technologies for Fruit and Vegetable Juices and Beverages: Overview and Advances. *Compr. Rev. Food Sci. Food Saf.* **2018**, *17*, 2–62. [CrossRef] [PubMed]

14. Nieva, S.G.; Jagus, R.J.; Agüero, M.V.; Fernandez, M. V Fruit and Vegetable Smoothies Preservation with Natural Antimicrobials for the Assurance of Safety and Quality. *LWT* **2022**, *154*, 112663. [CrossRef]
15. Salar, F.J.; Periago, P.M.; Agulló, V.; García-Viguera, C.; Fernández, P.S. High Hydrostatic Pressure vs. Thermal Pasteurization: The Effect on the Bioactive Compound Profile of a Citrus Maqui Beverage. *Foods* **2021**, *10*, 2416. [CrossRef] [PubMed]
16. de Souza, V.R.; Popović, V.; Bissonnette, S.; Ros, I.; Mats, L.; Duizer, L.; Warriner, K.; Koutchma, T. Quality Changes in Cold Pressed Juices after Processing by High Hydrostatic Pressure, Ultraviolet-c Light and Thermal Treatment at Commercial Regimes. *Innov. Food Sci. Emerg. Technol.* **2020**, *64*, 102398. [CrossRef]
17. Gupta, R.; Balasubramaniam, V.M. Chapter 5—High-Pressure Processing of Fluid Foods. In *Novel Thermal and Non-Thermal Technologies for Fluid Foods*; Cullen, P.J., Tiwari, B.K., Valdramidis, V.P., Eds.; Academic Press: San Diego, CA, USA, 2012; pp. 109–133. ISBN 978-0-12-381470-8.
18. Koutchma, T. Advances in Ultraviolet Light Technology for Non-Thermal Processing of Liquid Foods. *Food Bioprocess Technol.* **2009**, *2*, 138–155. [CrossRef]
19. Chiozzi, V.; Agriopoulou, S.; Varzakas, T. Advances, Applications, and Comparison of Thermal (Pasteurization, Sterilization, and Aseptic Packaging) against Non-Thermal (Ultrasounds, UV Radiation, Ozonation, High Hydrostatic Pressure) Technologies in Food Processing. *Appl. Sci.* **2022**, *12*, 2202. [CrossRef]
20. Cano-Lamadrid, M.; Artés-Hernández, F. By-Products Revalorization with Non-Thermal Treatments to Enhance Phytochemical Compounds of Fruit and Vegetables Derived Products: A Review. *Foods* **2022**, *11*, 59. [CrossRef]
21. Matusheski, N.V.; Juvik, J.A.; Jeffery, E.H. Heating Decreases Epithiospecifier Protein Activity and Increases Sulforaphane Formation in Broccoli. *Phytochemistry* **2004**, *65*, 1273–1281. [CrossRef]
22. Song, Y.; Bi, X.; Zhou, M.; Zhou, Z.; Chen, L.; Wang, X.; Ma, Y. Effect of Combined Treatments of Ultrasound and High Hydrostatic Pressure Processing on the Physicochemical Properties, Microbial Quality and Shelf-Life of Cold Brew Tea. *Int. J. Food Sci. Technol.* **2021**, *56*, 5977–5988. [CrossRef]
23. Alvarez-Jubete, L.; Valverde, J.; Kehoe, K.; Reilly, K.; Rai, D.K.; Barry-Ryan, C. Development of a Novel Functional Soup Rich in Bioactive Sulforaphane Using Broccoli (*Brassica Oleracea* L. Ssp. Italica) Florets and Byproducts. *Food Bioprocess Technol.* **2014**, *7*, 1310–1321. [CrossRef]
24. Klug, T.V.; Martínez-Hernández, G.B.; Collado, E.; Artés, F.; Artés-Hernández, F. Effect of Microwave and High-Pressure Processing on Quality of an Innovative Broccoli Hummus. *Food Bioprocess Technol.* **2018**, *11*, 1464–1477. [CrossRef]
25. Castillejo, N.; Martínez-Hernández, G.B.; Gómez, P.A.; Artés, F.; Artés-Hernández, F. Red Fresh Vegetables Smoothies with Extended Shelf Life as an Innovative Source of Health-Promoting Compounds. *J. Food Sci. Technol.* **2016**, *53*, 1475–1486. [CrossRef] [PubMed]
26. Shen, Y.; Zhu, D.; Xi, P.; Cai, T.; Cao, X.; Liu, H.; Li, J. Effects of Temperature-Controlled Ultrasound Treatment on Sensory Properties, Physical Characteristics and Antioxidant Activity of Cloudy Apple Juice. *LWT* **2021**, *142*, 111030. [CrossRef]
27. Dereli, U.; Türkyilmaz, M.; Yemiş, O.; Özkan, M. Effects of Clarification and Pasteurization on the Phenolics, Antioxidant Capacity, Color Density and Polymeric Color of Black Carrot (*Daucus carota* L.) Juice. *J. Food Biochem.* **2015**, *39*, 528–537. [CrossRef]
28. Petruzzi, L.; Campaniello, D.; Speranza, B.; Corbo, M.R.; Sinigaglia, M.; Bevilacqua, A. Thermal Treatments for Fruit and Vegetable Juices and Beverages: A Literature Overview. *Compr. Rev. Food Sci. Food Saf.* **2017**, *16*, 668–691. [CrossRef]
29. Sattar, S.; Imran, M.; Mushtaq, Z.; Ahmad, M.H.; Arshad, M.S.; Holmes, M.; Maycock, J.; Nisar, M.F.; Khan, M.K. Retention and Stability of Bioactive Compounds in Functional Peach Beverage Using Pasteurization, Microwave and Ultrasound Technologies. *Food Sci. Biotechnol.* **2020**, *29*, 1381–1388. [CrossRef]
30. Mandelová, L.; Totušek, J. Broccoli Juice Treated by High Pressure: Chemoprotective Effects of Sulforaphane and Indole-3-Carbinol. *High. Press. Res.* **2007**, *27*, 151–156. [CrossRef]
31. Gomes, W.F.; Tiwari, B.K.; Rodriguez, Ó.; de Brito, E.S.; Fernandes, F.A.N.; Rodrigues, S. Effect of Ultrasound Followed by High Pressure Processing on Prebiotic Cranberry Juice. *Food Chem.* **2017**, *218*, 261–268. [CrossRef]
32. Ferrario, M.; Alzamora, S.M.; Guerrero, S. Study of the Inactivation of Spoilage Microorganisms in Apple Juice by Pulsed Light and Ultrasound. *Food Microbiol.* **2015**, *46*, 635–642. [CrossRef] [PubMed]
33. Singleton, V.L.; Orthofer, R.; Lamuela-Raventós, R.M. Analysis of Total Phenols and Other Oxidation Substrates and Antioxidants by Means of Folin-Ciocalteu Reagent. In *Oxidants and Antioxidants Part A*; Methods in Enzymology; Academic Press: Cambridge, MA, USA, 1999; Volume 299, pp. 152–178.
34. Benzie, I.F.F.; Strain, J.J. Ferric Reducing/Antioxidant Power Assay: Direct Measure of Total Antioxidant Activity of Biological Fluids and Modified Version for Simultaneous Measurement of Total Antioxidant Power and Ascorbic Acid Concentration. In *Oxidants and Antioxidants Part A*; Methods in Enzymology; Academic Press: Cambridge, MA, USA, 1999; Volume 299, pp. 15–27.
35. Martínez-Zamora, L.; Castillejo, N.; Artés-Hernández, F. Postharvest UV-B and Photoperiod with Blue + Red LEDs as Strategies to Stimulate Carotenogenesis in Bell Peppers. *Appl. Sci.* **2021**, *11*, 3736. [CrossRef]
36. Gupta, P.; Sreelakshmi, Y.; Sharma, R. A Rapid and Sensitive Method for Determination of Carotenoids in Plant Tissues by High Performance Liquid Chromatography. *Plant Methods* **2015**, *11*, 5. [CrossRef] [PubMed]
37. Martínez-Zamora, L.; Castillejo, N.; Artés-Hernández, F. Ultrasounds and a Postharvest Photoperiod to Enhance the Synthesis of Sulforaphane and Antioxidants in Rocket Sprouts. *Antioxidants* **2022**, *11*, 1490. [CrossRef] [PubMed]

38. Kiddle, G.; Bennett, R.N.; Botting, N.P.; Davidson, N.E.; Robertson, A.A.B.; Wallsgrove, R.M. High-Performance Liquid Chromatographic Separation of Natural and Synthetic Desulphoglucosinolates and Their Chemical Vadation by UV, NMR and Chemical Ionisation-MS Methods. *Phytochem. Anal.* **2001**, *12*, 226–242. [CrossRef] [PubMed]
39. *ISO-8586:2012*; Sensory Analysis—General Guidelines for the Selection, Training and Monitoring of Selected Assessors and Expert Sensory Assessors. International Organization for Standarization: Geneva, Switzerland, 2012.
40. Cano-Lamadrid, M.; Tkacz, K.; Turkiewicz, I.P.; Clemente-Villalba, J.; Sánchez-Rodríguez, L.; Lipan, L.; García-García, E.; Carbonell-Barrachina, Á.A.; Wojdyło, A. How a Spanish Group of Millennial Generation Perceives the Commercial Novel Smoothies? *Foods* **2020**, *9*, 1213. [CrossRef]
41. Cano-Lamadrid, M.; Martínez-Zamora, L.; Castillejo, N.; Cattaneo, C.; Pagliarini, E.; Artés-Hernández, F. How Does the Phytochemical Composition of Sprouts and Microgreens from Brassica Vegetables Affect the Sensory Profile and Consumer Acceptability? *Postharvest Biol. Technol.* **2023**, *203*, 112411. [CrossRef]
42. Issa-Issa, H.; Cano-Lamadrid, M.; Calín-Sánchez, Á.; Wojdyło, A.; Carbonell-Barrachina, Á.A. Volatile Composition and Sensory Attributes of Smoothies Based on Pomegranate Juice and Mediterranean Fruit Purées (Fig, Jujube and Quince). *Foods* **2020**, *9*, 926. [CrossRef]
43. Mokrzycki, W.S.; Tatol, M. Colour Difference ΔE—A Survey. *MGV* **2011**, *20*, 383–411.
44. Izadifar, Z.; Babyn, P.; Chapman, D. Ultrasound Cavitation/Microbubble Detection and Medical Applications. *J. Med. Biol. Eng.* **2019**, *39*, 259–276. [CrossRef]
45. Ahmad, K.; Imran, M.; Ahmad, T.; Ahmad, M.H.; Khan, M.K. Impact of Ultrasound Processing on Physicochemical and Bioactive Attributes of Grape Based Optimized Fruit Beverage. *Pak. J. Agric. Sci.* **2020**, *57*, 1117–1124.
46. Barth, M.; Hankinson, T.R.; Zhuang, H.; Breidt, F. Microbiological Spoilage of Fruits and Vegetables. In *Compendium of the Microbiological Spoilage of Foods and Beverages*; Springer: New York, NY, USA, 2009; pp. 135–183.
47. Barba, F.J.; Esteve, M.J.; Frigola, A. Ascorbic Acid Is the Only Bioactive That Is Better Preserved by High Hydrostatic Pressure than by Thermal Treatment of a Vegetable Beverage. *J. Agric. Food Chem.* **2010**, *58*, 10070–10075. [CrossRef] [PubMed]
48. Kapoor, S.; Aggarwal, P. Effect of Processing and Storage on Bioactive Compounds and Antioxidant Activity of Carrot Juice. *J. Appl. Hortic.* **2014**, *16*, 80–84. [CrossRef]
49. Ibrahim, H.M.; Abou-Arab, A.A.; Abu Salem, F.M. Antioxidant and Antimicrobial Effects of Some Natural Plant Extracts Added to Lamb Patties during Storage. *Grasas Y Aceites* **2011**, *62*, 139–148. [CrossRef]
50. de Albuquerque, J.G.; Escalona-Buendía, H.B.; de Magalhães Cordeiro, A.M.T.; dos Santos Lima, M.; de Souza Aquino, J.; da Silva Vasconcelos, M.A. Ultrasound Treatment for Improving the Bioactive Compounds and Quality Properties of a Brazilian Nopal (Opuntia Ficus-Indica) Beverage during Shelf-Life. *LWT* **2021**, *149*, 111814. [CrossRef]
51. Kedare, S.B.; Singh, R.P. Genesis and Development of DPPH Method of Antioxidant Assay. *J. Food Sci. Technol.* **2011**, *48*, 412–422. [CrossRef] [PubMed]
52. Müller, L.; Fröhlich, K.; Böhm, V. Comparative Antioxidant Activities of Carotenoids Measured by Ferric Reducing Antioxidant Power (FRAP), ABTS Bleaching Assay (ATEAC), DPPH Assay and Peroxyl Radical Scavenging Assay. *Food Chem.* **2011**, *129*, 139–148. [CrossRef]
53. Kim, J.H.; Lim, Y.J.; Duan, S.; Park, T.J.; Eom, S.H. Accumulation of Antioxidative Phenolics and Carotenoids Using Thermal Processing in Different Stages of Momordica Charantia Fruit. *Molecules* **2023**, *28*, 1500. [CrossRef]
54. Sánchez-Moreno, C.; Plaza, L.; Elez-Martínez, P.; De Ancos, B.; Martín-Belloso, O.; Cano, M.P. Impact of High Pressure and Pulsed Electric Fields on Bioactive Compounds and Antioxidant Activity of Orange Juice in Comparison with Traditional Thermal Processing. *J. Agric. Food Chem.* **2005**, *53*, 4403–4409. [CrossRef]
55. Sánchez-Moreno, C.; Plaza, L.; de Ancos, B.; Cano, M.P. Nutritional Characterisation of Commercial Traditional Pasteurised Tomato Juices: Carotenoids, Vitamin C and Radical-Scavenging Capacity. *Food Chem.* **2006**, *98*, 749–756. [CrossRef]
56. Dallagi, W.; Rguez, S.; Hammami, M.; Bettaieb Rebey, I.; Bourgou, S.; Hamrouni Sellami, I. Optimization of Processing Conditions to Enhance Antioxidant and Carotenoid Contents of Carrot Juice. *J. Food Meas. Charact.* **2023**, *17*, 4384–4393. [CrossRef]
57. Meléndez-Martínez, A.J.; Mandić, A.I.; Bantis, F.; Böhm, V.; Borge, G.I.A.; Brnčić, M.; Bysted, A.; Cano, M.P.; Dias, M.G.; Elgersma, A.; et al. A Comprehensive Review on Carotenoids in Foods and Feeds: Status Quo, Applications, Patents, and Research Needs. *Crit. Rev. Food Sci. Nutr.* **2022**, *62*, 1999–2049. [CrossRef] [PubMed]
58. EFSA Panel on Dietetic Products and Nutrition and Allergies (NDA). Scientific Opinion on the Substantiation of Health Claims Related to: Flavonoids and Ascorbic Acid in Fruit Juices, Including Berry Juices; Flavonoids from Citrus; Flavonoids from Citrus Paradisi Macfad.; Flavonoids; Flavonoids in Cranerry Juice; Carotenoids; Polyphenols; among Others Pursuant to Article 13(1) of Regulation (EC) No 1924/2006. *EFSA J.* **2011**, *9*, 2082. [CrossRef]
59. Lafarga, T.; Bobo, G.; Viñas, I.; Collazo, C.; Aguiló-Aguayo, I. Effects of Thermal and Non-Thermal Processing of Cruciferous Vegetables on Glucosinolates and Its Derived Forms. *J. Food Sci. Technol.* **2018**, *55*, 1973–1981. [CrossRef] [PubMed]
60. Ramesh, M.N.; Wolf, W.; Tevini, D.; Bognár, A. Microwave Blanching of Vegetables. *J. Food Sci.* **2002**, *67*, 390–398. [CrossRef]
61. Parchem, K.; Piekarska, A.; Bartoszek, A. Enzymatic Activities behind Degradation of Glucosinolates. *Glucosinolates Prop. Recovery Appl.* **2020**, *2020*, 79–106. [CrossRef]
62. Halkier, B.A.; Gershenzon, J. Biology and Biochemistry of Glucosinolates. *Annu. Rev. Plant Biol.* **2006**, *57*, 303–333. [CrossRef] [PubMed]

53. EFSA Panel on Dietetic Products and Nutrition and Allergies (NDA). Scientific Opinion on the Substantiation of Health Claims Related to Various Foods/Food Constituents and "Immune Function/Immune System", "Contribution to Body Defences against External Agents", Stimulation of Immunological Responses, Reduction of Inflammation, among Others Pursuant to Article 13(1) of Regulation (EC) No 1924/2006. *EFSA J.* **2011**, *9*, 2061. [CrossRef]
54. EFSA Panel on Dietetic Products and Nutrition and Allergies (NDA). Scientific Opinion on the Substantiation of Health Claims Related to Various Food(s)/Food Constituent(s) and Protection of Cells from Premature Aging, Antioxidant Activity, Antioxidant Content and Antioxidant Properties, and Protection of DNA, Proteins and Lipids from Oxidative Damage Pursuant to Article 13(1) of Regulation (EC) No 1924/2006. *EFSA J.* **2010**, *8*, 1489. [CrossRef]
55. Pongmalai, P.; Devahastin, S.; Chiewchan, N.; Soponronnarit, S. Enhancing the Recovery of Cabbage Glucoraphanin through the Monitoring of Sulforaphane Content and Myrosinase Activity during Extraction by Different Methods. *Sep. Purif. Technol.* **2017**, *174*, 338–344. [CrossRef]
56. Sangkret, S.; Pongmalai, P.; Devahastin, S.; Chiewchan, N. Enhanced Production of Sulforaphane by Exogenous Glucoraphanin Hydrolysis Catalyzed by Myrosinase Extracted from Chinese Flowering Cabbage (*Brassica Rapa* Var. *Parachinensis*). *Sci. Rep.* **2019**, *9*, 9882. [CrossRef]
57. Mahn, A.; Quintero, J.; Castillo, N.; Comett, R. Effect of Ultrasound-Assisted Blanching on Myrosinase Activity and Sulforaphane Content in Broccoli Florets. *Catalysts* **2020**, *10*, 616. [CrossRef]

foods

MDPI

Article

Polymeric Compounds of Lingonberry Waste: Characterization of Antioxidant and Hypolipidemic Polysaccharides and Polyphenol-Polysaccharide Conjugates from *Vaccinium vitis-idaea* Press Cake

Daniil N. Olennikov [1,*], Vladimir V. Chemposov [2] and Nadezhda K. Chirikova [2]

1 Laboratory of Medical and Biological Research, Institute of General and Experimental Biology, Siberian Division, Russian Academy of Science, 6 Sakhyanovoy Street, 670047 Ulan-Ude, Russia
2 Department of Biology, Institute of Natural Sciences, North-Eastern Federal University, 58 Belinsky Street, 677027 Yakutsk, Russia
* Correspondence: olennikovdn@mail.ru; Tel.: +8-902-160-06-27

Citation: Olennikov, D.N.; Chemposov, V.V.; Chirikova, N.K. Polymeric Compounds of Lingonberry Waste: Characterization of Antioxidant and Hypolipidemic Polysaccharides and Polyphenol-Polysaccharide Conjugates from *Vaccinium vitis-idaea* Press Cake. *Foods* **2022**, *11*, 2801. https://doi.org/10.3390/foods11182801

Academic Editors: Noelia Castillejo Montoya and Lorena Martínez-Zamora

Received: 15 August 2022
Accepted: 6 September 2022
Published: 11 September 2022

Abstract: Lingonberry (*Vaccinium vitis-idaea* L.) fruits are important Ericaceous berries to include in a healthy diet of the Northern Hemisphere as a source of bioactive phenolics. The waste generated by the *V. vitis-idaea* processing industry is hard-skinned press cake that can be a potential source of dietary fiber and has not been studied thus far. In this study, water-soluble polysaccharides of *V. vitis-idaea* press cake were isolated, separated, and purified by ion-exchange and size-exclusion chromatography. The results of elemental composition, monosaccharide analysis, ultraviolet–visible and Fourier-transform infrared spectroscopy, molecular weight determination, linkage analysis, and alkaline destruction allowed us to characterize two polyphenol–polysaccharide conjugates (PPC) as neutral arabinogalactans cross-linked with monomeric and dimeric hydroxycinnamate residues with molecular weights of 108 and 157 kDa and two non-esterified galacturonans with molecular weights of 258 and 318 kDa. A combination of in vitro and in vivo assays confirmed that expressed antioxidant activity of PPC was due to phenolic-scavenged free radicals, nitrogen oxide, hydrogen peroxide, and chelate ferrous ions. Additionally, marked hypolipidemic potential of both PPC and acidic polymers bind bile acids, cholesterol, and fat, inhibit pancreatic lipase in the in vitro study, reduce body weight, serum level of cholesterol, triglycerides, low/high-density lipoprotein–cholesterol, and malondialdehyde, and increase the enzymatic activity of superoxide dismutase, glutathione peroxidase, and catalase in the livers of hamsters with a 1% cholesterol diet. Polysaccharides and PPC of *V. vitis-idaea* fruit press cake can be regarded as new antioxidants and hypolipidemic agents that can be potentially used to cure hyperlipidemic metabolic disorders.

Keywords: *Vaccinium vitis-idaea*; lingonberry; polysaccharides; polyphenol–polysaccharide conjugates; arabinogalactan; galacturonan; antioxidant activity; hypolipidemic activity

1. Introduction

The food industry produces more than one billion tons of waste per year, and the largest part is secondary products generated after mechanic procedures of cleaning, peeling, and pressing of fruits, vegetables, and berries [1]. Recycling of agro-food wastes is the focus of scientists who are interested in eco-friendly technologies for waste processing, which is an important problem of modern life [2]. The problem of fruit processing is of interest to many scientists, but insufficient attention is paid to berry industry waste [3]; at best, this type of waste is composted, and at worst, it is simply thrown away [4]. Among the processed berries, the least attention is paid to hard-skinned berries derived from *Vaccinium* and *Ribes* species, which is explained by the impossibility of its transformation into a puree or a homogeneous mass, despite the use of heat treatment methods [5].

Lingonberry (*Vaccinium vitis-idaea* L.) is one of the most common evergreen shrubs of the Ericaceous family in the Northern Hemisphere; it grows in green moss dry coniferous and mixed forests in the mountains, where it rises to the subalpine belt, and in the North, where it reaches subarctic light forests [6]. The lingonberry produces hard-skinned fruits that ripen in August–September as red multi-seeded spherical shiny berries, bearing a dried calyx at the top with a sweet and sour taste. After the first frost, they become soft and watery, and they can winter under snow until spring, but in spring, the berries fall off at the slightest touch. The berries are the reason for the human commercial interest in this plant, especially in the Siberian region, which offers a vast amount of land to allow the collection of more than 50,000 tons of possible annual fruit volume [7].

The focus of lingonberry processing on an industrial scale involves high-waste production as a juice, syrup, and jam, and the low-waste production as fruit drying and freezing [8]. The main residue of the high-waste production is a berry press cake, which may be more than half of the initial fruit weight, and most of the fruit processing factories are faced with some waste problems. A possible way to use the capacity of the secondary raw material is isolation of valuable compounds and the development of new products [9]. Isolation of phenolic-rich by-products is a good example of agri-food waste utilization for grapes [10], pomegranates [11], bananas [12], olives [13], oranges [14], and apples [15], which are rich in flavonoids, procyanidins, catechins, hydroxycinnamates, and chalcones [16]. The pomace of chokeberry, blueberry [17], strawberry [18], blackberry [19], and currants [20] is a good source of phenolic antioxidants and dietary fibers that can be used as bioactive products and food additives [21]. In addition to phenolics, agro-industrial fruit waste is a source of polymeric compounds as polysaccharides and polyphenol–polysaccharide conjugates with antioxidant, immunomodulatory, antiobesity and other bioactivities [22].

The phenolics of lingonberry pomace were previously characterized when anthocyanins, catechins, flavonols, and phenolic acids showed antioxidant activity in $ABTS^{\bullet+}$ scavenging capacity and ORAC assays [23], and the extract of lingonberry pomace showed anti-inflammatory properties by inhibition of the NF-κB pathway in LPS-stimulated THP-1 monocytes and COX-2 activity [24]. In the context of deep processing of lingonberry pomace, of special interest is the finding that unprocessed lingonberry pomace has hypolipidemic effects in in vitro assays [21], which may be explained by the high level of dietary fibers or their specific structure. The pomaces of other *Vaccinium* (e.g., blueberries and cranberries) showed in vitro binding of bile acids, which is caused, according to the authors, by the high fiber content [25]. Although there is considerable chemical information about metabolites of *V. vitis-idaea* for which flavonoids, catechins, procyanidins, simple phenolics, and anthocyanins are prevalent [26–28], to date, there are no scientific data on lingonberry fibers or polysaccharides. Only subspecies *V. vitis-idaea* ssp. *minus* was studied and showed the presence of neutral and acidic polysaccharides in the water-soluble fraction of polymers with antioxidant and α-glucosidase inhibitory activity [29]. Polysaccharides of other berries, such as goji (*Lycium barbarum*) [30], highland blackberry (*Rubus adenotrichos*) [31], and Cherokee rose (*Rosa laevigata*) [32] have proven their hypolipidemic potential; thus, the wastes of lingonberry processing industries might be a possible source of bioactive polymeric compounds.

As part of the ongoing study of Siberian lingonberry [27,28], the polymeric compounds (polysaccharides and polyphenol–polysaccharide conjugates) from *V. vitis-idaea* press cake produced after juice pressing were isolated, purified, characterized, and its antioxidant and hypolipidemic potential was studied in in vitro and in vivo experiments.

2. Materials and Methods

2.1. Plant Material and Chemicals

Samples of *V. vitis-idaea* press cake obtained after the juice pressing (moisture content of 88%) were purchased from Yagody Yakutii, Ltd. (Yakutsk, Russia). The plant material was dried in the ventilated heat oven at 40 °C within 7–10 days and stored at 3–4 °C before extraction. The reference standards were purchased from Sigma-Aldrich (St. Louis, MO,

USA): polysaccharides-pectin from citrus peel (galacturonic acid ≥ 74.0%; No. P9135), arabinogalactan from larch wood (No. 10830), starch from potato (No. 03967), and microcrystalline cellulose (No. Y0002021); hydroxycinnamates—ferulic acid (≥99%; No. 128708) and sinapic acid (≥98%; No. D7927); bioactivity standards—Trolox (≥97%; No. 238313), cholestyramine (No. 1133004), and simvastatin (≥97%; No. S6196).

2.2. Polysaccharide Fraction of V. vitis-idaea Press Cake (VVPS) Isolation

Dry and milled *V. vitis-idaea* press cake (55 kg) was extracted by hot water (90 °C, press cake:water ratio of 1:12; 10 h) in an industrial extractor (1000 L) coupled with an agitator, filtration unit, vacuum evaporator, condenser, and DNT-1.0 automatic control unit (SBN-Impex, Moscow, Russia) (Figure 1).

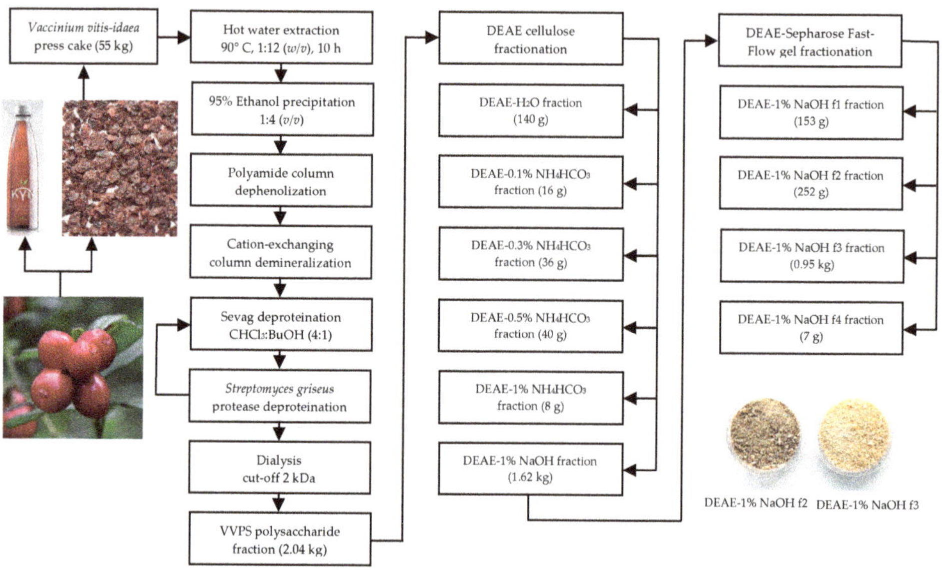

Figure 1. Flow chart for the extraction and fractionation of polysaccharides from *Vaccinium vitis-idaea* press cake. Abbreviation used: VVPS, *V. vitis-idaea* total polysaccharide fraction; DEAE, diethylaminoethyl cellulose; DEAE–sepharose fast-flow gel, diethylaminoethyl cellulose sepharose fast-flow gel.

The water extract was filtered through industrial filter CUNO™ Self Cleaning Metal Edge Filter Cartridge (200 µm; 3M Purification Inc., Meriden, CT, USA) and concentrated 20 times *in vacuo*, and the residue was mixed with 95% ethanol (1:4). The mixture was left for 24 h, and the crude precipitate was filtered through the cellulose membrane followed by drying, at 50 °C. The yield of crude precipitate was 3.2 kg (5.8% of press cake dry weight). Dry crude precipitate (3 kg) was suspended in 60 L of distilled water, the temperature of the mixture maintained at 60 °C for 1 h, and the resulting solution was filtered through two sequential columns with polyamide (10 kg; Sigma-Aldrich, cat. No. 02395) and a cation-exchanging column (KU-2-8, H^+-form; Eco-Vita, St. Petersburg, Russia; 20 kg) eluted with 100 L of distilled water (dephenolization and demineralization step). The water eluate was reduced in vacuum, at 30 °C, to 3 L, and the residue was vigorously mixed with a chloroform-*n*-butanol mixture (4:1) in the ratio of 1:1 and centrifuged (6000 rpm, 30 min) (Sevag deproteination step 1). Organic solvents were removed under vacuum, and the water residue was incubated with protease from *Streptomyces griseus* (type XIV, ≥3.5 units/mg; 1 unit per 1 mL of polysaccharide solution; Sigma-Aldrich, cat. No. P5147), at 35 °C, for 3 days (pronase deproteination step 2), and the Sevag deproteination step was repeated. The final solution was dialyzed in benzoylated dialysis tubes (cut-off of

2 kDa; Sigma-Aldrich, cat. No. D2272) against distilled water (48 h), and the non-dialyzed residue was freeze-dried to give *V. vitis-idaea* total polysaccharide fraction (VVPS) as a light-brownish powder in a yield of 2.04 kg (3.7% of press cake dry weight).

2.3. Diethylaminoethyl–Cellulose (DEAE–Cellulose) Fractionation of VVPS

The solution of VVPS (2 kg) in water (40 L) was passed through a DEAE–cellulose 52 column (10 kg; Sisco Research Laboratories Pvt. Ltd., New Delhi, India) in HCO_3^- form and eluted with water, NH_4HCO_3 solution (0.1%, 0.3%, 0.5%, 1%), and 1% NaOH until there was a negative reaction with phenol–sulfuric acid reagent. All eluates were dialyzed in benzoylated dialysis tubes (cut-off of 2 kDa; Sigma-Aldrich, cat. No. D2272) against distilled water (48 h), and non-dialyzed residues were freeze-dried. The yields of the DEAE–cellulose fractions were 140 g (DEAE–H_2O), 16 g (DEAE–0.1% NH_4HCO_3), 36 g (DEAE–0.3% NH_4HCO_3), 40 g (DEAE–0.5% NH_4HCO_3), 8 g (DEAE–1% NH_4HCO_3), and 1.62 kg (DEAE–1% NaOH).

2.4. DEAE–Sepharose Fast-Flow Gel Fractionation of the DEAE–1% NaOH Fraction

The solution of DEAE–1% NaOH (1.6 kg) in water (30 L) was passed through a DEAE–sepharose fast-flow gel column (2 L; GE Healthcare, Chicago, IL, USA) and eluted with water and NaCl solution (0.3%, 0.5%, 0.7%, 0.9%, 1.5%) for detecting the elution progress spectrophotometrically at 190 and 270 nm. All eluates were dialyzed in benzoylated dialysis tubes (cut-off of 2 kDa; Sigma-Aldrich, cat. No. D2272) against distilled water (48 h), and non-dialyzed residues were freeze-dried. The yields of the DEAE–sepharose fast-flow gel fractions were 153 g (DEAE–1% NaOH-f1; elution 0.5% NaCl), 252 g (DEAE–1% NaOH-f2; elution 0.7% NaCl), 0.95 kg (DEAE–1% NaOH-f3; elution 0.9% NaCl), and 7 g (DEAE–1% NaOH-f4; elution 1.5% NaCl).

2.5. Chemical Composition of VVPS and DEAE–Cellulose Fractions

Ready-to-use kits for spectrophotometric assays were applied to measure the total carbohydrate content (High Sensitivity Carbohydrate Assay Kit, BioVision, Inc., Milpitas, CA, USA; cat. No. K2049-100), uronic acids (D-Glucuronic/D-Galacturonic Acid Assay Kit, Megazyme, Bray, Ireland; cat. No. K-URONIC), starch (Total Starch Assay Kit, Megazyme; cat. No. K-TSTA-100A), protein (Pierce™ BCA Protein Assay Kit, Thermo Fisher Scientific, Waltham, MA, USA), and phenolics (Phenolic Compounds Assay Kit, Sigma-Aldrich; cat. MAK365). Arabinogalactan–protein complex content was estimated colorimetrically using Yariv reagent as described by Lamport et al. [33], and ash content was determined by the AOAC Official Method℠ 942.05 using muffle furnace ignition at 600 °C [34]. All analyses were performed five times, and the data were expressed as the mean value ± standard deviation (S.D.).

2.6. Elemental Composition

A 2400 Series II elemental analyzer (Perkin Elmer, Waltham, MA, USA) was used for analysis of carbon, hydrogen, oxygen, and nitrogen contents in the polysaccharides.

2.7. Monosaccharide Composition

The monosaccharide composition of polysaccharides was studied after trifluoroacetic acid (TFA) hydrolysis, followed by 1-phenyl-3-methyl-5-pyrazolone (PMP) labeling and HPLC with ultraviolet detection separation (HPLC–UV) as previously described with modifications [35]. Polysaccharide samples (10 mg) were subjected to the hydrolysis procedure with 1 mL of 2 M TFA using sealed ampoules incubated, at 120 °C, for 2 h. After cooling, the mixture was centrifuged (6000× *g*, 15 min) and evaporated in vacuo to remove TFA, and the residue was dissolved in 1 mL of distilled water. The hydrolyzed samples (60 µL) were mixed with 25 µL of 1.5 M NaOH (in water) and 80 µL of 0.5 M PMP (in methanol), incubated at 70 °C (2 h), cooled, and neutralized with 70 µL of 0.5 M HCl. The samples were purified by adding double chloroform (1 mL), followed by vigorous agitation (30 s) and centrifugation (3000× g, 10 min). Finally, the organic phase was removed, and the aqueous layer was analyzed by HPLC–UV. HPLC–UV separation of PMP-labeled sugars was performed using a MiLiChrom A-02 microcolumn chromatograph (Econova, Novosibirsk, Russia) coupled with a UV detector and microcolumn ProntoSIL-120-5-C18 AQ (75 mm × 1 mm × 1 μm; Metrohm AG, Herisau, Switzerland) eluted in gradient mode. A solution of 100 mM CH_3COONH_4 (pH 6.9) was eluent A, acetonitrile was eluent B, and the following gradient program was used: 0–20 min for 20–26% B. The parameters of column temperature, injection volume, and flow rate were 35 °C, 1 μL, and 150 μL/min, respectively. Chromatograms were recorded at 250 nm. Bidistilled water stock solutions (1 mg/mL) of reference monosaccharides of mannose, ribose, rhamnose, glucose, galactose, xylose, arabinose, fucose, galacturonic acid, and glucuronic acid were prepared and PMP-labeled in the same manner before analysis (Figure S1). The calibration curves were created by plotting the peak area vs. the concentration levels. All analyses were performed in triplicate.

2.8. Ultraviolet–Visible (UV–Vis) Spectroscopy

The UV–Vis spectra were obtained using a SF-2000 UV–Vis spectrophotometer (OKB Spectr, St. Petersburg, Russia) in the spectral range of 190–1000 nm using a quartz cell (10 mm). Polysaccharide solutions (5 mg/mL) were prepared using bidistilled water.

2.9. Fourier-Transform Infrared (FTIR) Spectroscopy

An FT-801 Fourier-transform infrared spectrometer (Simex, Novosibirsk, Russia; frequency 4000–600 cm^{-1}) with 200 scans and 2-cm^{-1} resolution was used to acquire FTIR spectra. Tablets with potassium bromide (spectroscopic grade) and sample in the ratio of 100:1 were produced by a GS15011 hydraulic press (Specac Ltd., Orpington, UK).

2.10. Molecular Weight Determination

A gel permeation–high performance liquid chromatography (GP–HPLC) procedure was used for the molecular weight determination. Experiments were performed on an LCMS 8050 liquid chromatograph coupled with a photodiode array detector (Shimadzu, Columbia, MD, USA) using a Shim-pack Diol-150 column (250 mm × 7.9 mm × 5 μm; Shimadzu) at a column temperature of 25 °C. The eluent was a 10 mM phosphate-buffered solution (pH 7.0). The injection volume was 1 μL, and the elution flow was 1 mL/min. Isocratic elution was applied, and chromatograms were integrated at 190 nm. A series of dextrans (10–410 kDa; Sigma-Aldrich) were used to create a calibration curve. The polysaccharide sample was dissolved in 10 mM phosphate-buffered solution (pH 7.0), centrifuged (6000× g), and filtered through a 0.22-μm PTFE syringe filter before injection into the HPLC system for analysis. All analyses were performed in duplicate.

2.11. Linkage Analysis

For linkage analysis, 10 mg of the polysaccharide was methylated by methyl iodide, followed by hydrolysis of the permethylated product using 90% formic acid and 2 M TFA, $NaBH_4$ reduction, and acetylation with acetic anhydride [36]. Partially methylated alditol acetates were analyzed by gas chromatography–mass spectrometry using a 5973N gas chromatograph mass spectrometer (Agilent Technologies, Santa-Clara, CA, USA) equipped with a 6890N mass selective detector, a diffusion pump, and an HP-Innowax capillary column (Agilent Technologies; 30 m × 250 μm × 0.50 μm) within a programmed temperature range of 150 to 250 °C, at a rate of 2 °C/min, with helium as the carrier gas (flow rate of 1 mL/min) [37]. The temperature of the transfer line and ion source was 280 °C. The sample injection volume was 1 μL with a split ratio of 50:1, and the scanning range was m/z 30–400. All analyses were performed in triplicate.

2.12. Alkaline Destruction and Analysis of Degradation Products by High-Performance Liquid Chromatography with Photodiode Array Detection and Electrospray Ionization Triple Quadrupole Mass Spectrometric Detection (HPLC–PDA–ESI–tQMS)

Alkaline hydrolysis of the polysaccharides was performed as previously described using 10% potassium hydroxide solution heating, 80% H_2SO_4 neutralization, and liquid–liquid extraction with ethyl acetate [38]. Degradation products were analyzed by high-performance liquid chromatography with photodiode array detection and electrospray ionization triple quadrupole mass spectrometric detection (HPLC–PDA–ESI–tQMS) using an LC-20 Prominence liquid chromatograph coupled with an SPD-M30A photodiode array detector (wavelength range 200–600 nm), and an LCMS 8050 triple quadrupole mass spectrometer (Shimadzu, Columbia, MD, USA) with two-eluent gradient elution [39]. The management of the LC–MS system was realized by LabSolution workstation software equipped with an internal LC–MS library. The final identification of metabolites was performed after an integrated analysis of retention time, ultraviolet spectra, and mass spectra in comparison with reference standards and literature data. The relative content of phenolic acids was calculated using calibration curves created using reference substances (ferulic acid, sinapic acid), methanolic solution (1–100 µg/mL) analysis, and by building concentration–peak area graphs. Contents of diferulic and triferulic acids were calculated as ferulic acid equivalents, and disinapic acid was measured as sinapic acid equivalents. All quantitative analyses were performed five times, and the data were expressed as the mean value ± standard deviation (S.D.).

2.13. Hydrolysis of DEAE–1% NaOH-f1 and DEAE–1% NaOH-f2 by 2% Oxalic Acid

For mild hydrolysis of polymers DEAE–1% NaOH-f1 and DEAE–1% NaOH-f2, a 500-mg sample was heated with 2% oxalic acid (100 mL), at 100 °C, for 1 h, and the hydrolysate was dialyzed in benzoylated dialysis tubes (cut-off of 2 kDa) against distilled water (48 h), and non-dialyzed residues were freeze-dried. The yields of degraded polymers were DEAE–1% NaOH-f1-d at 130.5 mg and DEAE–1% NaOH-f2-d at 146.5 mg.

2.14. Antioxidant Activity

In vitro microplate spectrophotometric assays of antioxidant activity were used to study the potential of samples to scavenge 2,2-diphenyl-1-picrylhydrazyl radicals ($DPPH^{\bullet}$) [40], 2,2′-azino-bis(3-ethylbenzothiazoline-6-sulfonic acid) cation radicals ($ABTS^{+\bullet}$) [35], superoxide radicals ($O_2^{\bullet-}$) [39], hydroxyl radicals ($OH^{\bullet}$) [39], and chlorine radicals ($Cl^{\bullet}$) [41], as well as nitric (II) oxide scavenging [42], hydrogen peroxide (H_2O_2) inactivating [43], and ferrous ion (Fe^{2+}) chelating activity [44]. All analyses were performed five times, and the data were expressed as the mean ± S.D.

2.15. Hypolipidemic Activity

2.15.1. In Vitro Assays

The cholesterol binding capacity of polysaccharides was measured by the method of Nagaoka et al. [45] and an enzymatic Amplex™ red cholesterol assay kit (Thermo Fisher Scientific, Waltham, MA, USA; No. A12216). The bile acid binding properties were studied using the Kim and White assay [46] with an enzymatic bile acid assay kit (Sigma-Aldrich; No. MAK309). The fat binding capacity of polysaccharides was estimated gravimetrically using the Jin et al. assay [47], and pancreatic lipase inhibition was studied spectrophotometrically using *p*-nitrophenol palmitate as a substrate [48]. All analyses were performed three times, and the data were expressed as the mean ± S.D.

2.15.2. In Vivo Assays

The experimental hyperlipidemia was reproduced using the recommendations of Cheng et al. [49] with modification. Fifty male Golden Syrian hamsters (weight of 70–72 g; BioNursery Stezar, Vladimir, Russia) were housed one per cage with a 12 h light/dark cycle (humidity of 50–55%), and regular rodent cholesterol-free chow (Asortiment-Agro

Company, Sergiev Posad, Russia) and free access to food and water were provided. After two weeks of adaptation, the animals were weighed and divided into five groups (n = 10): (1) the normal diet group; (2) the group with a 1% cholesterol supplementation diet; (3) the group with a 1% cholesterol supplementation diet + simvastatin; (4) the group with a 1% cholesterol supplementation diet + DEAE–1% NaOH-f2 polysaccharide; (5) the group with a 1% cholesterol supplementation diet + with 1% cholesterol diet + DEAE–1% NaOH-f3 polysaccharide. The diets of all groups were switched to a high-fat diet except for the normal diet group over a 3-month period. The general composition of the high-fat diet was 10% lard, 10% yoke powder, 1% cholesterol, and 79% regular rodent cholesterol-free chow. For the next six months, the animals were orally supplemented with simvastatin (10 mg/kg/day; group 3), DEAE–1% NaOH-f2 polysaccharide (250 mg/kg/day; group 4), and DEAE–1% NaOH-f3 polysaccharide (250 mg/kg/day; group 5). Animals in groups 1 and 2 received 0.9% NaCl solution. The experimental procedure was authorized by the Institute of General and Experimental Biology's Ethical Committee (protocol No. LM-0324, 27 January 2012) before starting the study and was conducted under the internationally accepted principles for laboratory animal use and care. Serum concentrations of total cholesterol, triglycerides, high-density lipoprotein–cholesterol, and low-density lipoprotein-cholesterol were measured enzymatically using Sigma-Aldrich cholesterol assay kit (No. MAK436), serum triglyceride determination kit (No. TR0100), and HDL and LDL/VLDL quantitation kit (No. MAK045), and the malondialdehyde level was measured using a malondialdehyde colorimetric assay kit (Elabscience Biotechnology, Inc., Houston, TX, USA; No E-BC-K025-S). Liver superoxide dismutase, glutathione peroxidase, and catalase were determined using Sigma-Aldrich SOD assay kit (No. 19160), glutathione peroxidase assay kit (No. MAK437), and catalase assay kit (No. MAK100). All analyses were performed five times, and the data were expressed as mean values ± S.D.

2.16. Statistical and Multivariate Analysis

Statistical analyses were performed by one-way analysis of variance, and the significance of the mean difference was determined by Duncan's multiple range test. Differences at $p < 0.05$ were considered statistically significant. The results are presented as the mean ± S.D. The linear regression analysis and generation of calibration graphs were conducted using Advanced Grapher 2.2 (Alentum Software, Inc., Ramat-Gan, Israel).

3. Results and Discussion

3.1. Yield and Chemical Composition of *V. vitis-idaea* Press Cake Polysaccharides (VVPS) and DEAE–Cellulose Fractions

Hot water extraction is able to isolate polysaccharides from the press cake *V. vitis-idaea*, which is also typical for other vacciniums such as blueberry, cranberry [25], and Manitoba lingonberry [29]. The total yield of *V. vitis-idaea* polysaccharide fraction (VVPS) after hot water extraction, ethanol precipitation, polyamide column dephenolization, cation-exchange demineralization, two steps of deproteination, and dialysis was 3.7% of dry press cake weight (Table 1). The total carbohydrate level in VVPS was measured to be 92.89%, which included uronic acids at 37.62% and starch at 2.04%. Non-carbohydrate constituents were proteins at 1.67%, phenolics at 2.96%, and ash at 1.56%. The positive test with Yariv reagent implied the presence of arabinogalactan–protein complexes (AGP) estimated as 0.29%. The early study of *Vaccinium* berry polysaccharide, indicating the presence of water-soluble polymers in *V. ashei*, yielded 2.7% of dry berry weight and contained 37% of uronic acids [50]. The purified water-soluble complex polysaccharide from the Northern manitoba lingonberry had 2.1% yield and showed 36% of total carbohydrates, 4.7% of proteins, and 3.5% of phenolics [29].

Table 1. Yields and chemical composition of VVPS and DEAE–cellulose fractions.

Polysaccharide Fraction	Yield	Total Carbohydrate, % [a]	Uronic Acids, % [a]	Starch, % [a]	Proteins, % [a]	AGP, % [a]	Phenolics, % [a]	Ash, % [a]
VVPS	3.7 [b]	92.89 ± 2.78 [i]	37.62 ± 0.75 [vii]	2.04 ± 0.04 [ix]	1.67 ± 0.05 [xiii]	0.29 ± 0.01 [xv]	2.96 ± 0.10 [xix]	1.56 ± 0.03 [xxii]
DEAE-H_2O	7.0 [c]	96.11 ± 2.89 [ii]	-	25.36 ± 0.52 [xi]	3.85 ± 0.12 [xiv]	3.44 ± 0.21 [xvii]	-	trace
DEAE-0.1% NH_4HCO_3	0.8 [c]	97.14 ± 2.99 [ii]	-	9.27 ± 0.18 [x]	1.86 ± 0.05 [xiii]	0.86 ± 0.03 [xvi]	-	trace
DEAE-0.3% NH_4HCO_3	1.8 [c]	98.06 ± 3.04 [iii]	3.22 ± 0.05 [iv]	0.53 ± 0.01 [viii]	0.53 ± 0.01 [xii]	-	-	trace
DEAE-0.5% NH_4HCO_3	2.0 [c]	98.15 ± 3.01 [iii]	29.53 ± 0.57 [v]	-	-	-	-	trace
DEAE-1% NH_4HCO_3	0.4 [c]	98.24 ± 3.11 [iii]	34.12 ± 0.48 [vi]	-	-	-	0.34 ± 0.01 [xviii]	0.52 ± 0.01 [xxi]
DEAE-1% NaOH	81.0 [c]	92.37 ± 2.76 [i]	39.67 ± 0.74 [vii]	-	0.61 ± 0.02 [xii]	-	3.54 ± 0.12 [xx]	1.90 ± 0.05 [xxiii]

[a] $n = 5$. [b] Yield—% of dry press cake weight. [c] Yield—% of VVPS weight. Values with different numbers (i–xxiii) indicate statistically significant differences among groups at $p < 0.05$ by one-way ANOVA. Abbreviation used: VVPS, *V. vitis-idaea* total polysaccharide fraction; DEAE, diethylaminoethyl cellulose.

DEAE–cellulose separation of VVPS resulted in the preparation of six polymer fractions eluted sequentially with water, ammonium bicarbonate (0.1–1%), and sodium hydroxide (1%) (Table 1). The neutral fraction DEAE–H_2O that eluted first with water showed 7% yield (of VVPS weight), high starch content (25.36%), the highest content of proteins (3.85%), and AGP (3.44%) and no phenolics.

The subsequent elution with NH_4HCO_3 resulted in isolation of four minor fractions with yields of 0.4–2.0%, and NaOH elution gave the dominant fraction DEAE–1% NaOH with 81% yield. Non-water eluted fractions (DEAE-0.1% NH_4HCO_3 → DEAE–1% NaOH) showed higher contents of uronic acids (0% → 39.67%), phenolics (0% → 3.54%), and ash (0% → 1.90%) and reduced contents of the starch (9.27% → 0%) and AGP (0.86% → 0%). Polysaccharide fractions of *V. vitis-idaea* press cake obtained through DEAE–cellulose separation were of various yields and chemical compositions.

3.2. Elemental Composition

The basic elements of polysaccharides are carbon, hydrogen, and oxygen, and the general formula is $C_xH_yO_z$. Fraction VVPS showed contents of C, H, and O at 39.08%, 6.14%, 54.52%, respectively, which is typical for carbohydrate polymers, and the composition of the DEAE–cellulose fractions varied in the ranges of 38.94–39.96% (C), 6.10–6.67% (H), and 52.81–54.57% (O) (Table 2).

Table 2. Elemental composition and O/C, H/C ratio of VVPS and DEAE–cellulose fractions, %.

Polysaccharide Fraction	C	H	O	N	O/C	H/C
VVPS	39.08	6.14	54.52	0.26	1.395	0.157
DEAE-H_2O	39.92	6.65	52.81	0.62	1.323	0.167
DEAE-0.1% NH_4HCO_3	39.96	6.67	53.08	0.29	1.328	0.167
DEAE-0.3% NH_4HCO_3	39.87	6.62	53.43	0.08	1.340	0.166
DEAE-0.5% NH_4HCO_3	39.22	6.21	54.57	-	1.391	0.158
DEAE-1% NH_4HCO_3	39.92	6.40	53.68	-	1.345	0.160
DEAE-1% NaOH	38.94	6.10	54.29	0.10	1.394	0.157
Hexose ($C_6H_{12}O_6$), pentose ($C_5H_{10}O_5$)	40.00	6.67	53.33	-	1.333	0.167
Desoxyhexose ($C_6H_{12}O_5$)	43.90	7.31	48.79	-	1.111	0.166
Hexuronic acid ($C_6H_{10}O_7$)	37.11	5.15	57.74	-	1.556	0.138

Nitrogen content was at a zero level (DEAE–0.5% NH_4HCO_3, DEAE–0.5% NH_4HCO_3), low (DEAE–0.3% NH_4HCO_3, DEAE–1% NaOH), or varied from 0.26% (VVPS) to 0.62% (DEAE–H_2O), which agreed with the previous chemical composition data. To better visualize the elemental composition results, we used a Van Krevelen diagram [51] to plot the atomic O/C ratio as a function of atomic H/C ratio (Figure 2).

Plant polysaccharides traditionally include hexoses, pentoses, hexuronic acids, and desoxyhexoses as structural blocks that are located in the Van Krevelen diagram in different places, creating a "monosaccharide triangle" reflecting extreme points of composition for the possible carbohydrate polymer. Because the monosaccharide composition of the polysaccharides varied in a wide range, the elemental composition data also varied. As shown, VVPS and DEAE–cellulose fractions were inside the triangle, whereas the DEAE–cellulose fraction eluted with water, 0.1%, and 0.3% NH_4HCO_3 were closer to the "neutral angle" of hexose (pentose), demonstrating low acidic monosaccharide content. The VVPS and DEAE–cellulose fractions eluted with 0.5% NH_4HCO_3 and 1% NaOH were in a lower position, reflecting the presence of uronic acids. The DEAE–cellulose fraction eluted with 1% NH_4HCO_3 was in the middle position away from the triangle side Hexose(Pentose)–Hexuronic Acid, which may be due to an increased content of desoxyhexoses.

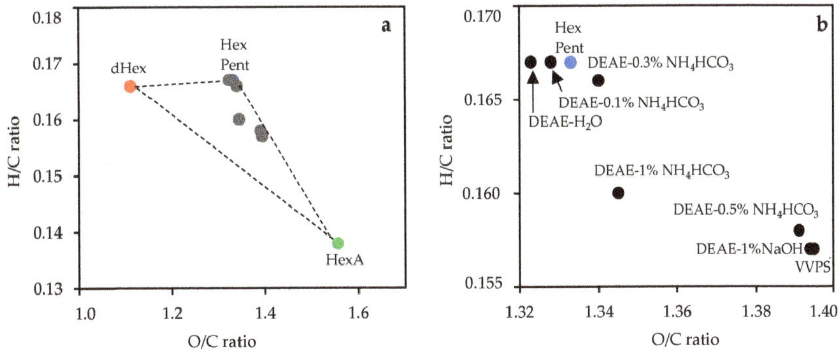

Figure 2. Van Krevelen diagram (correlation between O/C ratio and H/C ratio) for VVPS, DEAE–cellulose fractions (labeled as eluent type), hexose (Hex), pentose (Pent), desoxyhexose (dHex), and hexuronic acid (HexA) (**a**) and its enlarged fragment (**b**). Abbreviation used: VVPS, *V. vitis-idaea* total polysaccharide fraction; DEAE, diethylaminoethyl cellulose.

3.3. Monosaccharide Composition

Monosaccharides found in the total polysaccharide fraction of VVPS were neutral glucose, arabinose, and galactose in the ratio of 1.84:1.78:1.00 (58.2 mol% in sum) and minor rhamnose, fucose, xylose, and ribose, as well as acidic galacturonic acid (35.4 mol%) and glucuronic acid (0.6 mol%) (Table 3). The early data on *Vaccinium* polysaccharides showed that the mountain cranberry fruit (*V. vitis-idaea* subsp. *minus*) polysaccharides contain four basic neutral components of arabinose, glucose, galactose, and xylose in the ratio of 6.34:1.73:1.01:1.00 (58.5% in sum) and uronic acids at 39.1% [29]. Caucasian whortleberry fruit (*V. arctostaphylos*) polysaccharides were rich in arabinose (39.2 mol%), galactose (21.0 mol%), and xylose (13.7 mol%) with 18.3% uronic acids [52]. The fruits of rabbiteye blueberry (*V. ashei*) accumulate galacturonic acid in polysaccharides (37.3 mol%) and to a lesser degree arabinose (29.7 mol%), glucose (15.5 mol%), and galactose (10.9 mol%) [50]. Glucose (41.0 mol%) and xylose (33.0 mol%) were the main monosaccharides of European blueberries (*V. myrtillus*) [53], whereas arabinose (36.4 mol%), glucose (25.7 mol%), and galactose (24.5 mol%) dominated in bog blueberries (*V. uliginosum*) [54]. Probably, arabinose, galactose, glucose, and galacturonic acid are essential monosaccharides of *Vaccinium* polysaccharides.

Table 3. Monosaccharide composition of VVPS, DEAE–cellulose fractions and *Vaccinium* fruits polysaccharides (PS), mol%.

Polysaccharide Fraction, *Vaccinium* Species	Ara	Gal	Glc	Fuc	Man	Rib	Rha	Xyl	GalA	GlcA
VVPS	22.4	12.6	23.2	0.4	0.9	traces	4.6	traces	35.4	0.6
V. arctostaphylos PS [52]	39.2	21.0	6.3	-	-	-	1.5	13.7	18.3 *	
V. ashei PS [50]	29.7	10.9	15.5	-	1.2	-	1.6	3.6	37.3	-
V. myrtillus PS [53]	4.0	4.0	41.0	-	3.0	-	1.0	33.0	14.0	1.0
V. uliginosum PS [54]	36.4	24.5	25.7	-	3.1	-	-	-	10.3	-
V. vitis-idaea ssp. *minus* PS [29]	26.8	8.6	14.6	0.3	0.8	-	1.3	8.5	39.1 *	
DEAE-H_2O	29.3	22.5	45.5	-	2.6	-	-	-	-	-
DEAE-0.1% NH_4HCO_3	25.5	27.8	41.4	0.5	4.8	-	-	-	-	-
DEAE-0.3% NH_4HCO_3	31.2	33.0	28.3	-	4.7	-	-	-	2.8	-
DEAE-0.5% NH_4HCO_3	22.0	29.4	11.5	-	2.9	0.1	4.4	0.3	27.7	1.7
DEAE-1% NH_4HCO_3	24.0	19.2	5.0	-	3.6	-	17.6	-	30.6	-
DEAE-1% NaOH	26.2	13.9	0.9	-	0.7	-	3.6	-	54.8	-

* Calculated as a sum of GalA and GlcA.

DEAE–cellulose separation of VVPS resulted in isolation of polysaccharide fractions of various monosaccharide compositions (Table 3). Two neutral fractions of DEAE–H_2O and DEAE–0.1% NH_4HCO_3 were composed mainly of glucose, arabinose, and galactose in ratios of 2.02:1.30:1.00 and 1.49:0.92:1.00, respectively.

The subsequent fraction of DEAE–0.3% NH_4HCO_3 included 2.8 mol% of galacturonic acid and a reduced level of glucose (28.3 mol%). The decreasing of neutral monosaccharides and occurrence of rhamnose were detected in the DEAE–0.5% NH_4HCO_3 fraction with 29.4 mol% of uronic acids. The highest content of rhamnose was found in fraction DEAE–1% NH_4HCO_3 (17.6 mol%), and the most acidic fraction was DEAE–1% NaOH with 37.8 mol% of galacturonic acid. These findings were in good correlation with the elemental composition data and demonstrated the fine ability of sequential elution on the DEAE–cellulose to separate the polysaccharide mixture.

3.4. *Bioactivity of VVPS and DEAE–Cellulose Fractions of V. vitis-idaea Press Cake*

Polysaccharides of various *Vaccinium* species are known antioxidants with a potency to inactivate free radicals [52], chelate ferrous ions [50], and bind bile acids [25], making them good antioxidative and lipid-lowering agents. The primary cause of antioxidant potential for plant polysaccharides is phenolic constituents covalently bonded to carbohydrate chains [55], while uronic aids are capable of binding metal ions, bile acids, and cholesterol [56], providing the acidic polysaccharides with bioactivity. Given the diverse chemical properties of *V. vitis-idaea* press cake polysaccharides, we assumed potency for both VVPS and DEAE–cellulose fractions due to the high level of phenolics and uronic acids.

3.4.1. Antioxidant Activity

Eight in vitro assays were used to study the antioxidant properties of VVPS and DEAE–cellulose fractions against the following free radicals: 2,2-diphenyl-1-picrylhydrazyl radicals ($DPPH^{\bullet}$), 2,2′-azino-bis(3-ethylbenzothiazoline-6-sulfonic acid) cation radicals ($ABTS^{+\bullet}$), superoxide radicals ($O_2^{\bullet-}$), hydroxyl radicals ($OH^{\bullet}$), and chlorine radicals ($Cl^{\bullet}$); also, nitric (II) oxide scavenging, hydrogen peroxide (H_2O_2) inactivating, and ferrous ion (Fe^{2+}) chelating activity were examined (Table 4).

Comparative data of VVPS activity and Trolox used as a reference antioxidant demonstrated the better potential of Trolox in scavenging $DPPH^{\bullet}$, $ABTS^{+\bullet}$, $O_2^{\bullet-}$, $OH^{\bullet}$, $Cl^{\bullet}$, and H_2O_2 inactivation, but the total polysaccharide fraction VVPS was an effective scavenger of NO molecules and chelator of Fe^{2+} ions. Interestingly, three commercially available polysaccharides (i.e., pectin from citrus peel, starch, and arabinogalactan) were inactive in scavenging all free radicals, and only pectin demonstrated at least some activity in scavenging NO, H_2O_2 inactivation, and good Fe^{2+} chelation.

Analysis of DEAE–cellulose fractions showed DEAE–H_2O, DEAE–0.1% NH_4HCO_3, and DEAE–0.3% NH_4HCO_3 as inactive and DEAE–0.5% NH_4HCO_3 and DEAE–1% NH_4HCO_3 as poorly active. Only fraction DEAE–1% NaOH showed an antioxidant effect comparable to that of VVPS, which indicated the leading role of DEAE–1% NaOH in VVPS antioxidant activity. This fraction has the highest phenolic content, which may explain its excellent antioxidant activity.

The antioxidant activity of the *Vaccinium* polysaccharide fraction from *V. arctostaphylos* showed a 56% scavenging of $DPPH^{\bullet}$ and 49% scavenging of $OH^{\bullet}$ at a dose of 3 mg/mL [52], and the $ABTS^{+\bullet}$ scavenging potential of the *V. ashei* polysaccharide was 20–200 µmol Trolox eq./g [50]. Other berry polysaccharides also were effective scavengers of $DPPH^{\bullet}$ radicals in *Lycium barbarum* with an IC_{50} of 300–400 µg/mL [57], $ABTS^{+\bullet}$ radicals in sweet cherries (*Prunus avium*) and raspberries (*Rubus idaeus*) with an IC_{50} of 57–438 µmol Trolox eq./g [58], and $O_2^{\bullet-}$ and $OH^{\bullet}$ radicals in blackcurrant (*Ribes nigrum* L.) with IC_{50} values of 0.4–1.0 mg/mL and 0.2–0.5 mg/mL, respectively [59]. Polysaccharide fractions of *V. vitis-idaea* VVPS and DEAE–1% NaOH are excellent antioxidants that are effective against various damaging factors.

Table 4. Antioxidant activity of VVPS, DEAE–cellulose and DEAE–sepharose fast-flow gel fractions ($n = 5$).

Polysaccharide Fraction	DPPH [a]	ABTS^{+} [a]	$O_2^{\bullet -}$ [a]	OH• [a]	CI [b]	NO [a]	H2O$_2$ [c]	FeCA [d]
VVPS	35.69 ± 0.73 ^{v}	22.59 ± 0.45 ^{x}	144.17 ± 4.32 ^{xv}	32.60 ± 0.96 ^{xix}	27.56 ± 0.55 ^{xx}	92.75 ± 2.84 ^{xxvii}	0.36 ± 0.01 ^{xxxii}	4.71 ± 0.14 ^{xxxviii}
DEAE-H_2O	i.a.	i.a.	i.a.	i.a.	i.a.	i.a.	i.a.	i.a.
DEAE-0.1% NH_4HCO_3	i.a.	i.a.	i.a.	i.a.	i.a.	i.a.	i.a.	i.a.
DEAE-0.3% NH_4HCO_3	i.a.	i.a.	i.a.	i.a.	i.a.	i.a.	i.a.	i.a.
DEAE-0.5% NH_4HCO_3	i.a.	i.a.	i.a.	i.a.	i.a.	i.a.	0.25 ± 0.00 ^{xxxi}	0.31 ± 0.01 ^{xxxvii}
DEAE-1% NH_4HCO_3	i.a.	i.a.	i.a.	i.a.	i.a.	i.a.	0.18 ± 0.00 ^{xxxi}	0.24 ± 0.00 ^{xxxvi}
DEAE-1% NaOH	24.18 ± 0.48 ^{iv}	15.25 ± 0.31 ^{ix}	108.26 ± 2.07 ^{xiii}	22.86 ± 0.67 ^{xviii}	41.67 ± 0.83 ^{xxi}	72.11 ± 2.16 ^{xxvi}	0.27 ± 0.01	6.83 ± 0.20 ^{xxxxi}
DEAE-1% NaOH-f1	12.73 ± 0.25 ^{iii}	7.62 ± 0.15 ^{viii}	84.75 ± 1.76 ^{xii}	14.06 ± 0.42 ^{xvii}	126.79 ± 2.53 ^{xxii}	53.86 ± 1.61 ^{xxv}	0.10 ± 0.00 ^{xxx}	8.26 ± 0.34 ^{xxxxii}
DEAE-1% NaOH-f2	10.82 ± 0.21 ^{ii}	6.83 ± 0.12 ^{vii}	70.29 ± 1.70 ^{xi}	11.73 ± 0.39 ^{xvi}	157.11 ± 3.14 ^{xxiii}	41.09 ± 1.20 ^{xxiv}	0.08 ± 0.00 ^{xxx}	8.72 ± 0.35 ^{xxxxii}
DEAE-1% NaOH-f3	i.a.	i.a.	i.a.	i.a.	i.a.	i.a.	i.a.	6.14 ± 0.25 ^{xxxx}
DEAE-1% NaOH-f4	i.a.	i.a.	i.a.	i.a.	i.a.	i.a.	i.a.	6.09 ± 0.24 ^{xxxx}
Trolox [e]	8.94 ± 0.18 ^{i}	3.25 ± 0.06 ^{vi}	122.36 ± 2.44 ^{xiv}	15.23 ± 0.36 ^{xvii}	1000	125.11 ± 3.75 ^{xxviii}	0.59 ± 0.02 ^{xxxiii}	0.15 ± 0.00 ^{xxxv}
Pectin from citrus peel [e]	i.a.	i.a.	i.a.	i.a.	i.a.	265.82 ± 10.63 ^{xxix}	2.77 ± 0.11 ^{xxxiv}	5.26 ± 0.15 ^{xxxix}
Starch [e]	i.a.	i.a.	i.a.	i.a.	i.a.	i.a.	i.a.	i.a.
Arabinogalactan [e]	i.a.	i.a.	i.a.	i.a.	i.a.	i.a.	i.a.	0.22 ± 0.01 ^{xxxvi}

[a] IC_{50}, μg/mL; [b] Trolox-equivalents, mg/g; [c] IC_{50}, mg/mL; [d] mM Fe^{2+}/g; [e] Reference standards. Values with different numbers (i–xxxxii) indicate statistically significant differences among groups at $p < 0.05$ by one-way ANOVA. Abbreviation used: i.a., inactive; VVPS, *V. vitis-idaea* total polysaccharide fraction; DEAE, diethylaminoethyl cellulose.

3.4.2. In Vitro Hypolipidemic Activity

In four in vitro assays, polysaccharide VVPS showed the ability to bind bile acids, fat, and cholesterol and inhibit pancreatic lipase (Table 5). The bile acid binding capacity of the reference standard cholestyramine which is the styrene–divinylbenzene copolymer used as a bile acid sequestrant [60] was 10.29 μmol/100 g. Fraction VVPS showed activity value 5.73 μmol/100 g that was much higher than that of the reference polysaccharides such as microcrystalline cellulose (0.07 μmol/100 g), pectin (0.78 μmol/100 g), starch (0.02 μmol/100 g), and arabinogalactan (0.12 μmol/100 g). DEAE fractions of H_2O, 0.1% NH_4HCO_3, 0.3% NH_4HCO_3, and 0.5% NH_4HCO_3 showed weak activity, but DEAE–1% NaOH was most active (6.04 μmol/100 g). The fat binding levels of VVPS and of fractions DEAE–1% NH_4HCO_3 and DEAE–1% NaOH were the highest at 200.02, 231.02, and 252.37 g/100 g, respectively, which were significantly above the activities of microcrystalline cellulose (92.63 g/100 g) and starch (97.67 g/100 g).

Pectin and arabinogalactan demonstrated good fat binding potential with values of 186.85 and 156.14 g/100 g, respectively, and cholestyramine was inactive. A similar pattern was found for cholesterol binding by polysaccharides. Fractions VVPS and DEAE–1% NaOH were the most active (57.02 and 68.37 mg/g, respectively) but were less intensive binders than cholestyramine (93.11 mg/g). Inhibition of pancreatic lipase was detected only for VVPS and DEAE–1% NaOH polysaccharides that had IC_{50} values of 6.24 and 5.33 mg/mL, respectively, exceeding the activity of cholestyramine (IC_{50} 14.02 mg/mL). Thus, polysaccharide fraction VVPS from *V. vitis-idaea* press cake and its active constituent DEAE–1% NaOH showed good in vitro hypolipidemic potential.

Previously, freeze-dried dietary berries showed in vitro bile acid binding with a value ranging from 0.43 μmol/100 g (cranberries, *Vaccinium macrocarpon*) to 0.73 μmol/100 g (blueberry, *Vaccinium* spp.) owing to the high polysaccharide content (particularly dietary fibers) [25]. The bile acid binding potentials of dietary fruits were from 0.21 μmol/100 g for nectarines (*Prunus persica*) to 0.90 μmol/100 g bananas (*Musa paradisiaca*) [61]. Known polysaccharides with bile acid binding activity were also isolated from *Abelmoschus esculentus* [62], *Kadsura coccinea* [63], and *Laminaria japonica* [64], and in all cases, the presence of uronic acids was determined as a basic principle for effect manifestation [65]. The lipid lowering effect of *Inonotus obliquus* and *Volvariella volvacea* polysaccharides was associated with their fat and cholesterol binding [66,67], providing the hypolipidemic activity of polymers. In addition, the pancreatic lipase inhibition plays an important role in the lipid lowering effect of plant food and polysaccharide-derived products. Pectic polysaccharides with varied molecular weight and methoxylation degree are effective pancreatic lipase inhibitors, which is related to a reduction in the surface of the lipid droplet exposed to the enzyme [68], but neutral polysaccharides are also good inhibitors as in the case of *Dictyophora indusiata* polymers [69]. Summarizing the data, the polysaccharide fraction VVPS of *V. vitis-idaea* press cake and its component DEAE–1% NaOH are possible effective antioxidants and potential hypolipidemic agents that need further separation for the isolation of homogenous polymers and their chemical investigation, followed by an *in vivo* study of bioactivity.

Table 5. In vitro hypolipidemic activity of VVPS, DEAE–cellulose and DEAE–sepharose fast-flow gel fractions (n = 5).

Polysaccharide Fraction	Bile Acids Binding, μmole/100 g	Fat Binding, g/100 g	Cholesterol Binding, mg/g	Pancreatic Lipase Inhibition, IC_{50}, mg/mL
VVPS	5.73 ± 0.22 viii	200.02 ± 7.24 xvi	57.02 ± 2.56 xxiii	6.24 ± 0.18 xxx
DEAE-H_2O	0.10 ± 0.00 ii	103.75 ± 3.70 xii	15.37 ± 0.63 xx	i.a.
DEAE-0.1% NH_4HCO_3	0.14 ± 0.00 ii	127.80 ± 4.49 xiii	18.62 ± 0.80 xxi	i.a.
DEAE-0.3% NH_4HCO_3	0.27 ± 0.01 iii	131.29 ± 4.63 xiii	19.83 ± 0.93 xxi	i.a.
DEAE-0.5% NH_4HCO_3	0.92 ± 0.04 v	139.16 ± 4.90 xiii	37.10 ± 1.69 xxii	i.a.
DEAE-1% NH_4HCO_3	3.62 ± 0.16 vii	231.02 ± 8.14 xvii	59.22 ± 2.50 xxiii	i.a.
DEAE-1% NaOH	6.04 ± 0.29 viii	252.37 ± 8.97 xvii	68.37 ± 3.02 xxiv	5.33 ± 0.15 xix
DEAE-1% NaOH-f1	7.83 ± 0.39 ix	183.70 ± 6.40 xv	72.11 ± 3.24 xxv	4.27 ± 0.12 xxviii
DEAE-1% NaOH-f2	8.26 ± 0.44 ix	173.11 ± 6.04 xv	73.92 ± 3.36 xxv	3.86 ± 0.10 xxvii
DEAE-1% NaOH-f3	1.85 ± 0.09 vi	308.75 ± 10.83 xviii	59.27 ± 2.65 xxiii	i.a.
DEAE-1% NaOH-f4	1.90 ± 0.10 vi	315.61 ± 10.88 xviii	60.08 ± 2.72 xxiii	i.a.
Cholestyramine *	10.29 ± 0.40 x	i.a.	93.11 ± 4.15 xxvi	14.02 ± 0.42 xxxi
Microcrystalline cellulose *	0.07 ± 0.00 ii	92.63 ± 3.12 xi	10.33 ± 0.45 xix	i.a.
Pectin from citrus peel *	0.78 ± 0.03 iv	186.85 ± 6.51 xv	57.82 ± 2.69 xxiii	i.a.
Starch *	0.02 ± 0.00 i	97.67 ± 3.43 xi	9.63 ± 0.44 xix	i.a.
Arabinogalactan *	0.12 ± 0.00 ii	156.14 ± 5.46 xiv	21.16 ± 0.90 xxi	i.a.

* Reference standards. Values with different numbers (i–xxxi) indicate statistically significant differences among groups at $p < 0.05$ by one-way ANOVA. Abbreviation used: i.a., inactive; VVPS, *V. vitis-idaea* total polysaccharide fraction; DEAE, diethylaminoethyl cellulose.

3.5. Preparative Chromatography of Fraction DEAE-1% NaOH and Characterization of Homogenous Polymers

The elution curve for the DEAE–sepharose fast-flow gel of the DEAE–1% NaOH fraction demonstrated that it is composed of four polymers labeled as DEAE–1% NaOH-**f1**–DEAE–1% NaOH-f4 (Figure 3). Two polymers, DEAE–1% NaOH-f1 and DEAE–1% NaOH-f2, were UV-positive and detected at 270 nm, unlike DEAE–1% NaOH-f3 and DEAE–1% NaOH-f4, which showed no notable absorption in the UV region. The polymers were isolated, and after re-chromatography, the homogenous fractions were obtained in the yields of 9.5% (DEAE–1% NaOH-f1), 15.6% (DEAE–1% NaOH-f2), 58.4% (DEAE–1% NaOH-f3), and 4.2% (DEAE–1% NaOH-f4) of DEAE–1% NaOH weight. The molecular weights of the polymers were 157.6 kDa (DEAE–1% NaOH-f1), 108.2 kDa (DEAE–1% NaOH-f2), 258.3 kDa (DEAE–1% NaOH-f3), and 318.4 kDa (DEAE–1% NaOH-f4) (Figure 4a, Table 6).

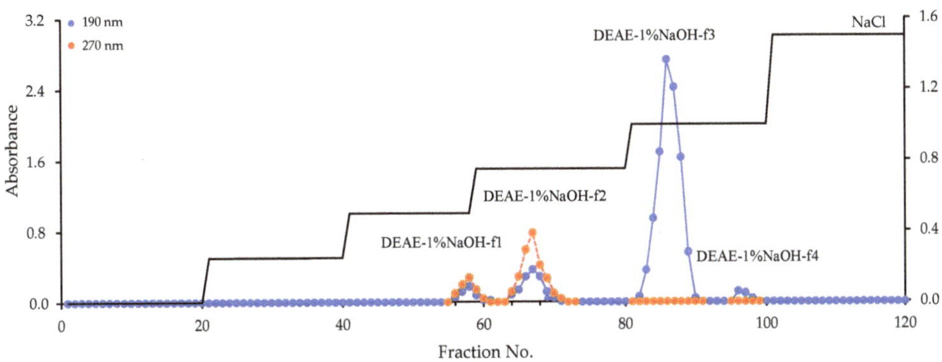

Figure 3. Elution curve of DEAE-1% NaOH fraction on the DEAE–sepharose fast-flow gel.

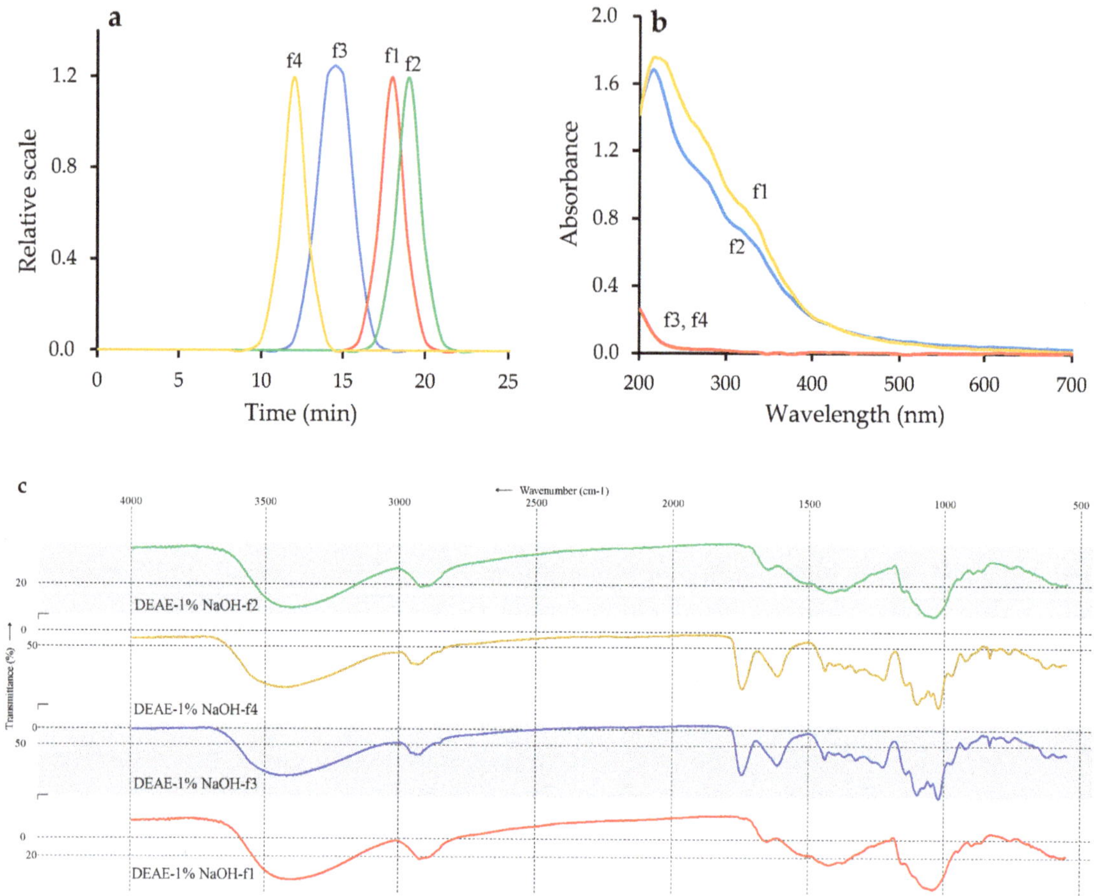

Figure 4. Characteristics of DEAE–sepharose fast-flow gel fractions of DEAE-1% NaOH-f1 (f1), DEAE-1% NaOH-f2 (f2), DEAE-1% NaOH-f3 (f3), and DEAE-1% NaOH-f4 (f4): (**a**) elution profiles on GP-HPLC; (**b**) UV–Vis spectra of 1 mg/mL water solutions; (**c**) FTIR spectra.

Table 6. Yield, molecular weight, monosaccharide composition, phenol content, and phenolic acid content after alkaline hydrolysis and linkage analysis of DEAE–sepharose fast-flow gel homogenic polymers and degradation polymers.

Parameter	DEAE-1% NaOH-f1	DEAE-1% NaOH-f1-d	DEAE-1% NaOH-f2	DEAE-1% NaOH-f2-d	DEAE-1% NaOH-f3	DEAE-1% NaOH-f4
Yield	9.5 [a]	26.1 [b]	15.6 [a]	29.3 [c]	58.4 [a]	4.2 [a]
M_w, kDa [d]	157.6 (±1.4%)	35.5 (±1.0%)	108.2 (±1.9%)	25.4 (±1.4%)	258.3 (±2.2%)	318.4 (±2.7%)
M_w/M_n [d]	1.56 (±2.9%)	1.64 (±3.7%)	1.48 (±2.1%)	1.42 (±2.6%)	1.71 (±3.7%)	1.62 (±2.4%)
Monosaccharide composition, mol%						
Ara	65.1	-	58.6	-	15.5	16.8
Gal	33.1	99.9	38.7	99.9	7.1	8.9
Glc	0.6	-	1.7	-	0.2	0.1
Man	1.2	-	1.0	-	2.7	2.1
Rha	-	-	-	-	6.7	4.1

Table 6. *Cont.*

Parameter	DEAE-1% NaOH-f1	DEAE-1% NaOH-f1-d	DEAE-1% NaOH-f2	DEAE-1% NaOH-f2-d	DEAE-1% NaOH-f3	DEAE-1% NaOH-f4
GalA	-	-	-	-	67.8	68.0
Linkage analysis, molar ratio						
Terminal Ara	12.6	-	12.4	-	3.7	3.2
1,5-Ara	38.4	-	31.6	-	10.3	10.9
1,3,5-Ara	14.1	-	15.0	-	1.2	2.8
Terminal Gal	7.1	1.0	10.2	1.4	8.2	7.9
1,3-Gal	10.5	98.9	9.1	98.5	4.2	5.7
1,3,6-Gal	15.7	-	19.4	-	2.6	3.0
1,4-Gal	-	-	-	-	60.9	61.3
1,2-Rha	-	-	-	-	4.6	3.6
1,2,4-Rha	-	-	-	-	2.0	0.6
Terminal Man	1.0	-	0.4	-	2.0	0.9
Terminal Glc	0.6	-	1.9	-	0.3	0.1
Phenols, % [e]	10.61 (±0.32)	-	14.52 (±0.44)	-	<0.1	<0.1

[a] Yield, % of DEAE-1%NaOH DW. [b] % of DEAE-1% NaOH-f1 DW. [c] % of DEAE-1% NaOH-f2 DW. [d] $n = 3$. [e] $n = 5$.

Polymers DEAE–1% NaOH-f1 and -f2 showed strong absorption at 290 ± 3, 310 ± 2 and 330 ± 3 nm caused by the high phenolic content (10.61 and 14.52%, respectively), unlike DEAE–1% NaOH-f3 and -f4, which were weakly absorbed in the UV region (Figure 4b). FTIR spectra of DEAE–1% NaOH-f1 and -f2 were similar, as well as the spectra of DEAE–1% NaOH-f3 and -f4 (Figure 4c). In the spectra of DEAE–1% NaOH-f1 and -f2, intensive bands at 3390 ± 5 and 2920 ± 3 cm^{-1} were attributed to vibrations of O–H and bending vibrations of C–H. Strong bands at 890 ± 2, 1040 ± 3, 1152 ± 2, 1415 ± 2, and 1650 ± 2 cm^{-1} were caused by the vibrations of C–OH, C–O–C, and C–C, and the "fingerprint" region included bands at 850 ± 2 and 910 ± 3 cm^{-1} due to α/β-glycosidic linkages [70]. Specific vibrations in region 1550–1530 cm^{-1} was related to phenolic fragments and phenyl hydroxyl structure [71]. The general profiles of FTIR spectra of DEAE–1% NaOH-f1 and -f2 were very consistent with neutral polysaccharide spectra, e.g., of arabinogalactans and galactans [72].

A more complex spectral pattern was found in the FTIR spectra of DEAE–1% NaOH-f3 and -f4 containing bands typical for the pectic polysaccharides [73]. The low-frequency (<1000 cm^{-1}) region of the FTIR spectra included bands of "breathing rings" (762 ± 1 cm^{-1}) and α/β-glycosidic linkages (850 ± 2, 890 ± 1, 917 ± 1 cm^{-1}). The specific and very intensive "pectic region" (1200–900 cm^{-1}) demonstrated vibrations of skeletal C–O and C–C of glycosidic bonds and pyranoid rings at 1025 ± 2, 1050 ± 2, 1075 ± 3, 1097 ± 2, and 1142 ± 2 cm^{-1} [74]. Two intense bans at 1740 ± 3 and 1610 ± 4 cm^{-1} were due to stretching C=O vibrations of esters, and carboxylate anion and symmetric vibrations of COO formed the band at 1415 ± 3 cm^{-1}.

Monosaccharide analysis indicated that DEAE–1% NaOH-f1 and -f2 were neutral polymers consisting mainly of arabinose and galactose in ratios of 1.97:1 and 1.51:1, respectively, with minor contents of glucose (0.6–1.7 mol%) and mannose (1.0–1.2 mol%). Polymers DEAE–1% NaOH-f3 and -f4 were characterized by a high content of galacturonic acid (67.8 and 68.0 mol%, respectively), medium levels of arabinose (15.5–16.8 mol%), rhamnose (4.1–6.7 mol%), and galactose (7.1–8.9 mol%), and low contents of mannose (4.1–6.7 mol%) and glucose (2.1–2.7 mol%).

Results of the monosaccharide composition study reflected the arabinogalactan and pectic nature of isolated polymers, as previously confirmed by FTIR and further verified by linkage analysis after methylation and GC-MS analysis. In DEAE–1% NaOH-f1 and -f2, eight peaks were identified as terminal Ara, (1 → 5)Ara, (1 → 3, 5)Ara, terminal Gal, (1 → 3)Gal, (1 → 3, 6)Gal, terminal Glc, and Man. Mild hydrolysis with 2% oxalic acid resulted in formation of degraded linear homopolymers of DEAE–1% NaOH-f1-d (from DEAE–1% NaOH-f1) and DEAE–1% NaOH-f2-d (from DEAE–1% NaOH-f2) consisting only of galactose, which were the core chains of the polymers. This result demonstrated that both neutral polymers were most likely (1 → 3)-linked galactans with branching points located at the *O*-6 positions containing (1 → 3)-linked arabinose chains branched at the *O*-5 positions. The linkage analysis results of DEAE–1% NaOH-f3 and -f4 indicated the dominance of (1 → 4)GalA fragments (60.9–61.3%), which were the building blocks in the construction of the polymer chains. The known data on pectin structure suggest that the rhamnose residues (1 → 2)Rha and (1 → 2, 4)Rha were probably incorporated in the basic polymer chains and that fragments of terminal Ara, (1 → 5)Ara, (1 → 3, 5)Ara, terminal Gal, (1 → 3)Gal, (1 → 3, 6)Gal, terminal Glc, and Man were the blocks for the side chains [75].

To understand the chemical reasons for DEAE–1% NaOH-f1 and -f2 absorption in the UV spectral region, we degraded both polymers in KOH media followed by HPLC-PDA-MS analysis of the cleavage products. Alkaline hydrolysis of DEAE–1% NaOH-f1 and -f2 resulted in liberation of phenolic acids identified as ferulic acid (*m/z* 195; 29.5–38.4%), sinapic acid (*m/z* 225; 9.6–15.8%), diferulic acids (*m/z* 387; 47.3–51.9%), and disinapic acids (*m/z* 447; 2.8–4.7%) (Figure 5, Table 7). Monomeric acids (degradation products 1 and 2) were identified using reference standards, and diferulic and disinapic acids gave a specific combination of daughter ions in the MS/MS spectra, which allowed us to predict their structure. Mass spectral patterns of three diferulic acids 3, 4, and 5 were similar to previously described data of 5-5′-diferulic acid, 8-*O*-4′-diferulic acid, and 8-5′-diferulic acid, respectively (Figure S2) [76]. The spectra of disinapic acids 6 and 7 gave mass spectral patterns that were analogous to those of diferulic acids 3 and 4, but the basic ions were 60 amu larger owing to two methoxy functional groups in the sinapoyl fragments. The predicted structures of 6 and 7 degradation products were 2-2′-disinapic acid and 8-*O*-4′-disinapic acid, respectively (Figure S2). The obtained data showed evidence of polyphenol–polysaccharide conjugates with cross-linked structures of polymers DEAE–1% NaOH-f1 and -f2 esterified by ferulic and sinapic acid and their dimers.

Table 7. Characterization and content of degradation products released after alkaline destruction of polymers DEAE-1% NaOH-f1 and DEAE-1% NaOH-f2.

Compound (No. Figure 5)	ESI-MS, $[M + H]^+$, *m/z*	ESI-MS/MS, *m/z*	Content after Alkaline Destruction, % [a]	
			DEAE-1% NaOH-f1	DEAE-1% NaOH-f2
Ferulic acid (1)	195	181	38.4 ± 1.1	29.5 ± 0.8
Sinapic acid (2)	225	211, 197	9.6 ± 0.2	15.8 ± 0.4
Diferulic acid (3)	387	369, 351, 325, 323, 319, 287	10.5 ± 0.1	12.7 ± 0.2
Diferulic acid (4)	387	369, 351, 326, 325, 263, 219, 204, 201, 193, 177, 149	28.6 ± 0.4	35.9 ± 0.7
Diferulic acid (5)	387	369, 351, 343, 325, 323, 307, 297, 293, 265, 237, 219, 201, 151	8.2 ± 0.2	3.3 ± 0.1
Disinapic acid (6)	447	429, 385, 383, 411, 379, 347	1.4 ± 0.0	0.8 ± 0.0
Disinapic acid (7)	447	429, 411, 401, 323, 279, 264, 261, 237, 223, 205, 179, 177	3.3 ± 0.1	2.0 ± 0.0

[a] Percentage of total peak area. $n = 3$.

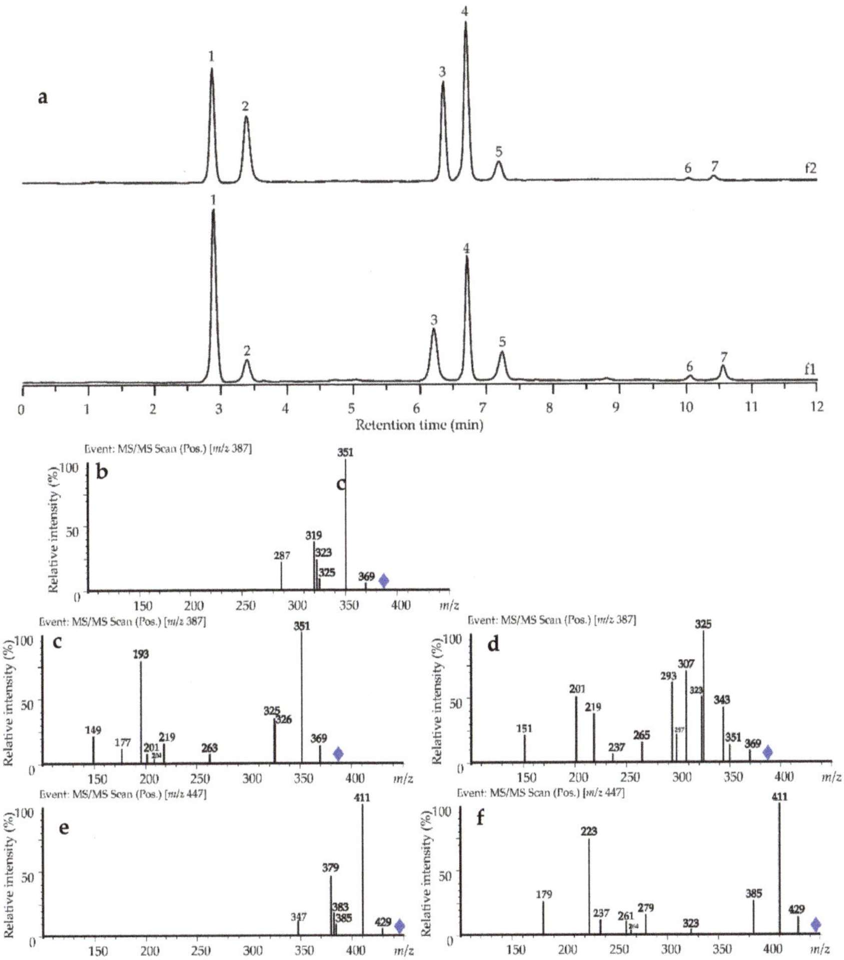

Figure 5. HPLC chromatogram of alkaline destruction products of DEAE–1% NaOH-f1 (f1) and DEAE–1% NaOH-f2 (f2) polymers (**a**) degradation products labeled 1–7 as described in Table 7 and ESI–MS/MS spectra (positive ionization) of degradation products 3 (**b**), 4 (**c**), 5 (**d**), 6 (**e**), and 7 (**f**).

Interestingly, the degradation polymers DEAE–1% NaOH-f1-d and -f2-d formed after elimination of arabinose from DEAE–1% NaOH-f1 and -f2, showed no hydroxycinnamates after alkaline cleavage and likely only occurred if phenolic acids were linked with arabinose residues. This is not unusual because feruloylated arabinose oligomer chains were previously found in the pectin fraction of spinach [77], diferuloyl fragments were found in *Zea mays* cell walls [78,79], and sinapoylated polysaccharides were detected in radish seedlings [80]. Acidic polymers DEAE–1% NaOH-f3 and -f4 released no phenolics after alkaline treatment.

The obtained data showed that homogenic components of the bioactive polysaccharide fraction of *V. vitis-idaea* press cake include four polymers, two of which are polyphenol–polysaccharide conjugates as neutral arabino-3,6-galactans esterified by hydroxycinnamoyl fragments and the other two are pectic-like polysaccharides. Although studies have been conducted on the primary structure of polymers, additional spectral studies need to identify the fine structure of polysaccharides isolated from *V. vitis-idaea*. Looking back at previous information about *Vaccinium* berry polysaccharides, only bilberry (*V. myrtillus*) [53] and rabbiteye blueberry (*V. ashei*) [50] have been mentioned as a source of bioactive polymers of pectic nature without fine structure determination; thus, it is still too early to speculate about genus features in the polysaccharide composition.

3.6. Antioxidant and Hypolipidemic Activity of Homogenic Polymers: In Vitro vs. In Vivo Assays

In a series of in vitro antioxidant assays, we found that the polyphenol–polysaccharide conjugates DEAE–1% NaOH-f1 and -f2 were the most active free radical scavengers compared with the parent fraction DEAE–1% NaOH and reference compound Trolox in some cases. The indices of 50% radicals inactivating (IC_{50}) were 10.73–12.73 μg/mL against DPPH$^{\bullet}$ (Trolox IC_{50} 8.94 μg/mL), 6.83–7.62 μg/mL against ABTS$^{+\bullet}$ (Trolox IC_{50} 3.25 μg/mL), 70.29–84.75 μg/mL against $O_2^{\bullet-}$ (Trolox IC_{50} 122.36 μg/mL), and 11.73–14.06 μg/mL against OH$^{\bullet}$ (Trolox IC_{50} 15.23 μg/mL), and the Cl$^{\bullet}$-scavenging potential was 126.79–157.11 mg/g vs. 1000 mg/g for Trolox. The values for NO scavenging and hydrogen peroxide inactivation were 41.09–53.86 μg/mL (Trolox IC_{50} 125.11 μg/mL) and 0.08–0.10 mg/mL (Trolox IC_{50} 0.59 μg/mL), respectively. The acidic polymers DEAE–1% NaOH-f3 and -f4 were inactive in the abovementioned assays. The ability to chelate ferrous ions was revealed for all homogenic polymers but to a greater degree for DEAE–1% NaOH-f1 (8.26 mM Fe^{2-}/g) and -f2 (8.72 mM Fe^{2-}/g) than for the DEAE–1% NaOH-f3 (6.14 mM Fe^{2-}/g) and -f4 (6.09 mM Fe^{2-}/g). Polymers with low phenolics content as DEAE–1% NaOH-f3, DEAE–1% NaOH-f3-f4 (Table 4), DEAE–1% NaOH-f1-d, and -f2-d (Table S1) showed low antioxidant potential.

Investigation of the in vitro hypolipidemic activity demonstrated higher bile acid binding potentials for DEAE–1% NaOH-f1 and -f2 polymers of 7.83 and 8.26 μmol/100 g, respectively, vs. 1.85 μmol/100 g for DEAE–1% NaOH-f3 and 1.90 μmol/100 g for DEAE–1% NaOH-f4 (Table 5). Both active polymers were also inhibitors of pancreatic lipase with IC_{50} values of 4.27 and 3.86 mg/mL for DEAE–1% NaOH-f1 and -f2, respectively, whereas DEAE–1% NaOH-f3 and -f4 were inactive. Acidic polysaccharides DEAE–1% NaOH-f3 and -f4 showed fat binding potentials (308.75 and 315.61 g/100 g, respectively), and the cholesterol binding activities of the four homogenic polymers were similar in the range of 59.27–73.92 mg/g. These results mean that the four components of active hypolipidemic fraction DEAE–1% NaOH also have potential to bind bile acids, fat, and cholesterol and inhibit pancreatic lipase, each in its own way. The polyphenol–polysaccharide conjugates showed good bile acid binding and inhibition of pancreatic lipase; however, the acidic non-phenolized polysaccharides were binders of fat, whereas the cholesterol binding activity was at a similar level. The removal of phenolic fragments from DEAE–1% NaOH-f1 and -f2 resulted in significant reduction in the in vitro hypolipidemic activity of the DEAE–1% NaOH-f1-d and -f2-d polymers (Table S2), which indicates the importance of phenolics as active sites of the neutral arabinogalactans.

To progress from in vitro to in vivo experiments, we chose polymer DEAE–1% NaOH-f2 as an example of high-yielded polyphenol–polysaccharide conjugates and DEAE–1% NaOH-f3 as high-yielded non-phenolized pectic polysaccharide from *V. vitis-idaea* press cake to treat experimental animals on the standard and high-fat diet (Figure 6). The reference substance was simvastatin, a known hypolipidemic drug, at a dose of 10 mg/kg/day [81]. The high-fat diet with 1% cholesterol resulted in animal body weight gain from 81 $\pm$ 4 g at the beginning of the test to a final weight of 195 $\pm$ 10 g (vs. 80 $\pm$ 4 g $\rightarrow$ 141 $\pm$ 7 g in the standard diet group). The changes in serum lipid profile involved a reduced level of cholesterol (6.29 mmol/L vs. 1.63 mmol/L in the standard diet group), triglycerides (2.97 mmol/L vs. 0.72 mmol/L in the standard diet group), low-density lipoprotein-cholesterol (3.28 mmol/L vs. 0.83 mmol/L in the standard diet group), and high-density lipoprotein–cholesterol (1.26 mmol/L vs. 0.52 mmol/L in the standard diet group). Experimental hyperlipidemia also affected antioxidant status of the animals caused by elevation of the serum malondialdehyde level (28.02 nmol/L vs. 3.02 nmol/L in the standard diet group) and reduction in liver enzymatic antioxidants as superoxide dismutase (SOD; 45.6 U/mg protein vs. 78.3 U/mg protein in the standard diet group), glutathione peroxidase (GPX; 3.83 U/mg protein vs. 9.63 U/mg protein in the standard diet group), and catalase (42 U/mg protein vs. 121 U/mg protein in the standard diet group).

This was an indication that the high-fat diet with 1% cholesterol leads to hyperlipidemia associated with antioxidant misbalance. Application of simvastatin lowered the animals' body weights and reduced serum lipid markers to sustainable levels close to those of the standard diet group, but the antioxidant effect was medium, which was not unexpected and has been previously demonstrated [82]. Both polysaccharides of *V. vitis-idaea* press cake DEAE–1% NaOH-f2 and -f3 demonstrated similar serum lipid lowering effects in the high-fat-diet animals against cholesterol (decreased by 33–38%), triglycerides (decreased by 29–35%), low-density lipoprotein–cholesterol (decreased by 48–53%), and high-density lipoprotein–cholesterol (decreased by 9–18%). This contrasted with the antioxidant potential when polymer DEAE–1% NaOH-f2 was more active than DEAE–1% NaOH-f3. The serum MDA value in the DEAE–1% NaOH-f2 group was 63% lower than in the high-fat diet group, and in the DEAE–1% NaOH-f3 group, we found a 36% reduction. The levels of SOD, GPX, and catalase after application of DEAE–1% NaOH-f2 increased by 60.9%, 186%, and 233%, respectively, compared with those in the high-fat diet group, while DEAE–1% NaOH-f3 resulted in 15%, 37%, and 128% boosts of enzymatic activity, respectively. This means that polysaccharides and polyphenol–polysaccharide conjugates of *V. vitis-idaea* are capable of being antioxidant and hypolipidemic agents in both in vitro and in vivo assays.

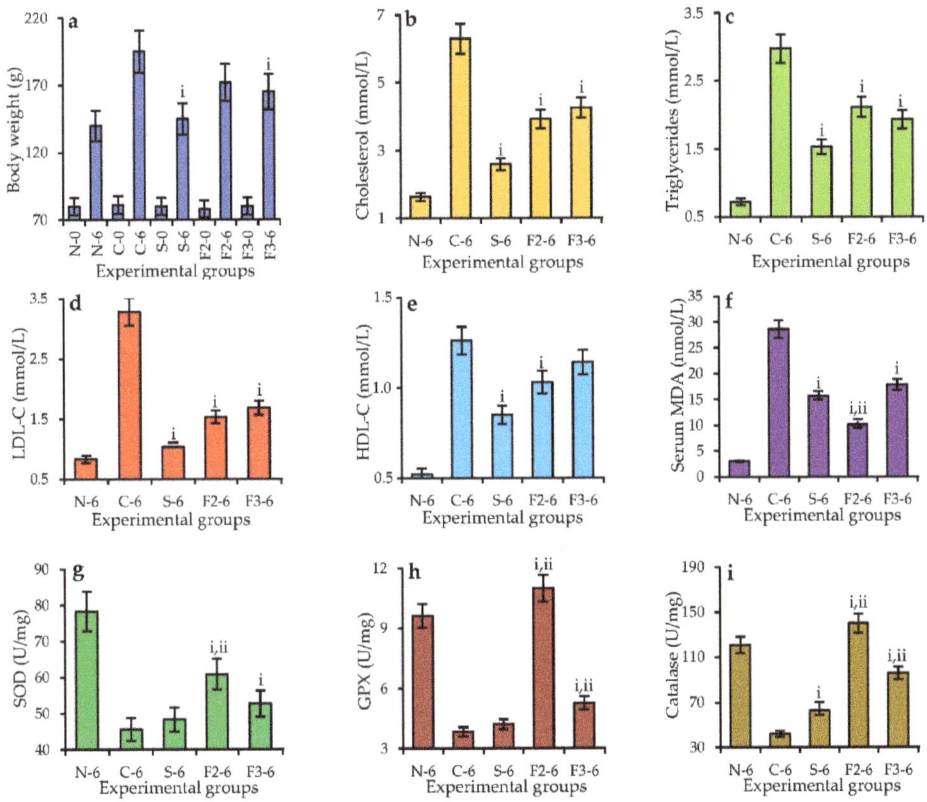

Figure 6. Changes in body weight (**a**), serum total cholesterol (**b**), serum total triglycerides (**c**), serum low-density lipoprotein–cholesterol (LDL–C) (**d**), serum high-density lipoprotein–cholesterol (HDL–C) (**e**),

serum malondialdehyde level (MDA) (**f**), liver superoxide dismutase (SOD) (**g**), liver glutathione peroxidase (GPX) (**h**) and liver catalase (**i**) in hamsters with a normal diet (N), with a 1% cholesterol diet (C), with a 1% cholesterol diet + simvastatin (10 mg/kg/day; S), with a 1% cholesterol diet + DEAE–1% NaOH-f2 polysaccharide (250 mg/kg/day; F2), with a 1% cholesterol diet + DEAE–1% NaOH-f3 polysaccharide (250 mg/kg/day; F3) before the feeding period (-0) and after a 6-month feeding period (-6). i—$p < 0.05$ vs. 1% cholesterol diet group (C-group); ii—$p < 0.05$ vs. 1% cholesterol diet + simvastatin group (S-group).

Polysaccharides of dietary origin are known hypolipidemic agents that normalize the lipid profile and antioxidant level of high-fat-diet animals [83]. The sources of bioactive polymers are fruits of pumpkin (*Cucurbita pepo*, *C. moschata*) [84,85], Chinese wolfberry (*Lycium barbarum*) [86], jujube or red date (*Ziziphus jujuba*) [87], Cherokee rose (*Rosa laevigata*) [32], and Japanese cornel (*Cornus officinalis*) [88]. In most cases, polysaccharides were polygalacturonates that reduced serum parameters (such as total cholesterol, high/low density lipoprotein cholesterol, and triglycerides) by inhibitory effects on the absorption of the bile acids and cholesterol [89], inhibition of the lipase activity [90], fat and cholesterol binding capacity and reduction in the accumulation of lipids and fecal fat and cholesterol contents [91], modulation of the gene expression of fatty acid synthesis [92], and increased formation of short chain fatty acids in the feces and regulation of lipid metabolism pathways [93]. At the same time, the regulation of observed antioxidant status and improvement of the oxidative stress [55] involves scavenging of free radicals and reduction in liver enzymes [88].

Antioxidant and hypolipidemic potential of polysaccharides and polyphenol– polysaccharide conjugates are markedly linked with structural specifics as molecular weight, elemental and monosaccharide composition, glycosidic linkage, and nature of conjugated polyphenolics [55,83]. The value of phenolic content in carbohydrate polymers is a crucial marker of radical-scavenging ability, nitric oxide (II) and hydrogen peroxide inactivating potential [94], as well as pancreatic lipase inhibition [95]. Additionally, high uronic content is an important factor of metal-chelating activity [96] and binding of bile, fat and cholesterol [83]. The homogenic water-soluble polymers of *V. vitis-idaea* press cake characterized by a high phenolics (polyphenol–polysaccharide conjugates DEAE-1% NaOH-f1 and DEAE-1% NaOH-f2) and uronic content (polysaccharides DEAE-1% NaOH-f3 and DEAE-1% NaOH-f4) and, therefore, the antioxidant and hypolipidemic activity of homogenic polymers, were caused by the various structural factors. Finally, considering the ever-increasing volumes of waste production by lingonberry processing factories, the press cake of *V. vitis-idaea* can become a promising feedstock for bioactive polymers manufacturing.

4. Conclusions

This is the first report of lingonberry (*Vaccinium vitis-idaea* L.) fruit press cake polymeric compounds characterization. The results of our study indicate the heterogeneity of *V. vitis-idaea* polysaccharides with a dominance of acidic polymers and polyphenol–polysaccharide conjugates which were neutral arabinogalactans esterified with hydroxycinnamates. This series of in vitro and in vivo studies suggest that polysaccharides normalize the lipid profile and antioxidant status of high-fat-diet hamsters. These findings support the idea of practical use of wastes from food processing as a source for antioxidant and hypolipidemic agent manufacturing in the pharmaceutical industry.

Supplementary Materials: The following are available online at https://www.mdpi.com/article/10.3390/foods11182801/s1: Table S1: Antioxidant activity of degradation polymers DEAE-1% NaOH-f1-d and DEAE-1% NaOH-f2-d; Table S2: In vitro hypolipidemic activity of degradation polymers DEAE-1% NaOH-f1-d and DEAE-1% NaOH-f2-d; Figure S1: HPLC-UV chromatograms of PMP-labeled samples of blank, standard monosaccharide mixture, and 2 M TFA hydrolysates of VVPS polysaccharide, fraction VVPS: DEAE-H_2O, and fraction VVPS: DEAE-1% NAOH; Figure S2: Possible ways of MS/MS cleavage of degradation products 3–7 released after alkaline destruction of DEAE-1% NaOH-f1 and DEAE-1% NaOH-f2 polymers.

Author Contributions: Conceptualization, D.N.O.; methodology, D.N.O. and N.K.C.; software, D.N.O.; validation, N.K.C.; formal analysis, N.K.C. and V.V.C.; investigation, D.N.O., N.K.C. and V.V.C.; resources, D.N.O. and N.K.C.; data curation, V.V.C.; writing—original draft preparation, D.N.O.; writing—review and editing, N.K.C.; visualization, D.N.O.; supervision, N.K.C.; project administration, N.K.C.; funding acquisition, D.N.O., N.K.C. and V.V.C. All authors have read and agreed to the published version of the manuscript.

Funding: This research was funded by the Scientific and Educational Foundation for the support of young scientists of Yakutia (project No 10); Ministry of Education and Science of Russia, grant numbers FSRG-2020-0019; 121030100227-7; and World-class Interregional Scientific and Educational Center (NOTs) "Baikal".

Institutional Review Board Statement: Not applicable.

Informed Consent Statement: Not applicable.

Data Availability Statement: Data is contained within the article.

Acknowledgments: The authors acknowledge the Buryat Research Resource Center for the technical support in chromatographic and mass-spectrometric research.

Conflicts of Interest: The authors declare no conflict of interest. The funders had no role in the design of the study, in the collection, analyses, or interpretation of data, in the writing of the manuscript, or in the decision to publish the results.

References

1. Fierascu, R.C.; Fierascu, I.; Avramescu, S.M.; Sieniawska, E. Recovery of natural antioxidants from agro-industrial side streams through advanced extraction techniques. *Molecules* **2019**, *24*, 4212. [CrossRef] [PubMed]
2. Maraveas, C. Production of sustainable and biodegradable polymers from agricultural waste. *Polymers* **2020**, *12*, 1127. [CrossRef] [PubMed]
3. Córdova, A.; Henríquez, P.; Nuñez, H.; Rico-Rodriguez, F.; Guerrero, C.; Astudillo-Castro, C.; Illanes, A. Recent advances in the application of enzyme processing assisted by ultrasound in agri-foods: A review. *Catalysts* **2022**, *12*, 107. [CrossRef]
4. Šelo, G.; Planinić, M.; Tišma, M.; Tomas, S.; Koceva Komlenić, D.; Bucić-Kojić, A. A comprehensive review on valorization of agro-food industrial residues by solid-state fermentation. *Foods* **2021**, *10*, 927. [CrossRef]
5. Bujor, O.-C.; Tanase, C.; Popa, M.E. Phenolic antioxidants in aerial parts of wild *Vaccinium* species: Towards pharmaceutical and biological properties. *Antioxidants* **2019**, *8*, 649. [CrossRef]
6. Malyshev, L.I. *Flora of Siberia: Pyrolaceae-Lamiaceae*; CRC Press: Boca Raton, FL, USA, 2006; pp. 25–28.
7. Timoshok, E.E.; Skorokhodov, S.N. Assessment of berry resources of Ericaceae family in Tomsk region, their rational use and protection. *Siberian Forest J.* **2019**, *4*, 80–88. [CrossRef]
8. Kowalska, K. Lingonberry (*Vaccinium vitis-idaea* L.) fruit as a source of bioactive compounds with health-promoting effects—A review. *Int. J. Mol. Sci.* **2021**, *22*, 5126. [CrossRef]
9. Vilas-Boas, A.A.; Pintado, M.; Oliveira, A.L.S. Natural bioactive compounds from food waste: Toxicity and safety concerns. *Foods* **2021**, *10*, 1564. [CrossRef]
10. Baroi, A.M.; Popitiu, M.; Fierascu, I.; Sărdărescu, I.-D.; Fierascu, R.C. Grapevine wastes: A rich source of antioxidants and other biologically active compounds. *Antioxidants* **2022**, *11*, 393. [CrossRef]
11. Moccia, F.; Flores-Gallegos, A.C.; Chávez-González, M.L.; Sepúlveda, L.; Marzorati, S.; Verotta, L. Ellagic acid recovery by solid state fermentation of pomegranate wastes by *Aspergillus niger* and *Saccharomyces cerevisiae*: A comparison. *Molecules* **2019**, *24*, 3689. [CrossRef]
12. Vu, H.T.; Scarlett, C.J.; Vuong, Q.V. Optimization of ultrasound-assisted extraction conditions for recovery of phenolic compounds and antioxidant capacity from banana (*Musa cavendish*) peel. *J. Food Process. Preserv.* **2017**, *41*, e13148. [CrossRef]
13. Markhali, F.S.; Teixeira, J.A.; Rocha, C.M.R. Olive tree leaves—A source of valuable active compounds. *Processes* **2020**, *8*, 1177. [CrossRef]
14. Boukroufa, M.; Boutekedjiret, C.; Petigny, L.; Rakotomanomana, N.; Chemat, F. Bio-refinery of orange peels waste: A new concept based on integrated green and solvent free extraction processes using ultrasound and microwave techniques to obtain essential oil, polyphenols and pectin. *Ultrason. Sonochem.* **2015**, *24*, 72–79. [CrossRef] [PubMed]
15. Ferrentino, G.; Morozova, K.; Mosibo, O.K.; Ramezani, M.; Scampicchio, M. Biorecovery of antioxidants from apple pomace by supercritical fluid extraction. *J. Clean Prod.* **2018**, *186*, 253–261. [CrossRef]
16. Panzella, L.; Moccia, F.; Nasti, R.; Marzorati, S.; Verotta, L.; Napolitano, A. Bioactive phenolic compounds from agri-food wastes: An update on green and sustainable extraction methodologies. *Front. Nutr.* **2020**, *7*, 60. [CrossRef] [PubMed]

17. Zokaityte, E.; Lele, V.; Starkute, V.; Zavistanaviciute, P.; Cernauskas, D.; Klupsaite, D.; Ruzauskas, M.; Alisauskaite, J.; Baltrusaitytė, A.; Dapsas, M.; et al. Antimicrobial, antioxidant, sensory properties, and emotions induced for the consumers of nutraceutical beverages developed from technological functionalised food industry by-products. *Foods* **2020**, *9*, 1620. [CrossRef] [PubMed]
18. Pukalskienė, M.; Pukalskas, A.; Dienaitė, L.; Revinytė, S.; Pereira, C.V.; Matias, A.A.; Venskutonis, P.R. Recovery of bioactive compounds from strawberry (*Fragaria* × *ananassa*) pomace by conventional and pressurized liquid extraction and assessment their bioactivity in human cell cultures. *Foods* **2021**, *10*, 1780. [CrossRef]
19. Mikulic-Petkovsek, M.; Veberic, R.; Hudina, M.; Zorenc, Z.; Koron, D.; Senica, M. Fruit quality characteristics and biochemical composition of fully ripe blackberries harvested at different times. *Foods* **2021**, *10*, 1581. [CrossRef]
20. Jurevičiūtė, I.; Keršienė, M.; Bašinskienė, L.; Leskauskaitė, D.; Jasutienė, I. Characterization of berry pomace powders as dietary fiber-rich food ingredients with functional properties. *Foods* **2022**, *11*, 716. [CrossRef]
21. Struck, S.; Plaza, M.; Turner, C.; Rohm, H. Berry pomace–A review of processing and chemical analysis of its polyphenols. *Int. J. Food Sci. Technol.* **2016**, *51*, 1305–1318. [CrossRef]
22. Souza, M.A.d.; Vilas-Boas, I.T.; Leite-da-Silva, J.M.; Abrahão, P.d.N.; Teixeira-Costa, B.E.; Veiga-Junior, V.F. Polysaccharides in agro-industrial biomass residues. *Polysaccharides* **2022**, *3*, 95–120. [CrossRef]
23. Kitrytė, V.; Kavaliauskaitė, A.; Tamkutė, L.; Pukalskienė, M.; Syrpas, M.; Rimantas Venskutonis, P. Zero waste biorefining of lingonberry (*Vaccinium vitis-idaea* L.) pomace into functional ingredients by consecutive high pressure and enzyme assisted extractions with green solvents. *Food Chem.* **2020**, *322*, 126767. [CrossRef] [PubMed]
24. Kunrade, L.; Rembergs, R.; Jēkabsons, K.; Kļaviņš, L.; Kļaviņš, M.; Muceniece, R.; Riekstiņa, U. Inhibition of NF-κB pathway in LPS-stimulated THP-1 monocytes and COX-2 activity in vitro by berry pomace extracts from five *Vaccinium* species. *J. Berry Res.* **2020**, *10*, 381–396. [CrossRef]
25. Kahlon, T.S.; Smith, G.E. In vitro binding of bile acids by blueberries (*Vaccinium* spp.), plums (*Prunus* spp.), prunes (*Prunus* spp.), strawberries (*Fragaria* × *ananassa*), cherries (*Malpighia punicifolia*), cranberries (*Vaccinium macrocarpon*) and apples (*Malus sylvestris*). *Food Chem.* **2007**, *100*, 1182–1187. [CrossRef]
26. Shamilov, A.A.; Bubenchikova, V.N.; Chernikov, M.V.; Pozdnyakov, D.I.; Garsiya, E.R. *Vaccinium vitis-idaea* L.: Chemical contents, pharmacological activities. *Pharm. Sci.* **2020**, *26*, 344–362. [CrossRef]
27. Olennikov, D.N.; Shamilov, A.A. Catechin-*O*-rhamnosides from *Vaccinium vitis-idaea* stems. *Chem. Nat. Comp.* **2022**, *58*, 269–273. [CrossRef]
28. Olennikov, D.N.; Shamilov, A.A. New compounds from *Vaccinium vitis-idaea*. *Chem. Nat. Comp.* **2022**, *58*, 240–244. [CrossRef]
29. Ross, K.A.; Godfrey, D.; Fukumoto, L. The chemical composition, antioxidant activity and α-glucosidase inhibitory activity of water-extractable polysaccharide conjugates from northern Manitoba lingonberry. *Cogent Food Agricult.* **2015**, *1*, 1109781. [CrossRef]
30. Luo, Q.; Cai, Y.; Yan, J.; Sun, M.; Corke, H. Hypoglycemic and hypolipidemic effects and antioxidant activity of fruit extracts from *Lycium barbarum*. *Life Sci.* **2004**, *76*, 137–149. [CrossRef]
31. Quesada-Morua, M.S.; Hidalgo, O.; Morera, J.; Rojas, G.; Pérez, A.M.; Vaillant, F.;.Fonseca, L. Hypolipidaemic, hypoglycaemic and antioxidant effects of a tropical highland blackberry beverage consumption in healthy individuals on a high-fat, high-carbohydrate diet challenge. *J. Berry Res.* **2020**, *10*, 459–474. [CrossRef]
32. Yu, C.H.; Dai, X.Y.; Chen, Q.; Zang, J.N.; Deng, L.L.; Liu, Y.H.; Ying, H.Z. Hypolipidemic and antioxidant activities of polysaccharides from Rosae Laevigatae Fructus in rats. *Carbohydr. Polym.* **2013**, *94*, 56–62. [CrossRef] [PubMed]
33. Lamport, D.T.A.; Kieliszewski, M.J.; Showalter, A.M. Salt stress upregulates periplasmic arabinogalactan proteins: Using salt stress to analyse AGP function. *New Phytol.* **2006**, *169*, 479–492. [CrossRef] [PubMed]
34. Thiex, N.; Novotny, L.; Crawford, A. Determination of ash in animal Feed: *AOAC Official Method* 942.05 revisited. *J. AOAC Int.* **2012**, *95*, 1392–1397. [CrossRef] [PubMed]
35. Olennikov, D.N.; Kashchenko, N.I.; Chirikova, N.K.; Gornostai, T.G.; Selyutina, I.Y.; Zilfikarov, I.N. Effect of low temperature cultivation on the phytochemical profile and bioactivity of Arctic plants: A case of *Dracocephalum palmatum*. *Int. J. Molec. Sci.* **2017**, *18*, 2579. [CrossRef]
36. Needs, P.W.; Selvendran, R.R. Avoiding oxidative degradation during sodium hydroxide/methyl iodide-mediated carbohydrate methylation in dimethyl sulfoxide. *Carbohydr. Res.* **1993**, *245*, 1–10. [CrossRef]
37. Olennikov, D.N.; Agafonova, S.V.; Rokhin, A.V.; Penzina, T.A.; Borovskii, G.B. Branched glucan from the fruiting bodies of *Piptoporus betulinus* (Bull.:Fr) Karst. *Appl. Biochem. Microbiol.* **2011**, *48*, 65–70. [CrossRef]
38. Olennikov, D.N.; Kirillina, C.S.; Chirikova, N.K. Water-soluble melanoidin pigment as a new antioxidant component of fermented willowherb leaves (*Epilobium angustifolium*). *Antioxidants* **2021**, *10*, 1300. [CrossRef]
39. Olennikov, D.N.; Chirikova, N.K.; Vasilieva, A.G.; Fedorov, I.A. LC-MS profile, gastrointestinal and gut microbiota stability and antioxidant activity of *Rhodiola rosea* herb metabolites: A comparative study with subterranean organs. *Antioxidants* **2020**, *9*, 526. [CrossRef]
40. Olennikov, D.N.; Chirikova, N.K.; Okhlopkova, Z.M.; Zulfugarov, I.S. Chemical composition and antioxidant activity of *Tánara Ótó* (*Dracocephalum palmatum* Stephan), a medicinal plant used by the North-Yakutian nomads. *Molecules* **2013**, *18*, 14105–14121. [CrossRef]
41. Olennikov, D.N.; Vasilieva, A.G.; Chirikova, N.K. *Fragaria viridis* fruit metabolites: Variation of LC-MS profile and antioxidant potential during ripening and storage. *Pharmaceuticals* **2020**, *13*, 262. [CrossRef]

42. Kumar, S.; Kumar, D.; Jusha, M.; Saroha, K.; Singif, N.; Vashishta, B. Antioxidant and free radical scavenging potential of *Citrullus colocynthis* (L.) Schrad. methanolic fruit extract. *Acta Pharm.* **2008**, *58*, 215–220. [CrossRef] [PubMed]
43. Fernando, C.D.; Soysa, P. Optimized enzymatic colorimetric assay for determination of hydrogen peroxide (H_2O_2) scavenging activity of plant extracts. *MethodsX* **2015**, *2*, 283–291. [CrossRef]
44. Olennikov, D.N.; Kashchenko, N.I.; Chirikova, N.K. A novel HPLC-assisted method for investigation of the Fe^{2+}-chelating activity of flavonoids and plant extracts. *Molecules* **2014**, *19*, 18296–18316. [CrossRef]
45. Nagaoka, S.; Futamura, Y.; Miwa, K.; Awano, T.; Yamauchi, K.; Kanamaru, Y.; Kuwata, T. Identification of novel hypocholesterolemic peptides derived from bovine milk β-lactoglobulin. *Biochem. Biophys. Res. Commun.* **2001**, *281*, 11–17. [CrossRef] [PubMed]
46. Kim, H.J.; White, P.J. In vitro bile-acid binding and fermentation of high, medium, and low molecular weight β-glucan. *J. Agricult. Food Chem.* **2010**, *58*, 628–634. [CrossRef] [PubMed]
47. Jin, Q.; Yu, H.; Wang, X.; Li, K.; Li, P. Effect of the molecular weight of water-soluble chitosan on its fat-/cholesterol-binding capacities and inhibitory activities to pancreatic lipase. *PeerJ* **2017**, *5*, e3279. [CrossRef] [PubMed]
48. Tan, Y.; Chang, S.K.C.; Zhang, Y. Comparison of α-amylase, α-glucosidase and lipase inhibitory activity of the phenolic substances in two black legumes of different genera. *Food Chem.* **2017**, *214*, 259–268. [CrossRef]
49. Cheng, Y.; Tang, K.; Wu, S.; Liu, L.; Qiang, C.; Lin, X.; Liu, B. *Astragalus* polysaccharides lowers plasma cholesterol through mechanisms distinct from statins. *PLoS ONE* **2011**, *6*, e27437. [CrossRef]
50. Deng, J.; Shi, Z.-J.; Li, X.-Z.; Liu, H.-M. Soluble polysaccharides isolation and characterization from rabbiteye blueberry (*Vaccinium ashei*) fruits. *BioRes* **2013**, *8*, 405–419. [CrossRef]
51. Rivas-Ubach, A.; Liu, Y.; Bianchi, T.S.; Tolić, N.; Jansson, C.; Paša-Tolić, L. Moving beyond the van Krevelen diagram: A new stoichiometric approach for compound classification in organisms. *Anal. Chem.* **2018**, *90*, 6152–6160. [CrossRef]
52. Jooyandeh, H.; Noshad, M.; Khamirian, R.A. Modeling of ultrasound-assisted extraction, characterization and in vitro pharmacological potential of polysaccharides from *Vaccinium arctostaphylos* L. *Int. J. Biol. Macromol.* **2018**, *107*, 938–948. [CrossRef] [PubMed]
53. Hilz, H.; Bakx, E.J.; Schols, H.A.; Voragen, A.G.J. Cell wall polysaccharides in black currants and bilberries—characterisation in berries, juice, and press cake. *Carbohydr. Polym.* **2005**, *59*, 477–488. [CrossRef]
54. Li, X.; Wang, X.; Guo, X.; Li, D.; Huo, J.; Yu, Z. Structural and biochemical characterization of a polysaccharide isolated from *Vaccinium uliginosum* L. *Starch* **2022**, *74*, 2100109. [CrossRef]
55. Wang, J.; Hu, S.; Nie, S.; Yu, Q.; Xie, M. Reviews on mechanisms of in vitro antioxidant activity of polysaccharides. *Oxidative Med. Cell. Longev.* **2016**, *2016*, 5692852. [CrossRef] [PubMed]
56. Brouns, F.; Theuwissen, E.; Adam, A.; Bell, M.; Berger, A.; Mensink, R.P. Cholesterol-lowering properties of different pectin types in mildly hyper-cholesterolemic men and women. *Eur. J. Clin. Nutr.* **2012**, *66*, 591–599. [CrossRef] [PubMed]
57. Liang, B.; Jin, M.; Liu, H. Water-soluble polysaccharide from dried *Lycium barbarum* fruits: Isolation, structural features and antioxidant activity. *Carbohydr. Polym.* **2011**, *83*, 1947–1951. [CrossRef]
58. Ross, K.; Siow, Y.; Brown, D.; Isaak, C.; Fukumoto, L.; Godfrey, D. Characterization of water extractable crude polysaccharides from cherry, raspberry, and ginseng berry fruits: Chemical composition and bioactivity. *Int. J. Food Prop.* **2015**, *18*, 670–689. [CrossRef]
59. Wu, C.; Zhao, M.; Bu, X.; Qing, Z.; Wang, L.; Xu, Y.; Bai, J. Preparation, characterization, antioxidant and antiglycation activities of selenized polysaccharides from blackcurrant. *RSC Adv.* **2020**, *10*, 32616–32627. [CrossRef]
60. Scaldaferri, F.; Pizzoferrato, M.; Ponziani, F.R.; Gasbarrini, G.; Gasbarrini, A. Use and indications of cholestyramine and bile acid sequestrants. *Intern. Emerg. Med.* **2013**, *8*, 205–210. [CrossRef]
61. Kahlon, T.S.; Smith, G.E. In vitro binding of bile acids by bananas, peaches, pineapple, grapes, pears, apricots and nectarines. *Food Chem.* **2007**, *101*, 1046–1051. [CrossRef]
62. Ren, D.; Chen, G. Extraction, purification and bile acid-binding capacity in vitro of polysaccharides from Okra. *Food Sci.* **2010**, *31*, 110–113. [CrossRef]
63. Long, H.; Xia, X.; Liao, S.; Wu, T.; Wang, L.; Chen, Q.; Wei, S.; Gu, X.; Zhu, Z. Physicochemical characterization and antioxidant and hypolipidaemic activities of a polysaccharide from the fruit of *Kadsura coccinea* (Lem.) A.C.Smith. *Front. Nutr.* **2022**, *9*, 903218. [CrossRef] [PubMed]
64. Gao, J.; Lin, L.; Sun, B.; Zhao, M. Comparison study on polysaccharide fractions from *Laminaria japonica*: Structural Characterization and bile acid binding capacity. *J. Agricult. Food Chem.* **2017**, *65*, 9790–9798. [CrossRef]
65. Gelissen, I.C.; Eastwood, M.A. Taurocholic acid adsorption during non-starch polysaccharide fermentation: An in vitro study. *Brit. J. Nutr.* **1995**, *74*, 221. [CrossRef] [PubMed]
66. Yang, M.; Hu, D.; Cui, Z.; Li, H.; Man, C.; Jiang, Y. Lipid-lowering effects of *Inonotus obliquus* polysaccharide in vivo and *in vitro*. *Foods* **2021**, *10*, 3085. [CrossRef]
67. Bai, J.; Li, J.; Pan, R.; Zhu, Y.; Xiao, X.; Li, Y.; Li, C. Polysaccharides from *Volvariella volvacea* inhibit fat accumulation in *C. elegans* dependent on the aak-2/nhr-49-mediated pathway. *J. Food Biochem.* **2021**, *45*, e13912. [CrossRef]
68. Aguilera-Angel, E.-Y.; Espinal-Ruiz, M.; Narváez-Cuenca, C.-E. Pectic polysaccharides with different structural characteristics as inhibitors of pancreatic lipase. *Food Hydrocoll.* **2018**, *83*, 229–238. [CrossRef]

69. Wu, D.-T.; Zhao, Y.-X.; Guo, H.; Gan, R.-Y.; Peng, L.-X.; Zhao, G.; Zou, L. Physicochemical and biological properties of polysaccharides from *Dictyophora indusiata* prepared by different extraction techniques. *Polymers* **2021**, *13*, 2357. [CrossRef]
70. Hong, Y.; Yin, J.; Nie, S.; Xie, M. Applications of infrared spectroscopy in polysaccharide structural analysis: Progress, challenge and perspective. *Food Chem. X* **2021**, *12*, 100168. [CrossRef]
71. Ho, T.C.; Kiddane, A.T.; Sivagnanam, S.P.; Park, J.-S.; Cho, Y.-J.; Getachew, A.T.; Nguyen, T.-T.T.; Kim, G.-D.; Chun, B.-S. Green extraction of polyphenolic-polysaccharide conjugates from *Pseuderanthemum palatiferum* (Nees) Radlk.: Chemical profile and anticoagulant activity. *Int. J. Biol. Macromol.* **2020**, *157*, 484–493. [CrossRef]
72. Kacuráková, M. FT-IR study of plant cell wall model compounds: Pectic polysaccharides and hemicelluloses. *Carbohydr. Polym* **2000**, *43*, 195–203. [CrossRef]
73. Čopíková, J.; Černá, M.; Novotná, M.; Kaasová, J.; Synytsya, A. Application of FT-IR spectroscopy in detection of food hydrocolloids in confectionery jellies and food supplements. *Czech J. Food Sci.* **2001**, *19*, 51–56. [CrossRef]
74. Synytsya, A. Fourier transform Raman and infrared spectroscopy of pectins. *Carbohydr. Polym.* **2003**, *54*, 97–106. [CrossRef]
75. Ridley, B.L.; O'Neill, M.A.; Mohnen, D. Pectins: Structure, biosynthesis, and oligogalacturonide-related signaling. *Phytochemistry* **2001**, *57*, 929–967. [CrossRef]
76. Vismeh, R.; Lu, F.; Chundawat, S.P.S.; Humpula, J.; Azarpira, A.; Balan, V.; Dale, B.E.; Ralph, J.; Jones, A.D. Profiling of diferulates (plant cell wall cross-linkers) using ultra high performance liquid chromatography-tandem mass spectrometry. *Analyst* **2013**, *138*, 1683. [CrossRef]
77. Fry, S.C. Feruloylated pectins from the primary cell wall: Their structure and possible functions. *Planta* **1983**, *157*, 111–123 [CrossRef]
78. Fry, S.; Willis, S.; Paterson, A. Intraprotoplasmic and wall-localised formation of arabinoxylan-bound diferulates and larger ferulate coupling-products in maize cell-suspension cultures. *Planta* **2000**, *211*, 679–692. [CrossRef]
79. Rouau, X.; Cheynier, V.; Surget, A.; Gloux, D.; Barron, C.; Meudec, E.; Criton, M. A dehydrotrimer of ferulic acid from maize bran *Phytochemistry* **2003**, *63*, 899–903. [CrossRef]
80. Chen, M.; Gitz, D.C.; McClure, J.W. Soluble sinapoyl esters are converted to wall-bound esters in phenylalanine ammonia-lyase-inhibited radish seedlings. *Phytochemistry* **1998**, *49*, 333–340. [CrossRef]
81. Pedersen, T.R.; Tobert, J.A. Simvastatin: A review. *Exp. Opin. Pharmacother.* **2004**, *5*, 2583–2596. [CrossRef]
82. Franzoni, F.; Quiñones-Galvan, A.; Regoli, F.; Ferrannini, E.; Galetta, F. A comparative study of the in vitro antioxidant activity of statins. *Int. J. Cardiol.* **2003**, *90*, 317–321. [CrossRef]
83. Kalita, P.; Ahmed, A.B.; Sen, S.; Chakraborty, R. A comprehensive review on polysaccharides with hypolipidemic activity: Occurrence, chemistry and molecular mechanism. *Int. J. Biol. Macromol.* **2022**, *206*, 681–698. [CrossRef]
84. Sedigheh, A.; Jamal, S.M.; Mahbubeh, S.; Somayeh, K.; Mahmoud, R.-K.; Azadeh, A. Hypoglycaemic and hypolipidemic effects of pumpkin (*Cucurbita pepo* L.) on alloxan-induced diabetic rats. *Afr. J. Pharm. Pharmacol.* **2011**, *5*, 2620–2626. [CrossRef]
85. Song, H.; Sun, Z. Hypolipidaemic and hypoglycaemic properties of pumpkin polysaccharides. *3Biotech* **2017**, *7*, 159. [CrossRef] [PubMed]
86. Ming, M.; Guanhua, L.; Zhanhai, Y.; Guang, C.; Xuan, Z. Effect of the *Lycium barbarum* polysaccharides administration on blood lipid metabolism and oxidative stress of mice fed high-fat diet in vivo. *Food Chem.* **2009**, *113*, 872–877. [CrossRef]
87. Ji, X.; Liu, F.; Peng, Q.; Wang, M. Purification, structural characterization, and hypolipidemic effects of a neutral polysaccharide from *Ziziphus jujuba* cv. Muzao. *Food Chem.* **2018**, *245*, 1124–1130. [CrossRef]
88. Wang, D.; Li, C.; Fan, W.; Yi, T.; Wei, A.; Ma, Y. Hypoglycemic and hypolipidemic effects of a polysaccharide from Fructus Corni in streptozotocin-induced diabetic rats. *Int. J. Biol. Macromol.* **2019**, *133*, 420–427. [CrossRef]
89. Zhao, X.H.; Qian, L.; Yina, D.; Zhou, Y. Hypolipidemic effect of the polysaccharides extracted from pumpkin by cellulase-assisted method on mice. *Int. J. Biol. Macromol.* **2014**, *64*, 137–138. [CrossRef]
90. Kolsi, R.B.A.; Jardak, N.; Hadjkacem, F.; Chaaben, R.; Jribi, I.; Feki, A.E.; Rebai, T.; Jamoussi, K.; Fki, L.; Belghith, H.; et al. Anti-obesity effect and protection of liver-kidney functions by *Codium fragile* sulphated polysaccharide on high fat diet induced obese rats. *Int. J. Biol. Macromol.* **2017**, *102*, 119–129. [CrossRef]
91. Teng, Z.; Qian, L.; Zhou, Y. Hypolipidemic activity of the polysaccharides from *Enteromorpha prolifera*. *Int. J. Biol. Macromol.* **2013**, *62*, 254–256. [CrossRef]
92. Rjeibi, I.; Feriani, A.; Hentati, F.; Hfaiedh, N.; Michaud, P.; Pierre, G. Structural characterization of water-soluble polysaccharides from *Nitraria retusa* fruits and their antioxidant and hypolipidemic activities. *Int. J. Biol. Macromol.* **2019**, *129*, 422–432. [CrossRef] [PubMed]
93. Nie, Q.; Hu, J.; Gao, H.; Fan, L.; Chen, H.; Nie, S. Polysaccharide from *Plantago asiatica* L. attenuates hyperglycemia, hyperlipidemia and affects colon microbiota in type 2 diabetic rats. *Food Hydrocoll.* **2019**, *86*, 34–42. [CrossRef]
94. Kolodziejczyk-Czepas, J.; Bijak, M.; Saluk, J.; Ponczek, M.B.; Zbikowska, H.M.; Nowak, P.; Tsirigotis-Maniecka, M.; Pawlaczyk, I. Radical scavenging and antioxidant effects of *Matricaria chamomilla* polyphenolic–polysaccharide conjugates. *Int. J. Biol. Macromol.* **2015**, *72*, 1152–1158. [CrossRef] [PubMed]
95. Campos, F.; Peixoto, A.F.; Fernandes, P.A.R.; Coimbra, M.A.; Mateus, N.; de Freitas, V.; Fernandes, I.; Fernandes, A. The antidiabetic effect of grape pomace polysaccharide-polyphenol complexes. *Nutrients* **2021**, *13*, 4495. [CrossRef] [PubMed]
96. Wang, R.; Liang, R.; Dai, T.; Chen, J.; Shuai, X.; Liu, C. Pectin-based adsorbents for heavy metal ions: A review. *Trends Food Sci. Technol.* **2019**, *91*, 319–329. [CrossRef]

foods

Article

Antialcohol and Hepatoprotective Effects of Tamarind Shell Extract on Ethanol-Induced Damage to HepG2 Cells and Animal Models

Shao-Cong Han [1], Rong-Ping Huang [1], Qiong-Yi Zhang [2], Chang-Yu Yan [2], Xi-You Li [1], Yi-Fang Li [2], Rong-Rong He [2] and Wei-Xi Li [1,*]

1 College of Traditional Chinese Medicine, Yunnan University of Chinese Medicine, Kunming 650500, China
2 Guangdong Engineering Research Centre of Chinese Medicine & Disease Susceptibility, Jinan University, Guangzhou 510632, China
* Correspondence: liweixi1001@163.com; Tel.: +86-158-0885-7436

Abstract: Alcohol liver disease (ALD) is one of the leading outcomes of acute and chronic liver injury. Accumulative evidence has confirmed that oxidative stress is involved in the development of ALD. In this study, we used chick embryos to establish ALD model to study the hepatoprotective effects of tamarind shell exttract (TSE). Chick embryos received 25% ethanol (75 µL) and TSE (250, 500, 750 µg/egg/75 µL) from embryonic development day (EDD) 5.5. Both ethanol and TSE were administrated every two days until EDD15. Ethanol-exposed zebrafish and HepG2 cell model were also employed. The results suggested that TSE effectively reversed the pathological changes, liver dysfunction and ethanol-metabolic enzyme disorder in ethanol-treated chick embryo liver, zebrafish and HepG2 cells. TSE suppressed the excessive reactive oxygen species (ROS) in zebrafish and HepG2 cells, as well as rebuilt the irrupted mitochondrial membrane potential. Meanwhile, the declined antioxidative activity of glutathione peroxidase (GPx) and superoxide dismutase (SOD), together with the content of total glutathione (T-GSH) were recovered by TSE. Moreover, TSE upregulated nuclear factor erythroid 2—related factor 2 (NRF2) and heme oxyense-1 (HO-1) expression in protein and mRNA level. All the phenomena suggested that TSE attenuated ALD through activating NRF2 to repress the oxidative stress induced by ethanol.

Keywords: alcohol liver disease; chick embryo; tamarind shell extract; oxidative stress; NRF2

Citation: Han, S.-C.; Huang, R.-P.; Zhang, Q.-Y.; Yan, C.-Y.; Li, X.-Y.; Li, Y.-F.; He, R.-R.; Li, W.-X. Antialcohol and Hepatoprotective Effects of Tamarind Shell Extract on Ethanol-Induced Damage to HepG2 Cells and Animal Models. *Foods* **2023**, *12*, 1078. https://doi.org/10.3390/foods12051078

Academic Editors: Noelia Castillejo Montoya and Lorena Martínez-Zamora

Received: 14 February 2023
Revised: 28 February 2023
Accepted: 28 February 2023
Published: 3 March 2023

1. Introduction

Alcohol abuse is becoming a global health problem. The immediate consequence of acute drinking is the hangover, with lethargy and cognitive impairment as the most prominent manifestations [1]. Because the liver is responsible for the metabolism of over 90% of consumed ethanol, alcohol liver disease (ALD) has been a main reason of acute and chronic liver injury [2]. For a start, ethanol is converted to acetaldehyde in hepatocytes, which is catalyzed by alcohol dehydrogenase (ADH) and microsomal ethanol-oxidizing system, particularly cytochrome P450 2E1 (CYP2E1) and CYP3A [3,4]. Acetaldehyde is subsequently transformed to acetate, under the catalysis of aldehyde dehydrogenase (ALDH) [5]. Excessive or chronic alcohol consumption causes alcohol metabolism disorder in hepatocytes decreasing ADH and ALDH and increasing CYP2E1. The oxidation of ethanol catalyzed by CYP2E1 accelerates the accumulation of acetaldehyde and produces large amounts of reactive oxygen species (ROS), such as singlet oxygen radicals, superoxide radicals, hydroxyl radicals and hydrogen peroxide [6]. Superfluous ROS combined with disrupted antioxidative defense systems, oxidative stress occurs and results in liver injury, including ethanolic hepatitis, liver fibrosis, cirrhosis, and even liver cancer. However, the treatment options of ALD are still limited. The demand of potential therapies for ALD is urgent.

Tamarind (*Tamarindus indica L.*), also known as Suanjiao, Suandou, Mahanghuang (Dai language), belongs to the subfamily Caesalpinioideae of the family Leguminosae (Fabaceae). It is widely cultured in Africa, Southeast Asia, and South America. Tamarind is high yielding, with an annual output of 300,000 tons in India alone [7]. Tamarind shells are leftovers of direct consumption and by-products of processing. With large amounts, tamarind shells waste resources and pollute the environment. Waste products such as peel and shell contain high levels of polyphenols, flavonoids, anthocyanins, vitamin C, and carotenoids [8], which are expected to be used as a natural source of medicinal antioxidants, cosmetics and food to increase economic benefits. Tamarind shell extract was also reported to attenuate carbon tetrachloride-induced liver injury in mice [9]. Our previous study has demonstrated that tamarind shell extract was rich in flavonoids. In our previous analysis by ultraperformance liquid chromatography-mass spectrometry (UPLC/MS), the major components of TSE were identified as naringenin, luteolin, myricetin, morin, eriocitrin, apigenin, (+) catechin, and taxifolin. We also found that TSE and the main flavonoids had remarkable antioxidant effects in vitro and in vivo [10]. However, limited research was found concerning the pharmacological effects of TSE on liver damage caused by alcohol. Therefore, in this study, the hepatoprotective effects of TSE were studied on ethanol-stimulated HepG2 cells in vitro and ethanol-induced liver injury in zebrafish and chicken embryos. The related mechanism was further explored. The results of this work would hopefully introduce new uses of tamarind shell and promote the comprehensive utilization of tamarind.

2. Materials and Methods

2.1. Chemicals and Reagents

From Jiangmen Yujun Trading Co., Ltd. (Jiangmen, China), 95% edible ethanol was. Then, 3-(4,5-dimethylthiazol-2-yl)-2,5-diphenyltetrazolium bromide (MTT) was purchased from Sigma (St. Louis, MO, USA). Rhodamine 123 and assay kits for the determination of reactive oxygen species (ROS), total glutathione, superoxide dismutase (SOD), catalase (CAT) and lipid peroxidation (MDA) were procured from Beyotime Biotechnology Company (Shanghai, China). Alcohol dehydrogenase (ADH) and acetaldehyde dehydrogenase (ALDH) assay kits were bought from Solarbio Biotechnology Company (Beijing, China). Triglyceride (TG) assay kits was obtained from Nanjing Jiancheng Bioengineering Research Institute (Nanjing, China). A chicken cytochrome P450 2E1 (CYP2E1) ELISA kit was purchased from Shanghai Jingkang Bioengineering Co., Ltd. (Shanghai, China). Primary antibodies of NRF2 and HO-1 were purchased from Proteintech Group (Chicago, IL, USA). HRP-labeled fluorescent secondary antibody (antimouse and antirabbit) and anti-β-actin were procured from Fude Biological Technology (Hangzhou, China). Dulbecco's modified Eagle's medium (DMEM), fetal bovine serum (FBS), and penicillin streptomycin (PS) were purchased from Life Technologies Corporation (Gibco, Grand Island, NY, USA).

2.2. Preparation of Tamarind Shell Extract (TSE)

Tamarind fruits (*Tamarindus indica* L.) were purchased in Xishuangbanna (China) and authenticated by Professor Ya-qiong Li, Yunnan University of Chinese Medicine. The fruit shells were separated, dried, and powdered. A total of 500 g of tamarind shell fine powder was refluxed twice with 1500 mL of 95% ethanol for 3 h. The extract was concentrated under reduced pressure, then freeze-dried and stored at 4 °C.

TSE was dissolved in ultrapure water (50 mg/mL) and stored at 4 °C. The solution was diluted to the experimental concentration with the corresponding culture medium in the following experiment.

2.3. Zebrafish Maintenance and Exposure

Zebrafish (wild-type AB) were bought from the China zebrafish resource center (Wuhan, China) and maintained following light and temperature standard conditions [11]. The embryos were collected from breeding groups of a 1:1 or 2:1 female-to-male ratio. The fertilized and normally developed eggs were selected and incubated in embryo medium. Zebrafish larvae of five days postfertilization (dpf) were used for subsequent experiments. All procedures were reviewed and approved by the Animal Care and Welfare Committee of Yunnan University of Chinese Medicine (approval No: R-062019009; approval date: 7 March 2019).

2.3.1. Toxicity

The toxicity of TSE to zebrafish larvae was determined by exposing zebrafish to TSE. Briefly, 5 dpf zebrafish larvae were divided to TSE exposure groups at different concentrations and, control group at random. TSE was added to the larvae (5, 10, 20, 40, and 60 μg/mL). The control group received embryo culture solution. Exposures were performed in triplicate. The survival rate was recorded every day until 9 dpf.

2.3.2. Exposure

The exposure of zebrafish to ethanol was following the protocol described earlier [12] with some modifications. In brief, at 5 dpf, zebrafish larvae were randomised into four groups, which were control group, model group, and two TSE treatment (10 μg/mL and 20 μg/mL) groups, and cultured in a six-well plate (50 fish per well, three multiple wells per group). All larvae except the control group were exposed to 1.5% ethanol or 1.5% ethanol with TSE prepared in embryo culture medium. Fish in control group were given embryo culture solution. The zebrafish were incubated for 32 h at 28.5 °C. Six zebrafish were selected from each group for phenotype analysis. The remaining fish were transferred to a 12-well plate (30 fish per well) and added 3 mL of appropriate solution: the experimental groups received TSE (10 μg/mL and 20 μg/mL, diluted in embryo culture solution), whereas control and model groups received embryo culture solution. After incubation for 48 h, the fish were collected for subsequent experiments. Triplicate wells were set for each group.

2.4. Chicken Embryos and Treatment

The fertilized Leghorn eggs required for the experiment were obtained from the poultry farm of South China Agriculture University (Guangzhou, China). The eggs were incubated in an incubator (ProCon automatic systems GmbH & Co. KG, Luebeck, Germany) at 38 °C under 70% humidity. At the embryonic development day (EDD) 5.5, live fertilized eggs were selected by an egg illuminator marked the air chamber, and a small hole was introduced into the shell for drug administration. Live eggs were randomly divided into a control group, an ethanol group and TSE (250, 500 and 750 μg per egg) groups with 24 eggs in each group. The eggs in control group were treated with 75 μL bird saline per egg, the eggs in ethanol group were administrated with 75 μL 25% ethanol per egg, and the eggs of TSE groups were dosed with 75 μL indicated dosage of TSE coadministration with 25% ethanol. The eggs were administered with appropriate solution every two days from EDD 5.5 until the EDD 15. The embryos were collected for subsequent experiments. A schematic illustration of the above protocol is presented in Figure 1.

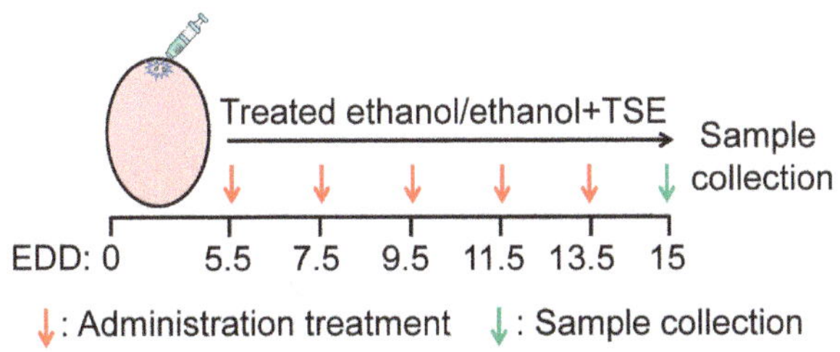

Figure 1. Schematic diagram of chicken embryo treatment protocol.

2.5. Cell Culture

Human hepatoma HepG2 cells (Procell CL-0103) were bought from Punuosai Life Technology (Wuhan, China). HepG2 cells were cultured in DMEM supplemented with 12% FBS and 1% PS in an incubator (Thermo, Waltham, MA, USA) at 37 °C, 5% CO_2 and 65% humidity. The following experiments were completed by using the third- to tenth-generation cells.

2.6. Cell Toxicity and Cell Experiment

The cytotoxicity of TSE and ethanol on HepG2 cells were determined by MTT assay. In brief, HepG2 cells were seeded into 96-well plates (2×10^4/well) and cultured overnight. Afterward, the cells were treated with TSE or ethanol for 24 h, and the cell toxicity was determined by MTT assay.

To assess the protective effect of TSE against ethanol-induced injury, HepG2 cells were grown in 48-well plates at a density of 6×10^4/well for 12 h. After treatment of TSE (5, 10, 20, and 40 μg/mL) for 24 h, cells were administrated with ethanol (700 mM) for 3 h. Cell viability was assessed by MTT assay. The DCFH-DA assay was carried out to quantified the generation of intracellular ROS on a flow cytometry (Beckman Coulter, Bria, CA, USA) as reported by Zhang et al. [13]. The mitochondrial membrane potential of HepG2 cells was measured by using rhodamine 123 as a fluorescent probe (Beyotime, Shanghai, China), following the method reported by Wang et al. [14].

2.7. Zebrafish Behavioral Test and Evaluation of ROS Level in Zebrafish

Zebrafish were treated as mentioned above. Five zebrafish were randomly selected from each group and placed in a 96-well plate individually. The swimming behavior, including total moving distance, trace, and total activity time, was monitored every 5 s for a total of 70 min by using a zebrafish behavior automatic analyzer (View Point Behavior Technology, Lyon, France) according to the manufacturer's guide.

For the measurement of ROS, zebrafish were homogenized (Beyotime, Shanghai, China) in PBS on ice, then centrifuged at 12,000 rpm for 15 min at 4 °C. A BCA protein assay kit was utilized to quantify the protein concentration. The protein was added to a black 96-well plate (55.84 μg in 50 μL/well) containing DCFH-DA fluorescent probe (50 μL/well, final concentration 10 μM). The intensity of fluorescence was recorded in a Synergy H1 fluorescence microplate reader (BioTek, Winooski, VT, USA), the excited and emitted wavelength were 488 and 525 nm.

2.8. Histological Analysis

Liver damage was evaluated by histological examination in the chick embryos, as the method described by Zhang et al. and with some modifications [13]. In brief, chicken embryo liver tissue was fixed in 4% paraformaldehyde solution for 10 days and then embedded in paraffin, then sectioned into 5 μm paraffin slices. After dewaxing and rehydrating, the slice was stained with hematoxylin and eosin (H&E) and sirius red (Servicebio, Wuhan, China). The morphology of the tissue sections was observed under an automatic scanning microscope (Pannormic, Budapest, Hungary).

2.9. Biochemical Analysis

The chicken embryo liver tissue homogenate was prepared on ice and then centrifuged at 12,000 rpm for 10–20 min at 4 °C. The protein content of supernatant was detected with BCA protein assay kit, and the supernatant was collected for subsequent experiments. The hepatic level of MDA, T-GSH and TG and the activities of GPx, SOD, CAT, ADH, ALDH and CYP2E1 were assessed by commercial kits according to the manufacturers' protocols.

2.10. Western Blot Analysis

Total proteins of HepG2 cells and chicken embryo liver tissues were extracted, respectively, as described before. The concentration of protein was determined with BCA protein assay kit. Protein samples (25 μg) were loaded on 10% SDS-PAGE to separate and blotted to a PVDF membrane. The transferred PVDF membrane was sealed in 5% skimmed milk solution at 25 °C for 2 h after being cleaned with TBST, then incubated with primary antibodies to detect NRF2 (1:1500) and HO-1 (1:1500) at 4 °C overnight; β-actin (1:5000) served as an internal standard. Subsequently, the membranes were incubated with HRP-labeled fluorescent secondary antibody, antimouse or antirabbit (1:5000) at room temperature for 2 h. Finally, the protein bands were imaged by using an ECL Detection Kit (Fdbio science, Hangzhou, China) and the Vilber FUSION FX 6 imaging system (Vilber, Collégien, France).

2.11. qRT-PCR

Total RNA was isolated from chick embryo liver tissues, HepG2 cells, and zebrafish by using TRIzol Reagent as described by the manufacturer's instructions. A NanoDrop ND-2000C spectrophotometer (Thermo, Waltham, MA, USA) was used to quantify the concentration of RNA. A total of 1000 ng of the total RNA was utilized to synthesize cDNA following the manufacturer's guide (TransGen Biotech, Beijing, China). The result of qRT-PCR was analyzed by LightCycler 96 system (Roche, Switzerland) by using the TransStart Top Green qPCR Supermix Kit (TransGen Biotech, Beijing, China). The target gene expression was assessed by using the $2^{-\Delta\Delta CT}$ method and GAPDH was used as the housekeeper gene and to a control group sample. The sequences of primers were listed in Table 1.

Table 1. Sequences of primers used for qRT-PCR [12,13,15,16].

Species	Gene	Forward Primer (5′–3′)	Reverse Primer (5′–3′)
Zebrafish	*cyp2y3*	TATTCCCATGCTGCACTCTG	AGGAGCGTTTACCTGCAGAA
Zebrafish	*cyp3a65*	AAACCCTGATGAGCATGGAC	CAAGTCTTTGGGGATGAGGA
Zebrafish	*adh8a*	CGAGTACACCGTCATCAAC	AGCACCGAGTCCGAATAC
Zebrafish	*adh8b*	ATTGATGATGATGCTCCTCTG	TAGACCAACCGCACCAAG
Zebrafish	*gapdh*	TGGTGCTGGTATTGCT	TTGCTGTAACCGAACTCA
Chicken	*Cyp3a4*	TCATAGTGTTGTTCCCCTT	GGTATCCTTCTTCCCGTTC
Chicken	*Cyp3a7*	GACTCCATGAACAACCCCAA	AAATCTACTCTGCCCGTGTG
Chicken	*Cyp2d6*	GAACCCTGCTTACATCCGAGA	CATGAACAGGAACGCCCAT
Chicken	*Cyp2c45*	CGGAGACAACAAGCACCACCA	TTCGTGATCGTCCTACTACCC
Chicken	*Nrf2*	CATAGAGCAAGTTTGGGAAGAG	GTTTCAGGGCTCGTGATTGT
Chicken	*Ho-1*	AACGCCACCAAGTTCAGTCTCC	AGCTTCTGCAGCGCCTCAA
Chicken	*Gapdh*	AGAACATCATCCCAGCGT	AGCCTTCACTACCCTCTTG
Human	*NRF2*	CCTCAACTATAGCGATGCTGAATCT	AGGAGTTGGGCATGAGTGAGTAG
Human	*HO-1*	CCAGTGCCACCAAGTTCAAG	CAGCTCCTGCAACTCCTCAA
Human	*GAPDH*	GCCTCAAGATCATCAGCAATGC	CCTTCCACGATACCAAAGTTGTCAT

2.12. Statistical Analysis

All values were expressed as mean ± SD. Data were analyzed by one-way variance (ANOVA) followed by Tukey's multiple comparison with SPSS 25 software (IBM SPSS Statistics 25.0, Armonk, NY, USA). The $p < 0.05$, $p < 0.01$ and $p < 0.001$ were set at the threshold for statistical significance, high statistical significance, and very high statistical significance, respectively. Graphs were constructed using GraphPad Prism 8 (GraphPad Prism 8.0, San Diego, CA, USA) and Adobe illustrator CS6 software (Adobe illustrator CS6, San Jose, CA, USA).

3. Results

3.1. TSE Protected Ethanol-Induced Damage in HepG2 Cells

The cell toxicity of TSE was tested on HepG2 cells firstly. The cell viability suggested that TSE had no significant toxicity to HepG2 cells with a concentration of less than or equal to 60 μg/mL (Figure 2A). When treated with ethanol, the cell viability dropped significantly (Figure 2B,C), accompanied by evident morphology changes, reflected as gradual shrinkage, increased intercellular gaps and complete loss of the original morphological characteristics (Figure 2D). We found that all doses of TSE could rescue the ethanol-treated cells from death and reverse the morphological alterations (Figure 2C,D). Numerous studies have confirmed that the oxidation of ethanol produces large amounts of ROS and results in the impairment of mitochondrial [17]. Using fluorescence probes, we measured the intracellular ROS level and the mitochondrial membrane potential. Similarly, TSE effectively inhibited the accumulation of ROS (Figure 2E), and restored the disrupted mitochondrial membrane potential in HepG2 cells induced by ethanol (Figure 2F). These findings indicated that TSE was effective on preventing ethanol-induced cell damage, eliminating intracellular ROS and reversing the depolarization of mitochondrial membrane.

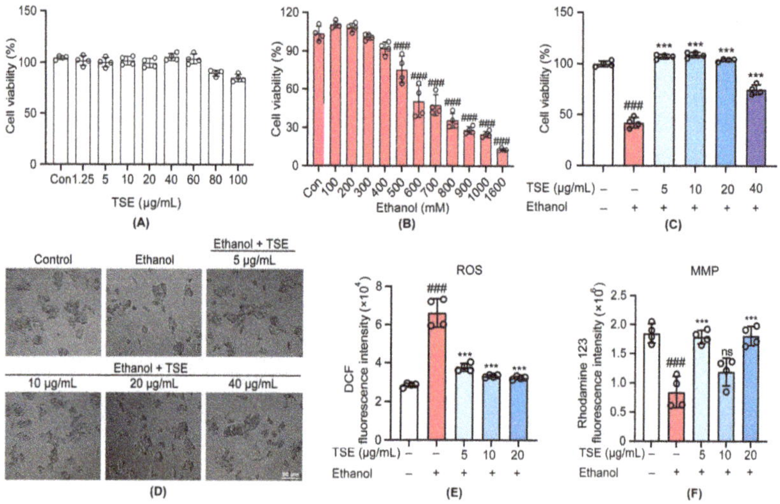

Figure 2. TSE protected HepG2 cells against ethanol-induced damage. (**A**) MTT assay of HepG2 cell viability following treatment with TSE for 24 h, $n = 4$ biologically independent experiments. (**B**) MTT assay of HepG2 cell viability following treatment with ethanol for 24 h, $n = 4$ biologically independent experiments. (**C**) MTT assay of HepG2 cell viability following pretreatment with or without TSE for 24 h before cotreatment with or without ethanol (700 mM) for 3 h, $n = 4$ biologically independent experiments. (**D**) Morphology of cells under microscope (magnification ×200). (**E**) Flow cytometric analysis of intracellular ROS levels detected by DCFH-DA probe, $n = 4$ biologically independent experiments. (**F**) Flow cytometric analysis of mitochondrial membrane potential by rhodamine 123 staining, $n = 4$ biologically independent experiments. Data represent the mean ± standard deviation and significant differences were analyzed by one-way ANOVA. ### $p < 0.001$ vs. control group; *** $p < 0.001$ vs. ethanol group. ns, not significant vs. ethanol group.

3.2. *TSE Ameliorated Ethanol-Induced Behavior Changes and Oxidative Stress in Zebrafish*

To further study the protective effect against ethanol of TSE in vivo, the ethanol-exposed zebrafish model was employed. At first, we evaluated the toxicity of TSE in zebrafish. The survival rate in all groups exceeded 91.67% (96 h after administration), indicating that TSE had negligible toxicity with the concentration of 60 μg/mL (Figure 3A). As shown in Figure 3B, ethanol produced severe developmental abnormality of zebrafish, presented as spinal curvature, pericardial edema, hepatomegaly yolk sac edema and lack of swim bladder. TSE prevented the incidence of malformation induced by ethanol. In the motor activity test, ethanol caused a dramatic reduction in total swimming distance and the amount of active time compared to the untreated fish ($p < 0.01$) (Figure 3C). TSE (20 μg/mL) greatly increased the swimming distance and the amount of active time. Furthermore, we detected the content of ROS in zebrafish. Consistently, TSE strikingly blocked the excessive

ROS generated by ethanol (Figure 3D). These data illustrated that TSE relieved alcoholism and suppressed ROS produced by ethanol.

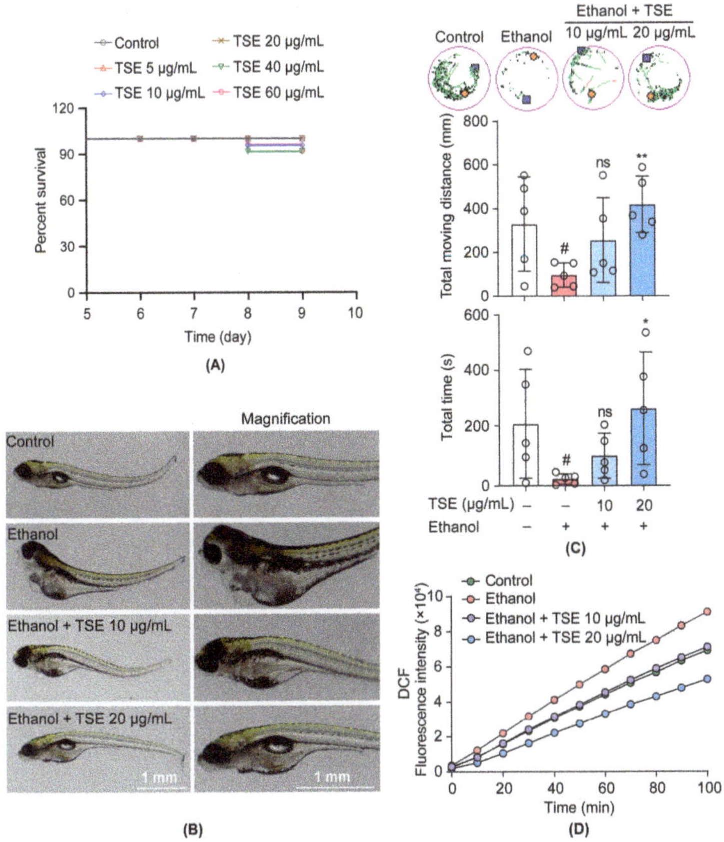

Figure 3. TSE ameliorated ethanol-induced behavior changes and oxidative stress in zebrafish. (**A**) TSE was used to treat 5 dpf zebrafish for 96 h to detect the toxicity of TSE. (**B**) TSE resisted ethanol-induced phenotypic changes in zebrafish, scale bar, 1 mm. (**C**) TSE ameliorated ethanol-induced behavior changes and increased the movement distance and activity time, $n = 5$ biologically independent samples. (**D**) The effect of TSE on the time-dependent changes of ethanol-induced DCF fluorescence intensity. Data represent the mean ± standard deviation and significant differences were analyzed by one-way ANOVA. $^{\#}$ $p < 0.05$ vs. control group; * $p < 0.05$, ** $p < 0.01$ vs. ethanol group; ns, not significant vs. ethanol group.

3.3. TSE Ameliorated Ethanol-Provoked Liver Dysfunction and Alleviated Liver Injury in Chicken Embryos

As the major site of detoxification, liver is susceptible to stimuli and pollutants. Furthermore, the lipase liberated by ethanol metabolism catalyzed the production of TG and the deposition of fat in liver cells [18]. Thus, TG is one of the biomarkers of ethanol-induced liver damage [19]. After ethanol administration, the chicken embryo survival rate dropped to 70.83%, whereas TSE treatment improved the survival rate (Figure 4A). In addition, TSE reversed the increased liver index caused by ethanol (Figure 4B). We also noted a significant increment of TG levels in embryo liver. The elevated TG levels could be remarkably decreased by TSE (Figure 4C). Ethanol consumption also gives rise to the accumulation of ROS disrupted metabolic enzymes. All of these changes trigger pathological responses, which contribute to the etiology of ethanol-induced liver injury [20,21]. Ethanol impeded the development of embryo liver, which was similar to the influence on zebrafish larvae In contrast with the normal embryo liver, the ethanol group showed obvious hemorrhage and fatty liver (Figure 4D). As presented in Figure 3F, hepatocytes in the control group were complete, with a neat, tight arrangement, intact nucleus, and clear gaps between hepatic sinuses. In contrast, ethanol led to evident hepatic pathological alterations. The liver sinuses dilated with disordered cell arrangement. Hemorrhage, inflammatory infiltration, fat deposition, and fibrosis were also found. TSE mitigated fat vacuole formation, inflammatory infiltration, and fibrosis. (Figure 4E,F).

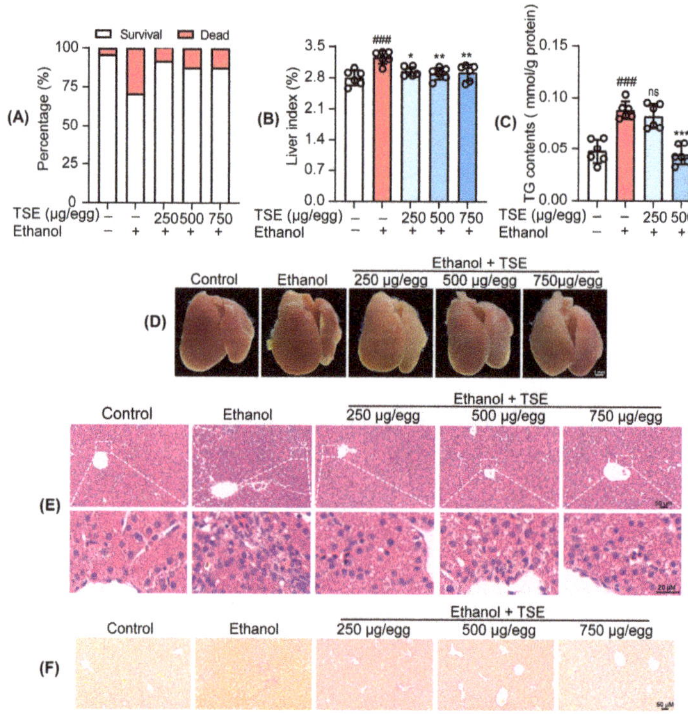

Figure 4. TSE ameliorated ethanol-induced chicken embryo death and liver damage. (**A**) TSE reversed ethanol-induced chick embryo death rates and (**B**) liver index, $n = 6$ biologically independent samples. (**C**) The effect of TSE on ethanol-induced changes of TG level in chicken embryo liver, $n = 6$ biologically independent samples. (**D**) Representative images of chicken embryo liver; scale bar, 1 mm. (**E**) H&E staining showing the histopathological changes of livers; scale bar, 50 μm (up), 20 μm (bottom). (**F**) Sirius red staining showing liver fibrosis; scale bar, 50 μm. Data represent the mean ± standard deviation and significant differences were analyzed by one-way ANOVA. ### $p < 0.001$ vs. control group; * $p < 0.05$, ** $p < 0.01$, *** $p < 0.001$ vs. ethanol group; ns, not significant vs. ethanol group.

3.4. TSE Corrected the Disorder of Ethanol-Metabolic Enzymes In Vivo

Disturbing metabolic enzymes are a typical feature of ethanol-induced liver damage. Ethanol is firstly transformed to acetaldehyde mainly by ADH, CYP2E1, and CYP3A. Acetaldehyde is subsequently metabolized by ALDH. Needless ROS generated during ethanol oxidation, especially through CYP2E1 [5]. Additionally, as an inducible enzyme, the expression of CYP2E1 is enhanced by alcohol ingestion, which exacerbates the production of ROS. Meanwhile, ethanol counteracts the effect of ALDH, leading to the accumulation of acetaldehyde, a robust hepatotoxic [12,18,22].

To evaluate the recovering effect of TSE on ethanol-evoked liver damage, we detected the expressions of ethanol metabolism-related genes by qPCR. In Figure 5A, we observed a slight rise in the mRNA expression of ethanol metabolism-related enzyme *cyp2y3* in ethanol-treated zebrafish. The mRNA expressions, *cyp3a65, adh8a,* and *adh8b* in zebrafish were markedly upregulated following ethanol exposure (Figure 5A). All doses of TSE could greatly abrogate the expressions of *cyp3a65, adh8a,* and *adh8b* in ethanol-stimulated zebrafish. In ethanol-exposed chick embryo liver, as shown in Figure 5B, the mRNA expression of *Cyp3a4, Cyp3a7,* and *Cyp2d6* were upregulated in the ethanol group, and these genes were involved in ethanol oxidation [23]. Similarly, the expression of *Cyp2c45* mRNA related to fat synthesis [24] was also upregulated by alcohol. TSE treatment decreased the expression of these genes. In addition, CYP2E1 enzyme activity was significantly increased. The activity of ALDH was drastically inhibited, accompanied by a mildly ADH. Likewise, TSE inhibited the overactive CYP2E1 and restored the inactive ALDH (Figure 5C–E). TSE was capable of correcting the disorder of ethanol-metabolic enzymes.

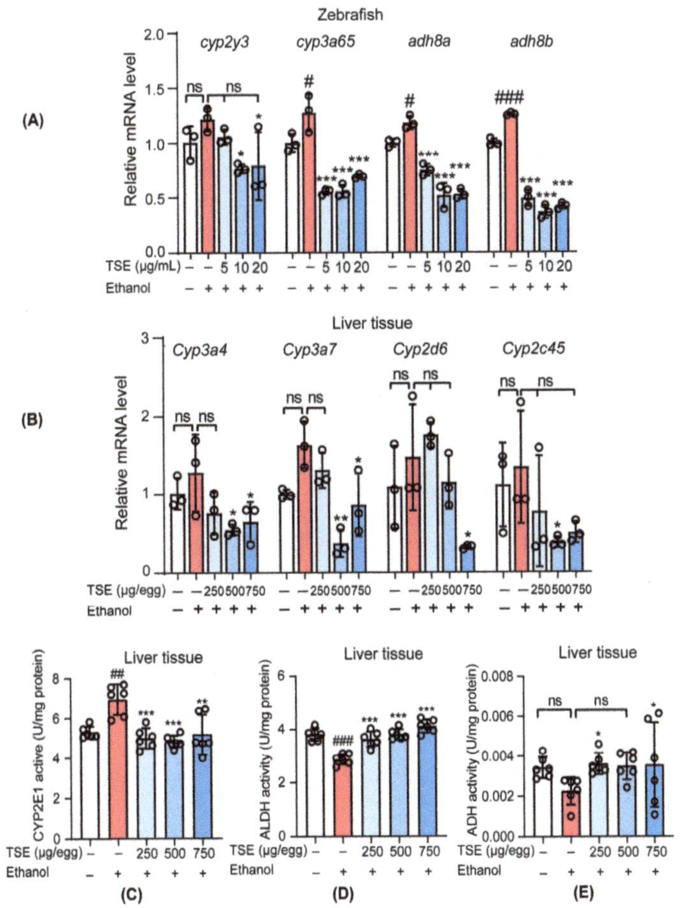

Figure 5. TSE ameliorated ethanol-induced metabolism dysregulation in vivo. (**A**) qRT-PCR analysis of the mRNA levels of *cyp2y3, cyp3a65, adh8a,* and *adh8b* in zebrafish, $n = 3$ biologically independent samples. (**B**) qRT-PCR analysis of the mRNA levels of *Cyp3a4, Cyp3a7, Cyp2d6,* and *Cyp2c45* in liver tissue, $n = 3$ biologically independent samples. Activity of (**C**) CYP2E1, (**D**) ALDH, and (**E**) ADH in the chicken embryo liver tissues, $n = 6$ biologically independent samples. Data represent the mean $\pm$ standard deviation and significant differences were analyzed by one-way ANOVA. $^{\#}$ $p < 0.05$, $^{\#\#}$ $p < 0.01$, $^{\#\#\#}$ $p < 0.001$ vs. control group; * $p < 0.05$, ** $p < 0.01$, *** $p < 0.001$ vs. ethanol group; ns, not significant vs. ethanol group.

3.5. TSE Attenuated Oxidative Stress in Ethanol-Induced Chick Embryo Livers

Ethanol exposure can interfere with the antioxidant system that protects hepatocytes against ROS damage. Antioxidant defense systems within cells include SOD, GPx, CAT and T-GSH, which are important in defending cells from oxidative damage and preventing lipid peroxidation. MDA is the final metabolite of lipid peroxidation and play an indicative function in oxidative damage. [25,26]. We noted that ethanol disrupted the antioxidant defense system of the liver, with declined SOD, GPx, and CAT activities as well as the T-GSH content (Figure 6A–D), although the change of CAT activity was not significant. In parallel, the MDA level in embryo liver was elevated after ethanol treatment (Figure 6E) Administration of TSE at 250, 500 or 750 μg/egg evidently recovered the activities of SOD Except for 250 μg/egg of TSE, the content of T-GSH was also evidently increased. Only 750 μg/egg of TSE can significantly restore the activity of GPx and CAT. Of course, the production of MDA was also inhibited by TSE.

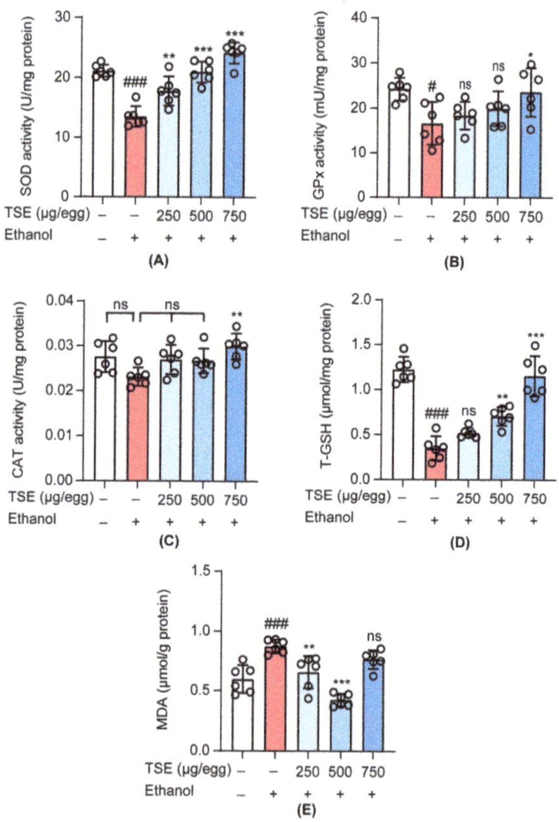

Figure 6. The antioxidant effect of TSE on ethanol-induced chicken embryo liver damage. (**A**–**E**) are the effects of TSE on the changes of ethanol-induced (**A**) SOD, (**B**) GPx, (**C**) CAT, (**D**) T-GSH, and (**E**) MDA levels in ethanol-induced chicken embryo liver, $n = 6$ biologically independent samples. Data represent the mean ± standard deviation and significant differences were analyzed by one-way ANOVA. $^{\#}$ $p < 0.05$, $^{\#\#\#}$ $p < 0.001$ vs. control group; * $p < 0.05$, ** $p < 0.01$, *** $p < 0.001$ vs. ethanol group; ns, not significant vs. ethanol group.

3.6. TSE Activated the Nuclear Factor Erythrocyte-2-Related Factor 2 (NRF2)-Mediated Antioxidant Response

NRF2 is a vital player in regulating intracellular redox homeostasis in cells [27]. Normally, NRF2 exists in the cytoplasm. When oxidative stress occurs, NRF2 translocates to the nucleus and binds antioxidant response elements (ARE) to drive the expression of the downstream genes, such as NQO-1, HO-1, SOD, GPx, CAT, and GSH [20,28]. Improved antioxidant capacity can restrain ROS-induced liver damage. However, chronic alcohol ingestion inhibits the function of NRF2, exacerbating oxidative stress in the liver [18,29]. Therefore, we monitored the levels of NRF2 and HO-1 both in chicken embryo liver and HepG2 cells. In chicken embryo liver tissue, NRF2 and HO-1 were significantly down-regulated at protein and mRNA levels in ethanol group, whereas remarkable restoration was detected in the TSE treatment groups (Figure 7A,B). The protein levels of NRF2 and HO-1 were evidently suppressed in ethanol-stimulated HepG2 cells, accompanied by mRNA expressions of *NRF2* and *HO-1* significant alteration. Pretreated with TSE (5 μg/mL, 10 μg/mL and 20 μg/mL) significantly restored the protein levels of ethanol-inhibited NRF2 and HO-1 (Figure 7C). The mRNA levels of *NRF2* and *HO-1* were also apparently raised in HepG2 cells by pretreated with 5 μg/mL and 10 μg/mL TSE (Figure 7D). Those findings elucidated the fact that TSE alleviated ethanol-induced liver damage by modulating NRF2 and its downstream antioxidant enzymes.

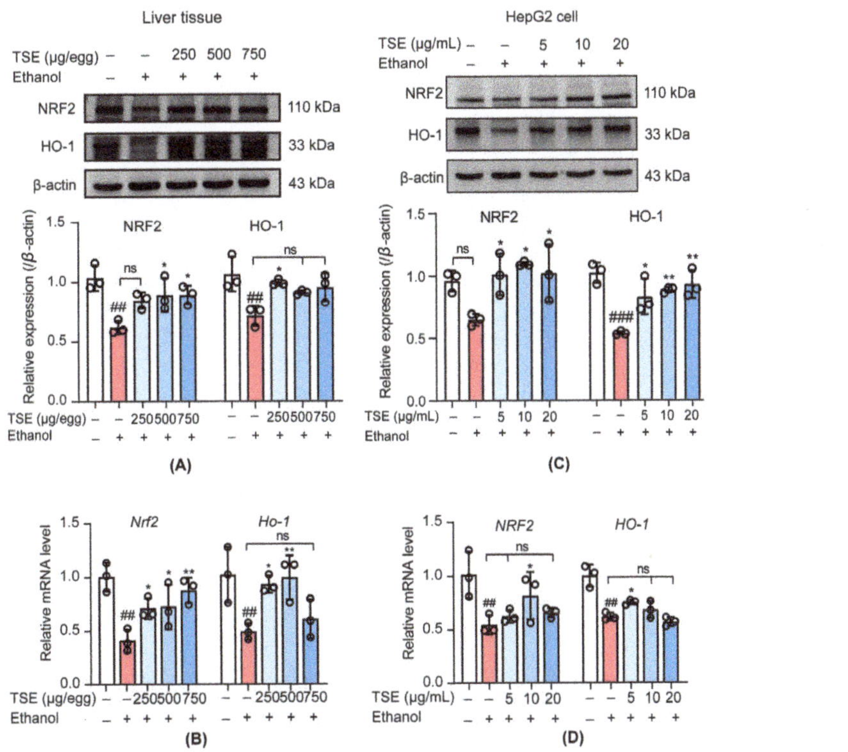

Figure 7. Effect of TSE on NRF2-mediated antioxidant signaling in ethanol-induced liver damage. (**A**) NRF2 and HO-1 protein expression levels in chicken embryo liver tissue, $n = 3$ biologically independent samples. (**B**) qRT-PCR analysis of the mRNA levels of *Nrf2* and *Ho-1* in chicken embryo liver tissue, $n = 3$ biologically independent samples. (**C**) NRF2 and HO-1 protein expression levels in HepG2 cells, $n = 3$ biologically independent experiments. (**D**) qRT-PCR analysis of the mRNA levels of *NRF2* and *HO-1* in HepG2 cells, $n = 3$ biologically independent experiments. Data represent the mean ± standard deviation and significant differences were analyzed by one-way ANOVA. ## $p < 0.01$, ### $p < 0.001$ vs. control group; * $p < 0.05$, ** $p < 0.01$ vs. ethanol group; ns, not significant vs. ethanol group.

4. Discussion

Until now, common treatments for ALD have been nutritional support and liver protection therapy, but we still lack efficacious and targeted treatment options. Suitable models for ALD are urgently needed to explore the pathogenesis and therapeutic strategies. Current experimental models of ALD are mainly based on rodents and cell lines. However, ALD related experiments on in vitro systems have limits in mimicking the symptoms of entire living organisms. In vivo ALD studies performed on rodents are of high cost, have long cycles, and have complicated operations with ethical problems. In contrast, chicken embryos are easy to operate, and the developmental process is easy to observe. In this work, we built an ethanol-induced liver injury model on chicken embryos for the first time. The embryonic liver of chicken has been formed at 20–22 somite stage (50–53 h after hatching) [30,31], and hemoblast cells have appeared in the hepatic sinus space at EDD 5 [32]. Therefore, we exposed chick embryos to 75 μL 25% (v/v) ethanol every two days from EDD5.5 until EDD15. There was a significant increase in the liver index due to ethanol. Severe hypertriglyceridemia is a prevalent complication of alcohol consumption, so TG is an important indicator of ALD [18]. A significant rise of TG in liver was noted after ethanol administration as well. The most important clinical pathological manifestations of ALD were steatosis, hepatocyte damage, and inflammatory infiltration. In ethanol-exposed embryo liver, we found obvious hepatic pathological alterations, presented as hemorrhage, inflammatory infiltration, fatty liver tissue and fibrosis. The disorder of metabolic enzymes was also noticed. Ethanol substantially inhibited the liver ALDH activity leading to excessive accumulation of acetaldehyde. ADH, another metabolic enzyme related to alcohol was mildly suppressed. Meanwhile, the activity of CYP2E1 was enhanced, and CYP3A7, the main CYP3A enzyme in fetal liver [33], was significantly upregulated by ethanol. These CYPs are associated with the generation of large amounts of ROS. Because oxidation stress is linked to alcohol-induced liver disease, we detected the change of hepatic antioxidant defense system in chick embryos subjected to ethanol. The activities of antioxidant enzymes including SOD, GPx, and CAT were decreased together with the content of low molecular weight antioxidant GSH. On the other hand, MDA, a final product and indicator of lipid peroxidation, remarkably elevated. The disrupted antioxidant defense system along with enhanced lipid peroxidation suggested the occurrence of oxidative stress. All the above indicated that the ALD model on chick embryos was successfully established.

Tamarind shells, wastes of consumption and processing, lack proper utilization for a long time. In our previous research, we found that tamarind shell was rich in functional chemicals. By using LC/MS, we identified and quantified eight flavonoids in TSE. These included naringenin, luteolin, myricetin, morin, eriodictyol, apigenin, catechin, and taxifolin. The antioxidative effects of TSE and the eight flavonoids were evaluated in vitro and in vivo [10]. Our research indicated that TSE possessed potent antioxidant capacity both in vitro and in vivo. The tamarind shell extract (60 μg/mL) was low in cytotoxicity, with no influence on the growth of HepG2 cells and zebrafish.

In the current work, we systematically evaluated the hepatoprotective effect of TSE on ethanol-induced liver damage in chick embryos, zebrafish, and HepG2 cells. TSE was found to rescued HepG2 cell from ethanol. In the presence of ethanol, there was a dramatic decrease in the motor activity of zebrafishes. Severe morphology changes were observed in ethanol-exposed zebrafish and HepG2 cells. Combined with the alteration in embryo liver described earlier, ethanol caused severe hepatic damage in vitro and in vivo. TSE could improve the pathological changes induced by alcohol in chick embryo liver, zebrafish, and cells, and increase in the amount of time and total distances of swimming in zebrafish. All the abovementioned results suggested that TSE relieved ethanol-provoked hepatic damage in vitro and in vivo.

Oxidation stress is an important feature of ethanol-induced liver damage. By using the fluorescence probe DCFH-DA, we found excessive ROS in zebrafish and HepG2 cells cultured with ethanol. Superfluous ROS contributes to mitochondrial impairment, including a reduction of the mitochondrial membrane potential ($\Delta\Psi m$) and damaged the permeability of membrane. The dysfunctional mitochondrial not only produces more ROS, but also releases cytochrome c, and then activates caspase 9 cascade which causes apoptosis of hepatocyte [34]. We noticed an evident reduction of ΔΨm and cell viability in HepG2 cells within ethanol incubation. Furthermore, the disruption of antioxidant defense system was found in ethanol-administrated embryo liver, as manifested by decreased SOD, GPx, CAT, and total GSH and increased MDA. In accordance with our previous research, TSE could efficiently scavenge the overproduced ROS in alcohol-exposed zebrafish and HepG2 cells. The ΔΨm of HepG2 cells was rebuilt by TSE. A salient remission of the malfunctional antioxidant defense system was also noted in chick embryo liver with the treatment of TSE. Taken together, TSE efficaciously combated the oxidation stress triggered by ethanol.

Disturbing metabolic enzymes re another outcome of alcohol intake. ADH, CYP2E1, and CYP3A are responsible for converting alcohol into acetaldehyde. However, CYP2E1-mediated oxidation of ethanol liberates large amounts ROS. Meanwhile, ethanol is a strong inducer of CYP2E1. The elevated CYP2E1 generates more ROS [35]. Additionally, ALDH, a major player in acetaldehyde metabolism, is severely inhibited by ethanol and results in the accumulated acetaldehyde, a powerful hepatotoxin. In chick embryo liver, the activities of ALDH and ADH were hampered by ethanol with the comparison of normal liver. TSE was capable of reversing the metabolic enzyme disorder by recovering ALDH and ADH and blocking CYP2E1 (Figure 8).

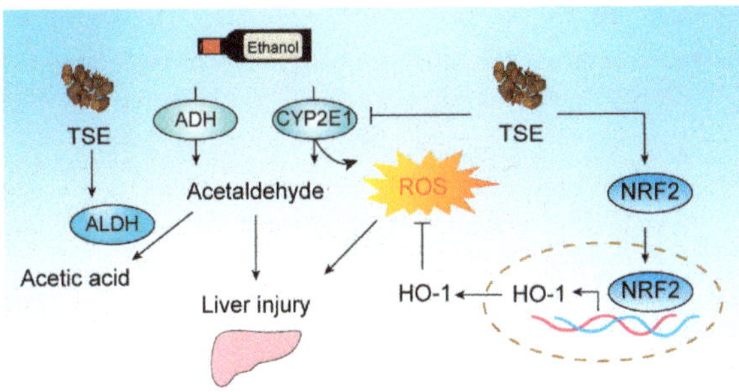

Figure 8. A schematic illustration of tamarind shell extract against ethanol-induced liver damage. Tamarind shell extract can improve alcohol-induced liver injury by improving alcohol metabolism and activating NRF2 pathway to inhibit oxidative stress.

NRF2/ARE pathway is essential in maintaining the redox homeostasis in humans. NRF2 mediates the transcription of antioxidation and detoxification enzyme genes, such as SOD, GPx, CAT, GSH, NQO-1, and HO-1 [36,37]. Thereafter, the oxidative homeostasis is reestablished. It has been reported that flavonoids such as luteolin, apigenin, naringenin and epicatechin are activators of NRF2/ARE in multiple cell lines. Therefore, we proposed that the protection of TSE on ethanol-evoked liver damage was related to the activation of NRF2. Western blot and PCR results indicated the apparent downregulation of NRF2 and HO-1 in ethanol exposed embryo liver. The levels of both genes were also decreased in HepG2 cells. These data combined with the inhibition of antioxidant defense system demonstrated that the NRF2 pathway was suppressed by ethanol. Consistently, TSE could significantly activate NRF2, following upregulation of HO-1 at the mRNA and protein level in embryo liver (Figure 8). Afterward, the oxidative homeostasis was rehabilitated.

5. Conclusions

In this study, we successfully built an ALD model on chick embryo to explore the hepatoprotective activity of TSE. TSE effectively attenuated ethanol-induced liver injury by relieving oxidative stress via activating the NRF2 pathway. These findings could facilitate the utilization of TSE, increase economic benefits, and provide raw material support for the study of natural ingredients of antialcohol and liver protection drugs. In addition, chicken embryo has the characteristics of simple and easy access, short experimental period and low experimental cost. This model of chicken embryo alcoholic liver injury can be used for high-throughput evaluation and screening of more active substances for antialcohol and liver protection.

Author Contributions: Conceptualization, S.-C.H., Q.-Y.Z. and W.-X.L.; methodology, S.-C.H., R.-P.H and X.-Y.L.; formal analysis, S.-C.H. and W.-X.L.; resources, W.-X.L.; writing—original draft preparation, S.-C.H.; writing—review and editing, C.-Y.Y., Y.-F.L. and W.-X.L.; supervision, Y.-F.L., R.-R.H and W.-X.L.; funding acquisition, W.-X.L. All authors have read and agreed to the published version of the manuscript.

Funding: This research was funded by Natural Science Foundation of China, grant number 81960780, U1801284, 81560661 and Yunnan Provincial Science and Technology Department-Applied Basic Research Joint Special Funds of Yunnan University of Chinese Medicine (2019FF002 (-007), 2017FF116 (-015)).

Institutional Review Board Statement: The animal study protocol was approved by the Animal Care and Welfare Committee of Yunnan University of Chinese Medicine (approval No: R-062019009; approval date: 7 March 2019).

Informed Consent Statement: Not applicable.

Data Availability Statement: The data presented in this study are available on request from the corresponding author.

Acknowledgments: We thank Ya-Qiong Li for authenticating the plant material.

Conflicts of Interest: The authors declare no conflict of interest.

References

1. Penning, R.; McKinney, A.; Verster, J.C. Alcohol hangover symptoms and their contribution to the overall hangover severity. *Alcohol. Alcohol.* **2012**, *47*, 248–252. [CrossRef] [PubMed]
2. Madushani Herath, K.; Bing, S.J.; Cho, J.; Kim, A.; Kim, G.; Kim, J.S.; Kim, J.B.; Doh, Y.H.; Jee, Y. Sasa quelpaertensis leaves ameliorate alcohol-induced liver injury by attenuating oxidative stress in HepG2 cells and mice. *Acta Histochem.* **2018**, *120*, 477–489. [CrossRef]
3. Niemelä, O.; Parkkila, S.; Juvonen, R.O.; Viitala, K.; Gelboin, H.V.; Pasanen, M. Cytochromes P450 2A6, 2E1, and 3A and production of protein-aldehyde adducts in the liver of patients with alcoholic and non-alcoholic liver diseases. *J. Hepatol.* **2000**, *33*, 893–901. [CrossRef] [PubMed]
4. Cardoso, J.L.; Lanchote, V.L.; Pereira, M.P.; Capela, J.M.; de Moraes, N.V.; Lepera, J.S. Impact of inhalational exposure to ethanol fuel on the pharmacokinetics of verapamil, ibuprofen and fluoxetine as in vivo probe drugs for CYP3A, CYP2C and CYP2D in rats. *Food Chem. Toxicol.* **2015**, *84*, 99–105. [CrossRef] [PubMed]
5. Teschke, R. Alcoholic Liver Disease: Alcohol Metabolism, Cascade of Molecular Mechanisms, Cellular Targets, and Clinical Aspects. *Biomedicines* **2018**, *6*, 106. [CrossRef]
6. Teschke, R.; Neuman, M.G.; Liangpunsakul, S.; Seitz, H.-K. Alcoholic Liver Disease and the co-triggering Role of MEOS with Its CYP 2E1 Catalytic Cycle and ROS. *Arch. Gastroenterol. Res.* **2021**, *2*, 9–25.
7. Yahia, E.M.; Salih, N.K.E. Tamarind (*Tamarindus indica* L.). In *Postharvest Biology and Technology of Tropical and Subtropical Fruits*; Yahia, E.M., Ed.; Woodhead Publishing: Cambridge, UK, 2011; pp. 442–458e.
8. Chimsah, F.A.; Nyarko, G.; Abubakari, A.H. A review of explored uses and study of nutritional potential of tamarind (*Tamarindus indica* L.) in Northern Ghana. *Afr. J. Food Sci.* **2020**, *14*, 285–294. [CrossRef]
9. Liman, M.; Atawodi, S. Hepatoprotective and Nephroprotective Effects of Methanolic Extract of Different Parts of Tamarindus Indica Linn in Rats Following Acute and Chronic Carbon Tetrachloride Intoxication. *Annu. Res. Rev. Biol.* **2015**, *5*, 109–123. [CrossRef]
10. Li, W.; Huang, R.; Han, S.; Li, X.; Gong, H.; Zhang, Q.; Yan, C.; Li, Y.; He, R. Potential of Tamarind Shell Extract against Oxidative Stress In Vivo and In Vitro. *Molecules* **2023**, *28*, 1885. [CrossRef]
11. Westerfield, M. *The Zebrafish Book: A Guide for the Laboratory Use of Zebrafish*; Inst of Neuro Science: Eugene, OR, USA, 2000; Available online: http://zfin.org/zf_info/zfbook/zfbk.html (accessed on 5 May 2022).
12. Liu, Y.S.; Yuan, M.H.; Zhang, C.Y.; Liu, H.M.; Liu, J.R.; Wei, A.L.; Ye, Q.; Zeng, B.; Li, M.F.; Guo, Y.P.; et al. Puerariae Lobatae radix flavonoids and puerarin alleviate alcoholic liver injury in zebrafish by regulating alcohol and lipid metabolism. *Biomed. Pharm.* **2021**, *134*, 111121. [CrossRef]
13. Zhang, Q.Y.; Han, S.C.; Huang, R.P.; Jiang, M.Y.; Yan, C.Y.; Li, X.Y.; Zhan, Y.J.; Li, X.M.; Li, Y.F.; Kurihara, H.; et al. Cyclo(-Phe-Phe) alleviates chick embryo liver injury via activating the Nrf2 pathway. *Food Funct.* **2022**, *13*, 6962–6974. [CrossRef] [PubMed]

4. Wang, M.; Ruan, Y.; Chen, Q.; Li, S.; Wang, Q.; Cai, J. Curcumin induced HepG2 cell apoptosis-associated mitochondrial membrane potential and intracellular free Ca^{2+} concentration. *Eur. J. Pharmacol.* **2011**, *650*, 41–47. [CrossRef] [PubMed]
5. Kluver, N.; Ortmann, J.; Paschke, H.; Renner, P.; Ritter, A.P.; Scholz, S. Transient overexpression of adh8a increases allyl alcohol toxicity in zebrafish embryos. *PLoS ONE* **2014**, *9*, e90619. [CrossRef] [PubMed]
6. Guo, K.; Ge, J.; Zhang, C.; Lv, M.W.; Zhang, Q.; Talukder, M.; Li, J.L. Cadmium induced cardiac inflammation in chicken (*Gallus gallus*) via modulating cytochrome P450 systems and Nrf2 mediated antioxidant defense. *Chemosphere* **2020**, *249*, 125858. [CrossRef]
7. Koch, O.R.; Pani, G.; Borrello, S.; Colavitti, R.; Cravero, A.; Farre, S.; Galeotti, T. Oxidative stress and antioxidant defenses in ethanol-induced cell injury. *Mol. Aspects Med.* **2004**, *25*, 191–198. [CrossRef]
8. Zheng, J.; Tian, X.; Zhang, W.; Zheng, P.; Huang, F.; Ding, G.; Yang, Z. Protective Effects of Fucoxanthin against Alcoholic Liver Injury by Activation of Nrf2-Mediated Antioxidant Defense and Inhibition of TLR4-Mediated Inflammation. *Mar. Drugs* **2019**, *17*, 552. [CrossRef]
9. Zhao, H.; Zhang, J.; Liu, X.; Yang, Q.; Dong, Y.; Jia, L. The antioxidant activities of alkalic-extractable polysaccharides from Coprinus comatus on alcohol-induced liver injury in mice. *Sci. Rep.* **2018**, *8*, 11695. [CrossRef] [PubMed]
20. Shu, G.; Qiu, Y.; Hao, J.; Fu, Q.; Deng, X. Nuciferine alleviates acute alcohol-induced liver injury in mice: Roles of suppressing hepatic oxidative stress and inflammation via modulating miR-144/Nrf2/HO-1 cascade. *J. Funct. Foods* **2019**, *58*, 105–113. [CrossRef]
21. Zhou, Y.; Tan, F.; Li, C.; Li, W.; Liao, W.; Li, Q.; Qin, G.; Liu, W.; Zhao, X. White Peony (Fermented *Camellia sinensis*) Polyphenols Help Prevent Alcoholic Liver Injury via Antioxidation. *Antioxidants* **2019**, *8*, 524. [CrossRef]
22. Lai, Y.; Zhou, C.; Huang, P.; Dong, Z.; Mo, C.; Xie, L.; Lin, H.; Zhou, Z.; Deng, G.; Liu, Y.; et al. Polydatin alleviated alcoholic liver injury in zebrafish larvae through ameliorating lipid metabolism and oxidative stress. *J. Pharmacol. Sci.* **2018**, *138*, 46–53. [CrossRef]
23. Kubiak-Tomaszewska, G.; Tomaszewski, P.; Pachecka, J.; Struga, M.; Olejarz, W.; Mielczarek-Puta, M.; Nowicka, G. Molecular mechanisms of ethanol biotransformation: Enzymes of oxidative and nonoxidative metabolic pathways in human. *Xenobiotica* **2020**, *50*, 1180–1201. [CrossRef] [PubMed]
24. Chen, Y.; Akhtar, M.; Ma, Z.; Hu, T.; Liu, Q.; Pan, H.; Zhang, X.; Nafady, A.A.; Ansari, A.R.; Abdel-Kafy, E.-S.M.; et al. Chicken cecal microbiota reduces abdominal fat deposition by regulating fat metabolism. *Res. Sq.* **2022**. [CrossRef]
25. Nagappan, A.; Jung, D.Y.; Kim, J.H.; Lee, H.; Jung, M.H. Gomisin N Alleviates Ethanol-Induced Liver Injury through Ameliorating Lipid Metabolism and Oxidative Stress. *Int. J. Mol. Sci.* **2018**, *19*, 2601. [CrossRef] [PubMed]
26. Wang, X.; Liu, M.; Zhang, C.; Li, S.; Yang, Q.; Zhang, J.; Gong, Z.; Han, J.; Jia, L. Antioxidant Activity and Protective Effects of Enzyme-Extracted Oudemansiella radiata Polysaccharides on Alcohol-Induced Liver Injury. *Molecules* **2018**, *23*, 481. [CrossRef]
27. Nagappan, A.; Kim, J.H.; Jung, D.Y.; Jung, M.H. Cryptotanshinone from the Salvia miltiorrhiza Bunge Attenuates Ethanol-Induced Liver Injury by Activation of AMPK/SIRT1 and Nrf2 Signaling Pathways. *Int. J. Mol. Sci.* **2019**, *21*, 265. [CrossRef]
28. Hou, R.; Liu, X.; Yan, J.; Xiang, K.; Wu, X.; Lin, W.; Chen, G.; Zheng, M.; Fu, J. Characterization of natural melanin from Auricularia auricula and its hepatoprotective effect on acute alcohol liver injury in mice. *Food Funct.* **2019**, *10*, 1017–1027. [CrossRef]
29. Jiang, X.; Zhou, Y.; Zhang, Y.; Tian, D.; Jiang, S.; Tang, Y. Hepatoprotective effect of pyrroloquinoline quinone against alcoholic liver injury through activating Nrf2-mediated antioxidant and inhibiting TLR4-mediated inflammation responses. *Process. Biochem.* **2020**, *92*, 303–312. [CrossRef]
30. Wong, G.K.; Cavey, M.J. Development of the liver in the chicken embryo. I. Hepatic cords and sinusoids. *Anat. Rec.* **1992**, *234*, 555–567. [CrossRef]
31. Yokouchi, Y. Establishment of a chick embryo model for analyzing liver development and a search for candidate genes. *Dev. Growth Differ.* **2005**, *47*, 357–366. [CrossRef]
32. Lin, L.P.S. Observation of Histological Development Process of Didestive Organs in Chicken Embryo. *Feed Rev.* **2015**, 6–10. [CrossRef]
33. Martínez-Jiménez, C.P.; Jover, R.; Teresa Donato, M.; Castell, J.V.; Jose Gomez-Lechon, M. Transcriptional regulation and expression of CYP3A4 in hepatocytes. *Curr. Drug Metab.* **2007**, *8*, 185–194. [CrossRef] [PubMed]
34. Hu, K.; Xiao, L.; Li, L.; Shen, Y.; Yang, Y.; Huang, J.; Wang, Y.; Zhang, L.; Wen, S.; Tang, L. The mitochondria-targeting antioxidant MitoQ alleviated lipopolysaccharide/ d-galactosamine-induced acute liver injury in mice. *Immunol. Lett.* **2021**, *240*, 24–30. [CrossRef]
35. Peter Guengerich, F.; Avadhani, N.G. Roles of Cytochrome P450 in Metabolism of Ethanol and Carcinogens. *Adv. Exp. Med. Biol.* **2018**, *1032*, 15–35. [CrossRef]
36. Rajput, S.A.; Shaukat, A.; Wu, K.; Rajput, I.R.; Baloch, D.M.; Akhtar, R.W.; Raza, M.A.; Najda, A.; Rafal, P.; Albrakati, A.; et al. Luteolin Alleviates AflatoxinB(1)-Induced Apoptosis and Oxidative Stress in the Liver of Mice through Activation of Nrf2 Signaling Pathway. *Antioxidants* **2021**, *10*, 1268. [CrossRef] [PubMed]
37. Ramezani, A.; Nahad, M.P.; Faghihloo, E. The role of Nrf2 transcription factor in viral infection. *J. Cell Biochem.* **2018**, *119*, 6366–6382. [CrossRef] [PubMed]

foods

MDPI

Article

Modulation of the Gut Microbiota by Tomato Flours Obtained after Conventional and Ohmic Heating Extraction and Its Prebiotic Properties †

Marta C. Coelho [1], Célia Costa [1], Dalila Roupar [2], Sara Silva [1], A. Sebastião Rodrigues [3], José A. Teixeira [2] and Manuela E. Pintado [1,*]

[1] CBQF—Centro de Biotecnologia e Química Fina—Laboratório Associado, Escola Superior de Biotecnologia, Universidade Católica Portuguesa, Rua Diogo Botelho 1327, 4169-005 Porto, Portugal
[2] CEB—Centre of Biological Engineering, University of Minho, 4710-057 Braga, Portugal
[3] Centre for Toxicogenomics and Human Health (ToxOmics), Genetics, Oncology and Human Toxicology, NOVA Medical School | Faculdade de Ciências Médicas, Universidade Nova de Lisboa, Campo dos Mártires da Pátria, 130, 1169-056 Lisbon, Portugal
* Correspondence: mpintado@ucp.pt

† This article is part of the PhD thesis of Marta C. Coelho.

Abstract: Several studies have supported the positive functional health effects of both prebiotics and probiotics on gut microbiota. Among these, the selective growth of beneficial bacteria due to the use of prebiotics and bioactive compounds as an energy and carbon source is critical to promote the development of healthy microbiota within the human gut. The present work aimed to assess the fermentability of tomato flour obtained after ohmic (SFOH) and conventional (SFCONV) extraction of phenolic compounds and carotenoids as well as their potential impact upon specific microbiota groups. To accomplish this, the attained bagasse flour was submitted to an in vitro simulation of gastrointestinal digestion before its potential fermentability and impact upon gut microbiota (using an in vitro fecal fermentation model). Different impacts on the probiotic strains studied were observed for SFCONV promoting the *B. animalis* growth, while SFOH promoted the *B. longum*, probably based on the different carbohydrate profiles of the flours. Overall, the flours used were capable of functioning as a direct substrate to support potential prebiotic growth for *Bifidus longum*. The fecal fermentation model results showed the highest Bacteroidetes growth with SFOH and the highest values of *Bacteroides* with SFCONV. A correlation between microorganisms' growth and short-chain fatty acids was also found. This by-product seems to promote beneficial effects on microbiota flora and could be a potential prebiotic ingredient, although more extensive in vivo trials would be necessary to confirm this.

Keywords: gut microbiota; short-chain fatty acids; prebiotic

Citation: Coelho, M.C.; Costa, C.; Roupar, D.; Silva, S.; Rodrigues, A.S.; Teixeira, J.A.; Pintado, M.E. Modulation of the Gut Microbiota by Tomato Flours Obtained after Conventional and Ohmic Heating Extraction and Its Prebiotic Properties. *Foods* **2023**, *12*, 1920. https://doi.org/10.3390/foods12091920

Academic Editors: Noelia Castillejo Montoya and Lorena Martínez-Zamora

Received: 13 February 2023
Revised: 3 May 2023
Accepted: 5 May 2023
Published: 8 May 2023

1. Introduction

The gut microbiota arrangement depends on individual intrinsic factors (e.g., age, ethnicity, genetic markers) and environmental factors (e.g., geographic area, lifestyle, diet, and drugs) [1,2], whereas the host intestine provides the necessary environmental conditions for the bacteria therein to survive and reproduce. The gut microbiota modulates various physical functions (e.g., nutrient processing and digestion, immune cell growth and immune response, and immunity towards pathogens, among others), hence representing a mutualistic relationship [2].

Most bacterial fermentation happens in the proximal colon, where there is higher substrate accessibility. Toward the distal colon, the convenience of substrates falls, and the recovery of available food reduces both substrate and microbial products' distribution. The fermentation results in the production of short-chain fatty acids (SCFAs) together with gas

(CO_2 and H_2) [3]. These compound molecules are mainly generated in the large intestine by gut microbiota fermentation of carbohydrates that had escaped digestion and absorption in the small intestine, although non-digested proteins or peptides are also essential upstream compounds for their production [4,5].

Bioactive compound sources are mainly found in plants, such as tomato fruit. While also used for fresh consumption, tomatoes are primarily used for processing into juice, pulp and sauces, hence originating many by-products whose potential valorization is still scarce [6,7]. A full and integrated recovery from tomato by-products with zero residues, in a context of a circular economy, could be used as a strategy. For instance, the final solid extraction by-product can be dried under controlled conditions, resulting in flour with a high fibre content combined with bonded bioactive compounds, such as phenolic compounds and carotenoids [8–10]. As such, the resulting material could be used with an ingredient that, given its characteristics, could have interesting biological potential, particularly in the modulation of the gut microbiota [11–13]. In addition, different extraction techniques have been tested to valorize these by-products, including "green techniques" like ohmic (OH) processing, which obtain bioactive extracts with significant differences from extracts obtained by conventional extraction [14].

Still, even though diet arrangement has been demonstrated to have a modulating effect on gut microbial communities, knowledge of the effects exerted by particular foods in driving gut microbial variety is limited, hence hampering their optimal use.

Therefore, the present work aimed to characterize the prebiotic potential of two tomato flours obtained after ohmic (SFOH) and conventional (SFCONV) extraction of phytochemicals from tomato bagasse. To accomplish this, both flours were subjected to an in vitro stimulation of the gastrointestinal tract. After the characterization of the impact of this process on each sample, the digested samples were evaluated upon pure probiotic cultures and on fresh human fecal samples to assess the prebiotic potential and the effect on the metabolic and population dynamics of gut microflora.

2. Materials and Methods

2.1. *Tomato Bagasse Flours*

Tomato Bagasse Flour Preparation

Two different tomato bagasse (peel and seeds) flours were used in the present work. The first (OH) was prepared from the solid by-product leftover after ohmic extraction (70 °C, 15 min, 70% ethanol). The tomato bagasse was subjected to ohmic extraction (peel and seeds) as described elsewhere [14]. The second (CONV) was prepared using the solid by-product of a conventional solid–liquid extraction described in the literature using hexane as solvent [15]. In both cases, after extraction, the leftover solid by-product fraction (SF) was dried at 55 °C overnight and stored in a desiccator at room temperature until use. The phytochemical properties of tomato flours were described in previous studies [7,10,14]. The SF for OH tomato by-products is shortened as SFOH to simplify the nomenclature, while the SF for CONV samples is SFCONV.

2.2. *In Vitro Digestion Simulation (GID)*

2.2.1. Sample Preparation

The tomato SF was suspended in water (10%) and homogenized using an Ultra-Turrax (IKA Ultra-turrax T18, Wilmington, NC, USA) at 13,000× *g* for 1 min. The tomato bagasse solution was set up at 10% (*w*/*v*), as the composition showed that the dried sample contained ca. 50% fiber and, as per the European Food Safety Agency (EFSA), 6 g of fiber for each 100 g of the item [16]. Results obtained before showed the presence of 48.06 ± 0.11 g/100 g DW of insoluble dietary fiber and 46.01 ± 0.13 g/100 g DW for SFCONV [10].

2.2.2. In Vitro Digestion Simulation

Before executing fecal fermentation assays, samples were subjected to an in vitro simulation of the GI tract (including dialysis) to better mimic in vivo conditions. The tomato bagasse mixture's pH value was adjusted to 5.6–6.9, utilizing 1 M HCl. Mouth digestion was simulated by adding α-amylase from human saliva (100 U/mL in 1 mM aqueous $CaCl_2$), homogenizing the mixture for 2 min, and incubating at 37 °C and 200 rpm Afterward, to emulate stomach conditions, the mixture's pH value was lowered to 2.0 (utilizing 1 M HCl) and pepsin from gastric juice was added (12.5 mg/mL in HCl 1 M) at a ratio of 0.05 mL/mL of sample. The mixture was then incubated in a water bath for 2 h at 37 °C and 130 rpm. To simulate small intestine conditions, the mixture's pH was adjusted to 6.0 utilizing 1 M $NaHCO_3$. Pancreatin and bile salts (0.4 g pancreatin and 1.2 g bile salts in 200 mL of $NaHCO_3$ 1 M) were added to the mixture, at a ratio of 0.25 mL/mL of sample Finally, the obtained solution was maintained at 37 °C and 45 rpm for 2 h. Afterward, the dialysis was performed with 12 kDa membranes for 24 h, with known volume (to simulate the blood circulation). At the end of the dialysis process, the solution within the dialysis tubing (OUT) represented the non-absorbable sample (colon-available). This fraction was then freeze-dried and stored in a desiccator for later use in the fecal fermentation.

2.3. *Preliminary Evaluation of the Prebiotic Potential of Tomato SF*

2.3.1. Microorganisms

Probiotic bacteria species were selected for the present work, namely *Lactobacillus casei* 01 and *Bifidobacterium animalis subsp lacties* BO.

2.3.2. Selection of the Best Tomato Flour Concentration

To evaluate the effect of tomato flour on the growth of the target microorganisms, tomato SF (before and after digestion) at 2, 4, and 6% (*w*/*v*) was suspended in basal media, inoculated using a 24 h inoculum at 10% (*v*/*v*), and incubated for 24 h at 37 °C in an anaerobic environment. After this period, the viable cell numbers were determined by plating, using the spread plate method, in de Mann, Rogosa, and Sharpe agar (MRS) enhanced with 0.5 g/L of L-cysteine hydrochloride. After 48 h incubation at 37 °C, under anaerobiosis, the bifidobacterial and lactobacilli colonies were enumerated, and the outcomes plotted as log CFU/mL, in accordance with [17]. All inoculations were performed in triplicate, and plain inoculated cells were determined using decimal dilutions and plating through the spread plate technique, in MRS agar enhanced with cysteine and bromophenol blue [18]. In addition, pH values were measured using a 52-02 Crison electrode, and the organic acid production was assessed through HPLC-IR, as described elsewhere [17].

2.4. *In Vitro Fecal Fermentations*

2.4.1. Collection and Preparation of Fecal Inocula

Fresh fecal samples were provided by five healthy donors (A–E, three men and two women, between the ages of 23 and 39 years old), whose selection was based on established criteria regarding health status and dietary habits; namely, to assess the existence of chronic diseases, allergies, and probiotic ingestion, among others. Moreover, an informed consent form was distributed among donors to provide the participants with information about the study, with a consent certificate assigned to each. Donors were healthy unrelated anonymous volunteers, ≥18 and <50 years of age, who had not received antibiotics in the preceding six months or consumed any prebiotic supplement. The fecal samples were maintained under anaerobic conditions for a maximum of 2 h before being used. The fecal inocula (FI) were then prepared by diluting the fecal matter in Reduced Physiological Salt solution (RPS) (constituted by 0.5 g/L cysteine-HCl (Merck, Darmstadt, Germany) and 8.5 g/L NaCl (LabChem, Zelienople, PA, USA), with a final pH value of 6.8, at 100 g/L in an anaerobic workstation (Don Whitley Scientific, West Yorkshire, UK) (10% CO_2, 5% H_2, and 85% N_2) [2].

2.4.2. Nutrient Base Medium Preparation

Fecal fermentations were performed with Nutrient Base Medium. The medium comprised 5.0 g/L trypticase soy broth without dextrose (Fluka Analytical, St. Louis, MO, EUA), 5.0 g/L bactopeptone (Becton Dickinson Biosciences, New Jersey, NJ, USA), 0.5 g/L cysteine-HCl (Merck, Darmstadt, Germany), 1.0% (*v*/*v*) of salt solution A [100.0 g/L NH_4Cl (Merck, Darmstadt, Germany), 10.0 g/L $MgCl_2{\cdot}6H_2O$ (Merck, Darmstadt, Germany), 10.0 g/L $CaCl_2{\cdot}2H_2O$ (Carlo Erba, Chaussée du Vexin, France)], 1.0% (*v*/*v*) of trace mineral solution (ATCC, Manassas, VA, USA), 0.2% (*v*/*v*) of salt solution B [200.0 g/L $K_2HPO_4{\cdot}3H_2O$ (Merck, Darmstadt, Germany)], and 0.2% (*v*/*v*) of a 0.5 g/L resazurin solution (Sigma-Aldrich Chemistry, St. Louis, MO, USA). The medium final pH value was adjusted to 6.8 and was then bubbled with N_2 until it presented a translucent/yellowish color. Following this, 50 mL parts were then distributed into several containers. Fructooligosaccharides (FOS) from inulin-Raftilose® P95 (Beneo-Orafti, Oreye, Belgium) with a molecular weight of 0.6 Kda—3.3 DP were used as positive controls and then freeze-dried digested [2]. Tomato residue flours were added to the respective vessels at a final concentration of 2%. The bottles were capped and autoclaved. Following sterilization, and before adding the fecal inocula, the atmosphere of each flask was refluxed with a sterile gas mixture (10% CO_2, 5% H_2, and 85% N_2) [2].

2.4.3. Fecal Fermentations

The flasks prepared before (Section 2.4.2) were inoculated at 2% (*v*/*v*) with fecal inocula (Section 2.4.1) and incubated for 48 h at 37 °C under anaerobic atmosphere (10% CO_2, 5% H_2, and 85% N_2). Samples were collected after 0, 12, 24, and 48 h of incubation, and the pH values were measured using a MicropH 2002 pH meter (Crison, Barcelona, Spain) equipped with a 52-07 pH electrode (Crison, Barcelona, Spain). The positive and negative controls were, respectively, designated as C+ (FOS) and C- (plain media), while the digested biomass tomato flours were named OH and CONV, which are under flours. Afterward, the samples were stored at −30 °C until analysis. All the steps considered in this section were carried out inside an anaerobic workstation (Don Whitley Scientific, West Yorkshire, UK) [2].

2.4.4. Fecal Fermentation Sample Processing

Aliquots of each sample (4 mL) were centrifuged for 6 min at 4000× *g*. The resulting supernatants were used to evaluate sugars and short-chain fatty acids (SCFAs), according to Section 2.6, and the pellet was used to extract the genomic DNA.

2.5. Sugars and SCFA Analysis

Sugar consumption and organic acid production during fecal fermentation were analyzed using an HPLC system composed of a Knauer K-1001 pump (Berlin, Germany), an ion exchange Aminex HPX87H (300 × 7.8 mm) (Bio-Rad, Hercules, CA, USA) column, and two detectors assembled in series, namely a UV-vis detector (220 nm) and a refractive index detector, both from Knauer (Berlin, Germany,) at a temperature of 65 °C. An isocratic gradient was used (13 mM H_2SO_4 Merck, Darmstadt, Germany), at a flow rate of 0.6 mL/min. The injection volume was 40 µL and the running time was 30 min. Fermentation supernatants were filtered through a 0.22 µm syringe filter and each sample was injected in duplicate.

2.6. Bacterial Population Analysis

2.6.1. DNA Extraction

An NZY Tissue gDNA Isolation kit (NZYTech, Lisbon, Portugal) was used to extract DNA from the fecal samples with slight modifications. Briefly, pellets were washed with TE (pH 8.0; Tris EDTA buffer), vortexed, and centrifuged at 4000× *g* for 10 min. Then, 180 µL of a freshly prepared lysozyme solution (10 mg/mL lysozyme in a NaCl-EDTA (30 mM:10 mM) solution) was added and incubated for a period of 1 h at 37 °C, with periodic shaking. Afterward, 350 µL of NT1- buffer was added to the samples, which were then vortexed and incubated at 95 °C. After 10 min, samples were centrifuged

(11,000× *g*, 10 min, 4 °C), and supernatants (200 µL) were mixed with 25 µL of proteinase K and incubated at 70 °C for 10 min. The remaining steps were performed according to the manufacturer's instructions. After extraction, the DNA's purity and concentration (20 ng/µL) were assessed using a Thermo Scientific™ µDrop™ Plate coupled with a Thermo Scientific™ Multiskan™ FC Microplate Photometer (Thermo Fisher Scientifc, Waltham, MA, USA).

2.6.2. Real-Time Quantitative Polymerase Chain Reaction—Gut Composition Analysis

Real-time PCR was performed using a CFX96 Touch™ Real-Time PCR Detection System (Bio-Rad Laboratories, Inc., Hercules, CA, USA), under the conditions described in Table S1, to detect and amplify the purified bacterial gDNA [2]. The PCR reaction mixture comprised 5 µL of 2x iQTM SYBR® Green Supermix (Bio-Rad Laboratories, Inc., Hercules, CA, USA), 2 µL of sterile ultrapure water, 1 µL of sample DNA (equilibrated to 20 ng/µL), and 1 µL of forward and reverse primers (100 nM) targeting the 16S rRNA gene. The primers used were obtained from STABvida (Lisbon, Portugal) and are listed in Table S2. Standard curves were constructed using tenfold dilutions (from 2 log to 6 log of several copies of 16S rRNA gene/µL) of bacterial genomic DNA standards (DSMZ, Braunschweig, Germany); the primer sequences for qRT-PCR required in this experiment are present in Table S2. The amplification schedule included one initial activation cycle at 95 °C for 10 min, 45 cycles at 95 °C for 10 s, an annealing step at 45, 50, or 55 °C for 60 s depending on the primer, and an extension step at 72 °C for 15 s. Melting curve analysis was performed for each PCR to evaluate the specificity of the amplification, considering a temperature interval from 60 to 97 °C, with an increase of 0.1 °C (per 0.01 min). All assays were performed in quadruplicate. Data were processed and analyzed using LightCycler software obtained from Roche Applied Science. The target groups were chosen from among the most numerous phyla and genera in the healthy human gut microbiota (Firmicutes, *Clostridium leptum* subgroup, Bacteroidetes, and Bacteroides), as well as known probiotics (*Bifidobacterium* and *Lactobacillus*). The standard curves were calculated using tenfold bacterial dilution gDNA standards of *Clostridium leptum* (ATCC 29065), *Bacteroides vulgatus* (ATCC 8482), *Bifidobacterium longum* subsp. *infantis* (ATCC 15697) (DSMZ, Braunschweig, Germany), and *Lactobacillus gasseri* (ATCC 33323) (Table S2). The NCBI Genome database was utilized in this work to acquire the genome size and copy number of the 16S rRNA gene for each bacterial strain used as a benchmark.

2.7. Statistical Analysis

Statistical analysis of the data was done using IBM SPSS Statistics v21.0 (IBM, Chicago, IL, USA). The normality of the data's distribution was evaluated through Shapiro–Wilk's test. As the data proved to follow a normal distribution, one-way ANOVA, coupled with Tukey's post hoc test, was used to determine the significance of the effect of tomato bagasse biomass on bacterial populations at each time point. Repeated measures ANOVA was used to evaluate the effect of tomato bagasse biomass on the bacterial population over time. Differences were considered significant for p-values ≤ 0.05.

3. Results

3.1. Probiotic Effect

The most used probiotic microorganisms belong to the *Lactobacillus* and *Bifidobacterium* genera. Thus, at the first stage, these microorganisms were used to understand the potential prebiotic effects of tomato bagasse flours after extraction. For this, the bacteria were inoculated into a growth medium with different concentrations of tomato flours (0%, 1%, 2%, 4%) to select the minimum concentration exerting a prebiotic effect. The results (Figure 1) showed that the flours had little impact, with differences observed between the SFCONV and SFOH samples of 2 and 4% by-product concentration. The viable cells number for *Lactobacillus* was ca. 10^8 CFU/mL for the various percentages of tomato samples, while for *Bifidobacterium*, the observed values were lower.

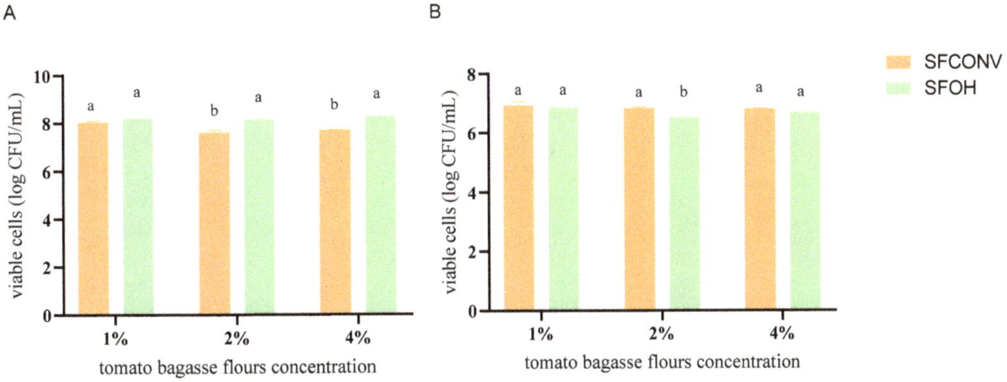

Figure 1. Impact of different concentrations of digested tomato by-products (after carotene extraction, OH, and CONV) on the growth of Lactobacillus (**A**) and *Bifidobacterium* (**B**) after 24 h anaerobic incubation. Letters mean the significant difference between methods for each tomato flours' concentration $p < 0.05$.

For *Lactobacillus casei*, when comparing the viable cells of the positive control and the sample at 2%, the by-product appeared to allow for some growth of this bacteria; i.e., the total viable counts were above those registered for the control and allowed for more prolonged survival of the bacterial cells, indicating that tomato biomass may be used as a source of nutrients by this microorganism. The SFOH extraction had a significant impact when compared with SFCONV ($p < 0.05$). The results suggest better bacteria accessibility to nutrients, such as carbohydrates, promoting their growth [2,18–20]. In addition, according to the chemical flours profile, SFOH has a greater amount of galactose, arabinose, and uronic acids than CONV, which could promote the growth of these bacteria, as described in the literature [21,22]. The results are according to the literature, given the recognized metabolic diversity of *Lactobacillus*, as previous results also reported strain-specific effects of tomato flours [23,24]. Thus, these flours can also be used as a medium for probiotic growth.

It was reported that XOS is not fermented by most of the lactobacilli tested, whereas arabinoxylan was not used by any of the strains examined [25]. However, [22] verified a *L. casei* growth in arabinoxylan. The authors also demonstrated that another strain of *Lactobacillus*, *L. brevis* DSM 20054, was genetically equipped with functional arabinoxylan-oligosaccharide-degrading hydrolases, which could explain the use of arabinose for growth.

In addition, tomato by-products could promote *L. casei* growth, suggesting that this sample may be used as a source of essential nutrients by bacteria, as has been the case in some studies that utilized tomato juice as material for the manufacture of a probiotic drink. One study showed that tomato juice enriched with *Lactobacillus plantarym ST III* strain positively affected fermented skimmed milk's taste and health-promoting activity. A fermented tomato juice with *L. casei* and *L. plantarum* was used to create a high-bioactivity probiotic drink [26]. Phenolic compounds, lycopene, and other carotenoids are related as they cause a positive correlation between antioxidant activity and prebiotic impact [17,26,27]. No differences were found in antioxidant activity (Table S5) between SFOH and SFCONV used in this study (87.50 ± 1.26 and 90.49 ± 2.54 g trolox eq./100 g DW, respectively), which indicated that both flours could be used as a probiotic.

Relatively to *B. animalis*, significant differences were observed at 24 h between CONV and OH extracts with 2% of tomato by-products. CONV seemed to promote the growth of this microorganism as there were viable cell numbers after 24 h incubation. Nevertheless, OH-extracted flour led to a slight decrease in bacterial growth over time. These differences between results may originate from the fact that different *Bifidobacterium* strains may have distinctive carbohydrate metabolic abilities, as has been observed in several previous studies [28–31]. Studies revealed that the *B. longum* encodes ABC transporters, PEP-PTS systems, and secondary transporters required to carry mono- and disaccharides. In comparison, *B. animalis* has a significantly smaller genome than *B. longum*, with a lower number of metabolic pathways to take advantage of carbon sources, does not encode PEP-PTS frameworks, and contains just two qualities determining sugar-specific ATP-binding proteins characteristic of ABC transporters [28]. Therefore, since previous studies [7] showed that CONV has more disaccharides and monosaccharides (glucose, fructose, and mannose) than OH (which contains more polysaccharides, such as cellulose, hemicellulose, and pectins), it is plausible that more metabolically limited bacteria, such as *B. animalis*, cannot use them as a carbon and energy source to grow.

Since there were no marginal gains in growth or the death of the target microorganism provided by the different concentrations of tomato pomace, subsequent experiments used the 2% concentration as there was limited sample availability, and in the future, it will be easy to justify as a commercial ingredient to minimize interference in final food products [32].

Impact of the Digested Tomato SF on Organic Acid Production

Probiotic bacteria can produce a variety of organic acids. The primary fermentation product from the breakdown of complex dietary carbohydrates is lactic acid, specially synthesized by *Lactobacillus* and *Bifidobacterium*. In Figures 2 and 3, it can be seen that at time 0 h, lactic acid is detected in the tomato SF. Greater concentrations of lactic acid concentrations were obtained, both in SFOH and in SFCONV at 2%, after 24 h. Moreover, the tomato by-product's presence promoted an overall increase in the production/accumulation of lactic acid.

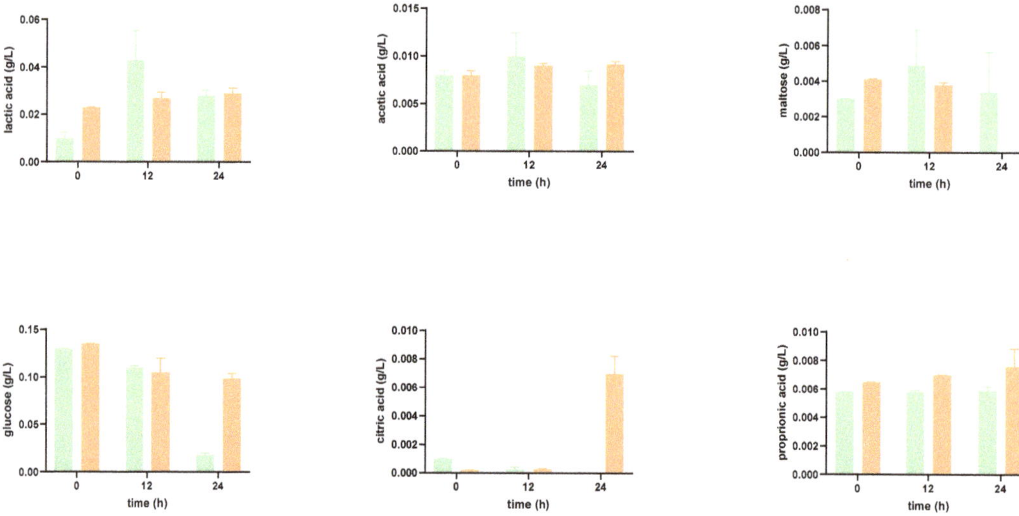

Figure 2. Concentrations of organic acids and sugars during the 24 h of growth of prebiotic bacteria Lactobacillus, incubated in the presence of SFOH (green) and SFCONV (orange) of tomato SF 2%, after simulation of the gastrointestinal tract.

A slight increase in the production of lactic acid in *Bifidobacterium* species can be noted. This result corroborates the literature: *Bifidobacterium* produce lactic and acetic acids in large amounts, that is, larger than the amounts secreted by *Lactobacillus*, even though the latter is known to be very acid-tolerant [33].

Bifidobacterium and *Lactobacillus* fermentation also result in the production of acetic acid. Acetic acid was not identified in the tomato by-product control at time 0 h. Moreover, during the graphical execution of the acetic acid concentrations for the function of time, it would be expected that the concentration of this product would increase over time. However, this condition was not verified for *L. casei*, in the presence of tomato by-products, at 24 h, which presents a high standard deviation, and for the mixture of *Lactobacillus* and *Bifidobacterium*, in the absence of tomato, at 12 h. Bifidobacteria produced acetic and lactic acids at proportions of 3:2, which were not comprised by analyzing the fermentation end products: the concentration of acetic acid was approximately three times inferior to the concentration of lactic acid obtained [34].

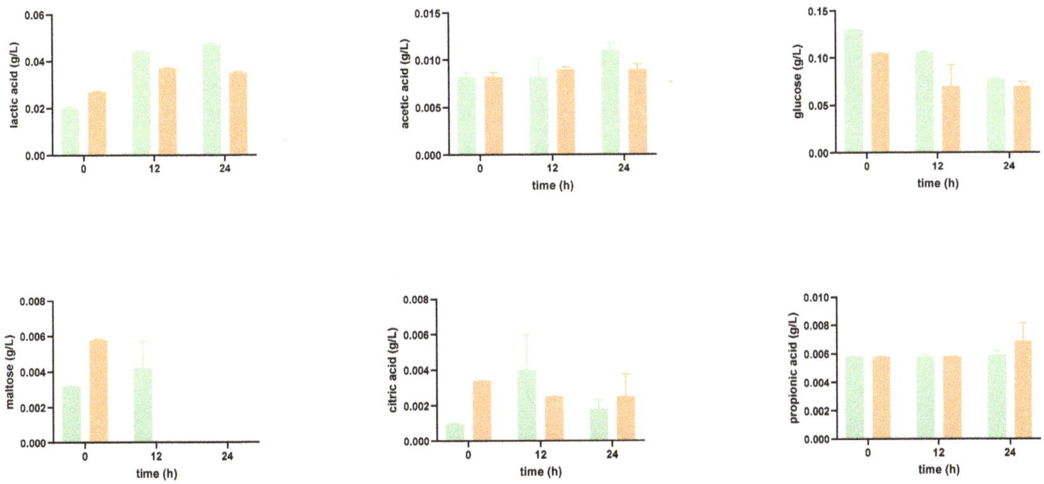

Figure 3. Concentrations of organic acids and sugars during the 24 h of growth of prebiotic bacteria *Bifidobacterium*, incubated in the presence of SFOH (green) and SFCONV (orange) of tomato SF 2%, after simulation of the gastrointestinal tract. Citric acid, in the control, increased throughout the incubation time for all probiotic bacteria as well as in the mixture of prebiotics. In addition, a significant concentration of citric acid was produced by *L. casei*. Towards the presence of tomato by-products, an increase in citric acid is visible in the first 12 h; however, after this time, the concentration decreases substantially, with *L. casei* production reaching null values. It is possible to conclude that *B. animalis* and *B. longum* achieved the highest concentrations in the presence of these flours.

Lactobacillus and *Bifidobacterium* can break down and metabolize a variety of substrates. The glucose concentration is initially high due to large amounts of this monosaccharide (Figures 2 and 3). After 12 h, there is a decrease in its concentration since the bacterial strains consume this substrate. The metabolic capacity to consume sugars did not differ significantly between the various probiotic bacteria, except for *L. casei* in the presence of tomato by-products, which degraded glucose more sharply after 24 h, though the production of lactic acid as well as of acetic acid did not increase.

Besides glucose, the disaccharide maltose was also found in the SF but at a lower concentration. During the incubation and in SFCONV, the overall amount of maltose decreased. As bacterial enzymes hydrolyze maltose in two glucose molecules, the bacteria are likely consuming it. However, in SFOH, consumption of maltose by the microorganisms was, overall, significantly lower, with the amount of maltose in the media increasing after 12 h (possibly released from the matrix), with a subsequent reduction in the concentration after 24 h (Figures 2 and 3). In SFOH, only *B animalis* was capable of consuming present maltose.

3.2. *Impact of Tomato Flour after Extraction on Gut Microbiota*

3.2.1. Microbial Population Modulation

The gut microbiota assay mimics our organism's complexity, which goes far beyond *Lactobacillus* and *Bifidobacterium*. There is a set of microorganisms that interact with each other and with different preferences for substrates.

A simulated gut microbiota fermentation was made through an in vitro model to evaluate the potential prebiotic impact of tomato flours (promotion of positive microorganism growth and metabolite production) obtained after CONV and OH extraction.

The phyla Bacteroidetes, which are Gram-negative, and Firmicutes, which are Gram-positive, are the most plentiful in the human gut. Bacteroidetes and Bacteroides presented significant differences between SFCONV and the controls. For the Bacteroidetes cluster, there was an increase in the presence of SFCONV at 12 h, with significant differences compared with C^-. In addition, according to Figure 4, there is a greater dispersion in the number of gene copies for SFOH than for SFCONV. The latter population is less spread out, concentrating in the 5 log of number of copies of 16S rRNA/ng DNA). It is also possible to verify that for SFOH, about 25% of the population presented a 6 log 16S rRNA gene copies/ng of DNA, similar in behavior to FOS. The results are in agreement with the literature, as Bacteroidetes may metabolize complex nutrient polymers, many of which are molecules in the plant cell wall (e.g., cellulose, pectin, and xylan), which, through the of action human digestive enzymes' cleavage activity, are released and may reach the colon intact [35]. Studies reveal that dietary habits and lifestyle turn into determinants and play a critical part in gut microbiota variations. High-fiber and animal protein foods increase Bacteroidetes, whereas the presence of high-fiber and carbohydrate foods increases Firmicutes and Prevotella [36,37]. This information reinforces our results since SFOH presents more protein and insoluble fiber than SFCONV and, consequently, more Bacteroidetes than SFOH ($p < 0.05$) [10].

In addition, SFCONV samples contain more bound phenolics than SFOH. Xue and colleagues (2016) showed that phenolic compounds, namely quercetin and catechin, inhibit the growth of Bacteroidetes and Firmicutes. Nevertheless, other microorganisms maintain the ability for carbohydrate and energy metabolism in each group. It is still unknown how other bacteria use FOS and their metabolites [35]. The presence of multiple FOS transport systems with different specificities in each strain may also explain the selective metabolism of particular oligosaccharide components observed here. Moreover, bound phenolic compounds may be responsible for altering the metabolism pathway, inhibiting Bacteroidetes growth. The increase of Bacteroidetes leads to an increase of acidic compounds such as pyruvic, citric, fumaric, and malic acids, indicators of higher energy metabolism, and thus contributes to the healthy metabolome [38–40]. For Bacteroides, the results also showed the same tendency as with Bacteroidetes genera. SFOH presented a more heterogeneous distribution for Bacteroides than SFCONV, which presents similar behavior for different donors (Figure 4).

Furthermore, the results showed that 25% of SFOH and SFCONV have a gene copy number higher than FOS. Additionally, there is a significant increase in the Bacteroides population caused by SFOH and SFCONV at 12 h when compared to controls ($p < 0.05$) (Figure 2). The SFCONV have more rutin than SFOH, a polyphenol compound, as described in [7]. The authors claimed that polyphenols might modify the microbiota balance through biased effects on Bacteroides [41].

Regarding the Firmicutes results, a slight increase of 16S rRNA at 24 h for SFCONV samples' exposure was observed compared to control samples, with significant differences ($p < 0.05$). According to Figure 2, while the SFOH samples presented similar results to FOS, the bacterial population is more dispersed than the observed SFCONV bacterial population. SFOH contains more fatty acids and fiber than SFCONV (Table S3). Studies have demonstrated that a diet rich in fiber and low in fat promotes Firmicutes' growth, which metabolizes dietary plant-derived polysaccharides to SCFAs [36,42].

The ratio of Firmicutes to Bacteroidetes (F/B) was also analyzed at 1:1 for all samples during the experience. Commonly, healthy individuals display a nearly 1:1 ratio of F/B [35,43], and the ratio's increase (e.g., to 20:1, F/B) or decrease has been associated with obesity and weight loss, respectively [44]. In addition, dietary intake, such as fiber, and phytochemicals have a higher impact on microbiota. An example is a diet rich in fiber, which increases Bacteroidetes; diets rich in calories increase Firmicutes. The maintenance of the F/B ratio during all experiences corroborates the higher amounts of fatty acids and dietary fiber present in samples, contributing to the ratio's equilibrium. Thus, this is a good indicator of the use of SFOH in diets to contribute to the health of individuals. Other studies with obese or malnourished individuals could be interesting to understand the alterations caused by these samples.

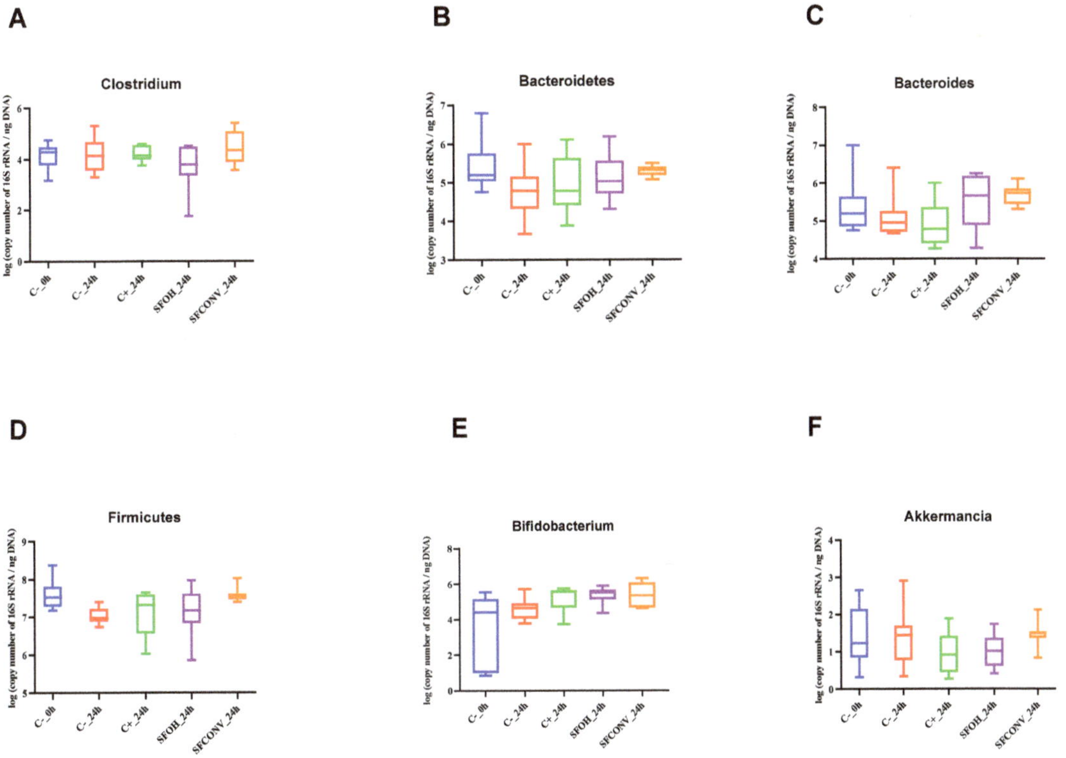

Figure 4. Distribution of gut bacterial populations (log 16S rRNA gene copies/ng of DNA, means ± SD) detected by PCR in fecal samples. The used probes: *Clostridium leptum* (**A**), Bacteroidetes (**B**), Bacteroides (**C**), Firmicutes (**D**), *Bifidobacterium* (**E**), and *Akkermansia* (**F**).

In general, the *Bifidobacterium* showed a slight increase over time for both tomato SFs tested. According to Figure 4, about 50% of the initial population of *Bifidobacterium* is between 1 and 4 log of copies of the number of 16S rRNA/ng DNA. At 6 h fermentation, a significant increase was observed in the number of copies with time ($p < 0.05$) for SFOH. In addition, CONV fermentations showed higher levels of gene copies at 12 h compared to the controls. In addition, at 24 h, 25% of the bacteria presented more copies than FOS. Given the results showed by [7], it appears that SFCONV has more soluble fiber than SFOH and FOS, being a good carbon source and promoting a greater growth of this bacteria. Studies have shown that, depending on the strains, they use different substrates for growth. While not all strains are capable of using (most of the components of) galactooligosaccharides, the capacity of intestinal communities for the metabolization of galactooligosaccharides would not be excluded [45]. The combined activity of multiple bacteria is responsible for the fermentation of complex carbohydrates in the gut [21]. Researchers have investigated several strains and revealed that 11 of the bifidobacterial strains were significantly growing on polydextrose (soluble fiber), final $OD_{600} > 05$, whereas 34 of the bifidobacterial strains exhibited positive growth FOS. This is an essential result to deduce the prebiotic potential of different carbohydrates to increase the diversity of the gut microbiota. McLaughlin et al. (2015) verified superior growth with inulin to *B. longum* subsp. CCUG 18157 when compared to the other strains tested. It is possible that this strain produces a specific enzyme, such as β-fructofuranosidase, with specificity for FOS or inulin.

One intestinal bacterium naturally existing in the gut microbiota of healthy people is *Akkermansia.* When present in the intestinal flora, this group also produces propionate and acetate; however, the fecal samples contained lower gene copies of *Akkermansia.* The fermentation of both samples induced a significant reduction in *Akkermansia* levels from 0 to 12 h. Nonetheless, no differences were found between samples and controls ($p > 0.05$). The lower concentrations of this bacteria, when compared with other groups of bacteria, could be associated with its human intestinal colonization at a very young age (it is found in breast milk and infant formula) (Lukovac et al., 2014); the median age of the donors was 40 years old. Additionally, recent studies have shown that diets rich in fiber and protein decrease the *Akkermansia* population [46,47]. Our results agree with previous reports, where SFOH presents more fat and soluble fiber than SFCONV, resulting in a population distribution with less 16 rRNA gene copies of *Akkermansia.*

C. leptum belongs to the group of anaerobic bacteria that mainly produce propionate and butyrate in gut microbiota and use amino acids as the primary energy source. No differences were observed in cell numbers ($p > 0.05$); nevertheless, interesting results were observed in Figure 4. The first quartile population (25%) of SFOH presented a lower number of gene copies than other samples, while the second and third quartiles, 50% of the *Clostridium leptum* population, presented similar 16 rRNA gene copies with FOS. Regarding SFCONV, this seems to promote the growth of this bacteria better. As seen with *Akkermansia*, a diet rich in fermentable fiber can promote the growth of clostridium; however, the presence of lipids can also inhibit it, thus verifying the discrepancies in the growth of this microorganism [36,48].

The differences obtained for microorganisms agree with recent research, suggesting that food changes may drastically modify endogenous microbial communities' total composition and organization in the gut.

3.2.2. SCFA Analysis

As referred to previously, some of the welfare benefits attributed to fiber fermentation by the colonic bacteria are related to the metabolites generated.

Relative to butyrate, SCFAs (Figure 5) are the primary energy source for normal, healthy colon cells. In addition, they safeguard against colon cancer and inflammation due to their capacity to help defend the gene-expression structure that discourages the development and proliferation of cancer cells. Results suggest a significant increase of n-butyrate at 6 h in samples fermented with SFOH- compared with control samples. At 24 h, the SFCONV-fermented samples also showed an increase of n-butyrate higher than SFOH samples, but no statistical differences were found ($p > 0.05$).

The results showed a significantly higher acetate concentration for tomato flour CONV than for OH ($p < 0.05$). In addition, there was an increase in acetate concentration over time. The results align with previously reported observations since SCFAs are, for the most part, created by enteric microorganisms as Bacteroidetes and *Bifidobacterium* because of carbohydrate fermentation through the hydrolysis of acetyl-CoA. A little part is synthesized by acetogenic microorganisms that use hydrogen, carbon dioxide, or formic acid through the Wood–Ljungdahl pathway [49,50]. The results are aligned with the literature, since the observed formic acid concentration decreases as acetate concentration increases.

There was a slight decrease in propionate concentration at 12 h in the SFOH sample compared with the positive control, with a significant difference ($p < 0.05$). Nevertheless, increased propionate concentration was found at 24 h for both positive control and tomato bagasse flours (SFOH, SFCONV), with significant differences compared to the negative control. The different propionate pathways may explain these results: succinate, acrylate, and propanediol. The succinate pathway is related to the Firmicutes and Bacteroidetes [50,51]. The results are illustrated in Figure 5, where succinate concentration decreases over time, suggesting a production of propionate based on the succinate pathway.

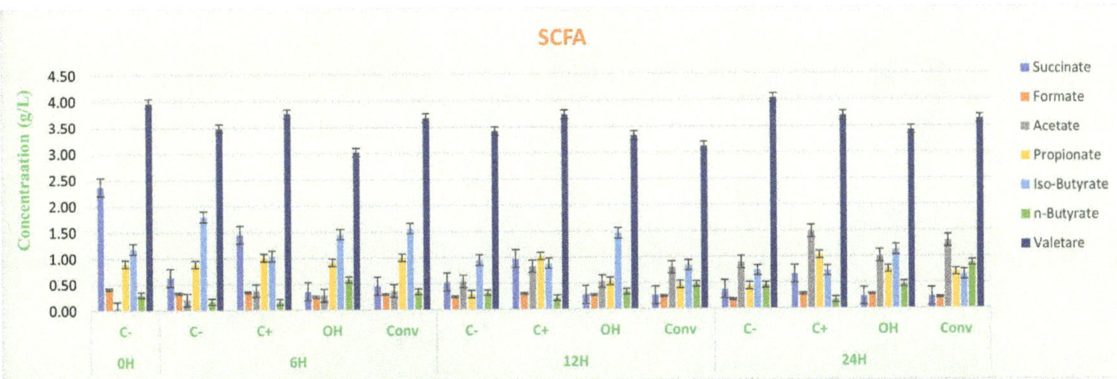

Figure 5. Concentration (mg/mL ± SD) of the SCFAs produced along with fermentation time in fecal samples. Negative control (C-), positive control (C+), tomato residue flour after OH extraction (OH), and tomato residue flour after conventional extraction (CONV). Different letters mark statistically significant ($p < 0.05$) differences.

The results showed higher production of butyrate at 24 h for SFOH and SFCONV samples than C-, which indicates a tomato bagasse flour fermentation stimulating the production of this acid.

Although acetate and propionate production for SFOH and SFCONV were almost the same, the CONV sample had higher butyrate production than the OH sample. In addition, acetate and propionate are related to the advancement of satiety, thus taking into account the phytochemical profile of the tomato bagasse flour and the results obtained for propionate and acetate production; they could be applied as a substitute for animal-derived proteins and fiber in foods.

According to the Pearson correlation (Figure 6), a significant impact is observed in some SCFAs on the expression of some microorganisms, which is considered in the evaluation of the previously described results. Propionate is correlated with the growth of Clostridium and Firmicutes, and the production of formate and succinate is correlated with Akkermancia. The last one also influences the Firmicutes.

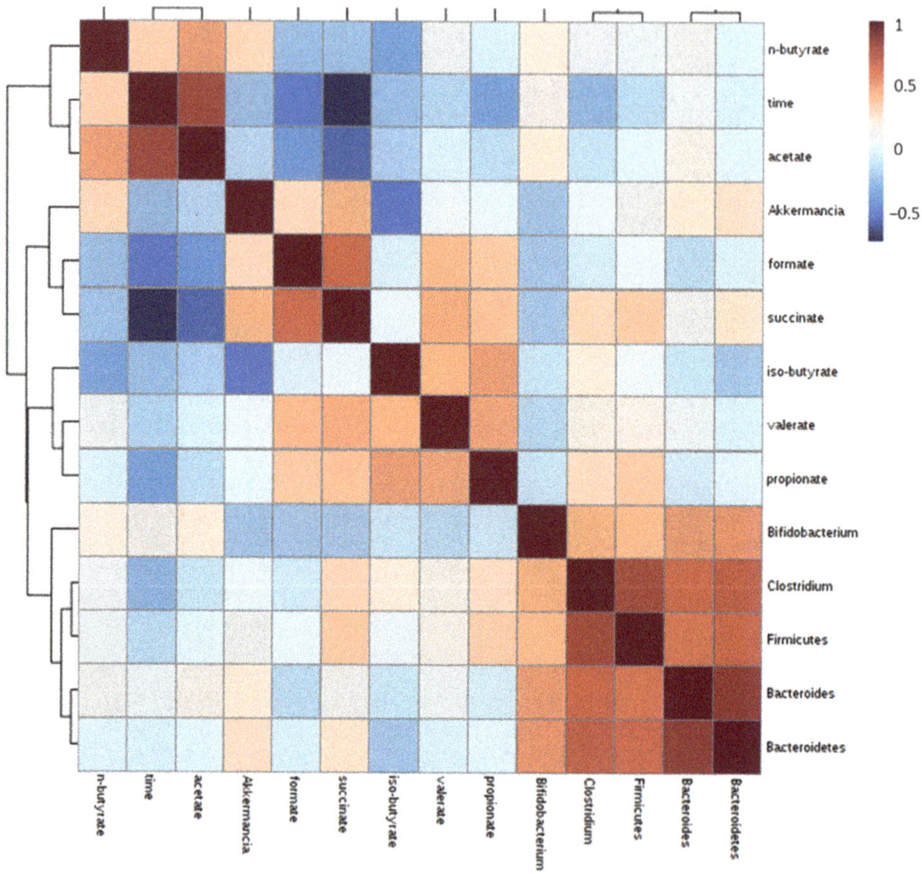

Figure 6. Pearson correlation between microorganisms and SCFAs produced during fermentations.

4. Conclusions

The screening of the prebiotic properties of SF obtained after OH and CONV extraction from tomato by-products was assessed using *Lactobacillus* and *Bifidobacterium* as probiotics. Differences in bacterial carbohydrate utilization patterns between species were identified, with the best results being obtained for Bifidobacterium animalis BO. The impact of SFOH and SFCONV on these probiotics was small, with differences observed for the *Lactobacillus* and Bifidobacteria strains. SFOH at 2% and 4% contributes to *L. casei* growth when compared to SFCONV, while for Bifidobacterium, SFCONV at 2% promotes it growth.

Regarding the fecal fermentation based on volunteers' feces, both flours' main groups are the *Bifidobacterium* and *Akkermansia*. In addition, 25% of SFOH and SFCONV samples presented gene copies higher than the positive control for Bacteroides. Regarding SFOH, this sample enhanced Bacteroidetes' growth. In addition, 50% of population tests presented similar results with FOS to *C. leptum*, while SFCONV presented a higher number of genes copies than the positive control for *Bifidobacterium*.

Concerning SCFA results, both flours increased the propionate, butyrate, and acetate concentration compared to the negative control, which indicates the production capacity of these acids by SFOH and SFCONV during fermentation. In addition, SFCONV produces more butyrate than the OH samples.

However, a relation was observed between certain bacterial groups and SCFA concentration. For the *Bifidobacterium*, both acetate and n-butyrate influence its growth, while *Clostridium* is influenced by iso-butyrate and propionate.

The outcomes propose that both tomato flours favor a potential modulatory impact upon the gut microbiota, thus giving a counteractive action for different diets. Moreover, SFOH comes from a cleaner extraction than SFCONV, making it a potential sustainable ingredient with a prebiotic impact through the growth enhancement of *Bifidobacterium animalis* and improvement of the generation of SCFA.

Nonetheless, initial and promising evidence of their potential prebiotic effect was demonstrated, raising the need for more extensive testing in vivo as part of future work.

Supplementary Materials: The following supporting information can be downloaded at: https://www.mdpi.com/article/10.3390/foods12091920/s1. Table S1: real-time PCR conditions; Table S2: primer sequences targeting bacterial groups, genomic DNA standards, and PCR product size. Adapted from Marques et al., 2016; Table S3: The proximate composition and total phenolic compounds of SFOH and SFCONV samples (g/100 g); Table S4: Constituents (g/100 g fibre DW) of SDF and IDF from SFOH and SFCONV samples; Table S5: Recovery index and bioaccessibility of total phenolic compounds, antioxidant activity, and individual phenolic compounds from SFCONV and SFOH samples throughout digestion; Table S6: Recover index and bioaccessibility percent of carotenoids identified by mass spectrometry LC-ESI-UHR-QqTOF-MS to OH and extracts; Figure S1: Bacterial populations (log (copy number of 16 S rRNA/ng of DNA), means $\pm$ SD) detected by PCR in Fecal samples. The used probes: *Clostridium leptum* (A), Bacteroidetes (B), Bacteroides (C), Firmicutes (D), *Bifidobacterium* (E), and *Akkermansia* (F) Different letters mark statistically significant ($p < 0.05$) differences between samples at each sampling point.

Author Contributions: Conceptualization, M.C.C. and M.E.P.; methodology, M.C.C., D.R. and C.C.; formal Analysis: M.C.C. and D.R.; investigation, M.C.C. and M.E.P.; writing—original draft preparation, M.C.C.; writing—review and editing, M.C.C., S.S., M.E.P., A.S.R. and J.A.T. All authors have read and agreed to the published version of the manuscript.

Funding: This work was supported by National Funds from FCT—Fundação para a Ciência e a Tecnologia through project UIDB/50016/2020.

Institutional Review Board Statement: Ethical review and approval were waived for this study, as it was conducted according to internal rules legally established, based on research ethics recommendations and with the informed consent of all subjects involved in the study.

Informed Consent Statement: Informed consent was obtained from all subjects in the study.

Data Availability Statement: The data presented in this study are available on request from the corresponding author. The data are not publicly available due to restrictions, e.g., privacy or ethics.

Acknowledgments: The authors would like to thank to "MultiBiorefinery: Estratégias multiuso para a valorização de uma gama alargada de subprodutos agroflorestais e das pescas: Um passo em frente na criação de uma biorrefinaria" financiado pelo Programa Operacional Competitividade e Internacionalização (POCI-01-0145-FEDER-016403) e pelo Programa Operacional Regional de Lisboa (LISBOA-01-0145-FEDER-016403), na sua componente FEDER e pela Fundação para a Ciência e Tecnologia, I.P. na componente nacional (SAICTPAC/0040/2015). The authors would also like to acknowledge the scientific collaboration under the FCT project UID/Multi/50016/2019, UID-BIM-00009-2020, and UID/BIO/04469/2020. Author Marta Coelho would like to acknowledge FCT for their PhD grant with the reference [grant number SFRH/BD/111884/2015].

Conflicts of Interest: The authors declare no conflict of interest.

References

1. Bell, V.; Ferrão, J.; Pimentel, L.; Pintado, M.; Fernandes, T. One Health, Fermented Foods, and Gut Microbiota. *Foods* **2018**, *7*, 195 [CrossRef] [PubMed]
2. Roupar, D.; Coelho, M.C.; Gonçalves, D.A.; Silva, S.P.; Coelho, E.; Silva, S.; Coimbra, M.A.; Pintado, M.; Teixeira, J.A.; Nobre, C. Evaluation of Microbial-Fructo-Oligosaccharides Metabolism by Human Gut Microbiota Fermentation as Compared to Commercial Inulin-Derived Oligosaccharides. *Foods* **2022**, *11*, 954. [CrossRef] [PubMed]
3. Cait, A.; Hughes, M.R.; Antignano, F.; Cait, J.; Dimitriu, P.A.; Maas, K.R.; Reynolds, L.A.; Hacker, L.; Mohr, J.; Finlay, B.B.; et al Microbiome-driven allergic lung inflammation is ameliorated by short-chain fatty acids. *Mucosal Immunol.* **2018**, *11*, 785–795. [CrossRef]
4. Sivaprakasam, S.; Prasad, P.D.; Singh, N. Benefits of short-chain fatty acids and their receptors in inflammation and carcinogenesis. *Pharmacol. Ther.* **2016**, *164*, 144–151. [CrossRef]
5. Fan, P.; Li, L.; Rezaei, A.; Eslamfam, S.; Che, D.; Ma, X. Metabolites of Dietary Protein and Peptides by Intestinal Microbes and their Impacts on Gut. *Curr. Protein Pept. Sci.* **2015**, *16*, 646–654. [CrossRef] [PubMed]
6. Coman, V.; Teleky, B.-E.; Mitrea, L.; Martău, G.A.; Szabo, K.; Călinoiu, L.-F.; Vodnar, D.C. Bioactive potential of fruit and vegetable wastes. *Adv. Food Nutr. Res.* **2019**, *91*, 157–225. [CrossRef] [PubMed]
7. Coelho, M.C.; Ribeiro, T.B.; Oliveira, C.; Batista, P.; Castro, P.; Monforte, A.R.; Rodrigues, A.S.; Teixeira, J.; Pintado, M. In Vitro Gastrointestinal Digestion Impact on the Bioaccessibility and Antioxidant Capacity of Bioactive Compounds from Tomato Flours Obtained after Conventional and Ohmic Heating Extraction. *Foods* **2021**, *10*, 554. [CrossRef]
8. Ramírez-Jiménez, A.K.; Rangel-Hernández, J.; Morales-Sánchez, E.; Loarca-Piña, G.; Gaytán-Martínez, M. Changes on the phytochemicals profile of instant corn flours obtained by traditional nixtamalization and ohmic heating process. *Food Chem.* **2018**, *276*, 57–62. [CrossRef]
9. Coelho, M.; Rodrigues, A.; Teixeira, J.; Pintado, M. Integral valorisation of tomato by-products towards bioactive compounds recovery: Human health benefits. *Food Chem.* **2023**, *410*, 135319. [CrossRef]
10. Coelho, M.C.; Ghalamara, S.; Campos, D.; Ribeiro, T.B.; Pereira, R.; Rodrigues, A.S.; Teixeira, J.A.; Pintado, M. Tomato Processing By-Products Valorisation through Ohmic Heating Approach. *Foods* **2023**, *12*, 818. [CrossRef]
11. Campos, D.A.; Coscueta, E.R.; Vilas-Boas, A.A.; Silva, S.; Teixeira, J.A.; Pastrana, L.M.; Pintado, M.M. Impact of functional flours from pineapple by-products on human intestinal microbiota. *J. Funct. Foods* **2020**, *67*, 103830. [CrossRef]
12. Tomás-Barberán, F.A.; Selma, M.V.; Espín, J.C. Interactions of gut microbiota with dietary polyphenols and consequences to human health. *Curr. Opin. Clin. Nutr. Metab. Care* **2016**, *19*, 471–476. [CrossRef]
13. Chen, T.; Long, W.; Zhang, C.; Liu, S.; Zhao, L.; Hamaker, B.R. Fiber-utilizing capacity varies in Prevotella- versus Bacteroides-dominated gut microbiota. *Sci. Rep.* **2017**, *7*, 2594. [CrossRef]
14. Coelho, M.; Pereira, R.; Rodrigues, A.S.; Teixeira, J.A.; Pintado, M.E. Extraction of tomato by-products' bioactive compounds using ohmic technology. *Food Bioprod. Process.* **2019**, *117*, 329–339. [CrossRef]
15. Oliveira, A.; Alexandre, E.M.; Coelho, M.; Barros, R.M.; Almeida, D.P.; Pintado, M. Peach polyphenol and carotenoid content as affected by frozen storage and pasteurization. *LWT—Food Sci. Technol.* **2016**, *66*, 361–368. [CrossRef]
16. Vilella, S.B.; Vaqué, L.G. The Commission Establishes the Specific Compositional and Information Requirements for "Total Diet Replacement for Weight Control Products": Commission Delegated Regulation (EU) 2017/1798. *Eur. Food Feed. Law Rev. EFFL* **2018**, *13*, 108–115.
17. Sousa, S.; Pinto, J.; Pereira, C.; Malcata, F.X.; Pacheco, M.B.; Gomes, A.M.; Pintado, M. In vitro evaluation of yacon (*Smallanthus sonchifolius*) tuber flour prebiotic potential. *Food Bioprod. Process.* **2015**, *95*, 96–105. [CrossRef]
18. Li, Y.; Qin, C.; Dong, L.; Zhang, X.; Wu, Z.; Liu, L.; Yang, J.; Liu, L. Whole grain benefit: Synergistic effect of oat phenolic compounds and β-glucan on hyperlipidemia via gut microbiota in high-fat-diet mice. *Food Funct.* **2022**, *13*, 12686–12696. [CrossRef] [PubMed]
19. Gullón, B.; Gullón, P.; Tavaria, F.; Alonso, J.L.; Pintado, M. In vitro assessment of the prebiotic potential of Aloe vera mucilage and its impact on the human microbiota. *Food Funct.* **2015**, *6*, 525–531. [CrossRef]
20. de Carvalho, N.M.; Walton, G.; Poveda, C.; Silva, S.; Amorim, M.; Madureira, A.R.; Pintado, M.M.; Gibson, G.; Jauregi, P. Study of in vitro digestion of Tenebrio molitor flour for evaluation of its impact on the human gut microbiota. *J. Funct. Foods* **2019**, *59*, 101–109. [CrossRef]
21. Watson, D.; Motherway, M.O.; Schoterman, M.; van Neerven, R.J.; Nauta, A.; van Sinderen, D. Selective carbohydrate utilization by lactobacilli and bifidobacteria. *J. Appl. Microbiol.* **2013**, *114*, 1132–1146. [CrossRef] [PubMed]
22. McLaughlin, H.P.; Motherway, M.O.; Lakshminarayanan, B.; Stanton, C.; Ross, R.P.; Brulc, J.; Menon, R.; O'Toole, P.W.; van Sinderen, D. Carbohydrate catabolic diversity of bifidobacteria and lactobacilli of human origin. *Int. J. Food Microbiol.* **2015**, *203*, 109–121. [CrossRef]
23. de Carvalho, N.M.; Teixeira, F.; Silva, S.; Madureira, A.R.; Pintado, M.E. Potential prebiotic activity of *Tenebrio molitor* insect flour using an optimized in vitro gut microbiota model. *Food Funct.* **2019**, *10*, 3909–3922. [CrossRef] [PubMed]
24. Gaglio, R.; Alfonzo, A.; Polizzotto, N.; Corona, O.; Francesca, N.; Russo, G.; Moschetti, G.; Settanni, L. Performances of Different Metabolic *Lactobacillus* Groups during the Fermentation of Pizza Doughs Processed from Semolina. *Fermentation* **2018**, *4*, 61. [CrossRef]

5. Crittenden, R.; Karppinen, S.; Ojanen, S.; Tenkanen, M.; Fagerström, R.; Mättö, J.; Saarela, M.; Mattila-Sandholm, T.; Poutanen, K. In vitro fermentation of cereal dietary fibre carbohydrates by probiotic and intestinal bacteria. *J. Sci. Food Agric.* **2002**, *82*, 781–789. [CrossRef]
6. Liu, Y.; Chen, H.; Chen, W.; Zhong, Q.; Zhang, G.; Chen, W. Beneficial Effects of Tomato Juice Fermented by Lactobacillus Plantarum and Lactobacillus Casei: Antioxidation, Antimicrobial Effect, and Volatile Profiles. *Molecules* **2018**, *23*, 2366. [CrossRef] [PubMed]
7. Zhang, Y.; Li, Y.; Ren, X.; Zhang, X.; Wu, Z.; Liu, L. The positive correlation of antioxidant activity and prebiotic effect about oat phenolic compounds. *Food Chem.* **2023**, *402*, 134231. [CrossRef]
8. Pokusaeva, K.; Fitzgerald, G.F.; van Sinderen, D. Carbohydrate metabolism in Bifidobacteria. *Genes Nutr.* **2011**, *6*, 285–306. [CrossRef]
9. Mazé, A.; O'Connell-Motherway, M.; Fitzgerald, G.F.; Deutscher, J.; van Sinderen, D. Identification and Characterization of a Fructose Phosphotransferase System in *Bifidobacterium breve* UCC2003. *Appl. Environ. Microbiol.* **2007**, *73*, 545–553. [CrossRef]
0. Viborg, A.H.; Katayama, T.; Hachem, M.A.; Andersen, M.C.; Nishimoto, M.; Clausen, M.H.; Urashima, T.; Svensson, B.; Kitaoka, M. Distinct substrate specificities of three glycoside hydrolase family 42 β-galactosidases from Bifidobacterium longum subsp. infantis ATCC 15697. *Glycobiology* **2014**, *24*, 208–216. [CrossRef]
1. Parche, S.; Beleut, M.; Rezzonico, E.; Jacobs, D.; Arigoni, F.; Titgemeyer, F.; Jankovic, I. Lactose-over-Glucose Preference in *Bifidobacterium longum* NCC2705: *glcP*, Encoding a Glucose Transporter, Is Subject to Lactose Repression. *J. Bacteriol.* **2006**, *188*, 1260–1265. [CrossRef]
32. Slavin, J.L.; Lloyd, B. Health Benefits of Fruits and Vegetables. *Adv. Nutr. Int. Rev. J.* **2012**, *3*, 506–516. [CrossRef] [PubMed]
33. Bermudez-Brito, M.; Plaza-Díaz, J.; Muñoz-Quezada, S.; Gómez-Llorente, C.; Gil, A. Probiotic Mechanisms of Action. *Ann. Nutr. Metab.* **2012**, *61*, 160–174. [CrossRef] [PubMed]
34. Amakiri, A.C.; Thantsha, M.S. Survival of Bifidobacterium longum LMG 13197 microencapsulated in Vegetal or Vegetal-inulin matrix in simulated gastrointestinal fluids and yoghurt. *Springerplus* **2016**, *5*, 1343. [CrossRef] [PubMed]
35. Xue, B.; Xie, J.; Huang, J.; Chen, L.; Gao, L.; Ou, S.; Wang, Y.; Peng, X. Plant polyphenols alter a pathway of energy metabolism by inhibiting fecal Bacteroidetes and Firmicutes in vitro. *Food Funct.* **2016**, *7*, 1501–1507. [CrossRef] [PubMed]
36. Simpson, H.L.; Campbell, B.J. Review article: Dietary fibre-microbiota interactions. *Aliment. Pharmacol. Ther.* **2015**, *42*, 158–179. [CrossRef]
37. Rinninella, E.; Raoul, P.; Cintoni, M.; Franceschi, F.; Miggiano, G.A.D.; Gasbarrini, A.; Mele, M.C. What Is the Healthy Gut Microbiota Composition? A Changing Ecosystem across Age, Environment, Diet, and Diseases. *Microorganisms* **2019**, *7*, 14. [CrossRef] [PubMed]
38. Jandhyala, S.M.; Talukdar, R.; Subramanyam, C.; Vuyyuru, H.; Sasikala, M.; Nageshwar Reddy, D. Role of the normal gut microbiota. *World J. Gastroenterol.* **2015**, *21*, 8787–8803. [CrossRef]
39. Kamada, N.; Seo, S.-U.; Chen, G.Y.; Núñez, G. Role of the gut microbiota in immunity and inflammatory disease. *Nat. Rev. Immunol.* **2013**, *13*, 321–335. [CrossRef]
40. Bäumler, A.J.; Sperandio, V. Interactions between the microbiota and pathogenic bacteria in the gut. *Nature* **2016**, *535*, 85–93. [CrossRef]
41. Rastmanesh, R. Use of Probiotics in Burn Patients to Improve Nutritional Status and Clinical Outcomes: A Hypothesis. *Int. J. Probiotics Prebiotics* **2011**, *6*, 159.
42. Di Paola, M.; De Filippo, C.; Cavalieri, D.; Ramazzotti, M.; Poullet, J.; Massart, S.; Collini, S.; Pieraccini, G.; Lionetti, P. PP90 impact of diet in shaping gut microbiota revealed by a comparative study in children from europe and rural africa. *Dig. Liver Dis.* **2011**, *43*, S445–S446. [CrossRef]
43. Parkar, S.G.; Trower, T.M.; Stevenson, D.E. Fecal microbial metabolism of polyphenols and its effects on human gut microbiota. *Anaerobe* **2013**, *23*, 12–19. [CrossRef]
44. Koliada, A.; Syzenko, G.; Moseiko, V.; Budovska, L.; Puchkov, K.; Perederiy, V.; Gavalko, Y.; Dorofeyev, A.; Romanenko, M.; Tkach, S.; et al. Association between body mass index and Firmicutes/Bacteroidetes ratio in an adult Ukrainian population. *BMC Microbiol.* **2017**, *17*, 120. [CrossRef] [PubMed]
45. Xu, J.; Mahowald, M.A.; Ley, R.E.; A Lozupone, C.; Hamady, M.; Martens, E.C.; Henrissat, B.; Coutinho, P.M.; Minx, P.; Latreille, P.; et al. Evolution of Symbiotic Bacteria in the Distal Human Intestine. *PLoS Biol.* **2007**, *5*, e156. [CrossRef]
46. Dao, M.C.; Everard, A.; Aron-Wisnewsky, J.; Sokolovska, N.; Prifti, E.; Verger, E.O.; Kayser, B.D.; Levenez, F.; Chilloux, J.; Hoyles, L.; et al. Akkermansia muciniphila and improved metabolic health during a dietary intervention in obesity: Relationship with gut microbiome richness and ecology. *Gut* **2016**, *65*, 426–436. [CrossRef] [PubMed]
47. Everard, A.; Belzer, C.; Geurts, L.; Ouwerkerk, J.P.; Druart, C.; Bindels, L.B.; Guiot, Y.; Derrien, M.; Muccioli, G.G.; Delzenne, N.M.; et al. Cross-talk between *Akkermansia muciniphila* and intestinal epithelium controls diet-induced obesity. *Proc. Natl. Acad. Sci. USA* **2013**, *110*, 9066–9071. [CrossRef]
48. Zhou, K. Strategies to promote abundance of *Akkermansia muciniphila*, an emerging probiotics in the gut, evidence from dietary intervention studies. *J. Funct. Foods* **2017**, *33*, 194–201. [CrossRef]
49. Schuchmann, K.; Müller, V. Energetics and Application of Heterotrophy in Acetogenic Bacteria. *Appl. Environ. Microbiol.* **2016**, *82*, 4056–4069. [CrossRef]

50. Ríos-Covián, D.; Ruas-Madiedo, P.; Margolles, A.; Gueimonde, M.; De Los Reyes-Gavilán, C.G.; Salazar, N. Intestinal Short Chain Fatty Acids and their Link with Diet and Human Health. *Front. Microbiol.* **2016**, *7*, 185. [CrossRef]
51. Morrison, D.J.; Preston, T. Formation of short chain fatty acids by the gut microbiota and their impact on human metabolism. *Gut Microbes* **2016**, *7*, 189–200. [CrossRef] [PubMed]

foods

Article

The Antioxidant Activities In Vitro and In Vivo and Extraction Conditions Optimization of Defatted Walnut Kernel Extract

Xiaomei Zhou [1,2,†], Xiaojian Gong [1,2,†], Xu Li [1,2], Ning An [3], Jiefang He [1], Xin Zhou [2,*] and Chao Zhao [1,*]

[1] Key Laboratory for Information System of Mountainous Areas and Protection of Ecological Environment, Guizhou Normal University, Guiyang 550001, China
[2] Guizhou Engineering Laboratory for Quality Control & Evaluation Technology of Medicine, Guizhou Normal University, Guiyang 550001, China
[3] Experimental Centre of Tropical Forestry, Chinese Academy of Forestry, Pingxiang 532600, China
* Correspondence: zhouxin@gznu.edu.cn (X.Z.); chaozhao@gznu.edu.cn (C.Z.)
† These authors contributed equally to this work.

Abstract: The objective of this study was to determine the antioxidant activities of defatted walnut kernel extract (DWE) and whole walnut kernel extract (WE) in vitro and in vivo. Three spectrophotometric methods, DPPH, ABTS, and FRAP, were used in in vitro experiments, and mice were used in in vivo experiments. In addition, response surface methodology (RSM) was used to optimize reflux-assisted ethanol extraction of DWE for maximum antioxidant activity and total phenolic content. The results of in vitro experiments showed that both extracts showed antioxidant activity; however, the antioxidant activity of DWE was higher than that of WE. Both extracts improved the mice's oxidative damage status in in vivo studies. An ethanol concentration of 58%, an extraction temperature of 48 °C, and an extraction time of 77 min were the ideal parameters for reflux-assisted ethanol extraction of DWE. The results may provide useful information for further applications of defatted walnut kernels and the development of functional foods.

Keywords: walnut kernel; antioxidant activity; in vitro; in vivo; response surface methodology

Citation: Zhou, X.; Gong, X.; Li, X.; An, N.; He, J.; Zhou, X.; Zhao, C. The Antioxidant Activities In Vitro and In Vivo and Extraction Conditions Optimization of Defatted Walnut Kernel Extract. *Foods* **2023**, *12*, 3417. https://doi.org/10.3390/foods12183417

Academic Editors: Noelia Castillejo Montoya and Lorena Martínez-Zamora

Received: 17 August 2023
Revised: 5 September 2023
Accepted: 7 September 2023
Published: 14 September 2023

1. Introduction

Originally from the Mediterranean Basin, walnut (*Juglans regia* L.) is a member of the Juglandaceae family. The major walnut producers in the world are China and the United States [1]. Walnut is one of the four most popular nuts in the world [2,3], favored for their nutritional and medicinal value. Walnut is widely distributed in various provinces of China, such as Heilongjiang, Hebei, Yunnan, and Guizhou. China has the largest planting area and production. Data shows that China's walnut planting area and production reached 631,000 hectares and more than 2.5 million tons in 2019, respectively [4]. Walnut kernels are rich in nutrients, containing large amounts of protein, dietary fiber, carbohydrates, vitamins, fat, and amino acids, as well as trace elements such as potassium, iron, and calcium [5–7]. Walnut kernels have anti-inflammatory and antiseptic effects, protecting blood vessels, lowering blood pressure, and preventing cancer. Walnut kernels can be eaten directly or processed into various food items for consumption, such as beverages, desserts, and snacks. The kernels, husks, green barks of walnuts, branches, and leaves of walnut trees are all treasures that have been extensively employed in medical research and cosmetics manufacturing [8].

The main cause of oxidative stress is inequality between prooxidants and antioxidants owing to the overproduction of reactive oxygen species (ROS) that exceeds the ability of cellular defenses [9,10]. The by-products of aerobic metabolism can damage the structure and function of cellular organelles. For example, reactive oxygen species (ROS), if not effectively removed, may harm proteins, lipids, and deoxyribonucleic acid (DNA) in cells, ultimately leading to cellular damage and cell death [11,12]. A variety of human diseases have been observed to have a bearing on oxidative stress, e.g., cancer [13], dementia [14], cardiovascular diseases, neurodegenerative diseases [15,16], hyponatremia [17], diabetes [18], uterine fibroid [19], and colorectal cancer [20]. Organisms contain natural defense systems against oxidative stress, including enzymatic antioxidant and non-enzymatic antioxidant systems. In addition, supplementation with exogenous antioxidants is also an efficacious way to effectively inhibit oxidative damage. Antioxidants can inhibit oxidative damage by scavenging or neutralizing free radicals, enhancing the activity of antioxidant enzymes in vivo, and reducing the products of lipid peroxidation [21,22]. When present in the medium in lower concentrations than oxidizable substrates, antioxidants prevent the oxidation of substrates [23,24]. In the last century, various synthetic antioxidants produced by chemical processes have been used in foods to prevent oxidation, such as butylhydroxytoluene (BHT) and butylhydroxyanisole (BHA). However, researchers have indicated that synthetic antioxidants can have negative consequences, like carcinogenesis [25–27]. Therefore, the search for natural alternatives from plants has generated great interest based on safety and health.

The presence of polyphenols, tocopherols, squalene, unsaturated fatty acids, phytosterols, and bioactive peptides in walnut kernels contributes to their nutritional and medicinal value [28,29]. Polyphenols, which have both antibacterial and antioxidant properties [30,31], are plentiful secondary metabolites in plants. The hydroxyl groups of polyphenolic compounds can react with free radicals, thus slowing down the harm of free radicals to cells and achieving antioxidant effects. Walnut kernels have the highest phenol content compared to other nuts, like pistachios, Brazil nuts, almonds, and peanuts [32,33]. A study even showed that walnut kernels have the greatest antioxidant ability among twenty-five common foods [34]. Therefore, walnut kernels are a good source of natural antioxidants.

Defatted walnut kernels are a by-product left over from the extraction of walnut oil. Defatted walnut kernels are rich in phenolics, which have antioxidant properties [35]. This property makes it a potential source for the development of functional foods and dietary products. The addition of substances with healthy properties can transform these products into healthier foods. Previous studies on the antioxidant activity of defatted walnut kernel have mainly focused on in vitro or in vivo studies [36–38]. However, many substances may not have consistent antioxidant activities both in vitro and in vivo. There are no reports on the use of in vitro and in vivo assays to comprehensively assess the antioxidant activity of defatted walnut kernel. One study used ultrasonic-assisted ethanol extraction of defatted walnut kernel extract to obtain optimal technical parameters for de-penalization treatments [39]. However, there are few studies on the optimization of defatted walnut kernel extract (DWE) based on the antioxidant effect as an indicator. Therefore, in the present study, the antioxidant activity of defatted walnut kernel extract and whole walnut kernel extract (WE) was comprehensively evaluated using in vitro and in vivo assays. In addition, response surface methodology was used to optimize extraction conditions for reflux-assisted ethanol extraction of defatted walnut kernel extracts. In this study, three spectrophotometric methods, DPPH, ABTS, and FRAP, were employed to measure the ability to eliminate free radicals in vitro. In in vivo experiments, malondialdehyde (MDA) content, total antioxidant activity (T-AOC), glutathione peroxidase (GSH-Px) activity, catalase (CAT) activity, and total superoxide dismutase (T-SOD) activity were determined in serum and tissues of mice.

2. Materials and Methods

2.1. Materials and Reagents

This test uses walnut kernels as the raw material from Shijiazhuang City, Hebei Province. BHA and ABTS were obtained from J&K Scientific (Beijing, China). Trolox was provided by Aldrich Co. (St. Louis, MO, USA). DPPH was supplied by TCI Development Co., Ltd. (Shanghai, China). Tripyridyltriazine and D-galactose were bought from Aladdin Co., Ltd. (Shanghai, China). The malondialdehyde (MDA), total antioxidant activity (T-

AOC), total superoxide dismutase (T-SOD), glutathione peroxidase (GSH-Px), and catalase (CAT) kits were provided by Nanjing Jiancheng Bioengineering Company (Nanjing, China). Other solvents utilized in the experiments were analytical- or HPLC-grade.

2.2. Preparation of Walnut Kernel Extracts

Weigh 0.30 g of the defatted walnut kernel (over 20 mesh) and walnut kernel dissolved in 70% ethanol at a ratio of 30:1. After being sonicated (HU-10260B, Tianjin Hengao Technology Development Co., Tianjin, China) for 40 min; the solution was centrifuged (TDL-5A, Changzhou Wanhe Instrument Manufacturing Co., Changzhou, China) to separate the supernatant, and then diluted with a 1.5% sodium carboxymethylcellulose solution to the desired concentration. Defatted walnut kernel extract (DWE) and walnut kernel extract (WE) were obtained.

The mixed fatty acids were prepared by saponifying walnut oil at 60 °C for 1 h. The saponified solution was then acidified with hydrochloric acid to pH 3–4, followed by diluting the acidified solution with anhydrous ethanol to the appropriate concentration.

2.3. In Vitro Antioxidant Activity Determination

Three complementary and common methods, DPPH, ABTS, and FRAP, were selected to determine the in vitro antioxidant activity of each walnut extract separately. As positive controls, the experiment employed vitamin C and butylhydroxyanisole (BHA).

2.3.1. DPPH Assay

The determination of DPPH free radical elimination activity was performed according to the literature with some modifications [40]. In total, 0.5 mL of 250 mmol/L DPPH (dissolved in anhydrous ethanol) working solution was added to 0.5 mL of the appropriate concentration of sample solutions or Trolox solutions. The mixture was reacted at room temperature and protected from light for 40 min. The absorbance values were subsequently determined at 517 nm using an enzyme marker (Spectra Max plus 383, MDC, Silicon Valley, CA, USA). A series of concentrations of Trolox (3.12–100 μmol/L) solutions were utilized as standard, and the calibration curve was plotted to compute DPPH radical scavenging activity. As follows: $Y_1 = 0.008X_1 + 0.1225$ ($R^2 = 0.9982$) (X_1, concentration of Trolox, μg/mL; Y_1, DPPH radical scavenging rate, %). DPPH scavenging activity can be shown as Trolox equivalent antioxidant capacity (TEAC).

2.3.2. ABTS Assay

The determination of ABTS free radical elimination activity was performed according to the literature with some modifications [41]. For the ABTS stock solution, ABTS solution and potassium persulfate solution were produced to final concentrations of 7 mmol/L and 2.45 mmol/L, individually. After being combined, the mixing solution was left at room temperature and shielded from light over 12 h. Anhydrous ethanol was used to dilute the stock solution into the working solution until the absorbance value at 734 nm was 0.80 ± 0.02. A total of 1.0 mL of ABTS working solution was mixed with 50 μL of proper sample concentration or Trolox solutions. The mixture was reacted at room temperature and protected from light for 20 min, and the absorbance values were determined at 734 nm using an enzyme marker (Spectra Max plus 383, MDC, Silicon Valley, CA, USA). A series of concentrations of Trolox (25–800 μmol/L) solutions were utilized as standard, and the calibration curve was drafted to compute ABTS radical scavenging activity as follows: $Y_2 = 0.0012X_2 - 0.0196$ ($R^2 = 0.9954$) (X_2, concentration of Trolox, μg/mL; Y_2, ABTS radical scavenging rate, %). ABTS scavenging activity can be shown as Trolox equivalent antioxidant capacity (TEAC).

2.3.3. FRAP Assay

The determination of ferric-reducing antioxidant power was performed according to the literature with some modifications [42]. Ferric chloride solution (20 mmol/L), acetate buffer solution (pH 3.6, 0.3 mol/L), and TPTZ stock solution (10 mmol/L) were mixed in a volume ratio of 1:1:10. This resulted in a fresh FRAP working solution. The samples were prepared with anhydrous ethanol to the appropriate concentration. In total, 50 μL of the sample solutions or Trolox solutions and 1 mL of FRAP reagent were well mixed at 37 °C After 10 min, the absorbance values were measured at 593 nm using an enzyme marker (Spectra Max plus 383, MDC, Silicon Valley, CA, USA). The calibration curve was used to compute the FRAP with Trolox as the standard: $Y_3 = 0.0011X_3 - 0.1113$ ($R^2 = 0.9996$) (X_3, concentration of Trolox, μg/mL; Y_3, FRAP). FRAP was signified as Trolox equivalent antioxidant capacity (TEAC).

2.4. *In Vivo Antioxidant Activity Determination*

2.4.1. Animals

Six to eight-week-old Kunming mice (SPF grade, Batch No. SCXK 2009-0012), weighing roughly 20 ± 2 g, were obtained from Changsha Tianqin Biotechnology Company (Changsha, China). The feed was purchased at Changsha Tianqin Biotechnology Company. Before the start of the experiment, the mice were fed with basic feed and water for one week to acclimatize them to the conditions of the animal room. Animal experiments in this study were conducted in accordance with current national and international laws and recommendations, and every effort was made to minimize suffering. The experiment was approved by the Animal Care and Use Committee of Guizhou Normal University (C00032801).

2.4.2. Mice Grouping and Experimental Design

After the mice were adapted, fifty mice were randomly divided into five groups of ten mice each as follows: the normal control group (Group I), the model control group (Group II), the positive control group (Group III); the defatted walnut kernel extract (DWE) group (Group IV), and the whole walnut kernel extract (WE) group (Group V). Vitamin C was the positive control that was employed.

The doses administered to the WE and DWE groups were set based on the defatted walnut powder extract group. To produce an oxidative damage model, mice in Groups II through V received an intraperitoneal injection of 0.3 mL of D-galactose solution (0.5 g/kg/d). Mice in Group I received a 0.3 mL physiological saline injection. In addition, Vitamin C (0.2 g/kg/d), DWE (0.2 g/kg/d), and WE (0.5 g/kg/d) were given daily to groups III, IV, and V, respectively. This was conducted for 42 consecutive days by gavage.

After the last gavage, all groups of mice were given a fast while still having access to water. The next day, blood was taken from the eyeballs of mice. Subsequently, after standing for a while, the collected blood was centrifuged (4 °C, 4000 rpm, 10 min) (TGL-16M, Changsha Meijiasen Instrument Co., Changsha, China) to collect the serum from the supernatant. The brain tissue, kidney, liver, and heart of mice were immediately dissected. The blood on the tissues was washed away separately with ice-cold physiologic saline and blotted dry with filter paper. The homogenization process involved combining the sample with iced saline at a ratio of 1:9 (mass to volume) to obtain a 10% homogenate. The resulting mixture was then subjected to centrifugation at 4 °C and 4000 revolutions per minute (rpm) for 5 min using a TGL-16M centrifuge (Changsha Meijiasen Instrument Co., Changsha, China). The supernatant was collected and preserved at a temperature of −20 °C for subsequent experiments.

2.4.3. Antioxidant Assays

The Technical Specification for Inspection and Evaluation of Health Food (2003 edition) stipulates that when any of the indicators of lipid peroxide content and any of the indicators of antioxidant function enzyme activity are positive, it can indicate that the tested extract has antioxidant health function. Therefore, we made a mouse model of oxidative damage using D-galactose. Subsequently, measurements of total antioxidant activity (T-AOC), catalase (CAT) activity, total superoxide dismutase (T-SOD) activity, glutathione peroxidase (GSH-PX) activity, and malondialdehyde (MDA) content were performed in the serum and tissues of mice. The operation procedure was regulated on the basis of the operating descriptions of the kit.

2.5. Optimization of Extraction Conditions of Defatted Walnut Kernel Extract

The extraction conditions for reflux-assisted ethanol extraction of defatted walnut kernel extract were optimized for maximum antioxidant activity and total phenolic content using single-factor experiments and response surface methodology.

2.5.1. Single-Factor Experimental Design

In this study, 0.30 g of defatted walnut powder was weighed. Firstly, the best extraction method was determined by a single-factor experiment as reflux extraction. The best extraction solvent was anhydrous ethanol, and the best liquid-to-material ratio was 30:1 (mL/g). Based on this, further single-factor experiments were carried out to determine the optimal extraction conditions for defatted walnut kernel extract (DWE). The extraction conditions included extraction time, ethanol concentration, and extraction temperature. Briefly, the extraction times were 40 min, 60 min, 80 min, 100 min, 120 min, and 140 min, respectively. The ethanol concentrations were 0%, 20%, 40%, 60%, 80%, and 100%, respectively. The extraction temperatures were 35 °C, 45 °C, 55 °C, 65 °C, and 75 °C, respectively. The common conditions were an extraction time of 80 min, an ethanol concentration of 50%, and an extraction temperature of 60 °C. When one of the independent variables is modified, the other parameters remain unchanged. The antioxidant activity was determined according to Sections 2.3.1–2.3.3. The total phenol content (TPC) was performed according to the literature with some modifications [43].

Add 2.5 mL of Folin–Ciocalteu phenol reagent (10%) to 0.1 mL of the appropriate concentration of DWE or 100 µg/mL gallic acid standard solution. Allow to stand for 5 min. Then, add Na_2CO_3 solution (10%, 2 mL) and distilled water to bring the volume to 10 mL. After another hour of keeping the mixture at room temperature and shielded from light, the absorbance values were gauged at 765 nm. The TPC of the samples was represented as gallic acid equivalents per gram of dry weight.

2.5.2. Response Surface Methodology (RSM) Experimental Design

Following the results of the single-factor experiments, the defatted walnut kernel extract (DWE) was optimized by response surface methodology (RSM) using Central Composite Design (CCD). Ethanol concentration, extraction time, and extraction temperature were the independent variables, and ABTS, DPPH, FRAP, and total phenol content (TPC) were the response variables. Five distinct levels (−1.68, −1, 0, 1, 1.68) of study were performed on each design variable (Table 1). A total of 19 randomized experimental cycles were run. Table 2 displays the response surface methodology run design and outcomes. Following the completion of the CDD matrix, an analysis of variance (ANOVA) was conducted, and the data were fitted to a second-order polynomial model.

$$y = \beta 0 + \sum_{i=1}^{k} \beta ixi + \sum_{i=1}^{k} \beta iixi2 + \sum_{i}^{k-1} \sum_{j}^{k} \beta ijxixj$$

where y denotes the response variable, xi and xj represent the independent variables, and k is the number of independent variables (k is 3 in this study). $\beta 0$, βi, βii, and βij indicate the intercept term coefficient, the linear term coefficient, the quadratic term coefficient, and the interaction term coefficient, respectively.

Table 1. Independent variables, real and coded values in Central Composite Design (CCD).

Variable	Levels				
	−1.68	−1	0	1	1.68
Ethanol concentration (A, %)	16.36	30	50	70	83.64
Extraction time (B, minutes)	46.36	60	80	100	113.64
Extraction temperature (C, °C)	28.18	35	45	55	61.82

Table 2. Central Composite Design (CCD) scheme and experimentally observed antioxidant activity (ABTS, DPPH, FRAP) and total phenolic content (TPC) response values.

Run	Variables			Response Values			
	A	B	C	ABTS	DPPH	FRAP	TPC
1	30	60	55	85.91	37.04	220.79	32.84
2	50	80	45	106.98	46.52	247.49	41.13
3	50	80	45	106.32	46.37	250.13	40.13
4	50	46.36	45	81.05	35.70	215.33	39.53
5	50	80	28.18	75.79	31.49	199.88	39.04
6	16.36	80	45	67.25	29.03	192.61	29.83
7	50	113.64	45	92.66	40.31	225.33	40.66
8	83.64	80	45	95.04	41.37	225.94	36.58
9	50	80	61.82	100.63	42.37	232.61	40.23
10	50	80	45	103.50	43.98	252.70	40.48
11	70	60	35	92.43	39.43	224.12	39.87
12	70	60	55	98.34	42.25	232.30	39.38
13	30	60	35	64.17	28.56	180.18	33.85
14	70	100	55	101.71	42.65	243.21	32.25
15	70	100	35	101.27	42.51	234.73	34.32
16	50	80	45	103.72	43.52	252.59	38.52
17	30	100	35	71.57	29.83	191.70	34.54
18	50	80	45	106.56	46.11	247.44	39.83
19	30	100	55	92.32	38.17	221.09	35.88

A, ethanol concentration; B, extraction time; C, extraction temperature; ABTS, ABTS radical scavenging activity of defatted walnut kernel extracts; DPPH, DPPH radical scavenging activity of defatted walnut kernel extracts; FRAP, ferric-reducing antioxidant power of defatted walnut kernel extracts; TPC, total phenolic content of defatted walnut kernel extracts.

2.6. Statistical Analysis

All experiments in vivo were performed in eight replicates and averaged. All other experiments were repeated three times and averaged. The experimental result values are expressed as mean ± SD. The data obtained from in vivo and in vitro experiments were statistically analyzed by SPSS.18.0 software. A one-way analysis of variance (ANOVA) was used, followed by Student's test to assess significance. Statistically, significant results were signified as $p < 0.05$, more significant as $p < 0.01$, and extremely significant as $p < 0.001$. The optimization experiments were statistically analyzed using Design-Expert 8.0.6 software.

3. Results and Discussion

3.1. In Vitro Antioxidant Activities of Walnut Kernel Extracts

Table 3 summarizes the antioxidant activities of walnut kernel extracts determined by DPPH, ABTS, and FRAP methods. Vitamin C and butylhydroxyanisole (BHA) were employed as positive controls. Walnut kernels contain up to 65–70% oil, while fatty acids make up more than 90% of walnut oil. Therefore, the antioxidant potential of fatty acids was evaluated and compared with defatted walnut kernel extract (DWE) and whole walnut kernel extract (WE). The results suggested that extracts from all parts of walnut kernels had certain in vitro antioxidant effects. After oil removal, the cake meal extract of walnut kernels had the strongest antioxidant capacity in vitro. Moreover, trends in results measured by the three methods were consistent, with vitamin C

> BHA > DWE > WE > fatty acids. Specifically, the results of DPPH, ABTS, and FRAP in DWE (0.025 mg/mL) were 1026.19 ± 52.95 μM TEAC/g, 2754.30 ± 42.79 μM TEAC/g, and 5829.09 ± 36.36 μM TEAC/g. The clearance rate measured by DPPH and ABTS at a concentration of 0.025 mg/mL of WE was small and not in the linear range. In contrast, the FRAP result was 4180.61 ± 41.99 μM TEAC/g. At a fatty acid concentration of 100 mg/mL, the clearance measured by the three methods was small and not in the linear range. This showed that DWE has significantly different ($p < 0.01$) antioxidant activity from WE and fatty acids. Compared to BHA, vitamin C showed the greatest antioxidant activity among the two positive controls. The result measured by the DPPH method was 4006.05 ± 4.03 μM TEAC/g. The result measured by the ABTS method was 9018.41 ± 85.58 μM TEAC/g. The result measured by the FRAP method was 20,018.69 ± 116.89 μM TEAC/g. Interestingly, the antioxidant activities of walnut kernel extracts showed high differences in DPPH, FRAP, and ABTS. Under the same conditions, the FRAP method showed high antioxidant capacity and large differences. For example, for the antioxidant-active substance DWE, the antioxidant activity determined through the FRAP method increased 5.68-fold and 2.12-fold compared to DPPH and ABTS, respectively. This result is due to the different detection principles of the methods utilized. Each method has its range of applications and features. At present, no method can comprehensively measure the antioxidant activity of a substance. Consequently, the antioxidant activities of food extracts should be evaluated using various methodologies simultaneously in experiments since different approaches frequently provide different findings [44].

Table 3. Comparison of antioxidant activities of different parts in walnut kernels.

Sample	Concentration (mg/mL)	DPPH (μM TEAC/g)	ABTS (μM TEAC/g)	FRAP (μM TEAC/g)
Vitamin C	0.025	4006.05 ± 4.03	9018.41 ± 85.58	20,018.69 ± 116.89
BHA	0.025	3660.22 ± 13.98	7301.33 ± 171.15	16,362.42 ± 575.34
DWE	0.025	1026.19 ± 52.95	2754.30 ± 42.79	5829.09 ± 36.36
WE	0.025	ND	ND	4180.61 ± 41.99 [a]
Fatty acid	100	ND	ND	ND

BHA: butylhydroxyanisole; DWE: defatted walnut kernel extract; WE: whole walnut kernel extract. Data are expressed as mean ± SD (n = 3). Superscript letters indicate a more significant difference ($p < 0.01$) compared to DWE. ND, not detected.

Oxidative damage is an important cause of aging in organisms. Antioxidant and anti-aging are closely related. Plant extracts are rich in bioactive substances. These bioactive substances are highly valuable in antioxidants and disease prevention. Therefore, antioxidant studies on plant extracts can provide many possibilities for the development of natural antioxidant products.

Walnut kernels are one of the most beloved nuts in the world, and they are widely grown for this reason. In addition to their good taste, walnuts have great benefits for human health. The reason for this is that walnut kernels are rich in natural active substances. Phenolic compounds are one of these naturally active substances. Phenolic compounds have been shown to have antioxidant properties [45–47]. Walnut kernels have a higher antioxidant potential than other nuts due to their high phenolic content. Defatted walnut kernels, a by-product released after obtaining walnut oil, also have antioxidant properties. Defatted walnuts are becoming an inexpensive source of phenolic compounds.

Our study determined the ability of defatted walnut kernel extract and whole walnut kernel extract to scavenge free radicals in vitro. The results of this study confirmed the antioxidant activity of both extracts. Previous studies have shown that most of the antioxidant activity in walnut kernels is in defatted walnut kernels [36,48,49]. Our results are consistent with them. Walnut kernels contain 65–70% oil. Regrettably, walnut kernel oil contains very low levels of phenolics, contributing less than 5% to the total antioxidant activity of walnut kernels [48]. The reason may lie in the low oil solubility of phenolic compounds [50] or be related to the distribution of phenolic compounds in walnut kernels.

Defatted walnut kernels consist primarily of the skin that encases the kernel. The highest phenolic content was found in the pericarp [51,52]. Therefore, the antioxidant activity of DWE is sensibly superior to that of WE, which may also be related to the interference of walnut oil.

3.2. *In Vivo Antioxidant Activities of Walnut Kernel Extracts*

The malondialdehyde (MDA) content, the catalase (CAT) activity, total superoxide dismutase (T-SOD) activity, glutathione peroxidase (GSH-Px) activity, and total antioxidant activity (T-AOC) in serum, brain tissue, liver, kidney, and heart of mice were examined to assess the in vivo antioxidant activity of walnut kernel extracts. Figure 1 displays the results.

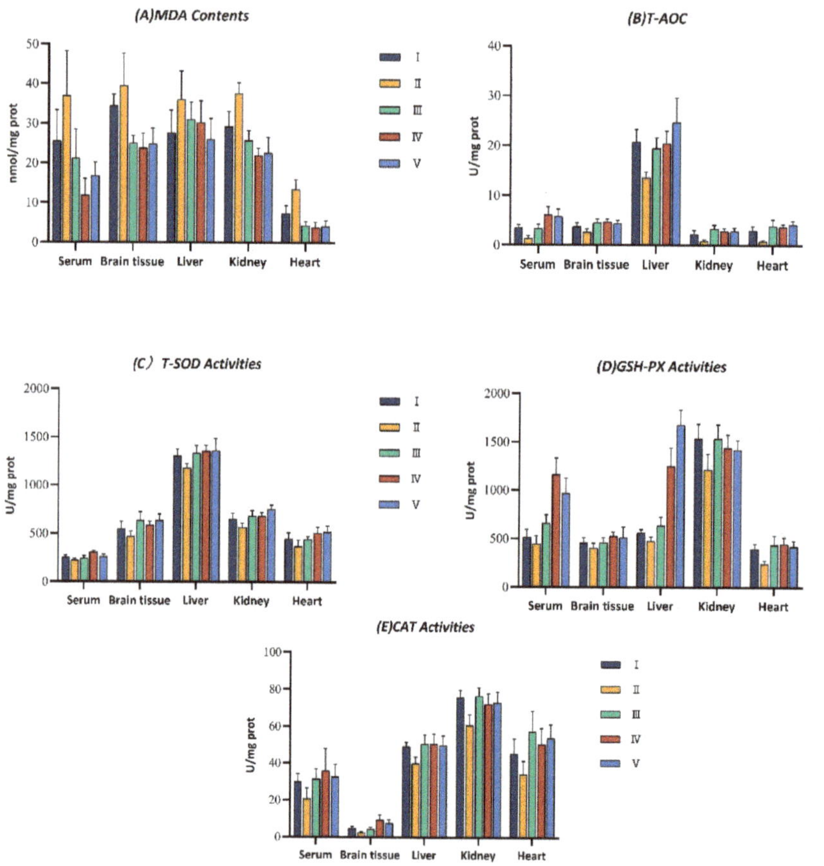

Figure 1. The changes of (**A**) malondialdehyde (MDA) contents; (**B**) total antioxidant activity (T-AOC); (**C**) total superoxide dismutase (T-SOD) activities; (**D**) glutathione peroxidase (GSH-Px) activities; and (**E**) catalase (CAT) activities. Data are expressed as mean ± SD (n = 8). Mice were divided into five groups as follows: I, normal control group (saline); II, model control group (D-gal solution); III, positive control group (VC solution, 0.2 g/kg/d); IV, DWE (walnut kernels extract, 0.2 g/kg/d); V, WE (walnut kernels extract, 0.5 g/kg/d). Compared with the model group, $p < 0.05$.

3.2.1. Effect on Lipid Peroxidation

As shown in Figure 1A, the malondialdehyde (MDA) content in serum and all tissues was obviously higher in the model control group (Group II) than in the normal control group (Group I). It indicates that mice injected intraperitoneally with D-galactose solutions could successfully create a model of oxidative damage. In comparison to the model group, the content of MDA in the serum and tissues of mice in the vitamin C, defatted walnut kernel extract (DWE), and whole walnut kernel extract (WE) groups were reduced and varied significantly ($p < 0.05$). It showed that two walnut kernel extracts and vitamin C may both lower the MDA content of aging mice due to D-galactose to varying degrees and protect mice from lipid peroxide damage. Walnut kernel extracts showed conspicuous antioxidant effects.

3.2.2. Effect on Antioxidant Enzyme Activities and Total Antioxidant Activities

As shown in Figure 1B–E, total superoxide dismutase (T-SOD) activities, glutathione peroxidase (GSH-Px) activities, catalase (CAT) activities, and total antioxidant activity (T-AOC) in the serum and various tissues were noticeably reduced in the model control group (Group II) compared with the normal control group (Group I). The ability to scavenge oxygen-free radicals in vivo was also correspondingly reduced. It indicates that intraperitoneal injection of D-galactose solution could successfully create a mouse model of oxidative damage. Figure 1B shows the effect of walnut kernel extracts on T-AOC in the serum and tissues of mice. The T-AOC in the serum and tissues of mice varied significantly among the vitamin C group, the defatted walnut kernel extract (DWE) group, and the whole walnut kernel extract (WE) group compared to the model group ($p < 0.05$). It notes that walnut kernel extracts and vitamin C could improve the T-AOC of D-galactose-aged mice to varying degrees and enhance their ability to scavenge free radicals. Figure 1C–E reflect the effects of walnut kernel extracts on antioxidant enzyme activity in the serum and tissues of mice. The activities of antioxidant enzymes in mice were reduced after the use of D-galactose. T-SOD activities, GSH-PX activities, and CAT activities in serum and tissues were improved after treatment with vitamin C, DWE, and WE. And all of them were significantly different from the model group ($p < 0.05$). Antioxidant enzymes can eliminate free radicals, maintain the redox balance of the organism, and play a certain antioxidant role in the organism.

Free radicals readily attack unsaturated fatty acids in biological membranes and cause lipid peroxidation. This action produces lipid peroxidation products, including MDA. MDA is a cytotoxic substance that can induce cross-linked polymeric of proteins, nucleic acids, and various biological macromolecules [53]. The MDA content is an indicator of the extent of lipid peroxidation occurring within the organism and may further indirectly reflect the level of cellular oxidative damage. The defense system of biological organisms includes both enzymatic and non-enzymatic antioxidant systems. T-AOC is a combination of enzymatic and non-enzymatic activity in a biological organism. The numerical value of T-AOC could objectively mirror the strength of the body's total antioxidant activity and further indirectly reflect the body's ability to defend against antioxidant damage. T-SOD, CAT, and GSH-Px are endogenous antioxidant enzymes that are crucial in the mechanism of antioxidant defense [54]. In the organism, these three antioxidant enzymes are the first line of defense against reactive oxygen species. T-SOD is essential for the oxidation and antioxidant balance of living organisms, converting superoxide anions into hydrogen peroxide (H_2O_2) and thus protecting cells from damage [55]. Hydroxyl radicals are chemically very active and can react with organic substances in living organisms, such as sugars, amino acids, organic acids, and phospholipids. The reaction speed is fast, and the damage to the cells of biological organisms is strong. However, it can be broken down into H_2O_2 by the highly destructive CAT under certain conditions to protect the cells of living organisms [56]. GSH-Px exclusively catalyzes the reduction reaction of reduced glutathione (GSH) to hydrogen peroxide, thus acting to defend the structure and functional completeness of the cytomembrane. These antioxidant indicators were therefore selected to assess the antioxidant activity of walnut kernel extracts.

Vitamin C is one of the most common and strong antioxidants that can perform antioxidant functions by reacting with free radicals and inhibiting lipid peroxidation damage. The selection of vitamin C as a positive control can distinctly elucidate the effect of antioxidants on antioxidant enzyme activity in oxidatively damaged mice. D-galactose is a well-stabilized reducing monosaccharide that increases tissue osmolarity and produces oxidative stress. The products of the enzyme-catalyzed reaction of D-galactose cannot be further metabolized. Instead, they accumulate in the body, damaging its antioxidant defense system and producing an overabundance of oxygen radicals. This results in a model of oxidative damage.

3.3. Single Factor Experimental Results

One of the primary factors impacting the total phenolic content (TPC) and antioxidant activity is the extraction variable. The results revealed that extraction time, ethanol concentration, and extraction temperature affected the antioxidant activity and TPC of defatted walnut kernel extract (DWE) (Figure 2). Specifically, the effect of extraction time on antioxidant activity manifested that the extraction time (from 40 min to 80 min) was positively correlated with antioxidant activity. The antioxidant activity, however, fell after 80 min of extraction time. Interestingly, the TPC reflects the same situation. The TPC was positively associated with the extraction time (from 40 min to 100 min). But sadly, the TPC started to progressively decline after the extraction time surpassed 100 min. Concerning the ethanol concentration, as the ethanol concentration rose from 0% to 40%, the antioxidant activity, as determined by the DPPH and ABTS techniques, increased. Subsequently, it gradually decreased. While the FRAP method showed that the antioxidant activity increased as ethanol concentration grew (from 0% to 80%), it gradually declined after ethanol concentration increased over 80%. The TPC was positively correlated with the ethanol concentration, within the range of 0–60% ethanol concentration, arriving at a maximum of 60%. After that, it gradually decreased with increasing ethanol concentrations. It is noteworthy that the extraction temperature showed the same trend as the extraction time. In general, the effects of these three variables on antioxidant activity and TPC followed a consistent trend: an early rise followed by a slow decline.

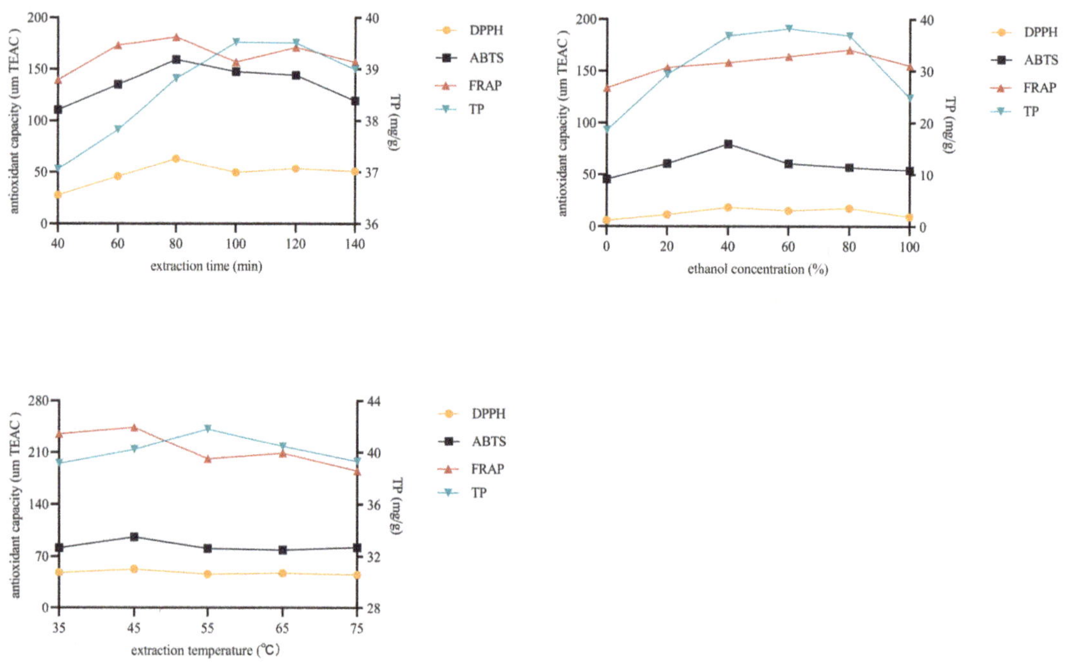

Figure 2. Effect of extraction variables on antioxidant activity (DPPH, ABTS, FRAP) and total phenolic content (TP) of defatted walnut kernel extract (DWE). Min, minutes.

Reflux extraction is a method of extracting active ingredients from plants with volatile organic solvents. The reflux extraction method is relatively simple to operate and has a high extraction efficiency. Ethanol, acetone, methanol, and water are organic solvents often used for the extraction of phenolic compounds [57], including the extraction of walnut kernel extract. Ethanol is a hydrophilic organic solvent with good solubility for all types of chemicals. Ethanol is safer compared to methanol and acetone. During the extraction process, the performance of the extraction is affected by some external factors, such as extraction temperature, feed/liquid ratio, extraction time, and solvent concentration [58].

The variable of extraction time may have a bearing on the instability of antioxidant active components in the DWE. Additionally, when extraction time lengthens, production costs rise. Taking all aspects into consideration, the extraction time was chosen to be controlled at about 80 min. For the variable of ethanol concentration, this may be due to the fact that polyphenols are not the only antioxidant active ingredient in the DWE. The content and type of antioxidant substances extracted differed at different ethanol concentrations. From the consideration of cost and subsequent operation, it is appropriate to control the ethanol concentration at about 50%. For the extraction temperature, this may be due to the poor thermal stability of polyphenols and the destruction of active ingredients due to oxidation reactions at high temperatures. In addition, the dissolution of impurities increases at high temperatures, and the subsequent separation and purification become more difficult. Taking all factors into consideration, the extraction temperature was controlled at about 45 °C as appropriate.

3.4. Correlation Analysis of Antioxidant Activity and Total Phenolic Content

The content of phenolic substances contained in nuts affects their antioxidant activity [59]. According to previous studies, phenolics and antioxidant activity are noticeably positively correlated [60]. Correlations were analyzed to evaluate the relevance of total phenolic content (TPC) and antioxidant activity of walnut kernel extracts. As can be seen in Table 4, using the DPPH, ABTS, and FRAP methods, the correlation coefficients between TPC and antioxidant activity were 0.517, 0.566, and 0.539, respectively. This indicated a significant relevance in antioxidants and TPC ($p < 0.05$). The results were in accordance with previous studies. There was an extremely significant relevance ($r > 0.97$; $p < 0.001$) among the three antioxidant tests in defatted walnut kernel extract (DWE). Therefore, the choice of any of these methods allows a reasonable one-way analysis of variance (ANOVA) of the influencing factors.

Table 4. Correlations between total phenolic content (TPC) and antioxidant activities (DPPH, ABTS, FRAP).

	FRAP	DPPH	ABTS
DPPH	0.971 **	1	–
ABTS	0.976 **	0.991 **	1
TPC	0.517 *	0.566 *	0.539 *

** $p < 0.01$; * $p < 0.05$.

3.5. Response Surface Methodology to Optimize the DWE Extraction Conditions

3.5.1. ANOVA and Quadratic Regression Analysis

Defatted walnut kernel extract (DWE) extraction conditions were optimized using the response surface method (RSM) based on the findings of single-factor experiments. The one-way analysis of variance (ANOVA) of influential variables was conducted using ABTS. As displayed in Table 5, the total phenolic content (TPC) of defatted walnut kernel extract (DWE) was more significantly correlated with extraction time ($p < 0.01$). It was extremely significantly correlated with ethanol concentration and extraction temperature ($p < 0.001$). The influencing variables were in the following order: ethanol concentration had the highest degree of influence, followed by extraction temperature and extraction time. Furthermore, the effect of extraction factors on the TPC was ranked in the following order: ethanol concentration > extraction time > extraction temperature. The quadratic multiple regression equations that were obtained to assess the ABTS and TPC outcomes are as follows: y (ABTS) = 105.25 + 9.26a + 3.33b + 6.64c − 0.20ab − 4.52ac − 0.81bc − 7.67a^2 − 5.65b^2 − 5.17c^2; y (TPC) = 40.10 + 1.47a − 0.52b − 0.017c − 2.05ab − 0.36ac + 0.096bc − 2.86a^2 − 0.43b^2 − 0.59c^2. It also shows that the effects of these three effects on antioxidant activity and TPC are not simply linear but quadratic. And there is an interaction between the three factors.

Table 5. ANOVA for CCD design.

Source	Sum of Squares	df	Mean Square	F-Value	p-Value	Significance
ABTS Model	3352.62	9	372.51	38.43	<0.0001	***
A	1172.04	1	1172.04	120.92	<0.0001	***
B	151.87	1	151.87	15.67	0.0033	**
C	601.23	1	601.23	62.03	<0.0001	***
AB	0.32	1	0.32	0.033	0.8597	
AC	163.12	1	163.12	16.83	0.0027	**
BC	5.22	1	5.22	0.54	0.4815	
A^2	803.14	1	803.14	82.86	<0.0001	***
B^2	435.69	1	435.69	44.95	<0.0001	***
C^2	364.90	1	364.90	37.65	0.0002	***
Residual	87.23	9	9.69			
Lack of Fit	76.09	5	15.22	5.46	0.0625	Not significant
Pure Error	11.14	4	2.79			
R^2	0.9746					
TPC						
Model	180.26	9	20.23	5.75	0.0078	**
A	29.47	1	29.47	8.46	0.0174	*
B	3.64	1	3.64	1.04	0.3334	
C	3.83×10^{-3}	1	3.83×10^{-3}	1.10×10^{-3}	0.9743	
AB	33.66	1	33.66	9.66	0.0125	*
AC	1.04	1	1.04	0.30	0.5974	
BC	0.074	1	0.074	0.021	0.8872	
A^2	111.96	1	111.96	32.14	0.0003	***
B^2	2.50	1	2.50	0.72	0.4189	
C^2	4.76	1	4.76	1.37	0.2724	
Residual	31.35	9	3.48			
Lack of Fit	27.61	5	5.52	5.90	0.0550	Not significant
Pure Error	3.74	4	0.94			
R^2	0.8518					

TPC, total phenolic content; A, ethanol concentration; B, extraction time; C, extraction temperature. NS, not significant; *, significant ($p < 0.05$); **, more significant ($p < 0.01$); ***, extremely significant ($p < 0.001$).

3.5.2. Response Surface Analysis

Three-dimensional response surface plots show the interactions between each of the two factors (Figure 3). Under the constant extraction temperature, the total polyphenol content (TPC) and ABTS scavenging activity increased and then decreased with the increase in extraction time and ethanol concentration. However, the magnitude of the change in ethanol concentration was more obvious. When the extraction time remains constant, the ABTS scavenging activity and TPC rise and then decline with the rise in extraction temperature and ethanol concentration. The impact of ethanol concentration on ABTS scavenging activity and TPC, however, was more remarkable. Under the constant ethanol concentration, the TPC and ABTS scavenging activities increased and then decreased with the increase in extraction time and extraction temperature. For ABTS scavenging capacity, the magnitude of the change in extraction temperature was more obvious. For TPC, the effect of extraction time on TPC was more prominent. The analysis of variance also supports these findings.

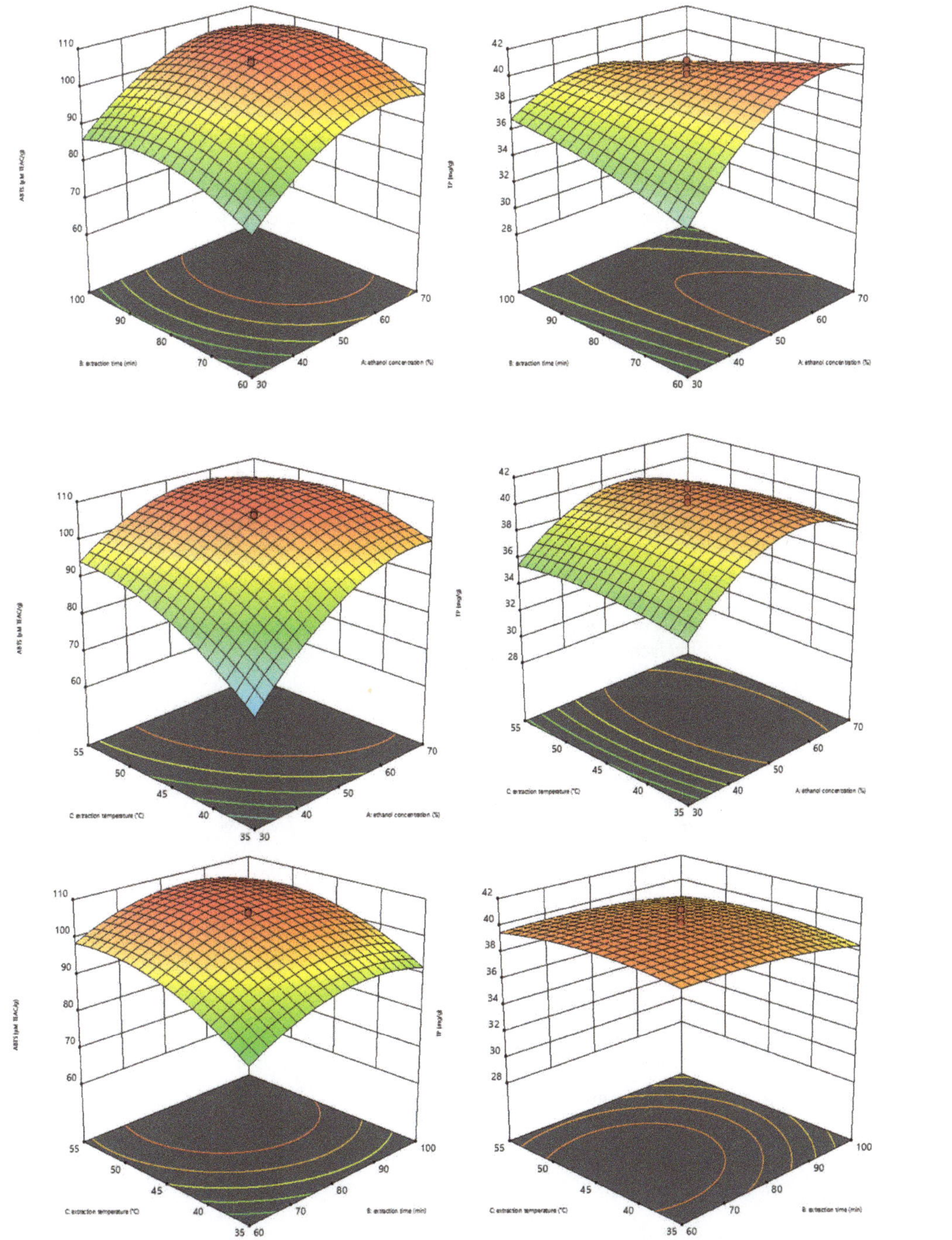

Figure 3. Response surface plots of antioxidant activity (ABTS) and total phenolic content (TP) of defatted walnut kernel extract (DWE) as influenced by the interaction of paired factors.

According to response surface methodology, the ethanol concentration of 57.67%, extraction temperature of 47.81 °C, and extraction time of 76.91 min were the optimal conditions for the extraction of DWE. The ABTS scavenging activity was predicted to be 108.04 μM TEAC, and the measured value was 108.93 μM TEAC. The predicted TPC was 40.33 mg/g, and the measured value was 40.76 mg/g. Considering the convenience of operation in practical production, the ethanol concentration of 58%, extraction temperature of 48 °C, and extraction time of 77 min were determined as the optimal extraction conditions.

In the past, the amount of walnut kernels needed for the oil extraction process was almost two times greater than the amount of oil that was produced. The remaining residue, without any economic value, is used as animal feed or plant fertilizer and may even be thrown away directly as garbage. Effective utilization of these by-products not only expands the processing industry chain of walnuts but also helps to reduce the waste of resources and protect the environment. The high antioxidant activity of DWE allows it to be used as a valuable ingredient in dietary supplements, nutraceuticals, and functional products. Or as a natural preservative to reduce rapid food spoilage. It can even work synergistically with other natural antioxidant extracts to enhance their performance. Therefore, the optimization of DWE provides a reference value for the application of defatted walnut kernels.

4. Conclusions

This study evaluated the antioxidant activity of defatted walnut kernel extract (DWE) and whole walnut kernel extract (WE) in vitro and in vivo experiments. In vitro experiments showed that DWE had better DPPH free radical scavenging, ABTS free radical scavenging, and ferric-reducing antioxidant abilities. In vivo tests revealed that both extracts increased the activity of antioxidant enzymes and decreased lipid peroxidation products in the serum and tissues of mice. In addition, the extraction conditions for reflux-assisted ethanol extraction of DWE were effectively optimized using response surface methodology. These extraction conditions include ethanol concentration, extraction temperature, and extraction time. DWE showed maximum antioxidant activity and total phenolic content under these optimized conditions. The results confirm that defatted walnut kernel extract is a natural antioxidant and can be a potential source for the development of functional foods. It is, however, necessary to conduct further research in order to determine the effectiveness of DWE added to foods.

Author Contributions: Conceptualization, X.Z. (Xiaomei Zhou), X.G. and C.Z.; Methodology, X.Z. (Xiaomei Zhou) and X.G.; Formal analysis, X.Z. (Xiaomei Zhou) and X.G.; Experimental activities and results, X.L.; Writing—original draft, X.Z. (Xiaomei Zhou) and X.G.; Writing—Review and Editing, N.A., J.H., C.Z. and X.Z. (Xin Zhou); Supervision, C.Z. All authors have read and agreed to the published version of the manuscript.

Funding: This work was funded by Natural Science Foundation of Guizhou [2020]1Y430, Guizhou Scholarship for Overseas Students ((2019) 12), and Innovation Program of Postgraduate Education of Guizhou Province (NO. Qian jiao he YJSCXJH (2020) 098).

Institutional Review Board Statement: Animal experiments in this study were conducted in accordance with current national and international laws and recommendations, and every effort was made to minimize suffering. The animal experiments were approved by the Animal Care and Use Committee of Guizhou Normal University (C00032801).

Data Availability Statement: The data used to support the findings of this study can be made available by the corresponding author upon request.

Conflicts of Interest: The authors declare no conflict of interest.

References

1. Masoodi, L.; Masoodi, F.A.; Gull, A.; Gani, A.; Muzaffer, S.; Sidiq, M. Effect of γ-Irradiation on the Physicochemical and Sensory Properties of Fresh Walnut Kernels (*Juglans regia*) during Storage. *Food Chem. Adv.* **2023**, *3*, 100301. [CrossRef]
2. Wei, F.; Chen, Q.; Du, Y.; Han, C.; Fu, M.; Jiang, H.; Chen, X. Effects of Hulling Methods on the Odor, Taste, Nutritional Compounds, and Antioxidant Activity of Walnut Fruit. *LWT* **2020**, *120*, 108938. [CrossRef]

3. Hu, J.; Shi, H.; Zhan, C.; Qiao, P.; He, Y.; Liu, Y. Study on the Identification and Detection of Walnut Quality Based on Terahertz Imaging. *Foods* **2022**, *11*, 3498. [CrossRef]
4. Wang, P.; Zhong, L.; Yang, H.; Zhu, F.; Hou, X.; Wu, C.; Zhang, R.; Cheng, Y. Comparative Analysis of Antioxidant Activities between Dried and Fresh Walnut Kernels by Metabolomic Approaches. *LWT* **2022**, *155*, 112875. [CrossRef]
5. Wang, W.; Wen, H.; Jin, Q.; Yu, W.; Li, G.; Wu, M.; Bai, H.; Shen, L.; Wu, C. Comparative Transcriptome Analysis on Candidate Genes Involved in Lipid Biosynthesis of Developing Kernels for Three Walnut Cultivars in Xinjiang. *Food Sci. Hum. Wellness* **2022**, *11*, 1201–1214. [CrossRef]
6. Xiao, H.; Zhang, S.; Xi, F.; Yang, W.; Zhou, L.; Zhang, G.; Zhu, H.; Zhang, Q. Preservation Effect of Plasma-Activated Water (PAW) Treatment on Fresh Walnut Kernels. *Innov. Food Sci. Emerg. Technol.* **2023**, *85*, 103304. [CrossRef]
7. Cintesun, S.; Ozman, Z.; Kocyigit, A.; Mansuroglu, B.; Kocacaliskan, I. Effects of Walnut (*Juglans regia* L.) Kernel Extract and Juglone on Dopamine Levels and Oxidative Stress in Rats. *Food Biosci.* **2023**, *51*, 102327. [CrossRef]
8. Jahanban-Esfahlan, A.; Ostadrahimi, A.; Tabibiazar, M.; Amarowicz, R. A Comparative Review on the Extraction, Antioxidant Content and Antioxidant Potential of Different Parts of Walnut (*Juglans regia* L.) Fruit and Tree. *Molecules* **2019**, *24*, 2133. [CrossRef] [PubMed]
9. Sun, Y.; Guo, F.; Peng, X.; Cheng, K.; Xiao, L.; Zhang, H.; Li, H.; Jiang, L.; Deng, Z. Metabolism of Phenolics of Tetrastigma Hemsleyanum Roots under In Vitro Digestion and Colonic Fermentation as Well as Their In Vivo Antioxidant Activity in Rats. *Foods* **2021**, *10*, 2123. [CrossRef]
10. Srisuksai, K.; Parunyakul, K.; Santativongchai, P.; Phaonakrop, N.; Roytrakul, S.; Tulayakul, P.; Fungfuang, W. Antioxidant Activity of Crocodile Oil (*Crocodylus siamensis*) on Cognitive Function in Rats. *Foods* **2023**, *12*, 791. [CrossRef] [PubMed]
11. Sridhar, K.; Charles, A.L. In Vitro Antioxidant Activity of Kyoho Grape Extracts in DPPH and ABTS Assays: Estimation Methods for EC50 Using Advanced Statistical Programs. *Food Chem.* **2019**, *275*, 41–49. [CrossRef]
12. Amidžić Klarić, D.; Klarić, I.; Mornar, A.; Velić, N.; Velić, D. Assessment of Bioactive Phenolic Compounds and Antioxidant Activity of Blackberry Wines. *Foods* **2020**, *9*, 1623. [CrossRef] [PubMed]
13. Kotha, R.R.; Tareq, F.S.; Yildiz, E.; Luthria, D.L. Oxidative Stress and Antioxidants—A Critical Review on In Vitro Antioxidant Assays. *Antioxidants* **2022**, *11*, 2388. [CrossRef]
14. Latif, F.; Imran, M. Antioxidants—A Combat against Oxidative Stress in Dementia. *Ann. Med. Surg.* **2022**, *82*, 104632. [CrossRef] [PubMed]
15. Mbah, C.; Orabueze, I.; Okorie, N. Antioxidants Properties of Natural and Synthetic Chemical Compounds: Therapeutic Effects on Biological System. *Acta Sci. Pharm. Sci.* **2019**, *3*, 28–42. [CrossRef]
16. Olufunmilayo, E.O.; Gerke-Duncan, M.B.; Holsinger, R.M.D. Oxidative Stress and Antioxidants in Neurodegenerative Disorders. *Antioxidants* **2023**, *12*, 517. [CrossRef]
17. Fibbi, B.; Marroncini, G.; Anceschi, C.; Naldi, L.; Peri, A. Hyponatremia and Oxidative Stress. *Antioxidants* **2021**, *10*, 1768. [CrossRef]
18. Muriach, M.; Flores-Bellver, M.; Romero, F.J.; Barcia, J.M. Diabetes and the Brain: Oxidative Stress, Inflammation, and Autophagy. *Oxid. Med. Cell. Longev.* **2014**, *2014*, 102158. [CrossRef]
19. AlAshqar, A.; Lulseged, B.; Mason-Otey, A.; Liang, J.; Begum, U.A.M.; Afrin, S.; Borahay, M.A. Oxidative Stress and Antioxidants in Uterine Fibroids: Pathophysiology and Clinical Implications. *Antioxidants* **2023**, *12*, 807. [CrossRef]
20. Bardelčíková, A.; Šoltys, J.; Mojžiš, J. Oxidative Stress, Inflammation and Colorectal Cancer: An Overview. *Antioxidants* **2023**, *12*, 901. [CrossRef]
21. Jian, F.; Zhang, Z.; Li, D.; Luo, F.; Wu, Q.; Lu, F.; Dai, Z.; Nie, M.; Xu, Y.; Feng, L.; et al. Evaluation of the Digestibility and Antioxidant Activity of Protein and Lipid after Mixing Nuts Based on in Vitro and in Vivo Models. *Food Chem.* **2023**, *414*, 135706. [CrossRef] [PubMed]
22. Moazzen, A.; Öztinen, N.; Ak-Sakalli, E.; Koşar, M. Structure-Antiradical Activity Relationships of 25 Natural Antioxidant Phenolic Compounds from Different Classes. *Heliyon* **2022**, *8*, e10467. [CrossRef] [PubMed]
23. Granato, D.; Shahidi, F.; Wrolstad, R.; Kilmartin, P.; Melton, L.D.; Hidalgo, F.J.; Miyashita, K.; van Camp, J.; Alasalvar, C.; Ismail, A.B.; et al. Antioxidant Activity, Total Phenolics and Flavonoids Contents: Should We Ban in Vitro Screening Methods? *Food Chem.* **2018**, *264*, 471–475. [CrossRef] [PubMed]
24. Nimal, R.; Selcuk, O.; Kurbanoglu, S.; Shah, A.; Siddiq, M.; Uslu, B. Trends in Electrochemical Nanosensors for the Analysis of Antioxidants. *TrAC Trends Anal. Chem.* **2022**, *153*, 116626. [CrossRef]
25. Caleja, C.; Barros, L.; Antonio, A.L.; Oliveira, M.B.P.P.; Ferreira, I.C.F.R. A Comparative Study between Natural and Synthetic Antioxidants: Evaluation of Their Performance after Incorporation into Biscuits. *Food Chem.* **2017**, *216*, 342–346. [CrossRef]
26. Blasi, F.; Cossignani, L. An Overview of Natural Extracts with Antioxidant Activity for the Improvement of the Oxidative Stability and Shelf Life of Edible Oils. *Processes* **2020**, *8*, 956. [CrossRef]
27. Bellucci, E.R.B.; Bis-Souza, C.V.; Domínguez, R.; Bermúdez, R.; da Barretto, A.C.S. Addition of Natural Extracts with Antioxidant Function to Preserve the Quality of Meat Products. *Biomolecules* **2022**, *12*, 1506. [CrossRef]
28. Ma, Y.; Wang, C.; Liu, C.; Tan, J.; Ma, H.; Wang, J. Physiochemical Responses of the Kernel Quality, Total Phenols and Antioxidant Enzymes of Walnut in Different Forms to the Low-Temperature Storage. *Foods* **2021**, *10*, 2027. [CrossRef]
29. Ampofo, J.; Grilo, F.S.; Langstaff, S.; Wang, S.C. Oxidative Stability of Walnut Kernel and Oil: Chemical Compositions and Sensory Aroma Compounds. *Foods* **2022**, *11*, 3151. [CrossRef]

30. Hama, J.R.; Omer, R.A.; Rashid, R.S.M.; Mohammad, N.-E.-A.; Thoss, V. The Diversity of Phenolic Compounds along Defatted Kernel, Green Husk and Leaves of Walnut (*Juglansregia* L.). *Anal. Chem. Lett.* **2016**, *6*, 35–46. [CrossRef]
31. Pop, O.L.; Suharoschi, R.; Socaci, S.A.; Ceresino, E.B.; Weber, A.; Gruber-Traub, C.; Vodnar, D.C.; Fărcaș, A.C.; Johansson, E. Polyphenols—Ensured Accessibility from Food to the Human Metabolism by Chemical and Biotechnological Treatments. *Antioxidants* **2023**, *12*, 865. [CrossRef]
32. Ruiz-Caro, P.; Espada-Bellido, E.; García-Guzmán, J.J.; Bellido-Milla, D.; Vázquez-González, M.; Cubillana-Aguilera, L.; Palacios-Santander, J.M. An Electrochemical Alternative for Evaluating the Antioxidant Capacity in Walnut Kernel Extracts. *Food Chem.* **2022**, *393*, 133417. [CrossRef] [PubMed]
33. Guo, X.; Gu, F.; Yang, T.; Shao, Z.; Zhang, Q.; Zhu, J.; Wang, F. Quantitative Conversion of Free, Acid-Hydrolyzable, and Bound Ellagic Acid in Walnut Kernels during Baking. *Food Chem.* **2023**, *400*, 134070. [CrossRef]
34. Wu, S.; Shen, D.; Wang, R.; Li, Q.; Mo, R.; Zheng, Y.; Zhou, Y.; Liu, Y. Phenolic Profiles and Antioxidant Activities of Free, Esterified and Bound Phenolic Compounds in Walnut Kernel. *Food Chem.* **2021**, *350*, 129217. [CrossRef] [PubMed]
35. Xu, X.; Ding, Y.; Liu, M.; Zhang, X.; Wang, D.; Pan, Y.; Ren, S.; Liu, X. Neuroprotective Mechanisms of Defatted Walnut Powder against Scopolamine-Induced Alzheimer's Disease in Mice Revealed through Metabolomics and Proteomics Analyses. *J. Ethnopharmacol.* **2023**, *319*, 117107. [CrossRef]
36. Trandafir, I.; Cosmulescu, S. Total Phenolic Content, Antioxidant Capacity and Individual Phenolic Compounds of Defatted Kernel from Different Cultivars of Walnut. *Erwerbs-Obstbau* **2020**, *62*, 309–314. [CrossRef]
37. Trandafir, I.; Cosmulescu, S.; Nour, V. Phenolic Profile and Antioxidant Capacity of Walnut Extract as Influenced by the Extraction Method and Solvent. *Int. J. Food Eng.* **2017**, *13*, 20150284. [CrossRef]
38. Tian, W.; Wu, B.; Sun, L.; Zhuang, Y. Protective Effect against D-Gal-Induced Aging Mice and Components of Polypeptides and Polyphenols in Defatted Walnut Kernel during Simulated Gastrointestinal Digestion. *J. Food Sci.* **2021**, *86*, 2736–2752. [CrossRef] [PubMed]
39. Wu, Q.; Lv, D.; Mao, X. Polyphenol Removal with Ultrasound-Assisted Ethanol Extraction from Defatted Walnut Powder: Optimization of Conditions and Effect on Functional Properties of Protein Isolates. *J. Sci. Food Agric.* **2023**. [CrossRef]
40. Culetu, A.; Fernandez-Gomez, B.; Ullate, M.; del Castillo, M.D.; Andlauer, W. Effect of Theanine and Polyphenols Enriched Fractions from Decaffeinated Tea Dust on the Formation of Maillard Reaction Products and Sensory Attributes of Breads. *Food Chem.* **2016**, *197*, 14–23. [CrossRef]
41. Re, R.; Pellegrini, N.; Proteggente, A.; Pannala, A.; Yang, M.; Rice-Evans, C. Antioxidant Activity Applying an Improved ABTS Radical Cation Decolorization Assay. *Free Radic. Biol. Med.* **1999**, *26*, 1231–1237. [CrossRef]
42. Thaipong, K.; Boonprakob, U.; Crosby, K.; Cisneros-Zevallos, L.; Hawkins Byrne, D. Comparison of ABTS, DPPH, FRAP, and ORAC Assays for Estimating Antioxidant Activity from Guava Fruit Extracts. *J. Food Compos. Anal.* **2006**, *19*, 669–675. [CrossRef]
43. Singleton, V.L.; Orthofer, R.; Lamuela-Raventós, R.M. Analysis of Total Phenols and Other Oxidation Substrates and Antioxidants by Means of Folin-Ciocalteu Reagent. *Methods Enzymol.* **1999**, *299*, 152–178. [CrossRef]
44. Müller, L.; Fröhlich, K.; Böhm, V. Comparative Antioxidant Activities of Carotenoids Measured by Ferric Reducing Antioxidant Power (FRAP), ABTS Bleaching Assay (ATEAC), DPPH Assay and Peroxyl Radical Scavenging Assay. *Food Chem.* **2011**, *129*, 139–148. [CrossRef]
45. Hilbig, J.; Alves, V.R.; Müller, C.M.O.; Micke, G.A.; Vitali, L.; Pedrosa, R.C.; Block, J.M. Ultrasonic-Assisted Extraction Combined with Sample Preparation and Analysis Using LC-ESI-MS/MS Allowed the Identification of 24 New Phenolic Compounds in Pecan Nut Shell [*Carya illinoinensis* (Wangenh) C. Koch] Extracts. *Food Res. Int.* **2018**, *106*, 549–557. [CrossRef] [PubMed]
46. Jia, X.; Luo, H.; Xu, M.; Zhai, M.; Guo, Z.; Qiao, Y.; Wang, L. Dynamic Changes in Phenolics and Antioxidant Capacity during Pecan (*Carya illinoinensis*) Kernel Ripening and Its Phenolics Profiles. *Molecules* **2018**, *23*, 435. [CrossRef]
47. Santos, J.; Alvarez-Ortí, M.; Sena-Moreno, E.; Rabadán, A.; Pardo, J.E.; Oliveira, M.B.P. Effect of Roasting Conditions on the Composition and Antioxidant Properties of Defatted Walnut Flour. *J. Sci. Food Agric.* **2018**, *98*, 1813–1820. [CrossRef]
48. Arranz, S.; Pérez-Jiménez, J.; Saura-Calixto, F. Antioxidant Capacity of Walnut (*Juglans regia* L.): Contribution of Oil and Defatted Matter. *Eur. Food Res. Technol.* **2008**, *227*, 425–431. [CrossRef]
49. Burbano, J.J.; Correa, M.J. Composition and Physicochemical Characterization of Walnut Flour, a By-Product of Oil Extraction. *Plant Foods Hum. Nutr.* **2021**, *76*, 233–239. [CrossRef]
50. Slatnar, A.; Mikulic-Petkovsek, M.; Stampar, F.; Veberic, R.; Solar, A. Identification and Quantification of Phenolic Compounds in Kernels, Oil and Bagasse Pellets of Common Walnut (*Juglans regia* L.). *Food Res. Int.* **2015**, *67*, 255–263. [CrossRef]
51. Labuckas, D.O.; Maestri, D.M.; Perelló, M.; Martínez, M.L.; Lamarque, A.L. Phenolics from Walnut (*Juglans regia* L.) Kernels: Antioxidant Activity and Interactions with Proteins. *Food Chem.* **2008**, *107*, 607–612. [CrossRef]
52. Colaric, M.; Veberic, R.; Solar, A.; Hudina, M.; Stampar, F. Phenolic Acids, Syringaldehyde, and Juglone in Fruits of Different Cultivars of *Juglans regia* L. *J. Agric. Food Chem.* **2005**, *53*, 6390–6396. [CrossRef] [PubMed]
53. Chen, L.; Long, R.; Huang, G.; Huang, H. Extraction and Antioxidant Activities in Vivo of Pumpkin Polysaccharide. *Ind. Crops Prod.* **2020**, *146*, 112199. [CrossRef]
54. Shang, H.-M.; Zhou, H.-Z.; Yang, J.-Y.; Li, R.; Song, H.; Wu, H.-X. In Vitro and in Vivo Antioxidant Activities of Inulin. *PLoS ONE* **2018**, *13*, e0192273. [CrossRef]
55. Shen, Y.; Zhang, H.; Cheng, L.; Wang, L.; Qian, H.; Qi, X. In Vitro and in Vivo Antioxidant Activity of Polyphenols Extracted from Black Highland Barley. *Food Chem.* **2016**, *194*, 1003–1012. [CrossRef] [PubMed]

56. Huang, H.; Chen, F.; Long, R.; Huang, G. The Antioxidant Activities in Vivo of Bitter Gourd Polysaccharide. *Int. J. Biol. Macromol.* **2020**, *145*, 141–144. [CrossRef]
57. de Mendes, M.K.A.; dos Oliveira, C.B.S.; Veras, M.D.A.; Araújo, B.Q.; Dantas, C.; Chaves, M.H.; Lopes Júnior, C.A.; Vieira, E.C. Application of Multivariate Optimization for the Selective Extraction of Phenolic Compounds in Cashew Nuts (*Anacardium occidentale* L.). *Talanta* **2019**, *205*, 120100. [CrossRef]
58. del Garcia-Mendoza, M.P.; Espinosa-Pardo, F.A.; Savoire, R.; Etchegoyen, C.; Harscoat-Schiavo, C.; Subra-Paternault, P. Recovery and Antioxidant Activity of Phenolic Compounds Extracted from Walnut Press-Cake Using Various Methods and Conditions. *Ind. Crops Prod.* **2021**, *167*, 113546. [CrossRef]
59. Ma, Y.; Gao, J.; Wei, Z.; Shahidi, F. Effect of in Vitro Digestion on Phenolics and Antioxidant Activity of Red and Yellow Colored Pea Hulls. *Food Chem.* **2021**, *337*, 127606. [CrossRef] [PubMed]
60. Lou, X.; Xu, H.; Hanna, M.; Yuan, L. Identification and Quantification of Free, Esterified, Glycosylated and Insoluble-Bound Phenolic Compounds in Hawthorn Berry Fruit (*Crataegus pinnatifida*) and Antioxidant Activity Evaluation. *LWT* **2020**, *130*, 109643. [CrossRef]

foods

MDPI

Article

Influence of *Benincasa hispida* Peel Extracts on Antioxidant and Anti-Aging Activities, including Molecular Docking Simulation

Pimpak Phumat [1], Siripat Chaichit [1], Siriporn Potprommanee [1], Weeraya Preedalikit [2], Mathukorn Sainakham [3], Worrapan Poomanee [3], Wantida Chaiyana [3] and Kanokwan Kiattisin [3,*]

[1] Faculty of Pharmacy, Chiang Mai University, Chiang Mai 50200, Thailand; pimpak.p@cmu.ac.th (P.P.); siripat.chaichit@cmu.ac.th (S.C.); siriporn_pot@cmu.ac.th (S.P.)

[2] Department of Cosmetic Sciences, School of Pharmaceutical Sciences, University of Phayao, Phayao 56000, Thailand; weeraya.pr@up.ac.th

[3] Department of Pharmaceutical Sciences, Faculty of Pharmacy, Chiang Mai University, Chiang Mai 50200, Thailand; mathukorn.s@cmu.ac.th (M.S.); worrapan.p@cmu.ac.th (W.P.); wantida.chaiyana@cmu.ac.th (W.C.)

* Correspondence: kanokwan.k@cmu.ac.th; Tel.: +66-899-603-699

Abstract: *Benincasa hispida* peel, a type of postconsumer waste, is considered a source of beneficial phytochemicals. Therefore, it was subjected to investigation for biological activities in this study. *B. hispida* peel was extracted using 95% v/v, 50% v/v ethanol and water. The obtained extracts were B95, B50 and BW. B95 had a high flavonoid content (212.88 ± 4.73 mg QE/g extract) and phenolic content (131.52 ± 0.38 mg GAE/g extract) and possessed high antioxidant activities as confirmed by DPPH, ABTS and lipid peroxidation inhibition assays. Moreover, B95 showed inhibitory effects against collagenase and hyaluronidase with values of 41.68 ± 0.92% and 29.17 ± 0.66%, which related to anti-aging activities. Via the HPLC analysis, one of the potential compounds found in B95 was rutin. Molecular docking has provided an understanding of the molecular mechanisms underlying the interaction of extracts with collagenase and hyaluronidase. All extracts were not toxic to fibroblast cells and did not irritate the hen's egg chorioallantoic membrane, which indicated its safe use. In conclusion, *B. hispida* peel extracts are promising potential candidates for further use as antioxidant and anti-aging agents in the food and cosmetic industries.

Keywords: *Benincasa hispida*; peel extract; rutin; antioxidant; anti-aging; anticollagenase; anti-hyaluronidase; HET-CAM; docking simulation

Citation: Phumat, P.; Chaichit, S.; Potprommanee, S.; Preedalikit, W.; Sainakham, M.; Poomanee, W.; Chaiyana, W.; Kiattisin, K. Influence of *Benincasa hispida* Peel Extracts on Antioxidant and Anti-Aging Activities, including Molecular Docking Simulation. *Foods* **2023**, *12*, 3555. https://doi.org/10.3390/foods12193555

Academic Editors: Noelia Castillejo Montoya and Lorena Martínez-Zamora

Received: 29 August 2023
Revised: 17 September 2023
Accepted: 18 September 2023
Published: 25 September 2023

1. Introduction

Ultraviolet (UV) radiation is a potent initiator of reactive oxygen species (ROS) generation in the skin, leading to oxidative imbalance [1]. ROS can impair the skin barrier functions by altering squalene, cholesterol and unsaturated lipids, which are essential components of the skin's structure. The effect of skin oxidative stress caused by excessive ROS leads to the initiation of several skin problems such as atypical pigmentation, skin inflammation, increasing sebum secretion and increased levels of oxidized lipids, including skin aging [2,3]. Collagen, elastin and hyaluronic acid (HA) are biological compounds that help promote healthy and youthful skin. Collagen and elastin, the proteins abundant in the dermal layer, play essential roles in maintaining skin flexibility, elasticity and integrity, whereas HA, a glucose-based polymer, in the dermis and the epidermis layers mainly promotes skin rejuvenation and moisture [4,5]. The over-accumulation of ROS leads to the activation of dermal enzymes including collagenase, elastase and hyaluronidase, which in turn degrade collagen, elastin and HA [6]. Thus, various antioxidants simplify skin protection against oxidative damage caused by ROS. Both chemical and natural antioxidants are investigated and incorporated into numerous skin care products to mitigate skin damage. Medicinal plants have been reported to provide potential antioxidation activity through various mechanisms. For example, *Thunbergia laurifolia* Lindl. leaf extracts have demonstrated antioxidation activities through radical scavenging and inhibiting peroxidation mechanisms [7], and *Acacia concinna* Linn. pod extracts exhibited scavenging activity against free radicals [8].

Benincasa hispida (Thunb.) Cogn, a winter melon also known as Fuk-Kiew in Thailand, is an edible plant in the family Cucurbitaceae, generally found in Asia, especially in northern Thailand. Scientific reports suggest that *B. hispida* is rich in nutrients, minerals, vitamins and phytochemical compounds [9]. The bioactive compounds in *B. hispida* extract include phenolics, triterpenoids, flavonoids, glycosides, carotenes and β-sitosterin [10,11]. Related studies have reported that *B. hispida* exhibits pharmacologic effects including antioxidation, anti-angiotensin-converting enzyme, anti-inflammation and antibacterial effects. [12,13]. In addition, the efficacy and phytochemical components of the pulp, fruit and seeds of *B. hispida* were determined [11]. Gallic acid, catechin and ascorbic acid, which are phenolic compounds found in the fruit of *B. hispida*, exhibit antioxidant activity by reducing free radicals [14–16].

The fruit of *B. hispida* is broadly cylindrical in shape, as shown in Figure 1A. Generally, the edible pulp of the fruit is used in cooking and the production of various food products such as tea and juice. Thus, the peel of the fruit then becomes a form of postconsumer waste, as shown in Figure 1B. Interestingly, *B. hispida* juice has gained popularity as a consumer product. This trend has resulted in an increased production of *B. hispida* peel, which is considered a waste product in the food industry. Notably, *B. hispida* fruit can be stored for several months, possibly because of the presence of a protective wax coating on the peel. Thus, investigating the constituents within the peel responsible for the ability to protect the pulp from external environmental influences as shown in Figure 1B would be intriguing. Therefore, the present study focuses on investigating the antioxidant and anti-aging properties of *B. hispida* peel, conducting simulations of the possible docking mechanisms of the potential compounds in the peel and performing safety tests. These research methods are designed to transform *B. hispida* peel, traditionally considered a waste product, into a high-value material for potential applications as an innovative cosmetic ingredient or another innovative material using its biological activities and the mechanisms of action determined through docking simulations.

(A) (B)

Figure 1. *B. hispida fruit* (**A**) and its peel (**B**) used in investigation.

2. Materials and Methods

2.1. Materials

The plant material used in this study was fresh *B. hispida* fruit, collected from Phrae Province, Thailand. The plant was identified, and its voucher specimen was stored in the Herbarium of the Faculty of Pharmacy, Chiang Mai University, Thailand.

The reagents and chemicals were of analytical grade. 2,2-Diphenyl-1-picrylhydrazyl (DPPH) and 2,2-azino-bis(3-ethylbenzthiazoline)-6-sulfonic acid (ABTS) were purchased from Fluka (Buchs, Switzerland). Folin–Ciocalteu reagent was purchased from Merck (Darmstadt, Germany). 6-Hydroxy-2,5,7,8-tetramethyl chroman-2-carboxylic acid (Trolox), gallic acid, ascorbic acid, kojic acid, epigallocatechin-3-gallate (EGCG), hyaluronic acid, linoleic acid, rutin, 3-(4,5-dimethylthiazolyl-2)-2,5-diphenyl tetrazolium bromide (MTT) and 2,4,6-Tris(2-pyridyl)-s-triazine (TPTZ) were purchased from Sigma (St. Louis, MO, USA). Dulbecco's Modified Eagle medium (DMEM) and penicillin–streptomycin were acquired from Invitrogen (Grand Island, NY, USA). Fetal bovine serum (FBS) and bovine serum albumin (BSA) were obtained from Biochrom AG (Berlin, Germany). Ethanol and dimethyl sulfoxide (DMSO) were purchased from Labscan Asia Co., Ltd. (Bangkok, Thailand). HPLC-grade acetonitrile was purchased from Merck (Darmstadt, Germany). Lipopolysaccharide (LPS) was purchased from Sigma-Aldrich (Darmstadt, Germany), and sodium carbonate, aluminum chloride and sodium nitrite were purchased from United Chemical & Trading Co., Ltd. (Chiang Mai, Thailand). Other chemicals and solvents were of the highest grade available.

2.2. Plant Extraction

The procedure was performed following a related study with modification [17]. Briefly, the fresh peel of *B. hispida* fruit was dried in a hot air oven (Memmert, WI, USA) at 50 °C for 24 to 48 h and then ground to a fine powder. The obtained plant powder was extracted with various solvents using ultrasound-assisted extraction (UAE). For extraction, 95% *v*/*v* ethanol, 50% *v*/*v* ethanol and deionized (DI) water were used as solvents. In each solvent extraction, an ultrasonic bath (Elma Schmidbauer GmbH, Singen, Germany) was used to generate the ultrasound effect for 30 min at 65.0 ± 1.0 °C. Each UAE was performed in triplicate. The extracted mixture (plant powder mixed with solvent) was filtered through a Whatman No. 1 filter paper. The filtrates obtained from 95% *v*/*v* ethanol and 50% *v*/*v* ethanol were subjected to a rotary vacuum evaporator (M.A-3S, Eyela, Tokyo, Japan) to remove the solvent, whereas the filtrate obtained from DI water was subjected to a lyophilizer (Christ Beta 2-8 LD plus, Osterode am Harz, Germany). After the solvents were completely removed, the extracts from 95% *v*/*v* ethanol (B95), 50% *v*/*v* ethanol (B50) and DI water (BW) were obtained, and all extracts were stored at 4 °C until use. The %yield of each extract was calculated using the following equation:

$$\%\text{yield} = (\text{E}/\text{P}) \times 100 \quad (1)$$

where E represents the weight of the extract and P represents the weight of the plant powder used for extraction.

2.3. Determination of Total Flavonoid Content

Total flavonoid content (TFC) was determined using aluminum chloride colorimetry following our related study [18]. All *B. hispida* extracts were prepared at a concentration of 1.0 mg/mL in DI water. The extract with 100 μL was mixed with 30 μL of 5% *v*/*v* sodium nitride. After 5 min, 50 μL of 2% *v*/*v* aluminum chloride was added. The solution mixture was left for 6 min, followed by adding 50 μL of 1N sodium hydroxide. The mixture was stored in the dark for 10 min at room temperature before analysis. The mixed solution was measured for absorbance at 510 nm using a UV spectrophotometer (UV-2600i, Shimadzu, Kyoto, Japan) against a blank without the extract. The standard curve for TFC determination was constructed using a quercetin (QE) solution (0.0 to 1.0 mg/mL) under the same procedure as the tested *B. hispida* extract. The TFC value was expressed as milligrams of QE equivalents per gram of extract (mg QE/g extract).

2.4. Determination of Total Phenolic Content

Total phenolic content (TPC) was analyzed using the Folin–Ciocalteu method [19]. First, 50 μL of each *B. hispida* extract (3 mg/mL in DI water) was mixed with 100 μL of 10% *v*/*v* Folin–Ciocalteu reagent. After 5 min, 50 μL of 10% *w*/*v* sodium carbonate was added. The mixture was vortexed and stored in the dark at ambient temperature for 2 h. Then, the absorbance of mixtures was measured using a UV spectrophotometer (UV-2600i, Shimadzu) at 765 nm against a blank without the extract. The TPC value was determined using a standard curve constructed with a gallic acid standard solution (0.0 to 1.0 mg/mL) and expressed as milligrams of gallic acid equivalents (GAEs) per gram of extract (mg GAE/g extract).

2.5. Antioxidant Activities

The *B. hispida* extracts were prepared to determine the antioxidant activities using various procedures. Trolox and ascorbic acid were used as standards in all assays.

2.5.1. DPPH Radical Scavenging Assay

This experiment was determined using the modified method [20]. In total, 20 µL of the extract solution (0.0 to 1.0 mg/mL) was mixed with 180 µL of DPPH reagent (167 µM) and then incubated in the dark at ambient temperature for 30 min. The absorbance of the extract–reagent mixture was investigated at 520 nm using a microplate reader (SpectraMax M3, Molecular Devices, San Jose, CA, USA). The efficacy against DPPH radicals was expressed as the percentage of inhibition (% inhibition), which was estimated using the following equation:

$$\% \text{ Inhibition} = [(Ac - Aw) - (Ae - Ab)/(Ac - Ab)] \times 100 \quad (2)$$

where Ac is the absorbance of positive control that used ethanol mixed with tested reagent, Aw is the absorbance of negative control (ethanol), Ae is the absorbance of tested extract, and Ab is the absorbance of blank that used extract solution mixed with ethanol.

2.5.2. ABTS Radical Scavenging Assay

The potential of *B. hispida* peel extracts was assessed in terms of scavenging activity using the ABTS assay following the method of Poomanee [21] with modifications. The ABTS reagent was prepared by mixing an ABTS solution in DI water (7 mM) with potassium persulfate solution in DI water (3 mM) in a ratio of 1:0.5 and incubated in a dark area for 18 h. The measurement was performed using a microplate reader (SpectraMax M3, Molecular Devices) at 734 nm. Before testing, the prepared ABTS reagent was diluted to provide an absorbance of 0.7 and was then used as a tested ABTS reagent. Altogether, 20 µL of the extract solution (0.0 to 1.0 mg/mL) and 180 µL of tested ABTS reagent were mixed and incubated in the dark at ambient temperature for 30 min. The absorbance of the extract–reagent mixture was measured at 520 nm using a microplate reader (SpectraMax M3, Molecular Devices). The % inhibition against ABTS radicals (% inhibition) was estimated using Equation (2) described above.

2.5.3. Ferric Reducing Antioxidant Power (FRAP) Assay

This experiment was modified from the related study [22]. The tested samples, which were 1.0 mg/mL of each of the *B. hispida* extracts, were used in this experiment. Briefly, 20 µL of the extract solution was added to the mixed solution, which was 0.3 M acetate buffer (pH 3.6) with 10 mM TPTZ in 40 mM of 37% *w*/*v* hydrochloric solution. The extract–reagent mixture was incubated under dark conditions at ambient temperature for 5 min and then analyzed using a microplate reader (SpectraMax M3, Molecular Devices) at 595 nm. FRAP values are expressed as mg ferrous sulfate ($FeSO_4$) per gram extract and were calculated from a linear regression equation constructed from various concentrations of $FeSO_4$ solutions (100 to 1000 µM).

2.5.4. Lipid Peroxidation Inhibition Assay

This investigation was based on the method in a related study [23]. The stock solution of *B. hispida* extracts was prepared at a concentration of 50.0 mg/mL in 50% DMSO. Then, 150 µL of extract solution was mixed with 100 µL of DI water, 350 µL of 20 mM phosphate buffer solution (PBS) (pH 7.0), 350 µL of 1.3% *v*/*v* linoleic acid in methanol and 50 µL of 25.14 mg/g 2,2′-azobis-(2-amidinopropane dihydrochloride) or (APPH) in 20 mM PBS (pH 7.0). The tested mixture was incubated at 50 °C for 4 h. Afterward, 5 µL of the mixture was added to a 96-well plate with 5 µL of 10% ammonium thiocyanate and 5 µL of 20 mM ferrous chloride, followed by adding 185 µL of 75% methanol. After 3 min, the absorbance of the mixed solution at 500 nm was determined using a Genios Pro microplate reader (Tecan, Crailsheim, Germany). The potential values expressed as % inhibition were calculated using Equation (2).

2.6. Anti-Aging Activities

2.6.1. Collagenase Inhibition Assay

The collagenase inhibition activity of *B. hispida* extracts was determined following the protocol of Thring [24] with some modification. The tested collagenase enzyme at a concentration of 2.0 mg/mL was prepared at 25 °C in PBS pH 7.5, containing 50 mM Tricine, 10 mM calcium chloride and 400 mM sodium chloride. As a substrate, 1.0 mM of N-[3- (2 furyl) acryloyl]-Leu-Gly-Pro-Ala, also known as FALGPA, in PBS pH 7.5 was used. *B. hispida* extracts (10 µL) at a concentration of 0.5 mg/mL were added to a 96-well plate, and then the tested collagen enzyme (10 µL) was added, followed by the substrate (10 µL). The reaction mixture was incubated for 30 min at 37 °C. After that, this mixture was analyzed using a microplate reader (SPECTROstar Nano, BMG Labtech, Aylesbury, Buckinghamshire, UK) at a wavelength of 340 nm. For comparison, 0.5 mg/mL of EGCG standard solution was used as a positive control. The results expressed as % collagenase enzyme inhibition were calculated according to the following equation:

$$\%\ \text{Collagenase enzyme inhibition} = [(C - C0) - (Cs - Cb)/(C - C0)] \times 100 \quad (3)$$

where C is the control absorbance (DI water mixed with collagen enzyme and substrate), C0 is the blank control absorbance (DI water mixed with PBS pH 7.5 and collagen enzyme), Cs is the tested extract absorbance and Cb is the extract blank absorbance (extract mixed with PBS and substrate).

2.6.2. Hyaluronidase Inhibition Assay

This assay was performed to determine the potential of *B. hispida* extract against the hyaluronidase enzyme following the sigma protocol with slight modification [25]. The diluted hyaluronidase enzyme (2.0 mg/mL) was prepared at 25 °C in the enzymatic diluent, comprising a mixture of PBS pH 7.0, 77 mM sodium chloride and 0.01% BSA. Then, 0.03% hyaluronic acid mixed with PBS pH 5.3, namely PHS, was prepared and incubated at 80 °C. The *B. hispida* extract at a concentration of 0.5 mg/mL in DI water was used for analysis by mixing 50 µL of the *B. hispida* extract with 100 µL of diluted hyaluronidase enzyme solution and incubating the mixture at 37.5 °C for 10 min; then, 100 µL of 0.03% hyaluronic acid was added and continuously incubated at 37.5 °C for 45 min. After that, acetic albumin solution pH 3.75, containing 24 mM sodium acetate, 79 mM acetic acid and 0.1% BSA, was added and mixed at 25 °C. This mixture was subjected to analysis at 600 nm using a microplate reader (SPECTROstar Nano, BMG Labtech). Tannic acid, at the same concentration of the tested extract, was used as a positive control. The activity was shown as % hyaluronidase enzyme inhibition and calculated using the following equation:

$$\%\ \text{Hyaluronidase enzyme inhibition} = [(H - H0) - (Hs - Hb)/(H - Hb)] \times 100 \quad (4)$$

where H is the control absorbance (tested mixture without the extract and using DI water as the tested sample), H0 is the blank control absorbance (DI water mixed with enzymatic diluent), Hs is the tested extract absorbance and Hb is the extract blank absorbance (extract mixed with PHS).

2.7. Chemical Marker Analysis by High-Performance Liquid Chromatography (HPLC)

Qualitative and quantitative analyses of the most promising *B. hispida* extract were performed using an HPLC system (Prominence LC2030C, Shimazu, Japan). A Knauer® vertex III reversed-phase HPLC column C18 (250 mm × 20 mm) (KNAUER Wissenschaftliche Geräte GmbH, Berlin, Germany) was used as the stationary phase. A gradient system was used as an analytical method with a detector at a wavelength of 360 nm. The mobile phase consisted of A (0.1% acetic acid in water) and B (acetonitrile) with a flow rate of 1.0 mL/min for 33 min. Gradient elution of the mobile phase was carried out as follows: 0.01 min, 95% A; 2 min, 95% A; 7 min, 80% A; 22 min, 70% A; 23 min, 95% A; and 33 min, 95% A. The analytical samples were dissolved in absolute ethanol and filtered through a 0.45 μm filter (Whatman, Marlborough, MA, USA) before analysis with an injection volume of 10 μL. *B. hispida* extract at 1.0 mg/mL was prepared for analysis. Rutin was used as a reference compound to analyze the active compounds in *B. hispida* extract. The retention time of rutin was approximately 17 min. The concentration of rutin in *B. hispida* extract was calculated from the peak area using a linear equation constructed from a standard curve of rutin (0.0 to 50 μg/mL).

2.8. Molecular Docking Simulation

The main ingredients of *B. hispida* peel, including ascorbic acid, quercetin and rutin, were submitted to molecular docking simulation. The 3D chemical structures of these compounds were retrieved from the PubChem database (https://pubchem.ncbi.nlm.nih.gov/ accessed on 12 June 2023). All structures were geometrically optimized using Gaussian09w [26] with the HF/6-31G(d,p) [27] basis set. The enzymes involved in the aging process such as collagenase (PDB ID: 1CGL) [28] and hyaluronidase (PDB ID: 2PE4) [29] were obtained from the RCSB Protein Data Bank (PDB) database (https://www.rcsb.org/ accessed on 12 June 2023). All co-crystallized ligands and water molecules were removed from the structures. Before molecular docking, the Gasteiger's and Kollman potential charges were assigned to ligands and proteins, respectively, using AutoDockTools 1.5.7 [30]. Molecular docking was carried out using the AutoDock Vina [31] program. The grid center of each protein was set according to its active site coordinates. The compound with the highest binding affinity was chosen as the representative structure for further analysis. Hydrogen bonds and hydrophobic interactions were identified using LigandScout 4.4.8 [32] and Discovery Studio Visualizer 2021 [33]. The molecular visualization was performed using the UCSF ChimeraX [34] Program.

2.9. Cell Cytotoxicity Test

To investigate the skin toxicity of *B. hispida* extracts, human dermal fibroblast (HDF) cells, obtained from the American Type Culture Collection (ATCC), were employed in this experiment. The test was performed using the MTT assay that was described in a related study with modifications [35]. The HDF cells (1×10^4 cells/well) were cultured and treated in a 96-well plate with a medium culture containing 10% DMEM, 10% FBS and antibiotic solution (100 U/mL penicillin and 100 μg/mL streptomycin) and incubated at 37 °C in a 5% CO_2 humidified atmosphere for 24 h. Then, the tested samples, *B. hispida* extracts at the concentration of 0.001 to 1.0 mg/mL, were added and continuously incubated in the same condition. After 24 h, the supernatant was removed, and 100 μL of 0.05 mg/mL MTT reagent was added to the well, followed by further incubation at 37 °C in a 5% CO_2 humidified atmosphere for 6 h. The MTT reagent was then removed, and 100 μL of DMSO was added to dissolve the formazan crystals. After 15 min, the absorbance was measured at 560 nm using a microplate reader (Bio Tek Instruments, Winooski, VT, USA). The results were calculated and expressed as the percentage of cell viability using the following equation:

$$\% \text{ Cell viability} = (A/A0) \times 100 \quad (5)$$

where A is the absorbance of the tested sample and A0 is the absorbance of the control, namely the tested well without *B. hispida* extract.

2.10. In Vitro Irritation Test Using Hen's Egg Chorioallantoic Membrane (HET-CAM) Assay

The irritation potential of *B. hispida* extracts was appraised using the HET-CAM assay according to a protocol described by Chaiyana [35]. This experiment serves as an alternative in vitro method for assessing skin irritation. For investigation using the HET-CAM assay, the placental membrane of a chicken embryo, providing a vascular network of capillaries, was used in the study. The fertilized hen eggs used in this study were obtained from the Faculty of Agriculture at Chiang Mai University, Thailand. Eggs aged between seven and nine days were employed. These eggs underwent incubation at a temperature of 37.5 ± 0.5 °C with a relative humidity of 62.5 ± 7.5%. The shell above the air cavitation of the egg was carefully removed using a rotating cutting blade attached to a Marathon champion 3 micromotor (Saeyang, Seoul, South Korea). A normal saline solution was then applied to an inner membrane that came into direct contact with the chorioallantoic membrane (CAM), followed by incubation under the same conditions as described above for 15 min. After this incubation period, the inner membrane was meticulously removed using forceps. Then, 30 μL of *B. hispida* extract at a concentration of 10 mg/mL was dropped onto the prepared CAM. In this study, the positive and negative controls were represented by 30 μL of 1% (*w*/*v*) sodium lauryl sulfate (SLS) and 30 μL of normal saline solution, respectively.

The assessment of irritation effects involved continuous observation using a stereomicroscope (Olympus, Tokyo, Japan) over a 5 min (300 s) period and expressed the short-term irritation effect. After 60 min of testing, observations were once again conducted to determine long-term irritation. The observed signs of irritation included vascular hemorrhage, vascular lysis and vascular irritation. The time of the first appearance of each of these irritating signs was recorded in seconds and was subsequently used to calculate an irritation score (IS) according to the following equation:

$$IS = [((301 - Ht)/300) \times 5] + [((301 - Lt)/300) \times 7] + [((301 - Ct)/300) \times 9] \quad (6)$$

where Ht is the time of the initial appearance of vascular hemorrhage, Lt is the time of the initial appearance of vascular lysis and Ct is the time point of the initial appearance of vascular coagulation. Based on the calculated IS, IS = 0.0 to 0.9 was classified as no irritation, IS = 1.0 to 4.9 was classified as mild irritation, IS = 5.0 to 8.9 was classified as moderate irritation and IS = 9.0 to 21.0 was classified as severe irritation.

2.11. Statistical Analysis

All measurements were performed independently in triplicate. The results are expressed as mean ± S.D. Statistical analysis was conducted using SPSS Software, Version 17.0 for Windows, and differences between groups were determined using one-way analysis of variance (ANOVA) followed by Tukey's test. The differences were considered significant at a p-value ≤ 0.05.

3. Results

3.1. Plant Extraction

From the extraction using various solvents, the obtained extracts from *B. hispida* peels and the percentage yields of each extract are shown in Table 1. The highest extract yield was found for BW, followed by B95 and B50. The extracts appeared viscous and light brown and had specific odors.

Table 1. Percentage yields of *B. hispida* peel extracts from various solvents.

Extract	Solvent	Yield (%)
B95	95% ethanol	29.6 ± 1.28 ^{b}
B50	50% ethanol	15.2 ± 0.98 ^{c}
BW	DI water	32.7 ± 0.87 ^{a}

The results show the average yield values of extracts ± S.D., and the different letters indicate significance level at $p < 0.05$.

3.2. Total Flavonoid and Total Phenolic Contents

The values of total flavonoid content (TFC) and total phenolic content (TPC) in *B. hispida* extracts are shown in Table 2. Flavonoid contents in different extracts were calculated from the regression equation of the quercetin standard curve ($y = 0.3502x + 0.0565$, $r^2 = 0.993$) and expressed as mg QE/g extract. Phenolic contents in different extracts were calculated from the regression equation of the gallic acid standard curve ($y = 8.2221x + 0.006$, $r^2 = 0.998$) and expressed as mg GAE/g extract.

Table 2. Total phenolic and total flavonoid contents of *B. hispida* peel extracts.

Extract	Total Flavonoid Content (mg QE/g Extract)	Total Phenolic Content (mg GAE/g Extract)
B95	212.88 ± 4.73 [a]	131.52 ± 0.38 [b]
B50	54.18 ± 0.11 [c]	140.43 ± 0.77 [a]
BW	60.83 ± 0.86 [b]	115.91 ± 0.78 [c]

The results show the average values ± S.D. Different letters indicate significance level at $p < 0.05$.

The highest TFC was found in B95 with a value of 212.88 ± 4.73 mg QE/g extract and was followed by the TFCs of BW and B50, consecutively. However, the highest TPC was found in B50 with the value of 140.43 ± 0.77 mg GAE/g extract and was followed by the TPCs of B95 and BW, consecutively. These results might be related to the extracted solvents; some flavonoid and phenolic compounds are well extracted in semi-polar solvents such as 95% ethanol, which was presented in the study as having a high value of TFC and TPC in B95.

3.3. Antioxidant Activities

The antioxidative activities of *B. hispida* peel extracts were investigated using various methods to evaluate various mechanisms of action. The results of *B. hispida* extracts from these procedures are shown in Table 3.

Table 3. Antioxidant activities with various assays.

Tested Sample	DPPH Assay * (IC_{50}; mg/mL)	ABTS Assay * (IC_{50}; mg/mL)	FRAP Assay * (mg $FeSO_4$/g Extract)	Lipid Peroxidation Assay * (% Inhibition)
B95	2.91 ± 0.07 [c]	0.17± 3.13 [c]	0.75 ± 0.09 [e]	53.42 ± 0.02 [c]
B50	1.73 ± 0.13 [b]	0.16 ± 3.95 [c]	2.46 ± 0.18 [d]	40.31 ± 0.01 [e]
BW	23.63 ± 0.60 [d]	0.27 ± 12.09 [d]	17.33 ± 0.19 [c]	45.39 ± 0.86 [d]
Ascorbic acid	0.01 ± 0.04 [a]	0.00002 ± 0.02 [a]	1926.16 ± 21.40 [a]	94.17 ± 3.74 [a]
Trolox	0.013 ± 0.26 [a]	0.00544 ± 1.71 [b]	566.65 ± 27.41 [b]	78.82 ± 3.36 [b]

* Results show the average antioxidant values of extracts ± S.D. from different assays. Different letters indicate significance level at $p < 0.05$.

DPPH and ABTS assays were performed to determine the radical scavenging activity of the extracts. From the DPPH assay, the results were expressed as the concentration of extract that could be active against radicals where the % inhibition is equal to 50 (IC_{50}). The IC_{50} values of extracts in the DPPH assay were calculated from the regression equations of each extract, namely $y = 0.0128x + 12.75$, $r^2 = 0.986$ for B95; $y = 0.0308x + 2.6124$, $r^2 = 0.9895$ for B50; and $y = 1.3828x + 16.589$, $r^2 = 0.925$ for BW. The standard compounds that were considered in comparison were ascorbic acid and Trolox. The IC_{50} values of these standard compounds were also calculated from the regression equations of each compound, namely $y = 6.1021 + 20.349$, $r^2 = 0.936$ for ascorbic acid and $y = 3.031 + 14.631$, $r^2 = 0.989$ for Trolox. The results demonstrated that B50 exhibited the significantly highest scavenging activity with an IC_{50} value of 1.73 ± 0.13 mg/mL, followed by B95 with an IC_{50} of 2.91 ± 0.07 mg/mL and BW with an IC_{50} value of 23.63 ± 0.60 mg/mL ($p < 0.05$).

The ABTS assay results were expressed as IC_{50} values, the same as the results obtained from the DPPH assay. IC_{50} values of the *B. hispida* extracts were calculated using the equations of linear regression of each extract. The obtained equations used to calculate IC_{50} values of B95, B50 and BW were $y = 0.227x + 9.7036$ ($r^2 = 0.9907$), $y = 0.2387x + 10.612$ ($r^2 = 0.9708$) and $y = 0.1387x + 10.685$ ($r^2 = 0.9279$), respectively. Ascorbic acid and Trolox were used as the standard compounds, and IC_{50} values were also calculated from the regression equations. The calculated equations of ascorbic acid and Trolox were $y = 7.2955x + 6.7331$ ($r^2 = 0.9841$) and $y = 3.9231x + 2.1272$ ($r^2 = 0.9803$), respectively. The obtained IC_{50} values from ABTS assays showed that B95 and B50 provided the highest ABTS scavenging activity with IC_{50} values of 0.17 ± 3.13 mg/mL and 0.16 ± 3.95 mg/mL, followed by BW with an IC_{50} value of 0.27 ± 12.09 mg/mL ($p < 0.05$).

The FRAP assay was used to determine the ability of radical reduction. The results were expressed as the mg $FeSO_4$/g extract which was calculated from the regression equation of $FeSO_4$ ($y = 6.724x + 0.1818$, $r^2 = 0.995$). A high FRAP value indicated a high radical reduction activity. *B. hispida* extracts (B95, B50 and BW) at a concentration of 1.0 mg/mL were used in this experiment. The BW showed the highest reducing property with a FRAP value of 17.33 ± 0.19 mg $FeSO_4$/g extract. Nevertheless, B95 showed less activity in comparison with others ($p < 0.05$).

The lipid peroxidation inhibition assay was performed to determine the antioxidant ability against oxidative degradation of lipids by peroxide radicals. The results were expressed as % inhibition obtained from B95, B50 and BW extracts including the standard compounds (ascorbic acid and Trolox) at the same concentration of 0.1 mg/mL. B95 possessed high activity with the % inhibition of 53.42 ± 0.02, against peroxide radicals, followed by B50 and BW, consecutively.

Ascorbic acid and Trolox, the compounds that were used for comparing the antioxidant activity in all tested assays, showed the significantly strongest activities when compared with all *B. hispida* peel extracts ($p < 0.05$). Interestingly, the overall antioxidative assays illustrated that B95 obtained from *B. hispida* peel has potential antioxidant activity related to all antioxidant mechanisms.

3.4. Anti-Aging Activities

To determine anti-aging activities of *B. hispida* peel extracts, the ability to inhibit collagenase and hyaluronidase enzymes was investigated. The results were expressed as % collagenase enzyme inhibition and % hyaluronidase enzyme inhibition. The obtained inhibitory effects of *B. hispida* peel extracts are illustrated in Figure 2. Among the samples tested against the collagenase enzyme (Figure 2A), B95 showed an inhibition effect that does not significantly differ ($p > 0.05$) from EGCG, a standard compound, at the same concentration of 0.5 mg/mL. B95 showed an inhibition of the collagenase enzyme with the value of $41.68 \pm 0.92\%$. Furthermore, the % collagenase inhibition exhibited by B50 and BW did not demonstrate statistically significant differences when compared with that of B95 ($p > 0.05$). Concerning the activity against the hyaluronidase enzyme, the obtained results are demonstrated in Figure 2B. B95, with a value of $29.17 \pm 0.66\%$, showed obviously higher hyaluronidase inhibition activity than other extracts ($p < 0.05$). Nevertheless, B50 and BW demonstrated hyaluronidase inhibition of the same potential ($p > 0.05$). However, all *B. hispida* peel extracts possessed an inhibitory effect against the hyaluronidase enzyme that was less than that of tannic acid, the standard compound in this experiment ($p < 0.05$).

3.5. Chemical Marker Analysis by High-Performance Liquid Chromatography (HPLC)

The results obtained from antioxidant and anti-aging assays showed that B95 exhibited the most potent effects among all the tested extracts of *B. hispida*. Consequently, B95 was selected to determine its chemical marker using HPLC. Rutin, a reference flavonoid compound used to identify the active compound in B95, was also included in the analysis for comparison. The HPLC chromatograms of B95 and rutin are presented in Figure 3. Several components were detected in B95, as indicated by the observed HPLC chromatograms; one of these components was rutin, which exhibited a retention time of approximately 17 min. The quantification of rutin in B95 was calculated using the linear equation of rutin, namely $y = 8914.9x + 793.63$ ($r^2 = 0.9999$). The results demonstrated that 1 mg of B95 contained 4.81 ± 0.03 μg of rutin.

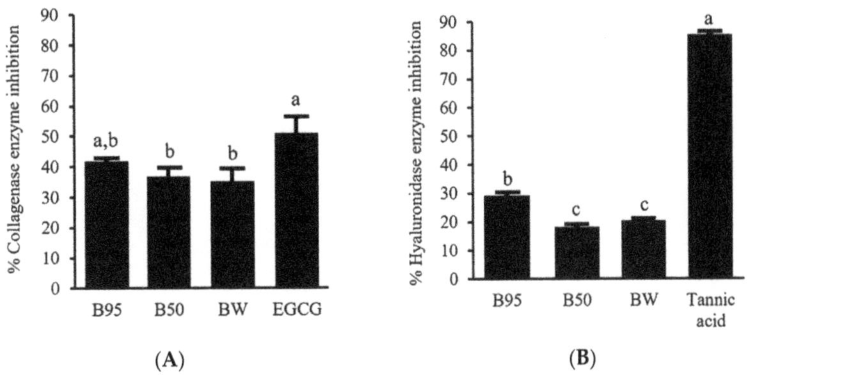

Figure 2. Potential of *B. hispida* extracts against the collagenase enzyme compared with EGCG (**A**) and the hyaluronidase enzyme compared with tannic acid (**B**). Different letters indicate significance level at $p < 0.05$.

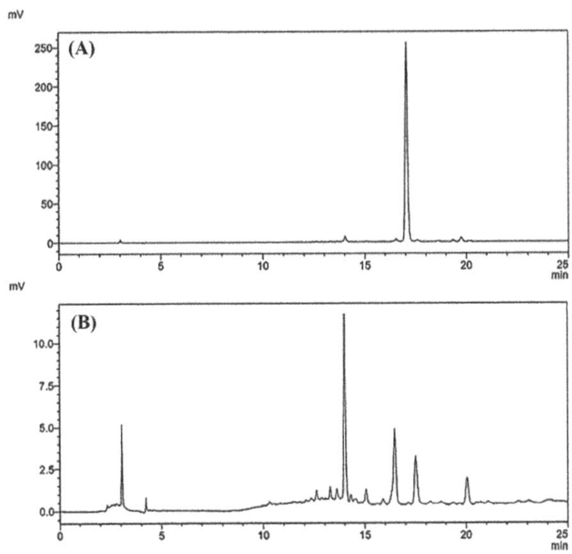

Figure 3. HPLC chromatograms of rutin at the concentration of 50 μg/mL (**A**) and the most promising *B. hispida* extract, B95 (**B**) at a concentration of 1.0 mg/mL, analysis at 360 nm.

3.6. Molecular Docking Simulation

The main ingredients of *B. hispida* peel extract, including ascorbic acid, quercetin and rutin, were submitted to molecular docking simulation. The molecular docking results indicated that quercetin exhibited the highest binding affinity against the collagenase enzyme (z8.9 kcal/mol), followed by rutin (−7.9 kcal/mol) and ascorbic acid (−6.3 kcal/mol), consecutively. The molecular recognition of these compounds against collagenase is depicted in Figure 4. All compounds fit well in the binding pocket of collagenase, as shown in Figure 4A. Ascorbic acid interacts with key amino acids, including Ala182, Arg214, His218, Glu219 and Pro238, via hydrogen bonding and also forms a metal interaction with Zn301 (Figure 4B). Quercetin forms hydrogen bonds with Arg214, Glu219 and Tyr237 and interacts with Leu181 and Val215 via hydrophobic interactions (Figure 4C). Rutin exhibits interactions with amino acids such as Asn180, Arg214, Tyr237 and Tyr240 through hydrogen bonds and forms hydrophobic interactions with Val215 (Figure 4D).

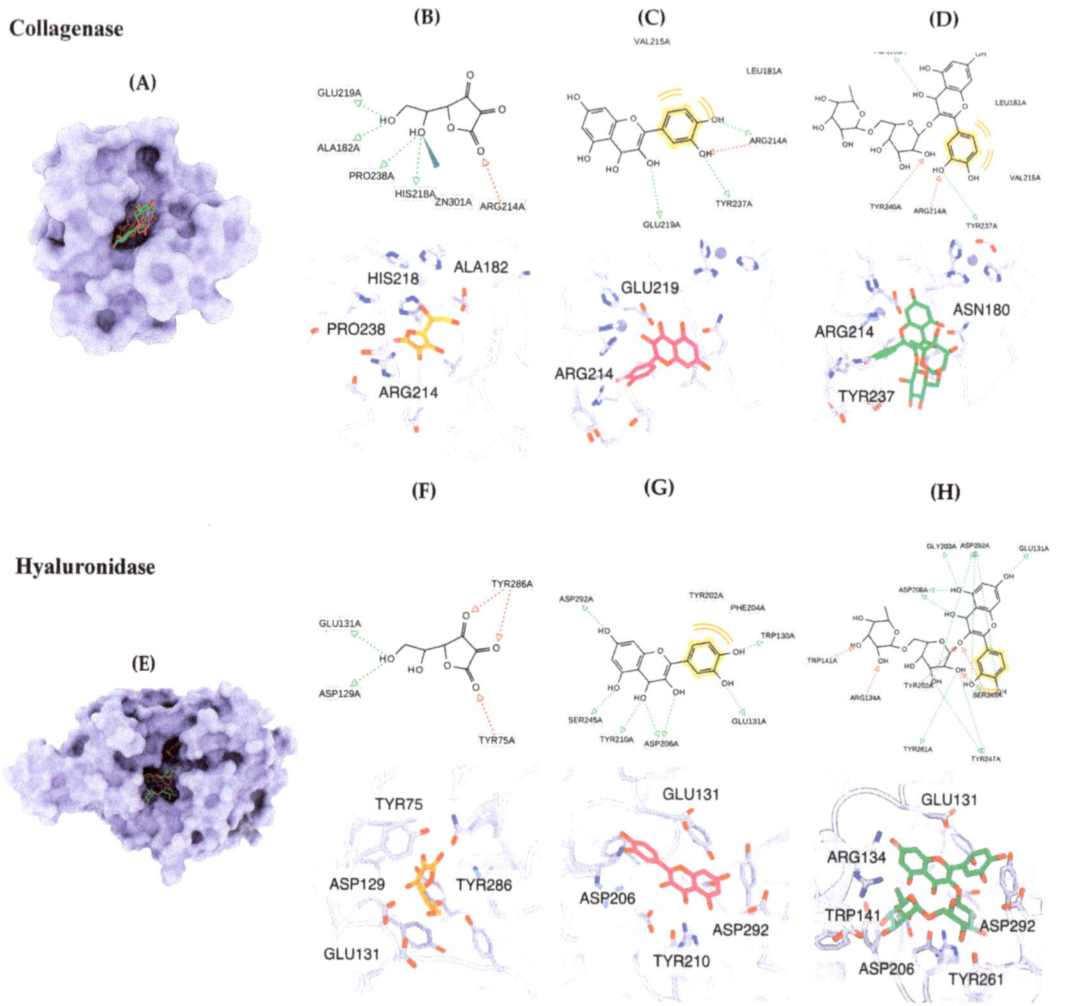

Figure 4. Binding modes of representative phytochemicals binding to collagenase (**A**) and hyaluronidase (**E**). The 2D and 3D interactions of ascorbic acid (**B**), quercetin (**C**) and rutin (**D**) complexed with collagenase, along with the 2D and 3D binding modes of ascorbic acid (**F**), quercetin (**G**) and rutin (**H**) in the binding site of hyaluronidase. The pharmacophore features include hydrogen bond acceptor (red arrow), hydrogen bond donor (green arrow) and hydrophobic property (yellow).

The molecular docking results for the hyaluronidase enzyme showed that rutin possessed the highest docking affinity (−9.8 kcal/mol), followed by quercetin (−8.1 kcal/mol) and ascorbic acid (−5.6 kcal/mol). The graphical illustration indicates that the molecular binding patterns of the compounds differ in terms of their positions within the binding pocket (Figure 4E). Ascorbic acid fits well in the narrow pocket to form hydrogen bonds with residues including Tyr75, Asp129, Glu131 and Tyr286 (Figure 4F). Quercetin interacts with Trp130, Glu131, Asp206, Tyr210, Ser245 and Asp292 via hydrogen bonds and forms hydrophobic interactions with Tyr202 and Phe204 (Figure 4G). Rutin is located on the same site as quercetin. The core structure of rutin forms hydrogen bonds with Glu131, Gly203, Tyr202, Asp206, Ser245 and Asp292, and the glycoside substituent forms hydrogen bonds with Arg134, Trp141, Tyr247 and Tyr261 (Figure 3).

3.7. Cell Cytotoxicity Test

A cytotoxicity test of B. hispida extracts (B95, B50 and BW) on HAF cells was performed using the MTT assay. The tested samples showing cell viability of above 80% were regarded as nontoxic to the cells. The obtained results after 24 h of exposure of the tested samples to the HAF cells are shown in Figure 5. All B. hispida extracts were not toxic to HAF cells at concentrations of 0.001 to 0.01 mg/mL, with cell viability of more than 90%. In addition, B50 and BW at a concentration of 1.0 mg/mL also exhibited nontoxicity to HAF cells with a % cell viability of 99.74 ± 8.63, and 92.35 ± 17.30, respectively. However, B95 at a concentration of 1.0 mg/mL showed slight toxicity to HAF cells with a % cell viability of 77.58 ± 2.82%.

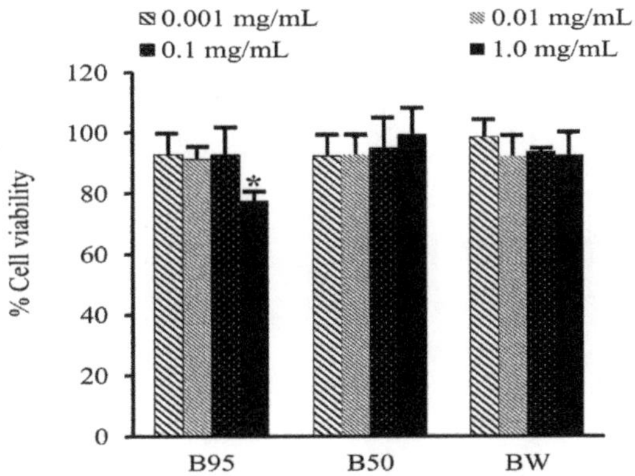

Figure 5. Percentage cell viability of each *B. hispida* extract at concentrations from 0.001 to 1.0 mg/mL. The values are presented mean ± S.D. (n = 3). Asterisk (*) represents a different value at $p < 0.05$.

3.8. In Vitro Irritation Test Using the HET-CAM Assay

To further investigate the effective activities of *B. hispida* peel extracts in vivo and in a clinical study, an in vitro irritation test using the HET-CAM assay was then employed. The irritation induced by *B. hispida* peel extracts (B95, B50 and BW) was evaluated and compared with that of the negative control, 0.9% *w/v* NaCl, and the positive control, 1% (*w/v*) SLS. The obtained results as irritation scores are shown in Table 4, and the observations of the alterations under a stereomicroscope are shown in Figure 6. None of the *B. hispida* peel extracts at a concentration of 10 mg/mL induced considerable alterations, namely vascular hemorrhage, vascular lysis and vascular irritation, in short-term testing (5 min) and long-term testing (60 min), similar to the result obtained from 0.9% *w/v* NaCl. However, 1% (*w/v*) SLS, the positive control, exhibited signs of severe irritation in 5 min of testing with vascular hemorrhages and vascular lysis.

Table 4. Irritation score of *B. hispida* peel extracts, 0.9% *w/v* NaCl and 1% *w/v* SLS.

Tested Compound	Irritation Score *	Irritation Category
B95	0.0 ± 0.0^{b}	No irritation
B50	0.0 ± 0.0^{b}	No irritation
BW	0.0 ± 0.0^{b}	No irritation
0.9% (*w/v*) NaCl	0.0 ± 0.0^{b}	No irritation
1% (*w/v*) SLS	11.1 ± 0.5^{a}	Severe irritation

* The results are shown as the average IS ± S.D. Different letters indicate significance level at $p < 0.05$.

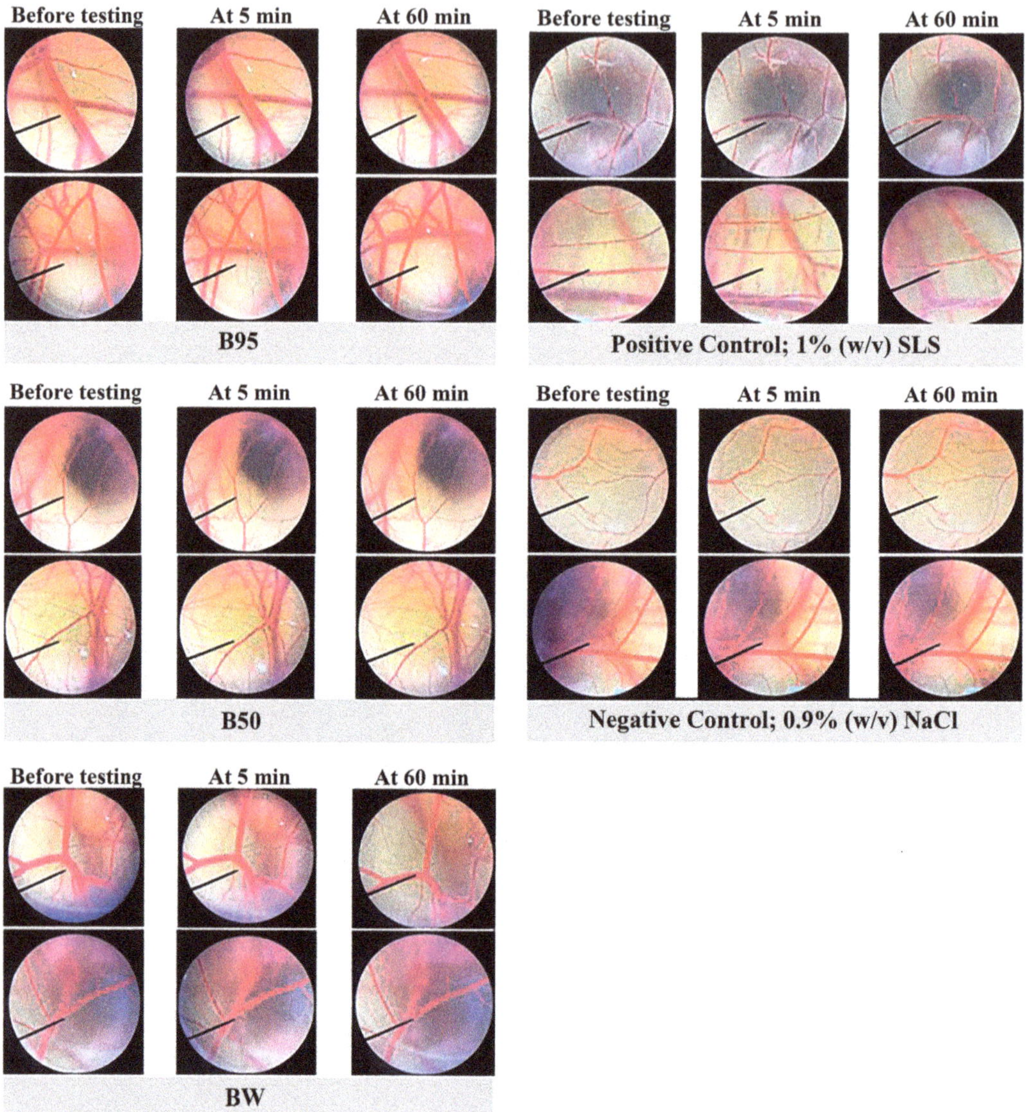

Figure 6. The stereomicroscope images of CAM vasculature before testing, at 5 min after treatment and at 60 min after treatment with 10 mg/mL of each extract (B95, B50 and BW), 0.9% *w*/*v* NaCl and 1% *w*/*v* SLS.

4. Discussion

It has been realized that the potential of antioxidants against free radical damage to the skin is a significant property that can decelerate skin aging and skin damage related to transduction pathways and epigenetic changes caused by oxidative stress [36]. Moreover, countless studies have explored the antioxidant activity from novel natural sources, including plants, animals, insects, algae and waste products, to protect the skin from damage caused by free radicals [37,38]. Phenolic compounds are powerful antioxidant agents with wide mechanisms of action including the capacity to scavenge free radicals such as hydroxyl radical ($^\bullet OH$) and superoxide anion ($O_2^{-\bullet}$) and chelate metal ions (Fe^{2+}, Fe^{3+},

Cu^{2+} and Cu^{+}) which react to hydrogen peroxide (H_2O_2). Highly reactive $^{\bullet}OH$ is formed by the dismutation of $O2^{-\bullet}$ by SOD, enhancing the activity or expression of intracellular antioxidant enzymes [39]. Flavonoids are a group of natural substances with variable phenolic structures. Flavones and catechins have been reported to be the most powerful flavonoids for protecting the body against ROS [40].

B. hispida is a potentially edible plant that has been used for remedies and protection against many diseases related to the phytochemicals in the used part [11]. The obtained extracts of *B. hispida* in this study showed a high yield in BW and B95, which could indicate that the chemical constituents in *B. hispida* peel are well solubilized in water and 95% ethanol. Related studies have reported the presence of phenols, alkaloids, saponins, steroids, carbohydrates and flavonoids in chloroform, alcoholic and aqueous extracts obtained from *B. hispida* peel [10,13]. This was consistent with the results of this study that found flavonoids in *B. hispida* peel extracts, especially in B95 (95% *v/v* ethanolic extract). Moreover, phenolic compounds were found in all extracts. The phytochemical profile of phenolic compounds in *B. hispida* has been reported by Islam et al., and the results revealed the presence of many bioactive chemicals, such as quercetin, rutin, astilbin, catechin, naringenin and hispidulin [11]. The antioxidant activities of *B. hispida* extracts were investigated with standard procedures to evaluate the different mechanisms of action. DPPH and ABTS methods have commonly been used to evaluate the antioxidant activity of compounds that act as scavengers of $O2^{-\bullet}$ and $^{\bullet}OH$ radicals [41]. The FRAP assay is a method that possesses the power of antioxidation by reducing ferric iron (Fe^{3+}) to ferrous iron (Fe^{2+}) [42]. In this study, B95 obviously possessed antioxidant activities with free radical scavenging mechanisms via DPPH and ABTS assays. Moreover, it could inhibit lipid peroxidation by stopping lipid chain reactions. Further, BW possessed good potential to reduce Fe^{3+} via the FRAP assay. These results are consistent with related studies reporting that the methanolic extracts of *B. hispida* peel provided antioxidant activity against free radicals when determined using DPPH and FRAP methods [10,43]. In addition, the high flavonoid and phenolic contents found in B95 possessed good potential antioxidant activities in all assays, referring to the synergistic action of both compounds [44].

Collagenase is a key enzyme that acts by degrading collagen in the extracellular matrix, promoting premature skin aging. Hyaluronidase is also a key enzyme that degrades hyaluronic acid, a critical component of the extracellular matrix in the skin for maintaining the normal hydration of the skin [45,46]. The inhibition of both enzymes provides the potential to prevent skin aging and its signs. Our study found that the strongest anticollagenase activity was provided by B95 (72.70 $\pm$ 1.24%), and the strongest antihyaluronidase activity was also exhibited by B95 (74.88 $\pm$ 4.84%). The results presumably recommend that flavonoid compounds that were found to possess the highest B95 levels may be the active compounds for these activities. In a related study, rutin and quercetin, which belong to the group of flavonoid compounds, were detected in the fruit of *B. hispida*, indicating good anti-aging activities [16,47]. In addition, the literature extensively supports the efficacy of flavonols as inhibitors of collagenase [48]. Quercetin and kaempferol exhibited a more significant inhibitory effect compared with flavones, isoflavones and flavanones, with the latter demonstrating a very negligible impact. The arrangement of hydroxyl groups in the B-ring of the flavonoid structure may play a crucial role in determining the inhibitory effect on enzyme activity [49]. Similarly, the antioxidative and anti-enzymatic activities of Thai plants were found to be influenced by their phenolic and flavonoid contents [6]. Thus, both compounds were investigated in a molecular docking simulation against collagenase and hyaluronidase enzymes.

Various phytochemicals have been identified in *B. hispida* extracts, primarily in the fruit portion [16]. However, limited information is available regarding the composition of phytochemicals obtained from *B. hispida* peel. In this study, the most promising *B. hispida* extract, namely B95, was used to determine the phytochemical active compounds through HPLC analysis. Rutin, a flavonoid that has been detected in various parts of *B. hispida*, was employed as a reference compound for the HPLC analysis of B95. This analysis revealed the presence of numerous compounds, which manifested as multiple HPLC peaks observed in the chromatogram. Rutin is one of the compounds identified in B95. This result aligns with related studies that have also detected rutin in other parts of *B. hispida* [50,51]. However, it is necessary to identify and characterize the other compounds in B95. This will provide a comprehensive understanding of B95′s composition and support the feasibility of its mechanism of action and efficacy in various applications. Conducting a broader analysis to identify and understand the various compounds in B95 can significantly enhance the ability to assess its potential benefits and applications. Molecular docking contributes to a better understanding of the molecular basis underlying the interaction between promising compounds from *B. hispida* and the collagenase and hyaluronidase enzymes, which are the key enzymes related to anti-aging properties. Ascorbic acid, quercetin and rutin were identified as the main bioactive compounds in the *B. hispida* peel extract and were further investigated for their molecular recognition within the binding sites of these enzymes. All phytochemicals are capable of binding to the collagenase active site cleft through hydrogen and hydrophobic interactions. The strength of the binding of these phytochemicals correlates to the number of hydrogen bonds and steric effects. Quercetin and rutin form interactions with Arg214 at the bottom of the S1′ pocket, which is suitable in size for an aromatic ring [52]. Moreover, the hydrogen bond with Glu219 also supports the binding affinity of quercetin within the binding pocket [28]. In hyaluronidase, rutin and quercetin form hydrogen bonds with Asp129 and Glu131, which are essential amino acids for the catalytic site of hyaluronidase [53,54]. The interaction of quercetin and rutin with Tyr202 has been observed, further supporting the binding strength [55]. This interaction information can provide insights into the potential use of the phytochemicals from *B. hispida* peel extract as anti-aging agents that inhibit the activity of collagenase and hyaluronidase, which are enzymes associated with the skin aging processes of collagen degradation and hyaluronic acid breakdown.

Regarding safety, the cytotoxicity and the irritation of *B. hispida* peel extracts need to be investigated before application and further development for food, cosmetics or cosmeceutical products. The cytotoxicity of all *B. hispida* peel extracts on human dermal fibroblast cells was evaluated using the MTT assay, which is a standard protocol for assessing in vitro cytotoxicity. This assay is based on the reduction of the yellow tetrazolium salt of MTT to purple formazan crystals by metabolically active cells. Thus, the high detection of purple formazan crystals in the experiment indicates a high density of living cells. A test sample showing cell viability above 80% is considered as safe or possessing non-cytotoxicity [56,57]. In the present study, *B. hispida* peel extracts (B95, B50 and BW) demonstrated cell viability of more than 80% after being exposed to the cells for 24 h. These results were consistent with one related study reporting that *B. hispida* extract is not toxic to normal cells [58]. Therefore, it could be concluded that *B. hispida* peel extracts are not harmful to dermal cells and can be safely applied on the skin.

As the irritation test, the HET-CAM assay was used to assess the irritation properties of *B. hispida* peel extracts. This protocol was developed by Lüepke [59] and is employed as an alternative to the Draize test for screening the potential of irritation and anti-irritation effect of cosmetic formulations and ingredients [60]. The HET-CAM test was compared with results obtained from the irritation test with human skin, and the results led to the conclusion that significant comparable efficacy patterns were found [61]. Thus, the HET-CAM assay has been used as a preliminary assessment of irritation before clinical trials. The results obtained in our study revealed that none of the *B. hispida* peel extracts showed any signs of irritation on the chorioallantoic membrane. The signs of irritation include hemorrhage, coagulation and vascular lysis. These results revealed that *B. hispida* peel extracts are suitable for further developing topical products or skin care products with safety similar to that of other extracts that have been reported to have no irritation potential after being tested using the HET-CAM assay [7,31,62]

5. Conclusions

In the present study, we have demonstrated that *B. hispida* peel extracts received from postconsumer waste, containing flavonoid and phenolic compounds, constitute a potential natural source. All the extracts possessed significant antioxidant and anti-aging activities, especially activities against enzymes that cause skin aging. The extract obtained from 95% ethanol as the extraction solvent showed the most promising antioxidant activities involving the mechanisms of radical scavenging of DPPH and ABTS radicals, reduced free radicals in the FRAP assay and inhibited peroxide radicals. Regarding the investigation of anti-aging activities, *B. hispida* peel extract, obtained from 95% ethanol, possesses good ability to inhibit the collagenase enzyme at the same potential as epigallocatechin gallate (standard control). Additionally, this extract also exhibits an inhibitory effect on the hyaluronidase enzyme. Based on the molecular docking results, the interaction of *B. hispida* peel extracts with collagenase and hyaluronidase suggests that they have the potential to be used as anti-aging agents. The in vitro toxicity against human dermal fibroblast cells and the in vitro irritation assay using hen's egg chorioallantoic membrane demonstrate that all *B. hispida* peel extracts possess no toxicity and produce no irritating effect. In conclusion, these valuable results provide scientific evidence to support further studies concerning the physiochemical properties of *B. hispida* extracts and their active compounds. This research serves as a preliminary study via in vitro investigation; the mechanisms of in vivo antioxidant and in vivo anti-aging effects could be clarified in further investigations. Additionally, developing products containing *B. hispida* extracts for use in food, cosmetics or cosmeceuticals, as well as clinical studies, could also be investigated. Expanding upon this research can open opportunities for a wide range of practical applications.

Author Contributions: Conceptualization, K.K. and M.S.; Method, K.K., M.S., W.C. and S.C.; Software, S.C.; Formal Analysis, P.P., S.C., W.C. and K.K.; Resources, W.C. and K.K.; Data Curation, K.K., S.P., W.P. (Weeraya Preedalikit), W.P. (Worrapan Poomanee), P.P., W.C. and S.C.; Writing—Original Draft Preparation, P.P., K.K. and S.C.; Writing—Review and Editing, K.K., P.P., W.P. (Weeraya Preedalikit), W.P. (Worrapan Poomanee), M.S. and W.C.; Supervision, K.K.; Project Administration, K.K.; Funding Acquisition, W.C., K.K., M.S. and W.P. All authors have read and agreed to the published version of the manuscript.

Funding: This research project was supported by the Fundamental Fund 2023, Chiang Mai University (funding number: 66A104000013).

Data Availability Statement: The data presented in this study are available from the corresponding author upon reasonable request.

Acknowledgments: The authors thank Chiang Mai University for supporting funding and are grateful to the Faculty of Pharmacy, Chiang Mai University, for providing us with the necessary resources. The authors would like to thank Thomas McManamon, Faculty of Pharmacy, Chiang Mai University, for English editing.

Conflicts of Interest: The authors declare no conflict of interest.

References

1. Poljšak, B.; Dahmane, R. Free radicals and extrinsic skin aging. *Dermatol. Res. Pract.* **2012**, *2012*, 135206. [CrossRef] [PubMed]
2. Masaki, H. Role of Antioxidants in the skin: Anti-aging effects. *J. Dermatol. Sci.* **2010**, *58*, 85–90. [CrossRef] [PubMed]
3. Jenkins, G. Molecular mechanisms of skin ageing. *Mech. Ageing Dev.* **2002**, *123*, 801–810. [CrossRef] [PubMed]
4. Jiratchayamaethasakul, C.; Ding, Y.; Hwang, O.; Im, S.T.; Jang, Y.; Myung, S.W.; Lee, J.M.; Kim, H.S.; Ko, S.C.; Lee, S.H. In Vitro screening of elastase, collagenase, hyaluronidase, and tyrosinase inhibitory and antioxidant activities of 22 halophyte plant extracts for novel cosmeceuticals. *Fish. Aquat. Sci.* **2020**, *23*, 6. [CrossRef]
5. Baier Leach, J.; Bivens, K.A.; Patrick, C.W.J.; Schmidt, C.E. Photocrosslinked hyaluronic acid hydrogels: Natural, biodegradable tissue engineering scaffolds. *Biotechnol. Bioeng.* **2003**, *82*, 578–589. [CrossRef]
6. Chatatikun, M.; Chiabchalard, A. Thai plants with high antioxidant levels, free radical scavenging activity, anti-tyrosinase and anti-collagenase activity. *BMC Complement. Altern. Med.* **2017**, *17*, 487. [CrossRef]

7. Chaiyana, W.; Chansakaow, S.; Intasai, N.; Kiattisin, K.; Lee, K.-H.; Lin, W.-C.; Lue, S.-C.; Leelapornpisid, P. Chemical constituents, antioxidant, anti-MMPs, and anti-hyaluronidase activities of *Thunbergia laurifolia* Lindl. leaf extracts for skin aging and skin damage prevention. *Molecules* **2020**, *25*, 1923. [CrossRef]
8. Poomanee, W.; Chaiyana, W.; Intasai, N. Leelapornpisid, Biological activities and charaterization of the pod extracts from Sompoi (*Acacia concinna* Linn.) grown in northern Thailand. *Int. J. Pharm. Pharm. Sci.* **2015**, *7*, 237–241.
9. Rayees, B.; Dorcus, M.; Chitra, S. Nutritional composition and oil fatty acids of Indian winter melon *Benincasa hispida* (Thunb.) seeds. *Int. Food Res. J.* **2013**, *20*, 1151–1155.
10. Rana, S.; Suttee, A. Phytochemical investigation and evaluation of free radical scavenging potential of *Benincasa hispida* peel extracts. *Int. J. Curr. Pharm. Rev. Res.* **2012**, *3*, 43–46.
11. Islam, M.T.; Quispe, C.; El-Kersh, D.M.; Shill, M.C.; Bhardwaj, K.; Bhardwaj, P.; Sharifi-Rad, J.; Martorell, M.; Hossain, R.; Al-Harrasi, A.; et al. A Literature-based update on *Benincasa hispida* (Thunb.) Cogn.: Traditional uses, nutraceutical, and phytopharmacological profiles. *Oxid. Med. Cell. Longev.* **2021**, *2021*, 6349041. [CrossRef] [PubMed]
12. Huang, H.-Y.; Huang, J.-J.; Tso, T.; Ying Chieh, T.; Chang, C.-K. Antioxidant and angiotension-converting enzyme inhibition capacities of various parts of *Benincasa hispida* (Wax Gourd). *Nahrung* **2004**, *48*, 230–233. [CrossRef] [PubMed]
13. Doharey, V.; Kumar, M.; Kumar Upadhyay, S.; Singh, R.; Kumari, B. Pharmacognostical, physicochemical and pharmaceutical paradigm of ash gourd, *Benincasa hispida* (Thunb.) Fruit. *Plant. Arch.* **2021**, *21*, 249–252. [CrossRef]
14. Fatariah, Z.; Zulkhairuazha, T.Y.T.; Rosli, W.W.I. Quantitative HPLC analysis of gallic acid in *Benincasa hispida* prepared with different extraction techniques. *Sains Malays.* **2014**, *43*, 1181–1187.
15. Fatariah, Z.; Zulkhairuazha, T.Y.T.; Rosli, W.W.I. Ascorbic acid quantification in *Benincasa hispida* fruit extracted using different solvents. *Int. Food Res. J.* **2015**, *22*, 208–212.
16. Busuioc, A.C.; Botezatu, A.V.D.; Furdui, B.; Vinatoru, C.; Maggi, F.; Caprioli, G.; Dinica, R.M. Comparative study of the chemical compositions and antioxidant activities of fresh juices from Romanian Cucurbitaceae varieties. *Molecules* **2020**, *25*, 5468. [CrossRef]
17. Phumat, P.; Khongkhunthian, S.; Wanachantararak, P.; Okonogi, S. Comparative inhibitory effects of 4-allylpyrocatechol isolated from *Piper betle* on *Streptococcus intermedius*, *Streptococcus mutans*, and *Candida albicans*. *Arch. Oral Biol.* **2020**, *113*, 104690. [CrossRef]
18. Phosri, S.; Kiattisin, K.; Intharuksa, A.; Janon, R.; Na Nongkhai, T.; Theansungnoen, T. Anti-aging, anti-acne, and cytotoxic activities of *Houttuynia cordata* extracts and phytochemicals analysis by LC-MS/MS. *Cosmetics* **2022**, *9*, 136. [CrossRef]
19. Ainsworth, E.A.; Gillespie, K.M. Estimation of total phenolic content and other oxidation substrates in plant tissues using Folin–Ciocalteu reagent. *Nat. Protoc.* **2007**, *2*, 875–877. [CrossRef]
20. Brem, B.; Seger, C.; Pacher, T.; Hartl, M.; Hadacek, F.; Hofer, O.; Vajrodaya, S.; Greger, H. Antioxidant dehydrotocopherols as a new chemical character of *Stemona* species. *Phytochemistry* **2004**, *65*, 2719–2729. [CrossRef]
21. Poomanee, W.; Wattananapakasem, I.; Panjan, W.; Kiattisin, K. Optimizing anthocyanins extraction and the effect of cold plasma treatment on the anti-aging potential of purple glutinous rice (*Oryza sativa* L.). *Cereal. Chem.* **2021**, *98*, 571–582. [CrossRef]
22. Payne, A.C.; Mazzer, A.; Clarkson, G.J.J.; Taylor, G. Antioxidant assays—Consistent findings from FRAP and ORAC reveal a negative impact of organic cultivation on antioxidant potential in Spinach but not watercress or rocket leaves. *Food Sci. Nutr.* **2013**, *1*, 439–444. [CrossRef] [PubMed]
23. Kiattisin, K.; Nitthikan, N.; Poomanee, W.; Leelapornpisid, P.; Viernstein, H.; Mueller, M. Anti-Inflammatory, Anti-inflammatory, antioxidant activities and safety of *Coffea arabica* leaf extract for alternative cosmetic ingredient. *Chiang Mai J. Sci.* **2019**, *46*, 284–294.
24. Thring, T.S.A.; Hili, P.; Naughton, D.P. Anti-collagenase, anti-elastase and anti-oxidant activities of extracts from 21 plants. *BMC Complement. Altern. Med.* **2009**, *9*, 27. [CrossRef] [PubMed]
25. Widowati, W.; Rani, A.; Hamzah, R.; Arumwardana, S.; Afifah, E.; Kusuma, H.; Rihibiha, D.; Nufus, H.; Amalia, A. Anti-collagenase, anti-elastase and anti-oxidant activities of extracts from 21 plants. *Nat. Prod. Sci.* **2017**, *23*, 192–200. [CrossRef]
26. Frisch, A. *Gaussian 09W Reference*; Gaussian, Inc.: Wallingford, CT, USA, 2009; 25p.
27. Hoher, W.; Ditchfield, R.; Stewart, R.F.; Pople, J.A. Self-consistent molecular orbital methods. Iv. use of Gaussian expansions of slater-type orbitals. Extension to second-row molecules. *J. Chem. Phys.* **1970**, *52*, 2769–2773.
28. Lovejoy, B.; Cleasby, A.; Hassell, A.M.; Longley, K.; Luther, M.A.; Weigl, D.; McGeehan, G.; McElroy, A.B.; Drewry, D.; Lambert, M.H.; et al. Structure of the catalytic domain of fibroblast collagenase complexed with an inhibitor. *Science* **1994**, *263*, 375–377. [CrossRef]
29. Chao, K.L.; Muthukumar, L.; Herzberg, O. Structure of Human Hyaluronidase-1, a hyaluronan hydrolyzing enzyme involved in tumor growth and angiogenesis. *Biochemistry* **2007**, *46*, 6911–6920. [CrossRef]
30. Morris, G.M.; Huey, R.; Lindstrom, W.; Sanner, M.F.; Belew, R.K.; Goodsell, D.S.; Olson, A.J. AutoDock4 and AutoDockTools4: Automated docking with selective receptor flexibility. *J. Comput. Chem.* **2009**, *30*, 2785–2791. [CrossRef]
31. Trott, O.; Olson, A.J. AutoDock Vina: Improving the speed and accuracy of docking with a new scoring function, efficient optimization, and multithreading. *J. Comput. Chem.* **2010**, *31*, 455–461. [CrossRef]
32. Wolber, G.; Langer, T. LigandScout: 3-D Pharmacophores derived from protein-bound ligands and their use as virtual screening filters. *J. Chem. Inf. Model.* **2005**, *45*, 160–169. [CrossRef] [PubMed]
33. BIOVIA Dassault Systèmes. *Discovery Studio Visualizer 2021*; Dassault Systèmes: San Diego, CA, USA, 2021.

34. Pettersen, E.F.; Goddard, T.D.; Huang, C.C.; Meng, E.C.; Couch, G.S.; Croll, T.I.; Morris, J.H.; Ferrin, T.E. UCSF ChimeraX: Structure visualization for researchers, educators, and developers. *Protein Sci.* **2021**, *30*, 70–82. [CrossRef] [PubMed]
35. Chaiyana, W.; Jiamphun, S.; Bezuidenhout, S.; Krueathanasing, N.; Thammasorn, P.; Jittasai, P.; Tanakitvanicharoen, S.; Tima, S.; Anuchapreeda, S. Enhanced cosmeceutical potentials of the oil from *Gryllus bimaculatus* de Geer by Nanoemulsions. *Int. J. Nanomed.* **2023**, *18*, 2955–2972. [CrossRef]
36. Fernandes, A.; Rodrigues, P.M.; Pintado, M.; Tavaria, F.K. A Systematic review of natural products for skin applications: Targeting inflammation, wound healing, and photo-aging. *Phytomedicine* **2023**, *115*, 154824. [CrossRef] [PubMed]
37. Coulombier, N.; Jauffrais, T.; Lebouvier, N. Antioxidant compounds from microalgae: A review. *Mar. Drugs* **2021**, *19*, 549. [CrossRef]
38. Lourenço, S.C.; Moldão-Martins, M.; Alves, V.D. Antioxidants of natural plant origins: From sources to food industry applications. *Molecules* **2019**, *24*, 4132. [CrossRef]
39. Mathew, S.; Abraham, T.E.; Zakaria, Z.A. Reactivity of phenolic compounds towards free radicals under In Vitro conditions. *J. Food Sci. Technol.* **2015**, *52*, 5790–5798. [CrossRef]
40. Panche, A.N.; Diwan, A.D.; Chandra, S.R. Flavonoids: An overview. *J. Nutr. Sci.* **2016**, *5*, e47. [CrossRef]
41. Tabrizi, L.; Dao, D.Q.; Vu, T.A. Experimental and theoretical evaluation on the antioxidant activity of a copper(Ii) complex based on lidocaine and ibuprofen amide-phenanthroline agents. *RSC Adv.* **2019**, *9*, 3320–3335. [CrossRef]
42. Antolovich, M.; Prenzler, P.D.; Patsalides, E.; McDonald, S.; Robards, K. Methods for Testing antioxidant activity. *Analyst* **2002**, *127*, 183–198. [CrossRef]
43. Abdullah, N.; Wan Saidatul, S.W.K.; Samicho, Z.; Zulkifli, K.S.; Aziman, N. Study on antioxidant capacity and phenolic content of various parts of wax gourd (*Benincasa hispida*). *World Appl. Sci. J.* **2012**, *19*, 1051–1056.
44. Hubner, A.; Sobreira, F.; Vetore Neto, A.; Pinto, C.A.S.d.O.; Dario, M.F.; Díaz, I.E.C.; Lourenço, F.R.; Rosado, C.; Baby, A.R.; Bacchi, E.M. The synergistic behavior of antioxidant phenolic compounds obtained from Winemaking Waste's valorization, increased the efficacy of a sunscreen system. *Antioxidants* **2019**, *8*, 530. [CrossRef] [PubMed]
45. Jung, H. Hyaluronidase: An overview of its properties, applications, and side effects. *Arch. Plast. Surg.* **2020**, *47*, 297–300. [CrossRef] [PubMed]
46. Ganceviciene, R.; Liakou, A.I.; Theodoridis, A.; Makrantonaki, E.; Zouboulis, C.C. Skin anti-aging strategies. *Dermatoendocrinology* **2012**, *4*, 308–319. [CrossRef]
47. Doshi, G.; Nalawade, V.; Mukadam, A.; Chaskar, P.; Zine, S.; Somani, R.; Une, H. Elucidation of flavonoids from *Carissa congesta*, *Polyalthia longifolia*, and *Benincasa hispida* plant extracts by hyphenated technique of liquid chromatography-mass spectroscopy. *Pharmacogn. Res.* **2016**, *8*, 281–286. [CrossRef]
48. Bo, Y.S.; Hyun, P.K. Inhibition of collagenase by naturally-occurring flavonoids. *Arch. Pharm. Res.* **2005**, *28*, 1152–1155.
49. Sim, G.-S.; Lee, B.-C.; Cho, H.S.; Lee, J.W.; Kim, J.-H.; Lee, D.-H.; Kim, J.-H.; Pyo, H.-B.; Moon, D.C.; Oh, K.-W.; et al. Structure activity relationship of antioxidative property of flavonoids and inhibitory effect on matrix metalloproteinase activity in UVA-irradiated human dermal fibroblast. *Arch. Pharm. Res.* **2007**, *30*, 290–298. [CrossRef]
50. Shakya, A.; Gogoi, N.; Chaudhary, S.K.; Bhat, H.R.; Ghosh, S.K. Development and validation of a high-performance thin-layer chromatography method for the quantification of rutin in the fruit pulp of *Benincasa hispida* (Thunb.) Cogniaux. *JPC—J. Planar. Chromat.* **2019**, *32*, 371–377. [CrossRef]
51. Doshi, G.M.; Une, H.D. Quantification of quercetin and rutin from *Benincasa hispida* seeds and *Carissa congesta* roots by high-performance thin layer chromatography and high-performance liquid chromatography. *Pharmacogn. Res.* **2016**, *8*, 37–42. [CrossRef]
52. Sung, E.; Kim, S.; Shin, W. Binary image representation of a ligand binding site: Its application to efficient sampling of a conformational ensemble. *BMC Bioinform.* **2010**, *11*, 256. [CrossRef]
53. Stern, R.; Jedrzejas, M.J. Hyaluronidases: Their genomics, structures, and mechanisms of action. *Chem. Rev.* **2006**, *106*, 818–839. [CrossRef] [PubMed]
54. Arming, S.; Strobl, B.; Wechselberger, C.; Kreil, G. In Vitro Mutagenesis of PH-20 hyaluronidase from human sperm. *Eur. J. Biochem.* **1997**, *247*, 810–814. [CrossRef]
55. Zhang, L.; Bharadwaj, A.G.; Casper, A.; Barkley, J.; Barycki, J.J.; Simpson, M.A. Hyaluronidase activity of human hyal1 requires active site acidic and tyrosine residues. *J. Biol. Chem.* **2009**, *284*, 9433–9442. [CrossRef]
56. Aslantürk, Ö.S. In Vitro cytotoxicity and cell viability assays: Principles, advantages, and disadvantages. In *Genotoxicity*; Larramendy, M.L., Soloneski, S., Eds.; IntechOpen: Rijeka, Croatia, 2017.
57. López-García, J.; Lehocký, M.; Humpolíček, P.; Sáha, P. HaCaT Keratinocytes response on antimicrobial atelocollagen substrates: Extent of cytotoxicity, cell vability and proliferation. *J. Funct. Biomater.* **2014**, *5*, 43–57. [CrossRef] [PubMed]
58. Choi, S.I.; Han, X.; Men, X.; Lee, S.J.; Oh, G.; Choi, Y.E.; Yang, J.M.; Cho, J.H.; Lee, O.H. *Benincasa hispida* extract prevents ovariectomy-induced osteoporosis in female ICR mice. *Appl. Sci.* **2023**, *13*, 832. [CrossRef]
59. Budai, P.; Lehel, J.; Tavaszi, J.; Kormos, É. HET-CAM test for determining the possible eye irritancy of pesticides. *Acta. Vet. Hung.* **2010**, *58*, 369–377. [CrossRef]
60. Rivero, M.N.; Lenze, M.; Izaguirre, M.; Pérez Damonte, S.H.; Aguilar, A.; Wikinski, S.; Gutiérrez, M.L. Comparison between HET-CAM protocols and a product use clinical study for eye irritation evaluation of personal care products including cosmetics according to their surfactant composition. *Food. Chem. Toxicol.* **2021**, *153*, 112229. [CrossRef]

61. Wilson, T.D.; Steck, W.F. A Modified HET–CAM assay approach to the assessment of anti-irritant properties of plant extracts. *Food Chem. Toxicol.* **2000**, *38*, 867–872. [CrossRef]
62. Hong-in, P.; Chaiyana, W. Potential cosmeceutical lamellar liquid crystals containing black longan (*Dimocarpus longan* Lour.) seed extract for MMP-1 and hyaluronidase inhibition. *Sci. Rep.* **2022**, *12*, 7683. [CrossRef]

foods

MDPI

Systematic Review

Flavonoid Extracts from Lemon By-Products as a Functional Ingredient for New Foods: A Systematic Review

Lorena Martínez-Zamora [1,2,*], Marina Cano-Lamadrid [2], Francisco Artés-Hernández [2] and Noelia Castillejo [2,3,*]

1 Department of Food Technology, Nutrition and Food Science, Faculty of Veterinary Sciences, University of Murcia, 30071 Espinardo, Murcia, Spain

2 Postharvest and Refrigeration Group, Department of Agronomical Engineering and Institute of Plant Biotechnology, Universidad Politécnica de Cartagena, 30203 Cartagena, Murcia, Spain; marina.cano@upct.es (M.C.-L.); fr.artes-hdez@upct.es (F.A.-H.)

3 Department of Agricultural Sciences, Food, Natural Resources and Engineering, University of Foggia, Via Napoli 25, I-71122 Foggia, Italy

* Correspondence: lorena.martinez23@um.es (L.M.-Z.); noelia.castillejomontoya@unifg.it (N.C.)

Abstract: This systematic review seeks to highlight, from the published literature about the extraction and application of lemon by-products rich in flavonoids, which works use environmentally friendly technologies and solvents and which ones propose a potentially functional food application, according to the Sustainable Development Goals (SDGs). WoS and SCOPUS were used as scientific databases for searching the documents, which were evaluated through 10 quality questions according to their adherence to our purpose (5 questions evaluating papers devoted to lemon flavonoid extraction and 5 concerning the application of such by-products in new foods). Each question was evaluated as "Yes", "No", or "does Not refer", according to its adherence to our aim. The analysis reported 39 manuscripts related to lemon flavonoid extraction; 89% of them used green technologies and solvents. On the other hand, 18 manuscripts were related to the incorporation of lemon by-products into new foods, of which 41% adhered to our purpose and only 35% evaluated the functionality of such incorporation. Conclusively, although the bibliography is extensive, there are still some gaps for further investigation concerning the extraction and application of lemon by-products to reduce food losses in an environmentally friendly way and the possible development of new functional foods, which must be performed following the SDGs.

Keywords: hesperidin; naringenin; food loss; lemon skin; co-products

Citation: Martínez-Zamora, L.; Cano-Lamadrid, M.; Artés-Hernández, F.; Castillejo, N. Flavonoid Extracts from Lemon By-Products as a Functional Ingredient for New Foods: A Systematic Review. *Foods* **2023**, *12*, 3687. https://doi.org/10.3390/foods12193687

Academic Editors: Mohamed Koubaa and Cornelia Witthöft

Received: 11 August 2023
Revised: 29 September 2023
Accepted: 6 October 2023
Published: 8 October 2023

1. Background

The Food and Agriculture Organization (FAO) reports that approximately one-third of the global production is lost or wasted at some stage of the food chain [1]. FAO's future challenges for 2050 are to reduce food waste by 50%, one of the SDGs. Circular economy has been seen as the principle of a society-driven and 'zero waste' economy, with waste as raw materials.

According to FAOSTAT, the global production of lemon was around 21.4 million tons in 2020 [2]. Although it depends on the variety, the juice yield of these citrus reaches values of 38–41% [3]. Peels, pulp, seeds, and pomace, which constitute approximately 50% of the fresh fruit, are some of the wastes generated by citrus processing and consumption [3]. Similar values are being obtained in our own laboratories (45%) in the Primofiori variety, while they decreased to 28% in varieties with a thicker albedo, such as the Verna variety (unpublished data). This means that, according to the variety, between 55 and 72% of a lemon is directly wasted after squeezing. In global figures, this leaves between 11.8 and 15.4 million tons of food losses per year with high nutritional value that can be recovered as sources of bioactive compounds, essential oils, and fiber.

In this sense, these food losses are currently used by flavor and extraction companies to obtain essential oils and fiber for their application as flavorings and odorizing or emulsifying agents. Linked to the rise in demand for a healthy diet and the pursuit of the SDGs, the extraction and purification of bioactive compounds from food by-products have exponentially increased in the last two decades.

In this perspective, sustainability, well-being, and health are currently popular topics in the food industry. 'Clean label' goods or components appeal to both consumers and food manufacturers [4–6]. This indicates that people are interested in a variety of green-processed foods and ingredients, including nutraceuticals, bioactive chemicals with health-promoting qualities, and non-thermal green solvents. The technical and functional properties of the bioactive compounds derived from fruit and vegetable by-products allow them to be incorporated into other food matrixes to improve their nutritional, functional, and sensory qualities [5,7]. Additionally, the use of bioactive chemicals from fruit and vegetable by-products has been previously categorized as a potential green element for the cosmetic and pharmaceutical sectors, generating several products aimed at niche markets such as athletes [8].

In fact, regarding European regulations, no health claim is yet authorized for 'antioxidants' and 'flavonoids' from lemon. The reason for the negative opinion from the EFSA is the non-compliance with the regulation based on the scientific evidence assessed. The claimed effect of this food has not been substantiated. There are currently no authorized health claims for lemon and its constituents in the European Union.

Consequently, several reviews have been published so far related to this topic, like the one recently published by Magalhães et al. [9], who widely exposed the major compounds found in lemons, their main extraction technologies, and their applications in food preservation. In this sense, as a review of the extensive bibliography on the topic has already been performed, the aim and novelty of the present systematic review is to account for which of the literature published on the topic is truly adapted to these SDGs, uses environmentally friendly technologies and solvents, and develops a potentially functional food application of flavonoids, the main bioactive compounds extracted from lemon by-products.

2. Lemon By-Products and Their Functional Quality

According to the structure of the lemon fruit, it is divided into the albedo, which is the main source of fiber (pectin and cellulose), the flavedo, which is rich in essential oils and pigments, and the pulp, where the juice, rich in water and nutritional and functional compounds (citric acid, ascorbic acid, minerals, and flavonoids), is obtained.

The albedo is the bitter white layer that surrounds the juicy pulp of the fruit. It contains pectin, fiber, and other nutrients [10]. Because of its bitterness, in the lemon processing industry, the albedo is usually removed from the fruit as a non-edible part. For lemon essential oil production, the processors use a method called cold-pressing, in which the lemon peels are soaked in water and then pressed to extract the oil. This method produces a high-quality oil with a fresh, citrusy fragrance, which is used in perfumes, cosmetics, and food flavorings.

In fact, this bitterness is caused mainly by the presence of phenolic compounds, which include phenolic acids and flavonoids. Hydroxycinnamic (chlorogenic, caffeic, ferulic, sinapic, and p-coumaric acids) and hydroxybenzoic acids (protocatechuic, p-hydroxybenzoic, vanillic, and gallic acids) have been identified in lemon peels [11]. Nevertheless, flavonoids such as hesperidin (59% of the flavanone content) and eriocitrin (35.6% of the flavanone content) are the most concentrated in the albedo. Furthermore, other minor flavanones such as didymin, naringin, neoeriocitrin, neohesperidin, narirutin, eriodictyol, and naringenin have been identified. Also, the favones diosmetin, diosmin, luteolin, vicenin, chrysoeriol, apigenin, and sinensetin and the flavonols quercetin, limocitrin, limocitrol, rutin, and kaempferol are also present in lemon by-products [9,10].

For this reason, and due to their bioactivity, lemon flavonoids have a good potential to be extracted and applied in new functional foods. Flavonoids are polyphenolic compounds with a broad range of biological activities, including antioxidant, anti-inflammatory, and anticancer effects. Hence, its consumption has been associated with the preventive effects of chronic diseases by avoiding inflammation and oxidative stress. The combination of these bioactive compounds and dietary fibers in lemon fruits makes them a valuable addition to a healthy diet and lifestyle.

In this respect, the dietary fiber contained in the lemon albedo includes gums, pectins, glucans, and some polysaccharides as insoluble fibers, while cellulose, hemicellulose, and lignin are soluble fibers. Particularly, pectin is the major component of such fiber, and although it cannot be digested by the human intestine, our microbiota is able to assimilate and convert it into beneficial metabolites [12].

In addition, the flavedo is rich in volatile compounds and essential oils. For instance, d-limonene is the main essential oil of the lemon flavedo, followed by β-pinene, γ-terpinene, α-pinene, sabinene, myrcene, and α-thujene, among others [10]. However, in the present review, we will focus on the bioactivity of the main compounds cited above.

3. Methods

WoS and SCOPUS were used as scientific databases for searching documents. The terms "lemon", "*Citrus limon*", "co-products", and "by-products" were used as keywords. Other search words used were "extraction", "flavonoids", "ultrasound-assisted extraction", "microwave-assisted extraction", and "enzymatic-assisted extraction" for manuscripts related to the extraction of lemon by-products. The terms "application", "food", "juice", and "beverage" were used for manuscripts related to the application of lemon by-products in new food matrixes. First, a description of the total literature found in reference to "lemon", "*Citrus limon*", "co-products", and "by-products" published in the last twenty years was carried out, including the number of reviews in reference to this topic (Figure 1).

From the bibliography found and described in the raw analysis, the inclusion criterion for our systematic review was "original studies included in JCR-SCI journals'". The exclusion criterion was "studies non-included in JCR-SCI journals, books, and reviews". The title and abstracts of the documents found were analyzed and classified depending on their significant interest using Microsoft Excel for the data curation. The potential scientific papers were subjected to a comprehensive analysis, in which all the papers were checked for the inclusion quality criteria. The 10 following questions were used as quality criteria (5 of the questions (from 1 to 5) were related to the evaluation of the quality of the manuscript related to the extraction of lemon by-products, and the other 5 (from 6 to 10) were related to the evaluation of the quality of the manuscript related to the application of such lemon by-product extracts): (Q1) Does the article include an experimental design (response surface methodology)?; (Q2) Does it use green solvents for the extraction?; (Q3) Does it use green technologies for the extraction?; (Q4) Does it extract flavonoids?; (Q5) Do the authors validate/characterize their extract?; (Q6) Do the authors incorporate these flavonoids into food?; (Q7) Do the authors encapsulate the extract?; (Q8) Do the authors conduct a shelf-life study with the new food?; (Q9) Do the authors compare it to a control sample?; (Q10) Do the authors evaluate the functionality of the enriched food?

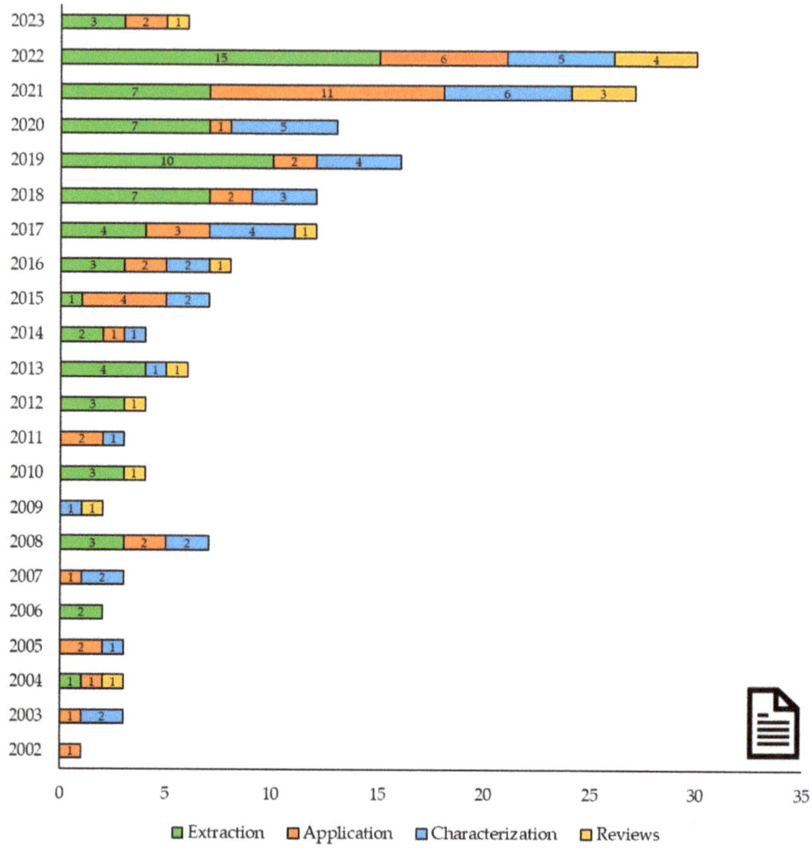

Figure 1. Number of research papers and reviews related to lemon by-products published in the last twenty years, according to WoS and SCOPUS (n = 176).

Each query was evaluated as "Yes", "No", or "does Not refer". The frequency of "Yes" responses for each one was used to determine the quality and reproducibility of this study. The works were arranged into three categories: excellent (>70% "Yes" responses), good (50–69% "Yes" responses), and bad (<50% "Yes" responses), according to their adherence to our purpose. The PRISMA flow diagram followed, and the results obtained in this systematic review are shown in Figure 2.

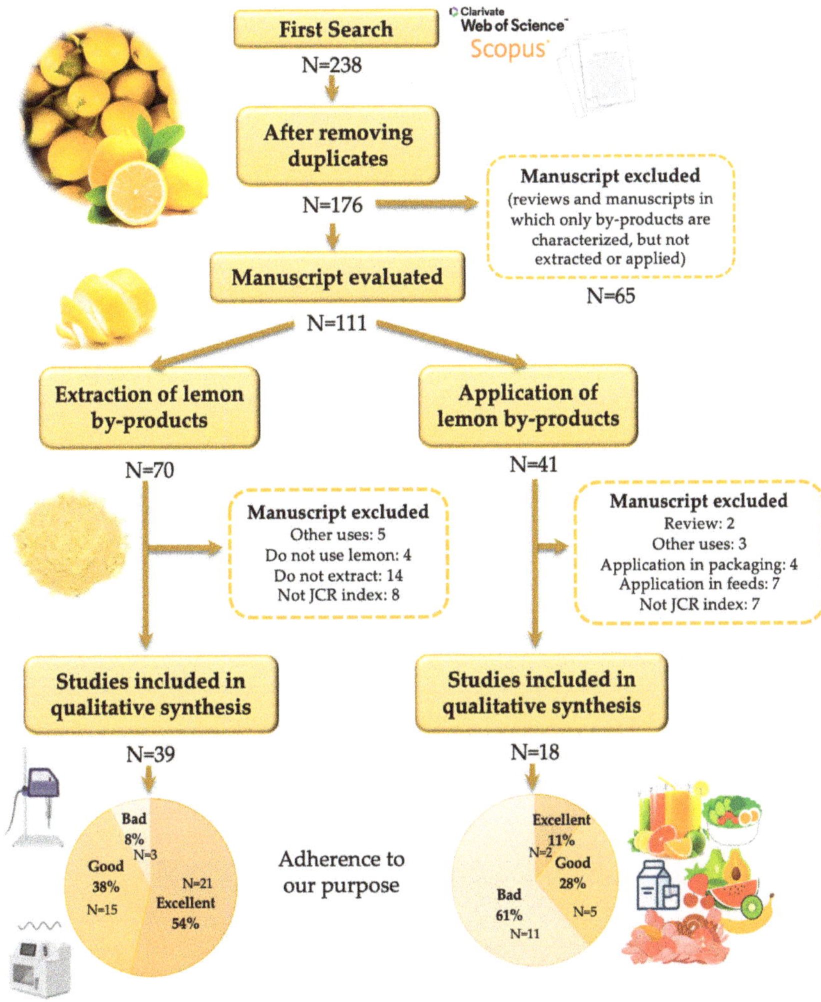

Figure 2. PRISMA flow diagram followed, studies selected, and classification in reference to the proposed quality criteria.

4. Results and Discussion

As shown in Figure 2, of all the studies found in the literature, 39 were included in the qualitative analysis for extraction, of which 92% suited well to our main goal, while the remaining 8% did not. Moreover, 18 scientific studies were included in the analysis related to the application of lemon by-products, of which 39% were in good compliance with our purpose.

4.1. Flavonoid Extraction

Table 1 shows the results obtained from the qualitative analysis carried out in the 39 scientific studies found.

Table 1. Non-thermal extractions of citrus by-products.

By-Product Characteristics	Q1	Q2	Q3	Q4	Q5	Optimum Conditions of Extraction	Adherence to Our Purpose	Ref.
Lemons (*Citrus limon* cv. *Meyer*), lemon (*C. limon* cv. *Yenben*) peels (epicarp and mesocarp), frozen and milled (Ø: <1 mm)	No	Yes (Water)	Yes (Enzymatic)	Yes	Yes	Celluzyme MX 1.5% at 50 °C	Excellent	[13]
Lemons (*Citrus limon* cv. *Meyer*), lemon (*C. limon* cv. *Yenben*) peels (epicarp and mesocarp), frozen and milled (Ø: <1 mm)	No	Yes (Water and ethanol)	Yes (Temp.)	Yes	Yes	85% ethanol and 80 °C	Excellent	[14]
Lemon (*Citrus limon* L.) peels and pulps, freeze-dried and milled (Ø: 0.25 mm)	Yes (Central composite design)	Yes (Ethanol and acidified date juice)	Yes (Stirring)	No	Yes	84.34 °C for 3 h 34 min, and pH 2.8	Excellent	[15]
Lemon (*Citrus limon* L.) peels and pulps, freeze-dried and milled (Ø: 0.25 mm)	No	Yes (Water with acidified date juice)	Yes (Stirring)	No	Yes	pH 3.5, 45% sucrose, and 0.1% calcium to improve the gelling properties of the extracted pectin	Good	[16]
Lemon (*Citrus limon* L.) peels and pulps, freeze-dried and milled (Ø: 0.25 mm)	No	Yes (Water with acidified date juice)	Yes (Stirring)	No	Yes	84.34 °C for 3 h 34 min, and pH 2.8	Good	[17]
Lemon peels (Ø: <3 mm)	No	Yes (Water)	Yes (Enzymatic)	No	Yes	Water at 160 °C in autohydrolysis with two membrane filtration stages (diafiltration and concentration)	Good	[18]
Lemon peels (Ø: <7 mm)	No	Yes (Water)	Yes (Enzymatic)	No	Yes	2:1 ratio (lemon peels:water) at 37 °C with 1.95 mg β-glucosidase, 2.21 mg pectinase, and 1.82 mg celullase using steam explosion	Good	[19]
Lemon peels	No	Yes (Water)	Yes (Enzymatic)	No	Yes	From 7.5 to 24 h enzymatic hydrolysis and membrane processing (filtration and concentration)	Good	[20]

Table 1. *Cont.*

By-Product Characteristics	Q1	Q2	Q3	Q4	Q5	Optimum Conditions of Extraction	Adherence to Our Purpose	Ref.
Lime peel (*Citrus limonium* cv. *Colima*) squares (5 mm × 5 mm)	No	Yes (Citric acid–sodium citrate)	Yes (Enzymatic)	No	No	1:2.5 ratio (peels:solvent) + 0.1% cellulase enzyme at 50 °C for 3 h	Bad	[21]
Lemon (*Citrus limon* L. var. Kütdiken) seeds	No	Yes –	Yes (Cold-pressed)	No	Yes	High-quality oils from seeds with a moisture content of 12% and cold-pressed (screw rotation speed of 30 rpm, outlet matrix of 10 mm, and oil outlet temp. of 40 °C)	Good	[22]
Yuzu (*Citrus junos*) peels and seeds, dried and milled (Ø: <0.71 mm)	Yes (Box–Behnken)	Yes (CO_2)	Yes (Supercritical fluids)	No	Yes	200.54 bar, 46.28 °C, and 34.98 g/min flow rate	Excellent	[23]
Cold-pressed EOs derived from lemon industrial processing	No	Yes (Water)	Yes (Cold-pressed)	No	Yes	Hydrodistillation for 3 h	Good	[24]
Lemon peels, membranes, and seeds, freeze-dried and milled (Ø: <1.4 mm)	No	Yes (Water)	Yes (MW)	Yes	Yes	360 W for 5 min (72 kJ/g)	Excellent	[25]
Lemon (*Citrus limon* L.) peels, membranes, and seeds, freeze-dried and milled (Ø: 1.4, 2, 2.8 mm)	Yes (Box–Behnken)	Yes (Water, hot water, and ethanol)	Yes (US, temp., and stirring)	Yes	Yes	US: 35–45 min, 48–50 °C, 150–250 W Temp.: 95 °C for 15 min	Excellent	[26]
Cold-pressed meals of lemon (*Citrus limon* L.) seeds, dried	No	Yes –	Yes (Cold-pressed)	Yes	Yes	150 °C for 30 min and cold-pressed (30 rpm, 10 mm die, and 40 °C)	Excellent	[27]
Domestic house solid waste lemon peel, milled and dried	No	No	Yes (HMIM)	Yes	Yes	80 °C for 3 h in a water bath, cooled, and filtered	Good	[28]
Lemon (*Citrus limon* L.) peels, dried and milled (Ø: 0.6–1.5 mm)	No	Yes (Water)	Yes (Subcritical fluids)	No	Yes	10 g at 1500 psi, 200 °C, 15 min	Good	[29]

Table 1. *Cont.*

By-Product Characteristics	Q1	Q2	Q3	Q4	Q5	Optimum Conditions of Extraction	Adherence to Our Purpose	Ref.
Lemon (*Citrus limon* L.) peels, chopped (Ø: 10–30 mm)	No	Yes –	Yes (Pulsed electric fields)	Yes	Yes	5 bars, 3.5 kV/cm, 30 pulses of 30 µs	Excellent	[30]
Lemon seeds, dried and milled (Ø: 0.25–0.425 mm)	No	Yes (CO_2)	Yes (Supercritical fluids)	No	Yes	First separator: 300 bar and 40 °C Second separator: 20 bar and 15 °C	Good	[31]
Lemon peels, dried and milled (Ø: 0.787 mm)	Yes (Box–Behnken)	Yes (Ethanol–water)	Yes (Homogenizer)	Yes	Yes	0.1 g mixed in 33.62% ethanol for 1.282 min at 5007 rpm	Excellent	[32]
Sweet lemon peels, dried and milled (Ø: 0.4 mm)	Yes (Box–Behnken)	Yes (Ethanol–water)	Yes (MW)	No	Yes	700 W, 3 min, and pH 1.5	Excellent	[33]
Lemon (*Citrus limon* L.) peels, dried and milled (Ø: <0.15 mm)	No	Yes (Water or ethanol)	Yes (Temp.)	Yes	Yes	10 g in 200 mL ethanol at 60 °C for 2 h	Excellent	[34]
Lemon (*Citrus limon* L.) peels, blanched and frozen	Yes (Box– Behnken)	Yes (Citric acid)	Yes (Ohmic heating)	No	Yes	8.7:1 (solvent:sample) ratio for 58.4 min and voltage gradient of 14.2 V/cm	Excellent	[35]
Persian lime (*Citrus latifolia*) seeds, dried	No	Yes (Phosphate buffer)	Yes (Enzymatic)	No	Yes	Alcalase, Protamex, and Neutrase mixed enzymes (1:1:1), pH 8.0, 50 °C	Good	[36]
Lemon peel pomace (Ø: 0.177 mm)	No	Yes (Citric acid and phosphate buffer)	Yes (Enzymatic)	No	Yes	0.45% xylanase, pH 5.0, rate (solid:liquid) 1:20 at 60 °C for 1.5 h at 30 rpm	Good	[37]
Frozen lemon peels and pulps	No	Yes (Ethanol)	Yes (HPH)	No	Yes	20 MPa to alcohol-insoluble residue	Good	[38]
Lemon (*Citrus limon* L. *Osbeck*) peels	No	Yes (Water and ethanol–water)	Yes (US and MW)	Yes	Yes	Ethanol:water (50:50), US at 70 °C for 30 min	Excellent	[39]
Lemon peels, dried and milled (Ø: ~0.1 mm)	Yes (Central composite design)	Yes (Deep eutectic solvents)	Yes (Stirring)	Yes	Yes	55% (sample/solvent), 13 mL/g and 36 min in deep eutectic solvents	Excellent	[40]

Table 1. *Cont.*

By-Product Characteristics	Q1	Q2	Q3	Q4	Q5	Optimum Conditions of Extraction	Adherence to Our Purpose	Ref.
Lemon peels	Yes (Box–Behnken)	Yes (Lactic acid-based automatic solvent)	Yes	Yes	Yes	1.5 h, 46% water, and 5 g of peel	Excellent	[41]
Lemon peels, dried and crushed (Ø: >1 mm)	No	Yes (Ethanol and water)	Yes (PLE-SPE)	Yes	Yes	Sepra™ C18-E columns, in water–ethanol, pH 6–7, temp. 50–70 °C	Excellent	[42]
Lemon peels, dried	Yes (Uncertainty analysis)	Yes (Water)	Yes (Steam distillation)	No	Yes	Lemon peel oils report better results compared to normal diesel in all aspects, except for NOx emissions. A content of 10% water in lemon peel oils improves the overall performance	Excellent	[43]
Cold-pressed essential oil (CPEO) from lemon and tangerine	No	No (Hexane and dichloromethane)	Yes (Distillation)	No	Yes	Higher carotenoid recovery through azeotropic condensation, adding isopropanol 3.5:1 (ratio) to the CPEO, evaporating under negative pressure, and heating for 2.5 h at 28–32 °C.	Bad	[44]
Lemon (*Citrus limon* cv. *Eureka*) peels, freeze-dried and milled (Ø: ~0.71 mm)	No	Yes (Ethanol–water)	Yes (US)	Yes	Yes	1:40 (*w:v*) 75% ethanol, US 550 W for 5 min	Excellent	[45]
Lemon peels, dried and milled (Ø: <0.4 mm)	Yes (Box–Behnken)	No (Hot acidic (HCl) water)	Yes (Temp.)	No	Yes	pH 1.5 at 90 °C for 120 min	Good	[46]
Citrus peel pomace, dried and milled (Ø: 0.177 mm)	No	No (Octenyl succinic anhydride)	Yes (Esterification)	No	Yes	Octenyl succinic anhydride:citrus fiber (1:5, *w:w*), pH 8.5 at 20 °C for 1.5 h	Bad	[47]
Lemon peels, dried and milled (Ø: 0.5–3.55 mm)	No	Yes (Deionized water)	Yes (Pressurized)	Yes	Yes	10.34 MPa, rinse volume (30%), purge for 90 s with N_2 gas, 160 °C for 5 or 30 min (depending on compound)	Excellent	[48]

Table 1. *Cont.*

By-Product Characteristics	Q1	Q2	Q3	Q4	Q5	Optimum Conditions of Extraction	Adherence to Our Purpose	Ref.
Lemon peels, freeze-dried and milled (Ø: 0.45 mm)	No	Yes (Sodium acetate buffer and ethanol–water)	Yes (Enzymatic and US)	Yes	Yes	Enzyme (5 U cellulase and pectinase in sodium acetate buffer (20 mM, pH 5.0) at 40 °C for 60 min) and US (400 W, 24 kHz, power level of 50%, 23 °C, 15 min in ethanol:water 50:50) treatments	Excellent	[49]
Lemon processing residues, dried and milled (Ø: <1 mm)	No	Yes (Water, ethanol, and ethanol–water)	Yes (Stirring)	No	Yes	Water as solvent at ratio 1:50, stirring for 30 min	Good	[50]
Lemon peel, dried and milled (Ø: <0.734 mm)	No	Yes (Water and ethanol)	Yes (Stirring)	Yes	Yes	50 g + 500 mL 98% ethanol stirring for 24 h at 25 °C, filtered and concentrated using the vacuum evaporator at 40 °C	Excellent	[51]

Q1: Does the article include an experimental design (response surface methodology)? Q2: Does it use green solvents for the extraction? Q3: Does it use green technologies for the extraction? Q4: Does it extract flavonoids? Q5: Do the authors validate their extract? Response surface methodology, solvent, and extractive technology used are specified in parentheses. Ø: diameter; Temp.: temperature; EOs: essential oils; CO_2: carbon dioxide; MW: microwave; US: ultrasound; HMIM: hybrid molecularly imprinted membrane; HPH: high-pressure homogenization; PLE-SPE: pressurized liquid extraction coupled in-line with solid-phase extraction.

The studies reviewed in this list explore various methods for the extraction and utilization of valuable components from citrus peels and by-products.

Due to the SDGs, petrochemical solvents have been so far replaced by green solvents in many recent studies. Green solvents must be environmentally sustainable and are characterized by high-quality products with fewer by-products produced during processing and low toxicity. The main green solvents are ionic liquids, deep eutectic solvents (DESs), polyethylene glycol (PEGs), ethyl lactate, water, supercritical fluids, alcohols (ethanol), esters (ethyl lactate and ethyl acetate), and terpenes [52–54]. Other solvents, such as xylenes, methanol, tetrahydrofuran, DMSO, chlorobenzene, thiophene, and diphenyl ether, are still widely used and sometimes considered to be green solvents, although little evidence of this has been found [55].

As previously described by Artés-Hernández et al. [4–6], some examples of green technologies are ultrasound-, microwave-, and enzymatic-assisted extraction, supercritical or subcritical fluids, and pressurized liquids, which are the most widely used in the studies reviewed. Another green technology is cold pressing, a fast, inexpensive, solvent-free, and environmentally friendly process, but its yield is often lower than that of solvent extraction [22].

Phenolics and pectins are commonly targeted for extraction, with enzyme-assisted extraction showing promise as an effective method. Lemon peels have been shown to be useful to produce pectin-derived oligosaccharides and polyphenol extracts, as well as bioethanol [19]. Other studies have explored the use of lemon peel waste for the removal of heavy metals from wastewater and the production of humic acid [29]. Novel approaches include the use of ultrasound- and microwave-assisted extractions, as well as the integration of pressurized liquid and in-line solid-phase extractions for the simultaneous extraction and concentration of phenolic compounds [42]. Overall, the studies suggest that utilizing citrus peels and by-products can be an effective way to reduce waste and extract valuable components for various applications that are going to be summarized below.

Li et al. [13,14] investigated the extraction of phenolics from citrus peels through two different methods: solvent extraction [14] and enzyme-assisted extraction [13]. Both methods resulted in high yields of phenolic compounds. However, the enzyme-assisted method was found to be more efficient and faster compared to the solvent extraction method [13]. Masmoudi et al. [15] studied the effect of different extraction methods on the antioxidant properties of citrus peels. They found that ethanol and water were the best solvents for extracting phenolics from citrus peels. The study also showed that microwave-assisted extraction had a higher extraction efficiency compared to traditional methods.

These studies highlight the potential value that can be generated from citrus peels and by-products. Efficient extraction methods can be used to obtain valuable bioactive compounds with potential health benefits and industrial applications using green solvents, mainly ethanol and water, and novel green technologies such as enzymatic-, ultrasound-, pressurized-, pulsed electric field-, or microwave-assisted extractions. Moreover, utilizing citrus peels can significantly decrease environmental pollution caused by the disposal of waste citrus materials and is an important source of flavonoids to be applied to new foods, as shown below.

4.2. Application

The results obtained from the qualitative analysis carried out for the 18 scientific studies found are shown in Table 2. The works listed in this collection focus on the use of citrus by-products, particularly lemon, in various food applications. These applications include the use of citrus fibers and albedo in meat products [56], the preparation and characterization of osmodehydrated fruits [57], the incorporation of citrus fibers in fermented milk containing probiotic bacteria [58], as well as cake or bakery products [51,59,60]. Other studies examine the potential of citrus by-products as fat replacers in chicken patties [61], as antioxidants in food flavorings, and as a means of improving the bio-accessibility of polyphenols in salad dressings [62]. Additionally, several works evaluate the efficacy of antioxidant extracts from lemon by-products in preserving the quality attributes of minimally processed radish [63].

In addition, lemon juice has been used as a natural acidifying agent, e.g., Banerjee et al. [64] used lemon juice instead of HCl in the valorization of mango peels to lower the pH to 2.5 and recover pectin. Furthermore, lemon peels have been studied as removers of heavy metal ions (Fe^{2+}, Zn^{2+}, and Mn^{2+}) in wastewater [65]. Lemon peels in a 0.1 M HCl solution were able to reach a value of 55.19% for Mn^{2+} desorption and 37.24% for Zn^{2+}, while for Fe^{2+}, the highest value of 25.82% was achieved in a 0.1 M HNO_3 solution.

Overall, these studies demonstrate the potential of citrus by-products as functional ingredients in various processed food products, such as fruits, vegetables, dairy, bakery, and meat products. For that reason, this topic must be further investigated with the goal of incorporating these new ingredients as food preservatives to recover part of the produced food discards with potential health benefits, which must be validated by international agencies (such as the EFSA in Europe or the FDA in the USA).

Table 2. Application of citrus by-products and their potential benefits.

Enriched Product	Q6 Q7 Q8 Q9 Q10	Main Findings	Adherence to Our Purpose	Ref.
Fresh British-style pork sausages	No No No Yes No	7% of citrus fiber extract reduced the shrinkage and the cooking loss, increased lightness (L*), and maintained their antioxidant effect as well as the overall acceptance of cooked sausages	Bad	[56]
Swedish-style beef meatballs	Yes No Yes Yes No	Citrus extracts reduced the rancidity of meat products by 50%	Good	[66]
Osmodehydrated fruits	No No Yes Yes No	Microbially stable for 3 months at 4 °C	Bad	[57]
Meat emulsion systems	No No No Yes No	Lemon albedo addition did not change the flow properties and improved the texture, acting as a source of fiber	Bad	[67]
Fermented milk	No No Yes Yes No	Enhanced survival of the tested probiotic bacteria and bacterial growth, maintaining the acceptability	Bad	[58]
Dough and Mantou (steamed bread)	Yes No No Yes No	3 or 6 g per 100 g of flour produced acceptable Mantou with higher antioxidant capacity and total phenolic content	Bad	[68]
Frankfurters	Yes No No Yes No	Incorporation of shaddock albedo increased hardness and decreased chewiness; hence, it can be a potential emulsifier	Bad	[69]
Oat–fruit juice mixed beverage	Yes No Yes Yes No	Antimicrobial effect against *Salmonella typhimurium* and *E. coli* of all the extracts at 5 °C	Good	[70]
Food flavoring	No No Yes No No	Extracts thermally stable and safe	Bad	[71]
Lemon oil	No No Yes Yes No	Nanoemulsions of lemon and fish oil by-products (8% lemon oil, 2% fish by-product oil, 10% surfactant, and 27.7% cosurfactants) inhibited 7 Gram-positive and 7 Gram-negative bacterial strains	Bad	[72]
Biscuits	Yes No No Yes No	Higher phenolic content and antioxidant activity with suitable acceptability	Bad	[59]
Cake	No Yes No Yes No	Greater acceptability by a 10% fat replacement, which presented an increase in dietary fiber	Bad	[60]
Sunflower oil	Yes No No Yes Yes	The antioxidant effects of citrus extracts (mandarin, orange, and lemon) were comparable to BHT, with lemon being the most antioxidant against lipid oxidation	Good	[73]
Ultra-low-fat chicken patties	Yes No No Yes Yes	Lemon albedo decreased fat and cholesterol content, increased cooking yield, and showed good acceptability	Good	[61]
Salad dressing	Yes No Yes Yes Yes	Increased the bioaccessibility of hydroxycinnamic acids and flavonols by 0.3- to 5.8-fold	Excellent	[62]
Fresh-cut radish	Yes No Yes Yes No	Lower color variation and mesophilic aerobic count, proving a shelf-life of 7 days at 3 °C	Good	[63]
Cake	Yes Yes Yes Yes No	Nanoencapsulated lemon reports lower antioxidant activity and yield compared to orange; no significant difference in sensory quality or acceptability	Excellent	[51]
Chicken emulsion	No No No Yes No	2% added citrus peel fiber reported the best quality (viscosity, cooking loss, and emulsion stability)	Bad	[74]

Q6: Do the authors incorporate these flavonoids into a food?; Q7: Do the authors encapsulate the extract?; Q8: Do the authors conduct a shelf-life study with the new food?; Q9: Do the authors compare to a control sample?; Q10: Do the authors evaluate the functionality of the enriched food?

5. Future Perspective and Main Conclusions

The main conclusion of the present systematic review is that almost 90% of the selected publications related to the extraction of bioactive compounds from lemon by-products used environmentally friendly technologies and solvents. They greatly contributed to the optimization of the extraction of the bio-compounds, which are mainly present in the flavedo and albedo of lemon peels, the main food discards of these citrus. Nevertheless, further research is still necessary relating to the incorporation of these lemon extracts into potential new functional foods, especially concerning the assessment of the functionality and direct benefits produced by the consumption of such new foods enriched in flavonoids from lemon by-products, which have been shown to be an important source of health-promoting compounds. In addition, further research is also needed regarding green technologies to reduce energy in the by-product's revalorization process by applying an efficient and environmentally friendly solvent extraction method.

Author Contributions: Conceptualization, L.M.-Z. and N.C.; Methodology, L.M.-Z. and N.C.; Validation, M.C.-L.; Formal Analysis, L.M.-Z., M.C.-L. and N.C.; Investigation, L.M.-Z. and N.C.; Data Curation, L.M.-Z.; Writing—Original Draft Preparation, L.M.-Z. and N.C.; Writing—Review and Editing, L.M.-Z., N.C., M.C.-L. and F.A.-H.; Visualization, L.M.-Z.; Supervision, L.M.-Z. and N.C.; Project Administration, F.A.-H.; Funding Acquisition, F.A.-H. All authors have read and agreed to the published version of the manuscript.

Funding: Project PID2021-123857OB-I00, financed by the Spanish Ministry of Science and Innovation, the Spanish State Research Agency/10.13039/501100011033/, and FEDER. This work has also been financed by the Autonomous Community of the Region of Murcia through the Seneca Foundation and the European program NextGenerationEU throughout the AGRO-ALNEXT project.

Acknowledgments: L.M.-Z.'s contract has been financed by the Program for the Re-qualification of the Spanish University System, Margarita Salas modality, by the University of Murcia. M.C.-L.'s contract has been co-financed by Juan de la Cierva-Formación (FJC2020-043764-I) from the Spanish Ministry of Education.

Conflicts of Interest: The authors declare no conflict of interest.

References

1. FAO. *The State of Food and Agriculture, 2019: Moving Forward on Food Loss and Waste Reduction*; FAO: Rome, Italy, 2019; ISBN 9789251317891.
2. FAO. FAOSTAT Statistics Database. Available online: http://www.fao.org/faostat/en/#data/QC/visualize (accessed on 8 March 2022).
3. Al-Jaleel, A.; Zekri, M.; Hammam, Y. Yield, Fruit Quality, and Tree Health of "Allen Eureka" Lemon on Seven Rootstocks in Saudi Arabia. *Sci. Hortic.* **2005**, *105*, 457–465. [CrossRef]
4. Artés-Hernández, F.; Martínez-Zamora, L.; Cano-Lamadrid, M.; Hashemi, S.; Castillejo, N. Genus Brassica By-Products Revalorization with Green Technologies to Fortify Innovative Foods: A Scoping Review. *Foods* **2023**, *12*, 561. [CrossRef]
5. Cano-Lamadrid, M.; Artés–Hernández, F. By-Products Revalorization with Non-Thermal Treatments to Enhance Phytochemical Compounds of Fruit and Vegetables Derived Products: A Review. *Foods* **2022**, *11*, 59. [CrossRef]
6. Cano-Lamadrid, M.; Martínez-Zamora, L.; Castillejo, N.; Artés-Hernández, F. From Pomegranate Byproducts Waste to Worth: A Review of Extraction Techniques and Potential Applications for Their Revalorization. *Foods* **2022**, *11*, 2596. [CrossRef]
7. Artés-Hernández, F.; Castillejo, N.; Martínez-Zamora, L.; Martínez-Hernández, G.B. Phytochemical Fortification in Fruit and Vegetable Beverages with Green Technologies. *Foods* **2021**, *10*, 2534. [CrossRef]
8. Carrillo, C.; Nieto, G.; Martínez-Zamora, L.; Ros, G.; Kamiloglu, S.; Munekata, P.E.S.; Pateiro, M.; Lorenzo, J.M.; Fernández-López, J.; Viuda-Martos, M.; et al. Novel Approaches for the Recovery of Natural Pigments with Potential Health Effects. *J. Agric. Food Chem.* **2022**, *70*, 6864–6883. [CrossRef]
9. Magalhães, D.; Vilas-Boas, A.A.; Teixeira, P.; Pintado, M. Functional Ingredients and Additives from Lemon By-Products and Their Applications in Food Preservation: A Review. *Foods* **2023**, *12*, 1095.
10. Liu, S.; Li, S.; Ho, C.T. Dietary Bioactives and Essential Oils of Lemon and Lime Fruits. *Food Sci. Hum. Wellness* **2022**, *11*, 753–764. [CrossRef]
11. Klimek-szczykutowicz, M.; Szopa, A.; Ekiert, H. Citrus limon (Lemon) Phenomenon—A Review of the Chemistry, Pharmacological Properties, Applications in the Modern Pharmaceutical, Food, and Cosmetics Industries, and Biotechnological Studies. *Plants* **2020**, *9*, 119. [CrossRef]
12. Minzanova, S.T.; Mironov, V.F.; Arkhipova, D.M.; Khabibullina, A.V.; Mironova, L.G.; Zakirova, Y.M.; Milyukov, V.A. Biological Activity and Pharmacological Application of Pectic Polysaccharides: A Review. *Polymers* **2018**, *10*, 1407. [CrossRef]

13. Li, B.B.; Smith, B.; Hossain, M.M. Extraction of Phenolics from Citrus Peels: II. Enzyme-Assisted Extraction Method. *Sep. Purif. Technol.* **2006**, *48*, 189–196. [CrossRef]
14. Li, B.B.; Smith, B.; Hossain, M.M. Extraction of Phenolics from Citrus Peels: I. *Solvent Extraction Method. Sep. Purif. Technol.* **2006**, *48*, 182–188. [CrossRef]
15. Masmoudi, M.; Besbes, S.; Chaabouni, M.; Robert, C.; Paquot, M.; Blecker, C.; Attia, H. Optimization of Pectin Extraction from Lemon By-Product with Acidified Date Juice Using Response Surface Methodology. *Carbohydr. Polym.* **2008**, *74*, 185–192. [CrossRef]
16. Masmoudi, M.; Besbes, S.; Ben Thabet, I.; Blecker, C.; Attia, H. Pectin Extraction from Lemon By-Product with Acidified Date Juice: Rheological Properties and Microstructure of Pure and Mixed Pectin Gels. *Food Sci. Technol. Int.* **2010**, *16*, 105–114. [CrossRef] [PubMed]
17. Masmoudi, M.; Besbes, S.; Abbes, F.; Robert, C.; Paquot, M.; Blecker, C.; Attia, H. Pectin Extraction from Lemon By-Product with Acidified Date Juice: Effect of Extraction Conditions on Chemical Composition of Pectins. *Food Bioprocess Technol.* **2012**, *5*, 687–695. [CrossRef]
18. Gómez, B.; Gullón, B.; Yáñez, R.; Parajó, J.C.; Alonso, J.L. Pectic Oligosacharides from Lemon Peel Wastes: Production, Purification, and Chemical Characterization. *J. Agric. Food Chem.* **2013**, *61*, 10043–10053. [CrossRef] [PubMed]
19. Boluda-Aguilar, M.; López-Gómez, A. Production of Bioethanol by Fermentation of Lemon (*Citrus limon* L.) Peel Wastes Pretreated with Steam Explosion. *Ind. Crops Prod.* **2013**, *41*, 188–197. [CrossRef]
20. Gómez, B.; Yáñez, R.; Parajó, J.C.; Alonso, J.L. Production of Pectin-Derived Oligosaccharides from Lemon Peels by Extraction, Enzymatic Hydrolysis and Membrane Filtration. *J. Chem. Technol. Biotechnol.* **2016**, *91*, 234–247. [CrossRef]
21. Chávez-González, M.L.; López-López, L.I.; Rodríguez-Herrera, R.; Contreras-Esquivel, J.C.; Aguilar, C.N. Enzyme-Assisted Extraction of Citrus Essential Oil. *Chem. Pap.* **2016**, *70*, 412–417. [CrossRef]
22. Yilmaz, E.; Güneşer, B.A. Cold Pressed versus Solvent Extracted Lemon (*Citrus limon* L.) Seed Oils: Yield and Properties. *J. Food Sci. Technol.* **2017**, *54*, 1891–1900. [CrossRef]
23. Ndayishimiye, J.; Lim, D.J.; Chun, B.S. Impact of Extraction Conditions on Bergapten Content and Antimicrobial Activity of Oils Obtained by a Co-Extraction of Citrus by-Products Using Supercritical Carbon Dioxide. *Biotechnol. Bioprocess Eng.* **2017**, *22*, 586–596. [CrossRef]
24. Kapsaski-Kanelli, V.N.; Evergetis, E.; Michaelakis, A.; Papachristos, D.P.; Myrtsi, E.D.; Koulocheri, S.D.; Haroutounian, S.A. "gold" Pressed Essential Oil: An Essay on the Volatile Fragment from Citrus Juice Industry By-Products Chemistry and Bioactivity. *Biomed. Res. Int.* **2017**, *2017*, 2761461. [CrossRef] [PubMed]
25. Papoutsis, K.; Vuong, Q.V.; Tesoriero, L.; Pristijono, P.; Stathopoulos, C.E.; Gkountina, S.; Lidbetter, F.; Bowyer, M.C.; Scarlett, C.J.; Golding, J.B. Microwave Irradiation Enhances the in Vitro Antifungal Activity of Citrus By-Product Aqueous Extracts against Alternaria Alternata. *Int. J. Food Sci. Technol.* **2018**, *53*, 1510–1517. [CrossRef]
26. Papoutsis, K.; Pristijono, P.; Golding, J.B.; Stathopoulos, C.E.; Bowyer, M.C.; Scarlett, C.J.; Vuong, Q.V. Optimizing a Sustainable Ultrasound-Assisted Extraction Method for the Recovery of Polyphenols from Lemon by-Products: Comparison with Hot Water and Organic Solvent Extractions. *Eur. Food Res. Technol.* **2018**, *244*, 1353–1365. [CrossRef]
27. Karaman, E.; Karabiber, E.B.; Yilmaz, E. Physicochemical and Functional Properties of the Cold Press Lemon, Orange, and Grapefruit Seed Meals. *Qual. Assur. Saf. Crops Foods* **2018**, *10*, 233–243. [CrossRef]
28. Mansour, M.S.M.; Abdel-Shafy, H.I.; Mehaya, F.M.S. Valorization of Food Solid Waste by Recovery of Polyphenols Using Hybrid Molecular Imprinted Membrane. *J. Environ. Chem. Eng.* **2018**, *6*, 4160–4170. [CrossRef]
29. Özkaynak Kanmaz, E. Humic Acid Formation during Subcritical Water Extraction of Food By-Products Using Accelerated Solvent Extractor. *Food Bioprod. Process.* **2019**, *115*, 118–125. [CrossRef]
30. Peiró, S.; Luengo, E.; Segovia, F.; Raso, J.; Almajano, M.P. Improving Polyphenol Extraction from Lemon Residues by Pulsed Electric Fields. *Waste Biomass Valorization* **2019**, *10*, 889–897. [CrossRef]
31. Rosa, A.; Era, B.; Masala, C.; Nieddu, M.; Scano, P.; Fais, A.; Porcedda, S.; Piras, A. Supercritical CO_2 Extraction of Waste Citrus Seeds: Chemical Composition, Nutritional and Biological Properties of Edible Fixed Oils. *Eur. J. Lipid Sci. Technol.* **2019**, *121*, 1800502. [CrossRef]
32. Kurtulbaş, E.; Yazar, S.; Makris, D.; Şahin, S. Optimization of Bioactive Substances in the Wastes of Some Selective Mediterranean Crops. *Beverages* **2019**, *5*, 42. [CrossRef]
33. Rahmani, Z.; Khodaiyan, F.; Kazemi, M.; Sharifan, A. Optimization of Microwave-Assisted Extraction and Structural Characterization of Pectin from Sweet Lemon Peel. *Int. J. Biol. Macromol.* **2020**, *147*, 1107–1115. [CrossRef]
34. Liu, Y.; Benohoud, M.; Galani Yamdeu, J.H.; Gong, Y.Y.; Orfila, C. Green Extraction of Polyphenols from Citrus Peel By-Products and Their Antifungal Activity against Aspergillus Flavus. *Food Chem. X* **2021**, *12*, 100144. [CrossRef]
35. Tunç, M.T.; Odabaş, H.İ. Single-Step Recovery of Pectin and Essential Oil from Lemon Waste by Ohmic Heating Assisted Extraction/Hydrodistillation: A Multi-Response Optimization Study. *Innov. Food Sci. Emerg. Technol.* **2021**, *74*, 102850. [CrossRef]
36. Fathollahy, I.; Farmani, J.; Kasaai, M.R.; Hamishehkar, H. Characteristics and Functional Properties of Persian Lime (Citrus latifolia) Seed Protein Isolate and Enzymatic Hydrolysates. *LWT* **2021**, *140*, 110765. [CrossRef]
37. Song, L.W.; Qi, J.R.; Liao, J.S.; Yang, X.Q. Enzymatic and Enzyme-Physical Modification of Citrus Fiber by Xylanase and Planetary Ball Milling Treatment. *Food Hydrocoll.* **2021**, *121*, 107015. [CrossRef]

38. Putri, N.I.; Celus, M.; Van Audenhove, J.; Nanseera, R.P.; Van Loey, A.; Hendrickx, M. Functionalization of Pectin-Depleted Residue from Different Citrus by-Products by High Pressure Homogenization. *Food Hydrocoll.* **2022**, *129*, 107638. [CrossRef]
39. Imeneo, V.; Romeo, R.; De Bruno, A.; Piscopo, A. Green-Sustainable Extraction Techniques for the Recovery of Antioxidant Compounds from "Citrus limon" by-Products. *J. Environ. Sci. Health Part B* **2022**, *57*, 220–232. [CrossRef]
40. Kalogiouri, N.P.; Palaiologou, E.; Papadakis, E.N.; Makris, D.P.; Biliaderis, C.G.; Mourtzinos, I. Insights on the Impact of Deep Eutectic Solvents on the Composition of the Extracts from Lemon (*Citrus limon* L.) Peels Analyzed by a Novel RP-LC–QTOF-MS/MS Method. *Eur. Food Res. Technol.* **2022**, *248*, 2913–2927. [CrossRef]
41. Toprakçı, G.; Toprakçı, İ.; Şahin, S. Highly Clean Recovery of Natural Antioxidants from Lemon Peels: Lactic Acid-Based Automatic Solvent Extraction. *Phytochem. Anal.* **2022**, *33*, 554–563. [CrossRef]
42. Chaves, J.O.; Sanches, V.L.; Viganó, J.; de Souza Mesquita, L.M.; de Souza, M.C.; da Silva, L.C.; Acunha, T.; Faccioli, L.H.; Rostagno, M.A. Integration of Pressurized Liquid Extraction and In-Line Solid-Phase Extraction to Simultaneously Extract and Concentrate Phenolic Compounds from Lemon Peel (*Citrus limon* L.). *Food Res. Int.* **2022**, *157*, 111252. [CrossRef]
43. Vellaiyan, S.; Kandasamy, M.; Subbiah, A.; Devarajan, Y. Energy, Environmental and Economic Assessment of Waste-Derived Lemon Peel Oil Intermingled with High Intense Water and Cetane Improver. *Sustain. Energy Technol. Assess.* **2022**, *53*, 102659. [CrossRef]
44. Myrtsi, E.D.; Koulocheri, S.D.; Evergetis, E.; Haroutounian, S.A. Agro-Industrial Co-Products Upcycling: Recovery of Carotenoids and Fine Chemicals from Citrus Sp. Juice Industry Co-Products. *Ind. Crops Prod.* **2022**, *186*, 115190. [CrossRef]
45. Gavahian, M.; Yang, Y.H.; Tsai, P.J. Power Ultrasound for Valorization of Citrus limon (Cv. Eureka) Waste: Effects of Maturity Stage and Drying Method on Bioactive Compounds, Antioxidant, and Anti-Diabetic Activity. *Innov. Food Sci. Emerg. Technol.* **2022**, *79*, 103052. [CrossRef]
46. El Fihry, N.; El Mabrouk, K.; Eeckhout, M.; Schols, H.A.; Filali-Zegzouti, Y.; Hajjaj, H. Physicochemical and Functional Characterization of Pectin Extracted from Moroccan Citrus Peels. *LWT* **2022**, *162*, 113508. [CrossRef]
47. Song, Y.T.; Qi, J.R.; Yang, X.Q.; Liao, J.S.; Liu, Z.W.; Ruan, C.W. Hydrophobic Surface Modification of Citrus Fiber Using Octenyl Succinic Anhydride (OSA): Preparation, Characterization and Emulsifying Properties. *Food Hydrocoll.* **2022**, *132*, 107832. [CrossRef]
48. Alasalvar, H.; Kaya, M.; Berktas, S.; Basyigit, B.; Cam, M. Pressurised Hot Water Extraction of Phenolic Compounds with a Focus on Eriocitrin and Hesperidin from Lemon Peel. *Int. J. Food Sci. Technol.* **2022**, *58*, 2060–2066. [CrossRef]
49. Durmus, N.; Kilic-Akyilmaz, M. Bioactivity of Non-Extractable Phenolics from Lemon Peel Obtained by Enzyme and Ultrasound Assisted Extractions. *Food Biosci.* **2023**, *53*, 102571. [CrossRef]
50. Al Chami, Z.; Alwanney, D.; De Pascali, S.A.; Cavoski, I.; Fanizzi, F.P. Extraction and Characterization of Bio-Effectors from Agro-Food Processing by-Products as Plant Growth Promoters. *Chem. Biol. Technol. Agric.* **2014**, *1*, 17. [CrossRef]
51. Mahmoud, K.F.; Ibrahim, M.A.; Mervat, E.D.; Shaaban, H.A.; Kamil, M.M.; Hegazy, N.A. Nano-Encapsulation Efficiency of Lemon and Orange Peels Extracts on Cake Shelf Life. *Am. J. Food Technol.* **2016**, *11*, 63–75. [CrossRef]
52. Nanda, B.; Sailaja, M.; Mohapatra, P.; Pradhan, R.K.; Nanda, B.B. Green Solvents: A Suitable Alternative for Sustainable Chemistry. *Mater. Today Proc.* **2021**, *47*, 1234–1240. [CrossRef]
53. Torres-Valenzuela, L.S.; Ballesteros-Gómez, A.; Rubio, S. Green Solvents for the Extraction of High Added-Value Compounds from Agri-Food Waste. *Food Eng. Rev.* **2020**, *12*, 83–100.
54. De los Ángeles Fernández, M.; Espino, M.; Gomez, F.J.V.; Silva, M.F. Novel Approaches Mediated by Tailor-Made Green Solvents for the Extraction of Phenolic Compounds from Agro-Food Industrial by-Products. *Food Chem.* **2018**, *239*, 671–678. [CrossRef]
55. Winterton, N. The Green Solvent: A Critical Perspective. *Clean Technol. Environ. Policy* **2021**, *23*, 2499–2522.
56. Aleson-Carbonell, L.; Fernández-López, J.; Pérez-Alvarez, J.A.; Kuri, V. Functional and Sensory Effects of Fibre-Rich Ingredients on Breakfast Fresh Sausages Manufacture. *Food Sci. Technol. Int.* **2005**, *11*, 89–97. [CrossRef]
57. Masmoudi, M.; Besbes, S.; Blecker, C.; Attia, H. Preparation and Characterization of Osmodehydrated Fruits from Lemon and Date By-Products. *Food Sci. Technol. Int.* **2007**, *13*, 405–412. [CrossRef]
58. Sendra, E.; Fayos, P.; Lario, Y.; Fernández-López, J.; Sayas-Barberá, E.; Pérez-Alvarez, J.A. Incorporation of Citrus Fibers in Fermented Milk Containing Probiotic Bacteria. *Food Microbiol.* **2008**, *25*, 13–21. [CrossRef]
59. Imeneo, V.; Romeo, R.; Gattuso, A.; De Bruno, A.; Piscopo, A. Functionalized Biscuits with Bioactive Ingredients Obtained by Citrus Lemon Pomace. *Foods* **2021**, *10*, 2460. [CrossRef]
60. Jiménez Nempeque, L.V.; Gómez Cabrera, Á.P.; Colina Moncayo, J.Y. Evaluation of Tahiti Lemon Shell Flour (*Citrus latifolia* Tanaka) as a Fat Mimetic. *J. Food Sci. Technol.* **2021**, *58*, 720–730. [CrossRef]
61. Chappalwar, A.M.; Pathak, V.; Goswami, M.; Verma, A.K.; Rajkumar, V. Efficacy of Lemon Albedo as Fat Replacer for Development of Ultra-Low-Fat Chicken Patties. *J. Food Process. Preserv.* **2021**, *45*, e15587. [CrossRef]
62. Kamiloglu, S.; Ozdal, T.; Tomas, M.; Capanoglu, E. Oil Matrix Modulates the Bioaccessibility of Polyphenols: A Study of Salad Dressing Formulation with Industrial Broccoli by-Products and Lemon Juice. *J. Sci. Food Agric.* **2022**, *102*, 5368–5377. [CrossRef]
63. Zappia, A.; Spanti, A.; Princi, R.; Imeneo, V.; Piscopo, A. Evaluation of the Efficacy of Antioxidant Extract from Lemon By-Products on Preservation of Quality Attributes of Minimally Processed Radish (*Raphanus sativus* L.). *Antioxidants* **2023**, *12*, 235. [CrossRef]
64. Banerjee, J.; Vijayaraghavan, R.; Arora, A.; MacFarlane, D.R.; Patti, A.F. Lemon Juice Based Extraction of Pectin from Mango Peels: Waste to Wealth by Sustainable Approaches. *ACS Sustain. Chem. Eng.* **2016**, *4*, 5915–5920. [CrossRef]

65. Meseldžija, S.; Petrovic, J.; Onjia, A.; Volkov-Husovic, T.; Nešic, A.; Vukelic, N. Removal of Fe2+, Zn2+and Mn2+from the Mining Wastewater by Lemon Peel Waste. *J. Serbian Chem. Soc.* **2020**, *85*, 1371–1382. [CrossRef]
66. Fernández-López, J.; Zhi, N.; Aleson-Carbonell, L.; Pérez-Alvarez, J.A.; Kuri, V. Antioxidant and Antibacterial Activities of Natural Extracts: Application in Beef Meatballs. *Meat Sci.* **2005**, *69*, 371–380. [CrossRef]
67. Sariçoban, C.; Özalp, B.; Yilmaz, M.T.; Özen, G.; Karakaya, M.; Akbulut, M. Characteristics of Meat Emulsion Systems as Influenced by Different Levels of Lemon Albedo. *Meat Sci.* **2008**, *80*, 599–606. [CrossRef]
68. Fu, J.T.; Chang, Y.H.; Shiau, S.Y. Rheological, Antioxidative and Sensory Properties of Dough and Mantou (Steamed Bread) Enriched with Lemon Fiber. *LWT* **2015**, *61*, 56–62. [CrossRef]
69. Shan, B.; Li, X.; Pan, T.; Zheng, L.; Zhang, H.; Guo, H.; Jiang, L.; Zhen, S.; Ren, F. Effect of Shaddock Albedo Addition on the Properties of Frankfurters. *J. Food Sci. Technol.* **2015**, *52*, 4572–4578. [CrossRef]
70. Sanz-Puig, M.; Pina-Pérez, M.C.; Martínez-López, A.; Rodrigo, D. Escherichia Coli O157:H7 and Salmonella typhimurium Inactivation by the Effect of Mandarin, Lemon, and Orange by-Products in Reference Medium and in Oat-Fruit Juice Mixed Beverage. *LWT* **2016**, *66*, 7–14. [CrossRef]
71. Long, J.M.; Mohan, A. Food Flavoring Prepared with Lemon By-Product. *J. Food Process. Preserv.* **2021**, *45*, e15462. [CrossRef]
72. Azmi, N.A.N.; Elgharbawy, A.A.M.; Salleh, H.M.; Moniruzzaman, M. Preparation, Characterization and Biological Activities of an Oil-in-Water Nanoemulsion from Fish By-Products and Lemon Oil by Ultrasonication Method. *Molecules* **2022**, *27*, 6725. [CrossRef]
73. Aydın, S.; Sayin, U.; Sezer, M.Ö.; Sayar, S. Antioxidant Efficiency of Citrus Peels on Oxidative Stability during Repetitive Deep-Fat Frying: Evaluation with EPR and Conventional Methods. *J. Food Process Preserv.* **2021**, *45*, e15584. [CrossRef]
74. Choi, Y.-S.; Kim, H.-W.; Hwang, K.-E.; Song, D.-H.; Kim, H.-Y.; Lee, M.-A.; Yoon, Y.-H.; Kim, C.-J. Effects of Dietary Fiber Extracted from Citrus (Citrus unshiu S. Marcoy) Peel on Physicochemical Properties of a Chicken Emulsion in Model Systems. *Korean J. Food Sci. Anim. Resour.* **2012**, *32*, 618–626. [CrossRef]

foods

MDPI

Review

Spent Grain: A Functional Ingredient for Food Applications

Ancuța Chetrariu and Adriana Dabija *

Faculty of Food Engineering, Stefan cel Mare University of Suceava, 720229 Suceava, Romania
* Correspondence: adriana.dabija@fia.usv.ro; Tel.: +40-748-845-567

Abstract: Spent grain is the solid fraction remaining after wort removal. It is nutritionally rich, composed of fibers—mainly hemicellulose, cellulose, and lignin—proteins, lipids, vitamins, and minerals, and must be managed properly. Spent grain is a by-product with high moisture, high protein and high fiber content and is susceptible to microbial contamination; thus, a suitable, cost-effective, and environmentally friendly valorization method of processing it is required. This by-product is used as a raw material in the production of many other food products—bakery products, pasta, cookies, muffins, wafers, snacks, yogurt or plant-based yogurt alternatives, Frankfurter sausages or fruit beverages—due to its nutritional values. The circular economy is built on waste reduction and the reuse of by-products, which find opportunities in the regeneration and recycling of waste materials and energy that become inputs in other processes and food products. Waste disposal in the food industry has become a major issue in recent years when attempting to maintain hygiene standards and avoid soil, air and water contamination. Fortifying food products with spent grain follows the precepts of the circular bio-economy and industrial symbiosis of strengthening sustainable development. The purpose of this review is to update information on the addition of spent grain to various foods and the influence of spent grain on these foods.

Keywords: spent grain; valuable by-product; beer industry; whisky; bioactive compound; valorization; circular economy

Citation: Chetrariu, A.; Dabija, A. Spent Grain: A Functional Ingredient for Food Applications. *Foods* **2023**, *12*, 1533. https://doi.org/10.3390/foods12071533

Academic Editors: Noelia Castillejo Montoya and Lorena Martínez-Zamora

Received: 2 March 2023
Revised: 25 March 2023
Accepted: 1 April 2023
Published: 4 April 2023

1. Introduction

Agro-industrial processes generate substantial amounts of by-products with increased contents of organic compounds that have a significant impact on the environment [1]. On the other hand, the increased demand for food globally drives the discovery of substitute raw materials with affordable pricing and superior nutritional value. By-product valorization has emerged as an important component of food research in recent years [2]. Spent grain (SG) is the main by-product of the beer and distillation industries [3,4]. During whisky and beer production, an average of 8–15 L of effluent and 2.5–3.0 kg of spent grain is generated for each liter of whisky produced, while 0.2 kg of spent grain is generated per liter of beer produced [4,5]. To support the circular economy and the environment, biomass conversion technologies and biorefineries must be developed so that the biomass-based economy grows. Spent grain is a valuable by-product that is rich in nutrients such as dietary fibers (hemicellulose, cellulose and lignin), proteins, monosaccharides (glucose, xylose and arabinosis), minerals, vitamins and lipids [6]. Products enriched with spent grain are called fortified foods [7,8]. Spent grain contains a substantial quantity of bioactive compounds with high antioxidant capacities, including hydroxycinamic acids (particularly ferulic acid and p-cumaric acid) [9,10]. There are several variations of this by-product due to the varied types of cereals, seasons and quality of crops, malting and lautering methods, and composition of the spent grain [11–13]. The roles of the functional components of spent grain in the human body are described in Table 1. Due to its high moisture, finding alternative sources of drying spent grain that consume less energy, produces a brighter color and with higher available and digestible protein, is challenging [14]. Spent grain is often used as animal feed due to its nutritional content and low cost; it is used either in wet or dry form [10,15]. Spent grain is also used for the generation of renewable energy to reduce the carbon footprint of alcohol production [16,17].

2. Nutritive Value of Spent Grains

Proteins in spent grain can be used as a substitute for fishmeal or soya flour in feed formulations [18]. Approximately 50–60% of dry matter in spent grain is carbohydrates, including glucans, starch, cellulose and arabinoxylans. These carbohydrates can be converted into various biochemical products and biofuels [19,20]. Spent grain can also be used in various biotechnological processes to produce lactic acid, xylitol, microbial enzymes and biopesticides [17,21,22]. The proteins in spent grain contain valuable amino acids, the most abundant of which are glutamine/glutamate and proline [20]. In addition, spent grain is a promising source of lipids, including triglycerides (67% of the total extract), a range of free fatty acids (18%) and lower quantities of monoglycerides (1.6%) and diglycerides (7.7%). Among the fatty acids are linoleic acid (18:2), palmitic acid (16:0) and oleic acid (18:1), as well as small quantities of stearic acid (18:0) and linolenic acid (18:3) [23–25]. It is necessary to add value to this by-product and to develop sustainable low-cost methods to make it economically attractive [26]. Harnessing this by-product and developing sustainable processes is an urgent need in food industry [27] (Table 1).

Table 1. Functional compounds in spent grain.

Functional Compounds in Spent Grain	Role in Human Body	Study
Non-cellulosic polysaccharides (β-glucan and arabinoxylans)	• β-glucan reduce the rise in blood glucose after meals; • Arabinoxylans reduce blood glucose levels.	[28,29]
Proteins	• Increase satiety; • Regulate long-term energy balance.	[30,31]
Polyphenols	• Anti-carcinogenic; • Anti-inflammatory and antioxidant activities.	[31–34]
Fiber	• Dietary fiber reduces cholesterol levels; • Increases fecal bulk.	[28,34–36]
Vitamins	• Vitamins have antioxidant properties.	[28]

Spent grain is a low cost by-product that is available throughout the year, is nutritionally valuable, rich in fiber, proteins and minerals and can be reused in both food and non-food sectors, including animal feed, compost preparations, biogas production, cultivation of microorganisms and production of biomaterials, biochemicals and bricks [33,37,38]. Spent grain extract is an approved ingredient in food supplements [39]. Spent grain undergoes pre-treatment to make it more accessible. It has a number of advantages, including open cell wall structure, decreased particle size and improved digestibility. Pre-treatments are used in acidic and alkaline environments and in ultrasonic or microwave extractions; however, simple, environmentally friendly, economical and efficient methods are desired. Spent grain is used in food products (Figure 1) for its health promoting effects against constipation, obesity, diabetes and cardiovascular diseases. Phenolics in spent grains are associated with the prevention of chronic cardiovascular and neurogenerative diseases, certain cancers and diabetes. The high fiber content helps in the elimination of cholesterol and fats and improves symptoms of ulcerative colitis [31,33–36]. Reusing spent grain as a value-added food source for human consumption is appealing because it increases the protein, fiber, vitamin and mineral contents while decreasing the starch and caloric content in grain-based products [40].

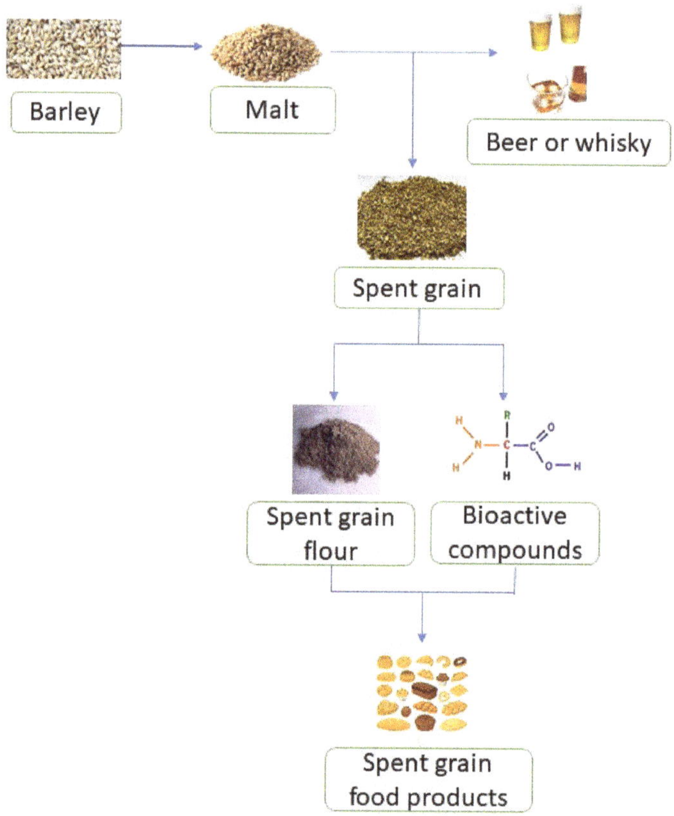

Figure 1. Synthetic process of generating and using spent grain to produce food products.

Spent grain is a by-product with high nutritional value, thus it is important to identify innovative solutions for returning waste and by-products into the production cycle to obtain innovative quality products [33,41]. A large number of companies have adopted the reduce-reuse-recycle approach, understanding that solving social and environmental issues require changing the strategy of organizations and introducing interdisciplinary actions and methods [38]. The circular economy is based on extending the life cycle of products by reusing, renovating and recycling them for as long as possible, thus reducing waste. In this respect, innovations are stimulated and solutions are found to meet the rising challenges [42,43]. A sustainable approach to the circular economy is necessary in order to use the circular economy structure for food by-products.

Spent grain is a good source of phenolic compounds (ferulic acid, p-coumaric acid, sinapic acid and caffeic acids), which are considered natural antioxidants [44]. There is a growing interest in the valorization of by-products and the circular bioeconomy, by developing alternatives to the conventional use of spent grain for animal feed and compost. The capitalization of food by-products is not only influenced by technical capabilities, but also by socio-economic, supply chain and regulatory factors [45].

The circular and sustainable bioeconomy is gaining traction as a means of addressing climate change and fossilization, increasing resource efficiency and creating new opportunities for long-term economic growth [33].

The aim of this review is to update information on the use of spent grain from the beer and whisky industries in the production of value-added food products.

3. Possible Uses of Spent Grain in Food Products

The food sector is continuously expanding, and consumers are becoming more and more interested in new recipes and healthy diets. The current global context challenges us to find low-cost, high-nutrition, healthy alternatives for food products, and to use industrial by-products. The principal characteristics of several food products enriched with spent grain are presented in Table 2.

Table 2. Characteristics of functional foods derived from spent grain.

Functional Foods derived from Spent Grain	Quantity of Spent Grain Added	Properties	Study
Bread	10–15% spent grain flour	Acceptable sensorial properties; High fiber content (health benefit); Increased mineral content; Influences the rheological and pasting properties of dough; The biaxial extensional viscosity is significantly higher; The strain-hardening index decreases with increasing quantities of flour substitution; Reduces the uniaxial extensibility, while the storage modulus, G″, increases; Addition of spent grain increases the composition/nutritional properties; The color of bread turned from light cream to brown; Water absorption increases with the quantity of spent grain; Increased crumb firmness; Increased antioxidant content.	[35,37,46–51]
Bread obtained from fermented spent grain	25%, 50%, 75%, 100% spent grain sourdough	Changes the porosity and acidity; Bacteriostatic function (the shelf life of bread increases).	[52]
Spent grain pasta	5–25% spent grain flour	Increased protein, fiber and β-glucan content; Increased antioxidant content; The higher the spent grain content, the darker the color of the pasta; A compact structure with higher firmness; Decreased cooking loss; Decreased degree of starch gelatinization; Reduced the optimal cooking time; Increased total organic matter.	[53–56]
Cookies	Max 30% spent grain added	Fiber and protein content increases; Dough development time and dough stability increases; Total antioxidant activity increases; Water absorption increases.	[57–61]
Shortbread	30%	Increase in fiber and protein content; Decrease in carbohydrate levels and energy value.	[29]
Muffins	15–30%	Increases the amount of fat, protein and total dietary fiber; Increases the viscosity of the batter.	[62,63]

Table 2. Characteristics of functional foods derived from spent grain.

Functional Foods derived from Spent Grain	Quantity of Spent Grain Added	Properties	Study
Wafers	5–20%	Gumminess, chewiness, springiness, firmness and cohesiveness increase.	[64–66]
Snacks	10–30%	Increase the total content of polyphenols, flavonoids, proteins, fats, dietary fiber and energy; Increases phytic acid and resistant starch content.	[48,67,68]
Yogurt and plant-based yogurt alternetives	5–20%	Yogurt's syneresis level was considerably reduced; Decreased fermentation time and increased viscosity and shear stress; Maintained textural and gelling formation.	[69,70]
Frankfurters sausages	1–5%	Total dietary fiber increases.	[71]
Tarhana	6%	Increase in protein and fiber content.	[72]
Fruit juice and smoothies	0–10%	Increased antioxidant activity.	[73]

3.1. Spent Grain in Bread

Bread is one of the most common foods. To obtain bread with acceptable sensorial properties, the amount of spent grain added is limited to 10–15%, as adding a higher quantity leads to a decrease in volume, affects the taste and aroma and changes the rheological properties [46]. Bread enriched with this by-product has a high fiber content, which is associated with health benefits, including increasing digestion and preventing some gastrointestinal diseases. [47]. Not only does the fiber content increase with the increase in spent grain content, the protein content also increases [48]. Furthermore, a high quantity of this by-product influenced the rheological and pasting properties of dough, significantly increased the biaxial extensional viscosity, decreased the strain hardening index as the quantity of substituted flour increased and reduced the uniaxial extensibility. On the other hand, the storage modulus, G″, increased, indicating changes in the structural properties of the dough. These properties negatively affected the baking quality of doughs, conduced to breads with low volumes and dense structures. On the other hand, adding spent grain increased the compositional/nutritional properties, e.g., protein and fiber content [49].

Ktenioudaki et al. [74] produced bread with 15% aded spent grain and sourdough with 15% added spent grain. The samples contained high fiber (11.9% in the spent grain flour added to bread and 12.1% in the spent grain added to the sourdough). As expected, sourdough bread had higher acidity (5.3 compared with 5.8 for breads with no sourdough). The mineral content in the sourdough spent grain samples was higher (107.9 $\pm$ 4.8 mg/100 g Ca, 12.7 $\pm$ 2.8 mg/100 g Mg, 100.4 $\pm$ 17.8 mg/100 g K) compared with bread with no sourdough (98.9 $\pm$ 12.5 8 mg/100 g Ca, 11.6 $\pm$ 2.0 mg/100 g Mg, 98.9 $\pm$ 12.5 mg/100 g K) [74]. Yitayew et al. found that the calcium, magnesium and potassium contents of the bread increased from 76.44 to 150.93 mg/100 g, 87.12 to 176.81 mg/100 g and 116.04 to 225.49 mg/100 g, respectively, as the spent grain quantity increased from 0 to 20% [37]. The in vitro antioxidant activity in the sourdough samples was also higher (132.7 $\pm$ 3.0 gallic acid equivalent (mg/100 g sample dwb) for total phenolic content and 83.1 $\pm$ 9.3 TEAC (IC50Trolox/IC50Sample) $\times 10^5$ for DPPH scavenging activity compared with samples with no spent grain sourdough (130.9 $\pm$ 4.3 gallic acid equivalent (mg/100 g sample dwb) of total phenolic content, 82.2 $\pm$ 9.0 TEAC (IC50Trolox/IC50Sample) $\times 10^5$ for DPPH scavenging activity) [74]. Similar observations were reported by Aprodu et al. [75] who found higher total phenolic content and antioxidant activity in bread prepared with sourdough. This can be attributed to the action of endogenous enzymes on cell walls. Phytic acid reduces mineral bioavailability through formation of insoluble complexes, but the sourdough fermentation reduces the phytic acid by approximatively 30%, as measured in the bread, hence potentially increasing the bioavailability of minerals in breads containing spent grain [74].

The increase in the concentration of spent grain added to bread decreases its specific volume due to the high amount of arabinoxylans found in spent grain, with an Arabinan:Xylan ratio of about 0.45, a ratio much lower than wheat bran and wheat endosperm (0.88 and 0.67). Sourdough fermentation increased sample volume, but this decreased once spent grain was added [75]. On the other hand, the bread crumb harness increased with addition of spent grain to the wheat flour; however, the bread crumb level was lower in samples with sourdough. This might be attributed to the effect of endoxylanases activity during sourdough fermentation [75]. Adding spent grain increased the starch gelatinization of wheat flour and affected the stability and retrogradation of the starch gels [75]. Shaitan et al. [52] made sourdough bread containing 25%, 50%, 75% and 100% spent grain by fermentation for 8 days. Bread samples prepared with 25% and 50% spent grain sourdough were characterized by higher porosity, acidity and corresponding moisture compared with samples prepared from 100% spent grain sourdough, which had lower porosity and acidity. The spent grain in the samples played a bacteriostatic function; the control sample was the first to show early signs of rope spoilage. The addition of spent grain increased the shelf life of the bread by 24–48 h, thus slowing down the development of rope spoilage in the bread.

Ginindza et al. [76] optimized spent grain in the wheat:maize:spent grain composite flour bread, up to 10%. A higher quantity of spent grain decreased the bread's specific volume but increased the volume and density. Fiber, protein and ash content increased with the increase in the quantity of spent grain added [76].

Czubaszek et al. [77] found that replacing wheat flour with spent grain reduced gluten yield and deteriorated its quality, leading to a decrease in the sedimentation value and stability, and increasing dough softening. These trends were attributed to the long mixing time and high shear force applied, in addition to the high fiber and protein content [37,50]. Bread with 10% spent grain did not differ significantly from wheat flour bread in terms of appearance, crust, and crumb properties; however, the color of the bread turned from light cream to brown as the spent grain replacement increased [77].

Water absorption ability of the bread increased significantly (from 58.40 to 66.67 mL/100 g) as the quantity of spent grain increased (0–20%), likely due to the increase in high protein and non-starchy polysaccharides in spent grain [37]. Similar results were obtained by Stojceska and Ainsworth [50]. The protein and fiber content of the spent grain also increased the dough development time (from 3.43 to 17.57 min) [37]. Yitayew et al. [37] showed that the loaf weight, volume and specific volume are modified. Increasing spent grain quantity increased the loaf weight (from 127.58 to 148.85 g), while the specific volume of bread loaf decreased (from 2.92 to 2.46 cm^3/g). The addition of fiber- and protein-rich ingredients increases the hardness of the bread, as fiber and protein absorb a lot of water, leading to a stronger structure. Sensory analysis showed that overall acceptability decreases with increase in spent grain content due to the cumulative effect of the darker color, the taste, the malt aroma and the crumb texture [37].

Sahin et al. [78] used two ingredients derived from spent grain (one rich in fiber and one rich in protein) to make bread. The bread containing additional fibers had high specific volume (3.72–4.66 mL/g), soft crumb texture (4.77–9.03 N) and crumb structure (4.77–9.03 N), whereas bread enriched with protein had increased dough resistance (+150% compared with control sample), which led to a lower specific volume (2.17–4.38 mL/g) and a harder crumb (6.25–36.36 N).

Steinmacher et al. [51] used enzyme-treated and untreated spent grain to produce bread. Enzymatic treatment did not affect the characteristics of the bread, but adding spent grain and enzymes (Pentopan Mono BG and Celluclast BG) directly to the dough improved the texture and volume. Stojceska and Ainsworth [50] used a wider range of enzymes (Maxlife 85, Lipopan Extra, Pentopan Mono BG and Celluclast) to evaluate the characteristics of bread with added spent grain. Increasing the level of fiber in bread has some advantages, including increasing dough development time, dough stability and crumb firmness; however, it also has some disadvantages, including decreased softening and loaf volume. Shelf life, loaf volume and textures improved when Lipopan Extra, Pentopan Mono BG and a mixture of Pentopan Mono and Celluclast were added.

Adding spent grain affected the rheological properties of the dough: the biaxial extensional viscosity was significantly higher in the supplemented doughs. Replacing wheat flour with spent grain significantly reduced the uniaxial extensibility, while the storage modulus (G″) increased, indicating changes in the structural properties of the dough. The strain hardening index decreased as the quantity of substituted spent grain increased [49]. The nutraceutical quality of bread enriched with spent grain is defined by the quantities of antioxidants and fiber. The antioxidant content of the bread increased with the increase in the quantity of spent grain in the bread formulation, as indicated by Baiano et al. [35]. Additionally, the fiber content increased without affecting the structural and sensory attributes of the bread.

3.2. Spent Grain in Pasta Products

Pasta is a ready-to-eat product made from durum wheat [79]; however, lately it is also obtained from other flours, flour mixtures with or without additional vegetables or other by-products, resulting in quality products that retain a good consistency after cooking. Several researchers have investigated the partial replacement of flour with ingredients from agro-industrial by-products to make pasta [80–84]. The increasing worldwide consumption of pasta is due to its high digestibility, slow carbohydrate release, relatively low glycemic index compared with bread, pizza or other cereal products, high shelf life, versatility and ease of cooking [56]. Foods that promote health by incorporating ingredients of plant or animal origin during manufacturing are considered nutritional products with added value [85].

In the case of pasta products, the addition of spent grain does not have a major influence on the functional properties, even at concentrations of 25%, which makes adding spent grain to these products in order to increase their nutrient content acceptable [46]. Nocente et al. conducted a study on pasta formulations by adding spent grain from two species of cereals (einkorn and tritordeum), resulting in pasta with noticeably higher protein, fiber and β-glucan content and, to a minor extent, increased antioxidant capacity and good sensory quality [53]. In another study, Nocente et al. [86] used a blend of semolina and spent grain to produce spent grain-enriched pasta characterized by high fiber and antioxidant content. Schettino et al. used bioprocessed spent grain to obtain fortified pasta labeled "High fiber" and "Source of protein" [55]. Cuomo et al. [54] used two fractions of spent grain (5–10% protein and 10–20% fiber) to obtain high fiber and high protein pasta. Several features of pasta were evaluated, including proximate composition, color, optimal cooking time, sensory features and texture. Protein-enriched pasta had a protein content of about 18% and a fiber content greater than 8%, meaning that it contained about 30% of the protein content recommended by EFSA [87]. Food products can be classified into "fiber source" products, which contain at least 3 g dietary fiber/100 g product, and "fiber-rich" products, which contain at least 6 g dietary fiber/100 g product according to Regulation (EC) No. 1924/2006 [80]. As expected, the protein- and fiber-enriched pasta had a darker color. The L* (brightness) parameter showed a significant reduction compared with the wholegrain semolina paste sample [46,54,86,88], as shown in Figure 2.

Optimal cooking time was between 11 min and 13 min 30 s. Values close to the blank samples and firmness of pasta were positively evaluated. The firmness characteristics, evaluated using instrumental analysis, were associated with sensorial features and were very close to each other [54]. Cappa and Alamprese [80] added egg white powder to spent grain pasta to improve the structural properties of fresh-egg pasta (lasagna) but the mechanical properties were poor.

Sahin et al. [15] developed enriched pasta containing protein and fiber fractions derived from spent grain, resulting products with stronger gluten networks and bonding properties, compact structure, higher firmness and higher tensile strength, but lower glycemic index. An important criterion for pasta quality is cooking loss, with low cooking loss (CL) being more desirable. Pasta labeled high in fiber had lower CL values ($3.47 \pm 0.86\%$). Increasing the quantity of ingredients derived from spent grain decreased cooking loss [15]. Starch gelatinization and protein coagulation are two processes responsible for the formation of the structure and quality of pasta during cooking. Adding ingredients derived from spent grain decreased the degree of starch gelatinization due to the high protein content [89]. Starch is physically captured in a protein matrix due to its interaction with proteins through molecular forces (ionic, hydrogen and covalent bonds), which leads to a reduction in the degree of gelatinization and an increase in resistance to shearing and heat. The increased quantities of starch and protein compete for water, leading to reduced swelling of the starch during gelatinization. Reducing the amount of total starch affects the degree of gelatinization and the cooking loss of pasta [15].

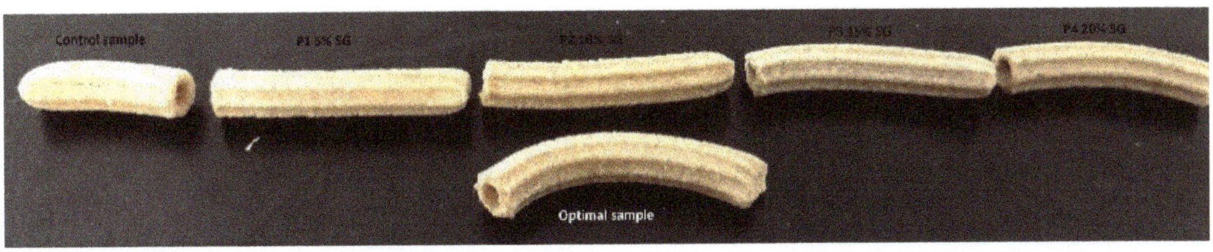

Figure 2. Spent grain pasta [84].

Nocente et al. [86] showed that the total antioxidant capacity in spent grain-enriched pasta increases as the quantity of spent grain increases. Most antioxidant compounds (phenolic acids and other polyphenols) are found in the outer layers of the barley grain and in the aleuronic layer of the kernels.

Optimal cooking time reduced after pasta fortification, probably due to the increase in dietary fiber content, which alters the structure of pasta and permits early starch gelatinization and accelerates water penetration [90]. These findings were in agreement with the findings by Nocente et al. [86]. Other quality parameters of the pasta include the amount of water absorbed (WA) by the pasta during the optimal cooking time—associated with the swelling and gelatinization of the starch. Good quality pasta has WA of 150–200 g water/100 g pasta. The swelling index (SI) gives us information about the integrity of the protein matrix, which restricts water penetration [89,91]. The increase in the swelling index is due to a weakened gluten network, which allows increased amounts of water to enter the starch granules, leading to faster gelatinization. Good quality pasta has SI values of approximately 1.8 [90]. These data are in agreement with those of the study by Chetrariu and Dabija [56].

Total organic matter increases with the increase in the spent grain content added to the recipe, possibly due to the high fiber content and the weakened gluten network, leading to swelling of the starch granules and release of a higher quantity of starch while cooking the pasta [86]. Very good quality pasta has total organic matter values lower than 1.4, good quality pasta has values between 1.4 and 2.1, and values higher than 2.1 represent poor quality pasta [91]. Microscopic analysis of the pasta showed a continuous protein network, with protein aggregates of different sizes derived from the spent grain. The degree of starch swelling also increased in the outer part of the sample but decreased towards the core [55].

Good quality pasta must have several attributes: moderate optimum cooking time, low cooking loss, water absorption and swelling index, moderate increase in volume with firmness and high chewiness and low adhesiveness, given by a consolidated and non-continuous protein matrix. This limits the swelling of starch and makes the diffusion of water to the core of the pasta difficult, leading to greater retention of amylose in the structure and less amylopectin on the surface [91]. All these studies show that spent grain can be used in a saturated market to produce innovative products; the ingredients used represent a stable and sustainable solution for spent grain upcycling.

3.3. Cookies and Shortbreads

Cookies are food products to which different flours can be added because they accommodate a wide variety of formulations and ingredients, are ready-to-eat, represent a good source of energy, have a long shelf life and are accepted by consumers of all ages [92]. Adding 20% spent grain to the cookie formulation increased the protein content by 55%, the lysine content by 90% and the fiber content by 220% compared with the control sample [93]. The addition of spent grain to cookies is proportional to the increase in dietary fiber content [13] and also depends on the particle size of the spent grain. Öztürk et al. [57] studied the influence of spent grain particle size on the quality of cookies. Medium and coarse particle sizes resulted in better properties in terms of spread ratio, hunter color values and overall sensory scores compared with cookies made with spent grain of fine particle size. Dough development time and dough stability increased with spent grain substitution level, resulting in higher energy costs. Adding spent grain increased the phenolic acid concentration and ferulic acid was predominant in all cookies. A 20% quantity of spent grain resulted in lower hydrolysis and glycemic index, and less total starch content compared with the control cookies [58]. Fat is needed for cookie production; as the quantity of substituted spent grain increased, fat levels in the cookies also increase. Additionally, the thickness and width of cookies increased with the addition of spent grain, while the spread ratio decreased in comparison with control samples [59]. Figure 3 shows cookies produced using 10% spent grain. Given that the replacement of wheat flour with spent grain in the cookie recipe increases the proportional fiber and protein content, the use of spent grain becomes promising for groups of consumers with nutritional deficiencies [60].

Figure 3. Spent grain cookies.

Petrović et al. [61] evaluated the effect of fresh spent grain (milled and non-milled) in cookie formulations on cookie quality parameters. Fresh spent grain had no negative effects on microbiological stability, and adding 25% of it produced the best sensory characteristics (appearance, hardness, grittiness and flavor). Fresh spent grain was susceptible to microbial attack and chemical damage due to its chemical composition, but the study shows that a proliferation of microbiological compounds (Yeast and molds, *Escherichia coli* and *Clostridium* spp.) did not occur. The high quantity of fiber and protein in spent grain increased water absorption, negatively impacting the hardness and chewiness.

Replacing 30% of wheat flour with spent grain flour in the shortbread recipe led to a significant increase in the fiber (particularly arabinoxylans) and protein content, and a decrease in carbohydrate levels and energy compared with the control samples. The replacement also showed acceptable sensory characteristics [29].

3.4. *Muffins*

Shih et al. looked into how two drying techniques (impingement and hot-air drying) affect the composition of spent grain and the quality of muffins made using spent grain. The study showed that impingement-dried spent grain may be used as a functional component in muffins to enhance the value to the food chain and to provide nutritional and environmental benefits. The study also found that adding spent grain flours to muffins (substituting 15% of the wheat flour) increased their protein and total dietary fiber contents by 23% and 13%, respectively, without influencing consumer acceptance of the products. In general, due to the higher concentrations of these nutrients in spent grain flour compared with wheat flour, the amount of fat, protein and total dietary fiber in fortified products is significantly higher [62]. The viscosity of the batter increases as the amount of spent grain in the muffin recipe increases. This may be because the spent grain has a high fiber content that acts as a thickening agent by absorbing water in the batter. A study of 18 participants who consumed muffins with 30% added spent grain daily for 8 weeks showed beneficial effects associated with reduced systolic blood pressure and insulin compared with the control group [63]. Another study on the use of spent grain in muffins found that 30% spent grain retains consumer approval and offers more chances of triggering biological reactions due to the higher levels of proteins, fiber and antioxidants. Furthermore, nutrient content could be mentioned on the labels of muffins with 20% and 30% spent grain, both of which are considered "good sources" of protein and fiber, because the portion size contains more than 10% of the recommended daily value of each nutrient [40].

3.5. *Wafers*

Wafers come in a wide variety of assortments and are obtained by baking special forms of fluid dough consisting of wheat flour, water, salt, aeration agents and other ingredients used to add taste and aroma, and are presented in the form of sheets or different alveoli formats [94] with high porosity and no filling (Figure 4). The disadvantages lie in the fact that parts of these products may have low nutritional values and poorly defined sensory characteristics.

Gumminess, chewiness, springiness, firmness and cohesiveness increased in the spent grain sample, while adhesiveness decreased with the addition of spent grain [64]. Water activity is an important instrumental measure of crispiness in wafers and should be between 0.387 and 0.52 [65]. Two regions of wafers with different porosities can be highlighted: a dense part called "skin" and a less porous part called "core", which is the central part of the baked sheet [66]. Analyzing the microstructure of the wafers is important for determining the quality of the products and involves measuring the size of the pores and cell wall sizes throughout the cross-section. This analysis shows that the distribution of the pores on the cross section is heterogeneous, with the center of the wafers having larger pores and the edges having smaller pores and denser skins [66].

Figure 4. Spent grain wafers.

3.6. Snacks

Nagy and Diósi obtained products with added spent grain that had positive nutritional values, including increased total content of polyphenols, flavonoids, proteins, fats, dietary fiber and energy, compared with the control samples [48]. Ktenioudaki et al. [95] obtained crispy snack slices containing 10% spent grain with a high crispiness index and low crispiness, indicating that this quantity of added spent grain did not negatively affect the crispiness of the finished product. Adding a higher quantity of spent grain modified the texture and crumb structure and altered the odor profile. Crispy slices containing 10% spent grain were highly acceptable to panelists compared with the control samples. This quantity of added spent grain almost doubled the fiber content of the baked snacks [95]. Crofton and Scanell conducted a study of four types of spent grain snacks; the crispy cracker snack was the most preferred, followed by the crispy sticks with dip, the fruity biscuits and finally the twisted breadsticks [96]. Stojceska et al. [67] studied the effects of spent grain on the textural and functional properties of extrudates and found that at between 10% and 30% reduced cell size, the expansion of the product reduced and the phytic acid content and bulk density increased. The optimal level of spent grain was set at 20% in order to obtain products similar to those available on the market, although a substitution of 30% still led to products with acceptable physico-chemical characteristics. The textural properties were adjusted by incorporating starch and specific mechanical energy and controlling extrusion parameters conduced to an acceptable expanded ready-to-eat snack. Ainsworth et al. [68] conducted a similar study on the effect of brewers spent grain and screw speed on the selected physical and nutritional properties of an extruded snack. They found that phytic acid and the resistant starch content of the samples increased significantly with the addition of up to 30% of spent grain to the formulation, although screw speed had no significant effect on total antioxidant capacity and total phenolic compounds. Kirjoranta et al. [97] conducted a study on the effects of spent grain on process parameters of snacks and concluded that 10% of spent grain increased hardness and caused a small expansion. The expansion increased with increasing screw speed and decreasing water content. Several ways of achieving a greater expansion of extruded snack include adding starch, enzymatic treatment of spent grain to solubilize part of the insoluble dietary fiber or grinding spent grain into small particles. The snack market is rapidly expanding with the frequent introduction of innovative bars fortified with proteins, fibers and other rich nutrients such as spent grain [98].

3.7. Extruded Spent Grain

Extrusion is a relatively new method of cooking that involves continuous mixing, cooking and extrusion of food products. The extrusion process takes place in an extruder in which thermo-mechanical processing takes place. The temperature inside the extruder is increased by subjecting the material to high compressive and shearing forces. This process leads to cooking of the product and has the advantage of immediate and efficient modification, giving a product with superior quality [66]. Extrusion is a thermo mechanical process that combines several unitary operations representing a viable, transferable opportunity with a beneficial impact on the functional, technological and food safety characteristics of the product [99]. The extrusion process can improve the balance between soluble and insoluble dietary fiber contents, breaking polysaccharide bonds under mechanical stress, and releasing the content of phenolic compounds trapped in the dietary fibers through shearing. Among the undesirable aspects of extrusion is the Maillard reaction, which favors the production of acrylamide or the reduction in the content of essential amino acids; aspects that depend on the required processing conditions [100]. Gutiérrez-Barrutia et al. studied the effect of extrusion on spent grain and found that extrusion had positive effects on spent grain, increasing the content of soluble dietary fibers, changes caused by the thermo mechanical process that can disrupt the cell wall matrix resulting in smaller and more soluble fragments. Extruded spent grain can be considered a suitable ingredient for human consumption from a microbiological point of view [100].

3.8. Yogurt and Plant-Based Yogurt Alternatives

Various quantities of spent grain were used as substitutes for yogurt fermentation, and the effects on microstructural characteristics such as surface chemical characteristics and confocal microstructures were investigated. Yogurt's syneresis level was considerably reduced when spent grain was added. Adding spent grain decreased fermentation time and increased viscosity and shear stress. As a result of amino acids being released, the proteolytic action of the additional microorganisms in yogurt manufacture shortens the fermentation process by enhancing microbial growth. The reduced fermentation period was also influenced by the inclusion of protein and fat substitutes. Yogurt's maximum quality, including its acidity, rheological behavior and lactic acid bacteria development improved with the increase in the added spent grain from 5% to 10%. While 15–20% of spent grain gave the lowest syneresis while producing the same amount of acidity and lactic acid bacteria, it reduced the yogurt's flow performance [69].

Spent grain has a great water-holding capacity due to its significant insoluble dietary fiber content (particularly arabinoxylans, which have a high capacity to bind water and play a potential prebiotic role). Due to these properties, spent grain may be able to control the behavior of semi-solid foods and hence replace the need for starch. In the study conducted by Naibaho et al., spent grain flour and three different protein extracts from spent grain added to plant-based yogurt-alternatives maintained textural and gelling formation, while increasing shear stress and viscosity [70].

3.9. Other Food Products

Spent grain can also be used as breadcrumbs for schnitzel, in fillings for vegetable burgers and in Frankfurter sausages [47]. Nine experimental Frankfurters were made using spent grain of three distinct particle sizes—fine (212 μm), medium (212–425 μm), and coarse (425–850 μm). As expected, the Frankfurters' total dietary fiber content improved as more spent grain was added. The water-holding capacity was correlated with the total dietary fiber in the experimental samples, and the total dietary fiber content of the samples generated with coarse-particle-size (425–850 μm) spent grain were higher than those of the other samples. The amount of spent grain and the low level of fat in the coarse, medium and small particle size groups appeared to have a negative impact on all textural criteria, with the exception of springiness. The study revealed that spent grain has great potential as a source of dietary fiber and might be used as a fat substitute to create meat products that are both high in dietary fiber and low in fat. Most of the textural and sensory criteria were substantially correlated with overall acceptance, based on statistical cluster analysis [71].

Özboy-Özbaş et al. showed that spent grain can be used as an ingredient in tarhana, a fermented wheat flour-yoghurt product, with acceptable results obtained using 6% spent grain. This quantity of spent grain increases the protein and fiber content while maintaining sensory qualities within acceptable limits [72].

McCarthy et al. used different methods to introduce phenolic extracts into fruit beverages (fruit juice and smoothies). The maximum concentration of phenolic extract in spent grain was 10%, considerably raising the FRAP activity of cranberry juice and demonstrating the possibility of using spent grain phenolic extracts as antioxidants in functional foods [73].

Spent grain has replaced flour/semolina in the traditional Herzegovinian product, Cupter. This product is made from wheat flour or semolina and grape must and is described as a sweet jelly. Spent grain influences the odor, color, texture and flavor profile of the traditional product, the taste of which is well known to consumers [101].

Spent grain can also be used to make edible coating composite films for fresh strawberries by immersing the strawberries in coating solution for 2 min before testing weight loss, pH, dry matter and anthocyanin content over 5 days. The carboxyl methylcellulose edible composite film positively affected the appearance of strawberries after the testing period, preserving their freshness for a longer period compared with uncoated strawberries. The strawberries had low weight loss because the films prevent moisture loss. No significant differences in anthocyanin levels were observed between coated and uncoated strawberries. The pH levels of the coated and uncoated strawberry samples also showed little variation over the course of storage [102].

4. Conclusions

The use of spent grain as an ingredient in finished food products is an opportunity to reduce by-products in the beer and whisky industries while improving the nutritional content of the food obtained. Studies showed that consumer acceptability limits for food products comparable to commercial ones fall within a 20% concentration of the spent grain, although 10–15% spent grain is considered optimal for sensory properties. By-products of the agro-industrial sector are important resources that can be used as raw materials to create food products with added value, supporting the circular economy. One of the basic tenets of the circular economy and one of the biggest problems in food engineering in recent years has been the sustainable use of organic waste and agri-food by-products. Due to its high quantity and low cost, spent grain is a source worth exploiting. The growing demand for products with stable ingredients obtained from food by-products is stimulating the identification of innovative alternatives. Additionally, food industry by-products are a good source of proteins, minerals, fatty acids, fiber and bioactive substances that can prevent nutrition-related disorders and improve consumers' physical and mental well-being. The global demand for food is rising, driving researchers to look for alternative raw materials with good nutritional value. There is a pressing need to utilize this by-product and to create sustainable methods of using it. The reuse of spent grain in food products (muffins, cakes, biscuits, etc.) brings both economic and environmental benefits, thus reducing pollution. In subsequent studies, we aim to determine the upper limit of spent grain that can be incorporated into food products, as well as to identify their advantages and disadvantages.

Author Contributions: Conceptualization, A.C. and A.D.; Methodology, A.C.; Formal Analysis, A.C. and A.D.; Investigation, A.C.; Resources, A.D.; Writing—Original Draft Preparation, A.C.; Writing—Review and Editing, A.C. and A.D. All authors have read and agreed to the published version of the manuscript.

Funding: This work was funded by Ministry of Research, Innovation and Digitalization within Program 1—Development of national research and development system, Subprogram 1.2—Institutional Performance—RDI excellence funding projects, under contract no. 10PFE/2021.

Institutional Review Board Statement: Not applicable.

Informed Consent Statement: Not applicable.

Data Availability Statement: No new data were created or analyzed in this study. Data sharing is not applicable to this article.

Acknowledgments: The authors acknowledge financial support from Stefan cel Mare University of Suceava, Romania.

Conflicts of Interest: The authors declare no conflict of interest.

References

1. Mitri, S.; Salameh, S.J.; Khelfa, A.; Leonard, E.; Maroun, R.G.; Louka, N.; Koubaa, M. Valorization of Brewers' Spent Grains: Pretreatments and Fermentation, a Review. *Fermentation* **2022**, *8*, 50. [CrossRef]
2. Neylon, E.; Nyhan, L.; Zannini, E.; Monin, T.; Münch, S.; Sahin, A.W.; Arendt, E.K. Food Ingredients for the Future: In-Depth Analysis of the Effects of Lactic Acid Bacteria Fermentation on Spent Barley Rootlets. *Fermentation* **2023**, *9*, 78. [CrossRef]
3. Naibaho, J.; Korzeniowska, M. The Variability of Physico-Chemical Properties of Brewery Spent Grain from 8 Different Breweries. *Heliyon* **2021**, *7*, e06583. [CrossRef] [PubMed]
4. Parchami, M.; Ferreira, J.A.; Taherzadeh, M.J. Starch and Protein Recovery from Brewer's Spent Grain Using Hydrothermal Pretreatment and Their Conversion to Edible Filamentous Fungi—A Brewery Biorefinery Concept. *Bioresour. Technol.* **2021**, *337*, 125409. [CrossRef] [PubMed]
5. Jackson, S.A.; Kang, X.; O'Shea, R.; O'Leary, N.; Murphy, J.D.; Dobson, A.D.W. Anaerobic Digestion Performance and Microbial Community Structures in Biogas Production from Whiskey Distillers Organic By-Products. *Bioresour. Technol. Rep.* **2020**, *12*, 100565. [CrossRef]
6. Fărcaș, A.C.; Socaci, S.A.; Chiș, M.S.; Martínez-Monzó, J.; García-Segovia, P.; Becze, A.; Török, A.I.; Cadar, O.; Coldea, T.E.; Igual, M. In Vitro Digestibility of Minerals and B Group Vitamins from Different Brewers' Spent Grains. *Nutrients* **2022**, *14*, 3512. [CrossRef]
7. Czubaszek, A.; Wojciechowicz-Budzisz, A.; Spychaj, R.; Kawa-Rygielska, J. Baking Properties of Flour and Nutritional Value of Rye Bread with Brewer's Spent Grain. *LWT* **2021**, *150*, 111955. [CrossRef]
8. Korcari, D.; Secchiero, R.; Laureati, M.; Marti, A.; Cardone, G.; Rabitti, N.S.; Ricci, G.; Fortina, M.G. Technological Properties, Shelf Life and Consumer Preference of Spelt-Based Sourdough Bread Using Novel, Selected Starter Cultures. *LWT* **2021**, *151*, 112097. [CrossRef]
9. Mussatto, S.I.; Dragone, G.; Roberto, I.C. Ferulic and P-Coumaric Acids Extraction by Alkaline Hydrolysis of Brewer's Spent Grain. *Ind. Crop. Prod.* **2007**, *25*, 231–237. [CrossRef]
10. Patrignani, M.; Brantsen, J.F.; Awika, J.M.; Conforti, P.A. Application of a Novel Microwave Energy Treatment on Brewers' Spent Grain (BSG): Effect on Its Functionality and Chemical Characteristics. *Food Chem.* **2021**, *346*, 128935. [CrossRef]
11. Kavalopoulos, M.; Stoumpou, V.; Christofi, A.; Mai, S.; Barampouti, E.M.; Moustakas, K.; Malamis, D.; Loizidou, M. Sustainable Valorisation Pathways Mitigating Environmental Pollution from Brewers' Spent Grains. *Environ. Pollut.* **2021**, *270*, 116069. [CrossRef]
12. Wagner, E.; Pería, M.E.; Ortiz, G.E.; Rojas, N.L.; Ghiringhelli, P.D. Valorization of Brewer's Spent Grain by Different Strategies of Structural Destabilization and Enzymatic Saccharification. *Ind. Crop. Prod.* **2021**, *163*, 113329. [CrossRef]
13. Lynch, K.M.; Steffen, E.J.; Arendt, E.K. Brewers' Spent Grain: A Review with an Emphasis on Food and Health. *J. Inst. Brew.* **2016**, *122*, 553–568. [CrossRef]
14. Malvandi, A.; Nicole Coleman, D.; Loor, J.J.; Feng, H. A Novel Sub-Pilot-Scale Direct-Contact Ultrasonic Dehydration Technology for Sustainable Production of Distillers Dried Grains (DDG). *Ultrason. Sonochem.* **2022**, *85*, 105982. [CrossRef] [PubMed]
15. Sahin, A.W.; Hardiman, K.; Atzler, J.J.; Vogelsang-O'Dwyer, M.; Valdeperez, D.; Münch, S.; Cattaneo, G.; O'Riordan, P.; Arendt, E.K. Rejuvenated Brewer's Spent Grain: The Impact of Two BSG-Derived Ingredients on Techno-Functional and Nutritional Characteristics of Fibre-Enriched Pasta. *Innov. Food Sci. Emerg. Technol.* **2021**, *68*, 102633. [CrossRef]
16. Gunes, B.; Stokes, J.; Davis, P.; Connolly, C.; Lawler, J. Optimisation of Anaerobic Digestion of Pot Ale after Thermochemical Pre-Treatment through Response Surface Methodology. *Biomass Bioenergy* **2021**, *144*, 105902. [CrossRef]
17. Leinonen, I.; MacLeod, M.; Bell, J. Effects of Alternative Uses of Distillery By-Products on the Greenhouse Gas Emissions of Scottish Malt Whisky Production: A System Expansion Approach. *Sustainability* **2018**, *10*, 1473. [CrossRef]
18. White, J.S.; Traub, J.E.; Maskell, D.L.; Hughes, P.S.; Harper, A.J.; Willoughby, N.A. *Recovery and Applications of Proteins from Distillery By-Products*; Elsevier Inc.: Amsterdam, The Netherlands, 2016; ISBN 9780128026113.
19. Foltanyi, F.; Hawkins, J.E.; Panovic, I.; Bird, E.J.; Gloster, T.M.; Lancefield, C.S.; Westwood, N.J. Analysis of the Product Streams Obtained on Butanosolv Pretreatment of Draff. *Biomass Bioenergy* **2020**, *141*, 105680. [CrossRef]
20. Kang, X.; Lin, R.; O'Shea, R.; Deng, C.; Li, L.; Sun, Y.; Murphy, J.D. A Perspective on Decarbonizing Whiskey Using Renewable Gaseous Biofuel in a Circular Bioeconomy Process. *J. Clean. Prod.* **2020**, *255*, 120211. [CrossRef]

21. Artola, A.; Sala, A.; Vittone, S.; Barrena, R.; Antoni, S. Scanning Agro-Industrial Wastes as Substrates for Fungal Biopesticide Production: Use of B Eauveria Bassiana and Trichoderma Harzianum in Solid-State Fermentation. *J. Environ. Manag.* **2021**, *295*, 113113. [CrossRef]
22. Gunes, B.; Carrié, M.; Benyounis, K.; Stokes, J.; Davis, P.; Connolly, C.; Lawler, J. Optimisation and Modelling of Anaerobic Digestion of Whiskey Distillery/Brewery Wastes after Combined Chemical and Mechanical Pre-Treatment. *Processes* **2020**, *8*, 492. [CrossRef]
23. Akunna, J.C.; Walker, G.M. Co-Products from Malt Whisky Production and Their Utilisation. *Alcohol Textb.* **2017**, *34*, 529–537.
24. Chetrariu, A.; Dabija, A. Spent Grain from Malt Whisky: Assessment of the Phenolic Compounds. *Molecules* **2021**, *26*, 3236. [CrossRef]
25. Nigam, P.S. An Overview: Recycling of Solid Barley Waste Generated as a by-Product in Distillery and Brewery. *Waste Manag.* **2017**, *62*, 255–261. [CrossRef]
26. Bower, J. Scotch Whisky: History, Heritage and the Stock Cycle. *Beverages* **2016**, *2*, 11. [CrossRef]
27. Joshi, I.; Truong, V.K.; Elbourne, A.; Chapman, J.; Cozzolino, D. Influence of the Scanning Temperature on the Classification of Whisky Samples Analysed by UV-VIS Spectroscopy. *Appl. Sci.* **2019**, *9*, 3254. [CrossRef]
28. Galanakis, C.M. Sustainable Applications for the Valorization of Cereal. *Foods* **2022**, *11*, 241. [CrossRef]
29. Sileoni, V.; Alfeo, V.; Bravi, E.; Belardi, I.; Marconi, O. Upcycling of a By-Product of the Brewing Production Chain as an Ingredient in the Formulation of Functional Shortbreads. *J. Funct. Foods* **2022**, *98*, 105292. [CrossRef]
30. Kamali Rousta, L.; Ghandehari Yazdi, A.P.; Amini, M. Optimization of Athletic Pasta Formulation by D-Optimal Mixture Design. *Food Sci. Nutr.* **2020**, *8*, 4546–4554. [CrossRef] [PubMed]
31. Thai, S.; Avena-Bustillos, R.J.; Alves, P.; Pan, J.; Osorio-Ruiz, A.; Miller, J.; Tam, C.; Rolston, M.R.; Teran-Cabanillas, E.; Yokoyama, W.H.; et al. Influence of Drying Methods on Health Indicators of Brewers Spent Grain for Potential Upcycling into Food Products. *Appl. Food Res.* **2022**, *2*, 100052. [CrossRef]
32. McCarthy, A.L.; O'Callaghan, Y.C.; Piggott, C.O.; FitzGerald, R.J.; O'Brien, N.M. Brewers' Spent Grain; Bioactivity of Phenolic Component, Its Role in Animal Nutrition and Potential for Incorporation in Functional Foods: A Review. *Proc. Nutr. Soc.* **2013**, *72*, 117–125. [CrossRef] [PubMed]
33. Zeko-Pivač, A.; Tisma, M.; Žnidaršič-Plazl, P.; Kulisic, B.; Sakellaris, G.; Hao, J. Mirela Planinić The Potential of Brewer's Spent Grain in the Circular Bioeconomy: State of the Art and Future Perspectives. *Front. Bioeng. Biotechnol.* **2022**, *10*, 870744. [CrossRef] [PubMed]
34. Pabbathi, N.P.P.; Velidandi, A.; Pogula, S.; Gandam, P.K.; Baadhe, R.R.; Sharma, M.; Sirohi, R.; Thakur, V.K.; Gupta, V.K. Brewer's Spent Grains-Based Biorefineries: A Critical Review. *Fuel* **2022**, *317*, 123435. [CrossRef]
35. Baiano, A.; Gatta, B.; Rutigliano, M.; Fiore, A. Functional Bread Produced in a Circular Economy Perspective: The Use of Brewers' Spent Grain. *Foods* **2023**, *12*, 834. [CrossRef]
36. Roth, M.; Jekle, M.; Becker, T. Opportunities for Upcycling Cereal Byproducts with Special Focus on Distiller's Grains. *Trends Food Sci. Technol.* **2019**, *91*, 282–293. [CrossRef]
37. Yitayew, T.; Moges, D.; Satheesh, N. Effect of Brewery Spent Grain Level and Fermentation Time on the Quality of Bread. *Int. J. Food Sci.* **2022**, *2022*, 8704684. [CrossRef]
38. Colpo, I.; Rabenschlag, D.R.; De Lima, M.S.; Martins, M.E.S.; Sellitto, M.A. Economic and Financial Feasibility of a Biorefinery for Conversion of Brewers' Spent Grain into a Special Flour. *J. Open Innov. Technol. Mark. Complex.* **2022**, *8*, 79. [CrossRef]
39. Ullah, H.; Esposito, C.; Piccinocchi, R.; De Lellis, L.F.; Santarcangelo, C.; Di Minno, A.; Baldi, A.; Buccato, D.G.; Khan, A.; Piccinocchi, G.; et al. Postprandial Glycemic and Insulinemic Response by a Brewer's Spent Grain Extract-Based Food Supplement in Subjects with Slightly Impaired Glucose Tolerance: A Monocentric, Randomized, Cross-Over, Double-Blind, Placebo-Controlled Clinical Trial. *Nutrients* **2022**, *14*, 3916. [CrossRef]
40. Combest, S.; Warren, C. The Effect of Upcycled Brewers' Spent Grain on Consumer Acceptance and Predictors of Overall Liking in Muffins. *J. Food Qual.* **2022**, *2022*, 6641904. [CrossRef]
41. Lech, M.; Labus, K. The Methods of Brewers' Spent Grain Treatment towards the Recovery of Valuable Ingredients Contained Therein and Comprehensive Management of Its Residues. *Chem. Eng. Res. Des.* **2022**, *183*, 494–511. [CrossRef]
42. Leite, P.; Silva, C.; Salgado, J.M.; Belo, I. Simultaneous Production of Lignocellulolytic Enzymes and Extraction of Antioxidant Compounds by Solid-State Fermentation of Agro-Industrial Wastes. *Ind. Crop. Prod.* **2019**, *137*, 315–322. [CrossRef]
43. Montes, J.A.; Rico, C. Biogas Potential of Wastes and By-Products of the Alcoholic Beverage Production Industries in the Spanish Region of Cantabria. *Appl. Sci.* **2020**, *10*, 7481. [CrossRef]
44. Ikram, S.; Huang, L.Y.; Zhang, H.; Wang, J.; Yin, M. Composition and Nutrient Value Proposition of Brewers Spent Grain. *J. Food Sci.* **2017**, *82*, 2232–2242. [CrossRef]
45. Bolwig, S.; Mark, M.S. Beyond Animal Feed? The Valorisation of Brewers' Spent Grain. In *From Waste to Value: Valorisation Pathways for Organic Waste Streams in Circular Bioeconomies*; Routledge: Abingdon, UK, 2019.
46. Maga, J.; Everen, K. van Chemical and Sensory Properties of Wholewheat Pasta Products Supplemented with Wheat-Derived Dried Distillers Grain (DDG). *J. Food Process. Preserv.* **1988**, *13*, 71–78. [CrossRef]

47. Nasim, J.; Bohn, T. Turning Apparent Waste into New Value: Up-Cycling Strategies Exemplified by Brewer's Spent Grains (BSG). *Curr. Nutraceut.* **2020**, *1*, 6–13. [CrossRef]
48. Nagy, V.; Diósi, G. Using Brewer's Spent Grain as a Byproduct of the Brewing Industry in the Bakery Industry. *Elelmvizsg. Kozl.* **2021**, *67*, 3327–3350. [CrossRef]
49. Ktenioudaki, A.; O'Shea, N.; Gallagher, E. Rheological Properties of Wheat Dough Supplemented with Functional By-Products of Food Processing: Brewer's Spent Grain and Apple Pomace. *J. Food Eng.* **2013**, *116*, 362–368. [CrossRef]
50. Stojceska, V.; Ainsworth, P. The Effect of Different Enzymes on the Quality of High-Fibre Enriched Brewer's Spent Grain Breads. *Food Chem.* **2008**, *110*, 865–872. [CrossRef] [PubMed]
51. Steinmacher, N.C.; Honna, F.A.; Gasparetto, A.V.; Anibal, D.; Grossmann, M.V.E. Bioconversion of Brewer's Spent Grains by Reactive Extrusion and Their Application in Bread-Making. *LWT-Food Sci. Technol.* **2012**, *46*, 542–547. [CrossRef]
52. Șaitan, O.; Țarnă, R. Aliona ghendov-mosanu influence of brewer's spent grain on qualitative indicators of bread from wheat flour. *MTFI* **2022**, *2022*, 80009.
53. Nocente, F.; Natale, C.; Galassi, E.; Taddei, F.; Gazza, L. Using Einkorn and Tritordeum Brewers' Spent Grain to Increase the Nutritional Potential of Durum Wheat Pasta. *Foods* **2021**, *10*, 502. [CrossRef] [PubMed]
54. Cuomo, F.; Trivisonno, M.C.; Iacovino, S.; Messia, M.C.; Marconi, E. Sustainable Re-Use of Brewer's Spent Grain for the Production of High Protein and Fibre Pasta. *Foods* **2022**, *11*, 642. [CrossRef] [PubMed]
55. Schettino, R.; Verni, M.; Acin-albiac, M.; Vincentini, O.; Krona, A.; Knaapila, A.; Di Cagno, R.; Gobbetti, M.; Rizzello, C.G.; Coda, R. Bioprocessed Brewers' Spent Grain Improves Nutritional and Antioxidant Properties of Pasta. *Antioxidants* **2021**, *10*, 742. [CrossRef]
56. Chetrariu, A.; Dabija, A. Quality Characteristics of Spelt Pasta Enriched with Spent Grain. *Agronomy* **2021**, *11*, 1824. [CrossRef]
57. Öztürk, S.; Özboy, Ö.; Ox, D.; Köksel, H.; Brew, J.I. Effects of Brewers Spent Grain on the Quality and Dietary Fibre Content of Cookies. *J. Inst. Brew.* **2002**, *108*, 23–27. [CrossRef]
58. Heredia-Sandoval, N.G.; del Carmen Granados-Nevárez, M.; de la Barca, A.M.C.; Vásquez-Lara, F.; Malunga, L.N.; Apea-Bah, F.B.; Beta, T.; Islas-Rubio, A.R. Phenolic Acids, Antioxidant Capacity, and Estimated Glycemic Index of Cookies Added with Brewer's Spent Grain. *Plant Foods Hum. Nutr.* **2020**, *75*, 41–47. [CrossRef]
59. Ajanaku, K.O.; Dawodu, F.A.; Ajanaku, C.O.; Nwinyi, O.C. Functional and Nutritive Properties of Spent Grain Enhanced Cookies. *Am. J. Food Technol.* **2011**, *6*, 763–771. [CrossRef]
60. Rigo, M.; Vidal Bezerra, J.R.M.; Rodrigues, D.D.; Moraes Teixeira, A. Avaliação Físico-Química e Sensorial de Biscoitos Tipo Cookie Adicionados de Farinha de Bagaço de Malte Como Fonte de Fibra. *Ambiência Guarapuava (PR)* **2017**, *13*, 47–57. [CrossRef]
61. Petrovic, J.; Pajin, B.; Tanackov-Kocic, S.; Pejin, J.; Fistes, A.; Bojanic, N.; Loncarevic, I. Quality Properties of Cookies Supplemented with Fresh Brewer's Spent Grain. *Food Feed Res.* **2017**, *44*, 57–63. [CrossRef]
62. Shih, Y.T.; Wang, W.; Hasenbeck, A.; Stone, D.; Zhao, Y. Investigation of Physicochemical, Nutritional, and Sensory Qualities of Muffins Incorporated with Dried Brewer's Spent Grain Flours as a Source of Dietary Fiber and Protein. *J. Food Sci.* **2020**, *85*, 3943–3953. [CrossRef]
63. Combest, S.; Warren, C.; Patterson, M. Upcycling Brewers' Spent Grain: The Development of Muffins and Biomarker Response After Consuming Muffins for 8-Weeks in Healthy Adults From Randomized-Controlled Trial. *Curr. Dev. Nutr.* **2020**, *4*, 745. [CrossRef]
64. Chetrariu, A.; Dabija, A. Study of the Utilization of Spent Grain from Malt Whisky on the Quality of Wafers. *Appl. Sci.* **2022**, *12*, 7163. [CrossRef]
65. Goerlitz, C.D.; Harper, W.J.; Delwiche, J.F. Relationship of Water Activity to Cone Crispness as Assessed by Positional Relative Rating. *J. Sens. Stud.* **2007**, *22*, 687–694. [CrossRef]
66. Butt, S.S. Geometrical and Material Effects on Sensory Properties of Confectionery Wafers and Similar Extruded Products. Ph.D. Thesis, Mechanics of Materials Department of Mechanical Engineering Imperial College London, London, UK, 2016.
67. Stojceska, V.; Ainsworth, P.; Plunkett, A.; Ibanoğlu, S. The Recycling of Brewer's Processing by-Product into Ready-to-Eat Snacks Using Extrusion Technology. *J. Cereal Sci.* **2008**, *47*, 469–479. [CrossRef]
68. Ainsworth, P.; Ibanoğlu, Ş.; Plunkett, A.; Ibanoğlu, E.; Stojceska, V. Effect of Brewers Spent Grain Addition and Screw Speed on the Selected Physical and Nutritional Properties of an Extruded Snack. *J. Food Eng.* **2007**, *81*, 702–709. [CrossRef]
69. Naibaho, J.; Butula, N.; Jonuzi, E.; Korzeniowska, M.; Laaksonen, O.; Föste, M.; Kütt, M.L.; Yang, B. Potential of Brewers' Spent Grain in Yogurt Fermentation and Evaluation of Its Impact in Rheological Behaviour, Consistency, Microstructural Properties and Acidity Profile during the Refrigerated Storage. *Food Hydrocoll.* **2022**, *125*, 107412. [CrossRef]
70. Naibaho, J.; Butula, N.; Jonuzi, E.; Korzeniowska, M.; Chodaczek, G.; Yang, B. The Roles of Brewers' Spent Grain Derivatives in Coconut-Based Yogurt-Alternatives: Microstructural Characteristic and the Evaluation of Physico-Chemical Properties during the Storage. *Curr. Res. Food Sci.* **2022**, *5*, 1195–1204. [CrossRef]
71. Ozvural, E.B.; Vural, H.; Gokbulut, I.; Ozboy-Ozbas, O. Utilization of Brewer's Spent Grain in the Production of Frankfurters. *Int. J. Food Sci. Technol.* **2009**, *44*, 1093–1099. [CrossRef]
72. Özboy-Özbaș, Ö.; Hançer, A.; Gökbulut, I. Utilization of Sugarbeet Fiber and Brewers' Spent Grain in the Production of Tarhana. *Zuckerindustrie* **2009**, *135*, 496–501. [CrossRef]

73. McCarthy, A.L.; O'Callaghan, Y.C.; Neugart, S.; Piggott, C.O.; Connolly, A.; Jansen, M.A.K.; Krumbein, A.; Schreiner, M.; FitzGerald, R.J.; O'Brien, N.M. The Hydroxycinnamic Acid Content of Barley and Brewers' Spent Grain (BSG) and the Potential to Incorporate Phenolic Extracts of BSG as Antioxidants into Fruit Beverages. *Food Chem.* **2013**, *141*, 2567–2574. [CrossRef] [PubMed]
74. Ktenioudaki, A.; Alvarez-Jubete, L.; Smyth, T.J.; Kilcawley, K.; Rai, D.K.; Gallagher, E. Application of Bioprocessing Techniques (Sourdough Fermentation and Technological Aids) for Brewer's Spent Grain Breads. *Food Res. Int.* **2015**, *73*, 107–116. [CrossRef]
75. Aprodu, I.; Simion, A.B.; Banu, I. Valorisation of the Brewers' Spent Grain Through Sourdough Bread Making. *Int. J. Food Eng.* **2017**, *13*, 1–9. [CrossRef]
76. Ginindza, A.; Solomon, W.K.; Shelembe, J.S.; Nkambule, T.P. Valorisation of Brewer's Spent Grain Flour (BSGF) through Wheat-Maize-BSGF Composite Flour Bread: Optimization Using D-Optimal Mixture Design. *Heliyon* **2022**, *8*, e09514. [CrossRef] [PubMed]
77. Czubaszek, A.; Wojciechowicz-Budzisz, A.; Spychaj, R.; Kawa-Rygielska, J. Effect of Added Brewer's Spent Grain on the Baking Value of Flour and the Quality of Wheat Bread. *Molecules* **2022**, *27*, 1624. [CrossRef]
78. Sahin, A.W.; Atzler, J.J.; Valdeperez, D.; Münch, S.; Cattaneo, G.; O'Riordan, P.; Arendt, E.K. Rejuvenated Brewer's Spent Grain: Evervita Ingredients as Game-Changers in Fibre-Enriched Bread. *Foods* **2021**, *10*, 1162. [CrossRef]
79. Kumar, G.S.; Sh, A.; Krishnan, R.; Mohammed, T. Pasta: Raw Materials, Processing and Quality Improvement. *Pharma Innov. J.* **2021**, *10*, 185–197. [CrossRef]
80. Cappa, C.; Cappa, C. Brewer's Spent Grain Valorization in Fiber-Enriched Fresh Egg Pasta Production: Modelling and Optimization Study. *LWT-Food Sci. Technol.* **2017**, *82*, 464–470. [CrossRef]
81. Iuga, M.; Mironeasa, S. Use of Grape Peels By-product for Wheat Pasta Manufacturing. *Plants* **2021**, *10*, 926. [CrossRef]
82. Khan, I.; Yousif, A.M.; Johnson, S.K.; Gamlath, S. Effect of Sorghum Flour Addition on in Vitro Starch Digestibility, Cooking Quality, and Consumer Acceptability of Durum Wheat Pasta. *J. Food Sci.* **2014**, *79*, S1560–S1567. [CrossRef]
83. Palavecino, P.M.; Bustos, M.C.; Alabí, M.B.H.; Nicolazzi, M.S.; Penci, M.C.; Ribotta, P.D. Effect of Ingredients on the Quality of Gluten-Free Sorghum Pasta. *J. Food Sci.* **2017**, *82*, 2085–2093. [CrossRef]
84. Chetrariu, A.; Dabija, A. Valorisation of Spent Grain from Malt Whisky in the Spelt Pasta Formulation: Modelling and Optimization Study. *Appl. Sci.* **2022**, *12*, 1441. [CrossRef]
85. Zarzycki, P.; Teterycz, D.; Wirkijowska, A.; Koz, K.; Miros, D. Use of Moldavian Dragonhead Seeds Residue for Pasta Production. *LWT* **2021**, *143*, 111099. [CrossRef]
86. Nocente, F.; Taddei, F.; Galassi, E.; Gazza, L. Upcycling of Brewers' Spent Grain by Production of Dry Pasta with Higher Nutritional Potential. *LWT* **2019**, *114*, 108421. [CrossRef]
87. Panel, E.; Nda, A. Scientific Opinion on Dietary Reference Values for protein. *EFSA J.* **2012**, *10*, 1–66. [CrossRef]
88. Kamali Rousta, L.; Pouya Ghandehari Yazdi, A.; Khorasani, S.; Tavakoli, M.; Ahmadi, Z.; Amini, M. Optimization of Novel Multigrain Pasta and Evaluation of Physicochemical Properties: Using D-Optimal Mixture Design. *Food Sci. Nutr.* **2021**, *9*, 5546–5556. [CrossRef]
89. Spinelli, S.; Padalino, L.; Costa, C.; Del Nobile, M.A.; Conte, A. Food By-Products to Fortified Pasta: A New Approach for Optimization. *J. Clean. Prod.* **2019**, *215*, 985–991. [CrossRef]
90. Bianchi, F.; Tolve, R.; Rainero, G.; Bordiga, M.; Brennan, C.S.; Simonato, B. Technological, Nutritional and Sensory Properties of Pasta Fortified with Agro-Industrial by-Products: A Review. *Int. J. Food Sci. Technol.* **2021**, *9*, 4356–4366. [CrossRef]
91. Bustos, M.C.; Perez, G.T.; Leon, A. Structure and Quality of Pasta Enriched with Functional Ingredients. *RSC Adv.* **2015**, *39*, 1–34. [CrossRef]
92. de Ávila Gonçalves, S.; Quiroga, F.; Vilaça, A.C.; Lancetti, R.; Blanco Canallis, M.S.; Caño de Andrade, M.H.; Ribotta, P.D. Physical–Chemical Evaluation of Flours from Brewery and Macauba Residues and Their Uses in the Elaboration of Cookies. *J. Food Process. Preserv.* **2021**, *45*, e15700. [CrossRef]
93. Kirssel, L.; Prentice, N. Protein and Fiber Enrichment of Cookie Flour with Brewers Spent Grain. *Cereal Chem.* **1979**, *56*, 261–266.
94. Raza, K.; Nadeem, M.; Hussain, S.; Jabbar, S.; Din, A.; Qureshi, T.M.; Ainee, A. Development and Physico-Chemical Characterization of Date Wafers. *J Agric. Res* **2016**, *54*, 368–1157.
95. Ktenioudaki, A.; Crofton, E.; Scannell, A.G.M.; Hannon, J.A.; Kilcawley, K.N.; Gallagher, E. Sensory Properties and Aromatic Composition of Baked Snacks Containing Brewer's Spent Grain. *J. Cereal Sci.* **2013**, *57*, 384–390. [CrossRef]
96. Crofton, E.C.; Scannell, A.G.M. Snack Foods from Brewing Waste: Consumer-Led Approach to Developing Sustainable Snack Options. *Br. Food J.* **2020**, *122*, 3899–3916. [CrossRef]
97. Kirjoranta, S.; Tenkanen, M.; Jouppila, K. Effects of Process Parameters on the Properties of Barley Containing Snacks Enriched with Brewer's Spent Grain. *J. Food Sci. Technol.* **2016**, *53*, 775–783. [CrossRef] [PubMed]
98. Stelick, A.; Sogari, G.; Rodolfi, M.; Dando, R.; Paciulli, M. Impact of Sustainability and Nutritional Messaging on Italian Consumers' Purchase Intent of Cereal Bars Made with Brewery Spent Grains. *J. Food Sci.* **2021**, *86*, 531–539. [CrossRef]
99. Leonard, W.; Zhang, P.; Ying, D.; Fang, Z. Application of Extrusion Technology in Plant Food Processing Byproducts: An Overview. *Compr. Rev. Food Sci. Food Saf.* **2020**, *19*, 218–246. [CrossRef]
100. Gutiérrez-Barrutia, M.B.; Del Castillo, M.D.; Arcia, P.; Cozzano, S. Feasibility of Extruded Brewer's Spent Grain as a Food Ingredient for a Healthy, Safe, and Sustainable Human Diet. *Foods* **2022**, *11*, 1403. [CrossRef]

101. Lali, A.; Karlovi, A.; Mari, M. Use of Brewers' Spent Grains as a Potential Functional Ingredient for the Production of Traditional Herzegovinian Product Cupter. *Fermentation* **2023**, *9*, 123. [CrossRef]
102. Oztuna Taner, O.; Ekici, L.; Akyuz, L. CMC-Based Edible Coating Composite Films from Brewer's Spent Grain Waste: A Novel Approach for the Fresh Strawberry Package. *Polym. Bull.* **2022**, *79*, 1–26. [CrossRef]

MDPI AG
Grosspeteranlage 5
4052 Basel
Switzerland
Tel.: +41 61 683 77 34

Foods Editorial Office
E-mail: foods@mdpi.com
www.mdpi.com/journal/foods

www.ingramcontent.com/pod-product-compliance
Lightning Source LLC
Chambersburg PA
CBHW040802250926
51850CB00007BA/1062